Birkhäuser

Progress in Mathematics

Volume 348

Progress in Mathematics is a series of books intended for professional mathematicians and scientists, encompassing all areas of pure mathematics. This distinguished series, which began in 1979, includes research level monographs, polished notes arising from seminars or lecture series, graduate level textbooks, and proceedings of focused and refereed conferences. It is designed as a vehicle for reporting ongoing research as well as expositions of particular subject areas.

Manuel Ritoré

Isoperimetric Inequalities in Riemannian Manifolds

Birkhäuser

Manuel Ritoré
Departamento de Geometría y Topología,
Facultad de Ciencias
Universidad de Granada
Granada, Spain

ISSN 0743-1643 ISSN 2296-505X (electronic)
Progress in Mathematics
ISBN 978-3-031-37900-0 ISBN 978-3-031-37901-7 (eBook)
https://doi.org/10.1007/978-3-031-37901-7

Mathematics Subject Classification: 49Q20

This book is published under the imprint Birkhäuser, www.birkhauser-science.com by the registered company Springer Nature Switzerland AG
The registered company address is: Gewerbestrasse 11, 6330 Cham, Switzerland

Paper in this product is recyclable.

In what nobler and more humanitarian enterprise could the intelligence be employed?

Santiago Ramón y Cajal
Recollections of my life

To Carmen and Álvaro

Preface

The purpose of this work is to give a coherent introduction to the theory and methods behind isoperimetric inequalities in Riemannian manifolds, including many of the results obtained in the last decades.

The original isoperimetric problem sought those compact regions in the Euclidean plane achieving maximal area among those regions of the same perimetric length or, equivalently, achieving the minimal perimetric length among those regions of the same area. The latter is the version of the problem we consider in this book, although naming it as isoperimetric is not completely correct from the etymological point of view. This isoperimetric problem is typically solved by means of an optimal inequality, an *isoperimetric inequality*, relating the volume and the perimeter of a set.

The literature on isoperimetric problems is vast. According to Knorr, we have three main sources of complete expositions on isoperimetric figures that survive from antiquity: a chapter of the *Introduction to the Syntaxis*, an anonymous compilation of edited extracts from commentaries on Ptolemy's Book I; the first section of Book V of the *Mathematical Collection* by Pappus; and Theon's Commentary on Ptolemy's Book I, see Chapter 10 in [249]. These sources clearly point to research developed by the Alexandrian school of Mathematics since the beginning of the third century BC. Only Theon cites a previous work, *On isoperimetric figures*, by a mathematician related to the Alexandrian school named Zenodorus, reveling a strong Archimedean influence, see pp. 272 ff. in [250]. Since then, the problem has attracted the interest of many scientists, including mathematicians. An overview of the large variety of techniques developed to face this problem, together with many references, can be found in Burago and Zalgaller's *Geometric inequalities* [83]. A more recent work treating this topic is Chavel's *Isoperimetric inequalities* [101], focused on different proofs of the classical isoperimetric inequality in Euclidean space and on obtaining coarse isoperimetric inequalities in noncompact manifolds. Chavel's *Riemannian geometry* also contains invaluable information on isoperimetric inequalities [102].

In contrast to these works, we are mostly interested in *optimal* inequalities, and in characterizing the minimizers of the problem, the *isoperimetric regions*, trying to obtain as much geometric and topological information as possible from their minimizing property. Our point of view will be mostly geometric, although we make use of techniques of geometric measure theory and of modern calculus of variations.

In this work, we do not consider some recent developments such as qualitative isoperimetric inequalities or isoperimetric inequalities in general metric measure spaces or sub-Riemannian manifolds, nor analytic applications to estimate physical quantities such as moment of inertia, capacity, or frequency of membranes. Neither do we consider multi-volume isoperimetric problems. We have not been exhaustive regarding regularity of isoperimetric boundaries.

This book should be of interest to students and researchers working on isoperimetric inequalities from either a geometric or analytic point of view. The results and, specially, the techniques are of application in many related problems in geometric variational theory. In this sense, the isoperimetric problems considered serve as a model for constrained variational problems.

This book assumes a reasonable background of the reader on Riemannian Geometry. The variety of techniques complicates the reading of the last part of the chapters, some of which are quite technical. At the end of every chapter, bibliographical notes have been included. I have attempted to be as exhaustive as possible when referring to the sources.

A brief summary of the chapters of this work is as follows.

In Chap. 1, we recall basic facts on Riemannian geometry, the variational formulas for the Riemannian volume of submanifolds, and the theory of sets of finite perimeter in Riemannian manifolds, as well as some other notions of boundary area: the Hausdorff measure and the Minkowski content. We show the equivalence between perimeter and relaxed Minkowski content obtained by Ambrosio et al. [20]. We also recall some boundary regularity properties of isoperimetric sets.

Chapter 2 treats isoperimetric inequalities on Riemannian surfaces. After a review of classical results, we focus on Benjamini and Cao's application of the geodesic curvature flow to obtain isoperimetric inequalities, and on later developments based on classical tools of the Calculus of Variations. The latter provide classifications of isoperimetric sets in surfaces where other methods were unsuccessful, such as on convex surfaces of revolution with monotonous Gauss curvature from the poles. In this chapter, we encounter the problem of existence of isoperimetric sets. We give an example of a complete surface without isoperimetric sets for any value of the area, and prove the existence of isoperimetric sets on complete convex surfaces for all values of the area.

In Chap. 3, we introduce the isoperimetric profile function I_M in a compact Riemannian manifold M, defined as

$$I_M(v) = \inf\{P(E) : E \subset M, |E| = v\},$$

where $P(E)$ is the perimeter of a measurable set E of volume $|E| = v \in (0, |M|)$. This is a key function which can be thought of as an optimal isoperimetric inequality in M. The main properties of the profile are obtained: continuity, positivity, and asymptotic behavior near 0. It is also shown that the profile satisfies, in a weak sense, a second order differential inequality involving a lower bound on the Ricci curvature of the manifold. A proof of the Lévy-Gromov isoperimetric inequality is obtained from the differential properties of the profile and, in particular, it is obtained in this work for the first time the characterization of the isoperimetric sets in the round sphere $\mathbb{S}^m$. The continuity of the normalized isoperimetric profile with respect to Gromov-Hausdorff convergence is also treated, as well as density estimates for isoperimetric sets.

Chapter 4 covers non-compact Riemannian manifolds, which isoperimetric behavior is in contrast to the one of compact manifolds. We provide a counterexample by Nardulli and Pansu of a complete Riemannian manifold with discontinuous isoperimetric profile, but give sufficient geometric conditions for the continuity of the profile: it is shown that the isoperimetric profile of a Cartan-Hadamard manifold and of a manifold with strictly positive sectional curvature is continuous. Then the problem of existence of isoperimetric sets in cocompact Riemannian manifolds is considered. Recall that in such a manifold, the action of the isometry group on M has a compact quotient. This provides a first characterization in this work of isoperimetric sets in Euclidean and hyperbolic spaces. A generalized existence result for isoperimetric regions is also proven. Although in general we cannot expect existence of isoperimetric sets in Riemannian manifolds, in the case of uniform Lipschitz geometry, we are able to obtain a generalized existence result, extending the space to include its asymptotic manifolds.

In Chap. 5, we obtain several classical symmetrizations from a general procedure on warped products. This result, obtained for some manifolds with density by Morgan et al. [309], is extremely powerful and allows the uniform treatment of many classical symmetrizations, from which new proofs of the isoperimetric inequalities in Euclidean, spherical, and hyperbolic spaces are obtained again. In the final part of the chapter, we consider the more recent Hsiang symmetrization and give several applications of this symmetrization to the classification of isoperimetric sets in some symmetric spaces.

In Chap. 6, exploiting the notion of stable hypersurfaces, first we provide again optimal isoperimetric inequalities in simply connected space forms, and then we focus on three-dimensional manifolds. Here all isoperimetric regions have smooth boundary and their genus can be estimated using techniques of complex variables theory. Spherical and toroidal isoperimetric boundaries can be easily characterized to provide a complete classification of isopenmetric sets in the three-dimensional real projective space and some lens spaces. Compactness results for the space of three-dimensional manifolds possessing isoperimetric regions with boundaries of genus higher than one are also obtained. Also an isoperimetric inequality by Hadwiger in the m-dimensional torus, as well as a description of isoperimetric candidates in quotients of $\mathbb{R}^3$ by orthogonal lattices, is given. An interesting example is the Riemannian manifold $\mathbb{S}^1 \times \mathbb{R}^m$, which exhibits isoperimetric sets which are neither spheres nor tubes for m large enough.

In Chap. 7, we consider the isoperimetric profile of a compact manifold for small volumes. They are invariant by the isometries fixing the center of mass of the isoperimetric set. This is a result by Nardulli [332] which implies a Taylor development of the isoperimetric profile for small volumes. In the second part of the chapter, isoperimetric sets of large volume in $M \times \mathbb{R}^k$, $k \geqslant 1$, are also characterized as $M \times B$, where $B \subset \mathbb{R}^k$ is a Euclidean ball (an interval in the case $k = 1$).

In Chap. 8, we review the known results related to the Cartan-Hadamard conjecture, that the Euclidean isoperimetric inequality holds on Cartan-Hadamard manifolds, and briefly discuss recent developments.

Finally, in Chap. 9, we briefly consider relative isoperimetric inequalities consisting of minimizing the relative perimeter under a volume constraint inside an open set of a Riemannian manifold. Variational formulas with boundary term are computed and a differential inequality for sets inside convex domains is obtained, proving that a certain power of the isoperimetric profile is a concave function. The characterization of stable sets in balls and properties of the isoperimetric profile in convex bodies, with non-necessarily smooth boundary, are obtained.

This work originated from the course *Mean curvature flow and isoperimetric inequalities*, published by Birkhäuser in the CRM series on advanced courses in mathematics [369]. The author wishes to thank the organizers of this course, Vicente Miquel and Joan Porti, for their interest in publishing the notes of the lectures.

In addition, there are several people I would like to warmly thank for their help and support. To all my colleagues at the Department of Geometry and Topology of the University of Granada, specially to Antonio Ros, Sebastián Montiel, Francisco J. López, and Francisco Urbano; to the members of my research group: Antonio Cañete, Ana Hurtado and César Rosales for discussions and the careful reading of large parts of the manuscript. To my collaborators Jaigyoung Choe, Mohammad Ghomi, Frank Morgan, and Gian Paolo Leonardi for sharing many ideas along time. To my students Matteo Galli, Efstratios Vernadakis, Julián Pozuelo, and Katherine Castro. To Joel Hass, Katrin Fässler, Nicola Fusco, and Benoît Kloeckner, for their encouragement and many interesting suggestions on earlier versions of this manuscript. Special thanks to Gioacchino Antonelli for his very careful reading of some parts of the manuscript.

Of course, any inaccuracy in the book is the sole responsibility of the author.

The author had support of the Junta de Andalucía grant P20_00164 and MEC-FEDER grant PID2020-118180GB-I00 during the elaboration of this work.

Granada, Spain
June 2023

Manuel Ritoré

Contents

Chapter 1
Introduction

In this chapter, we collect some basic material that will be widely used throughout this work. Some knowledge of well-known results in Riemannian geometry will be assumed. The reader will be referred to the appropriate sources when necessary, specially to the classical monographs by do Carmo [141], Warner [437], Petersen [345], Chavel [102], Lee [263], and Cheeger and Ebin [104].

After a brief introduction of the most basic concepts, we first present the Riemannian volume of a Riemannian manifold with some detail. Any submanifold of a Riemannian manifold inherits a Riemannian metric, which induces a Riemannian volume in the submanifold. This corresponds to the classical notions of length of a curve and area of a surface. Then we review the first and second variation formulas of the Riemannian volume for submanifolds, essential tools in calculus of variations. After that, we briefly review the notion and properties of sets of finite perimeter in Riemannian manifolds, which play a fundamental role in proving existence of minimizers of some variational problems including those of isoperimetric type. Some other notions of boundary area, as the Minkowski content and the Hausdorff measure, are also recalled. Finally, the notions of isoperimetric profile of a manifold and of isoperimetric set are introduced, together with some results on regularity of isoperimetric sets.

1.1 Riemannian Manifolds

A *Riemannian manifold* is a pair (M, g) consisting of a smooth (C^∞) manifold M and a tensor $g : TM \times TM \to C^\infty(M)$ giving a scalar product (symmetric and definitive positive) on every tangent space. Here TM is the tangent bundle to

M. Ritoré, *Isoperimetric Inequalities in Riemannian Manifolds*,
Progress in Mathematics 348, https://doi.org/10.1007/978-3-031-37901-7_1

M, and $C^\infty(M)$ is the set of smooth functions on M. If $p \in M$, then T_pM is the tangent space at p. We denote by m the dimension of M, and we assume, unless explicitly stated, that (M, g) is complete, connected, and without boundary. To ensure existence of partitions of unity, we also assume that M is second countable. We often refer to (M, g) simply by M and denote the Riemannian metric either by g or by $\langle \cdot, \cdot \rangle$, the latter preferred for calculations in a single manifold. The space of smooth vector fields on M, smooth sections $X : M \to TM$, is denoted by $\mathfrak{X}(M)$.

The *Levi-Civita connection* associated with g is denoted by ∇. The *exponential map* at some point $p \in M$ by $\exp_p$. The *Riemannian distance d* associated with g is computed as the infimum of the length of piecewise continuous curves connecting two given points in M. The open ball with respect to the Riemannian distance centered at $p \in M$ of radius $r > 0$ is denoted by $B(p, r)$, the closed ball by $\overline{B}(p, r)$, and the sphere by $S(p, r)$.

The *curvature operator R* is defined for vector fields X, Y, Z by

$$R(X, Y)Z = \nabla_X \nabla_Y Z - \nabla_Y \nabla_X Z - \nabla_{[X,Y]} Z.$$

The sectional curvature $K_{\text{sec}}(v, w)$ of a plane generated by the orthogonal unit vectors is $R(v, w, w, v)$. The *curvature tensor* is defined also by R but acts on four vector fields as

$$R(X, Y, Z, T) = \langle R(X, Y)Z, T \rangle.$$

The *Ricci tensor* Ric is the metric contraction of the curvature tensor

$$\operatorname{Ric}(X, Y) = \sum_{i=1}^{m} R(E_i, X, Y, E_i)$$

for any local orthonormal basis $(E_i)_i$. The *s*calar curvature of M is the trace of the Ricci tensor. For $p \in M$,

$$\operatorname{scal}(p) = \sum_{i=1}^{m} \operatorname{Ric}_p(e_i, e_i),$$

where (e_i) is any orthonormal basis of T_pM.

A *coordinate system* or *local chart* (U, φ) on M is given by an open set $U \subset M$ and a diffeomorphism φ from U onto an open set in $\mathbb{R}^m$. We say that U is a *coordinate neighborhood* and that φ is a *coordinate map*. The *coordinates* $x_1, \ldots, x_m$ associated with a coordinate system (U, φ) are the composition of φ followed by the corresponding coordinates of $\mathbb{R}^m$. A coordinate system (U, φ) is *centered* at $p \in U$ if $\varphi(p) = 0$.

For $m \geqslant 1$, the m-dimensional Euclidean space is denoted by $\mathbb{R}^m$. For $m \geqslant 2$, the m-dimensional sphere of curvature $\kappa > 0$ is the subset $\mathbb{S}^m(\kappa)$ of $\mathbb{R}^{m+1}$ defined by $\{x \in \mathbb{R}^{m+1} : ||x|| = \kappa^{-1/2}\}$, where $|| \cdot ||$ is the Euclidean norm in $\mathbb{R}^{m+1}$. When

$\kappa = 1$, we write $\mathbb{S}^m$ instead. The *circle* of radius $r > 0$ is the set $\mathbb{S}^1(r) = \{x \in \mathbb{R}^2 : ||x|| = r\}$. For $m \geqslant 2$, the m-dimensional hyperbolic space of curvature $\kappa < 0$ is denoted by $\mathbb{H}^m(\kappa)$. When $\kappa = -1$, we simply denote it by $\mathbb{H}^m$. For $m \geqslant 2$, the space $\mathbb{M}^m_\kappa$ is the m-dimensional Euclidean space if $\kappa = 0$, the sphere $\mathbb{S}^m(\kappa)$ if $\kappa > 0$, or the hyperbolic space $\mathbb{H}^m(\kappa)$ if $\kappa < 0$.

1.2 The Riemannian Volume

In this section, we introduce the Riemannian volume of Riemannian manifolds and submanifolds. This is done by transplanting the Lebesgue measure in Euclidean space to the Riemannian manifold using coordinate systems and partitions of unity.

1.2.1 Definition and Properties

Given a smooth map $F : M \to N$ between Riemannian manifolds (M, g_M), (N, g_N) of the same dimension m, its *Jacobian* at $p \in M$ is defined as the function

$$[\mathrm{Jac}(F)](p) = \sqrt{\det A(p)},$$

where $A(p)$ is the matrix whose (i, j) entry is $g_N\big(dF_p(e_i), dF_p(e_j)\big)$ and $e_1, \ldots, e_m$ is a g_M-orthonormal basis of T_pM. This definition does not depend on the choice of orthonormal basis in T_pM and coincides with the classical definition of Jacobian for smooth maps between Euclidean spaces.

Given a Riemannian manifold (M, g) and a continuous function f whose support $\mathrm{supp}(f)$ (the closure of the set $\{p \in M : f(p) \neq 0\}$) is compact and contained in the coordinate neighborhood U of a coordinate system (U, φ) in M, we can define the integral of f in M by the equality

$$\int_{\varphi(U)} (f \circ \varphi^{-1})\, \mathrm{Jac}(\varphi^{-1})\, d\mathcal{L}^m,$$

where $\mathcal{L}^m$ denotes Lebesgue's measure in $\mathbb{R}^m$ and the Jacobian is computed using the standard Riemannian metric in $\mathbb{R}^m$. The classical area formula in Euclidean space easily implies that this definition is independent of the coordinate system. If we express the Riemannian metric in the coordinate system as the matrix (g_{ij}), then we have

$$\mathrm{Jac}(\varphi^{-1}) = \sqrt{\det(g_{ij})}.$$

The integral of a continuous function $f : M \to \mathbb{R}$ with compact support can be defined by covering the support of f with a finite number of local charts (U_i, φ_i), $i = 1, \ldots, k$. Taking smooth non-negative functions $\rho_1, \ldots, \rho_k$ so that $\mathrm{supp}(\rho_i) \subset U_i$ and $\rho_1 + \cdots + \rho_k = 1$ on $\mathrm{supp}(f)$, we can define the integral of f on M as

$$\Lambda(f) = \sum_{i=1}^{n} \int_{\varphi_i(U_i)} (f\rho_i \circ \varphi^{-1}) \,\mathrm{Jac}(\varphi_i^{-1})\, d\mathcal{L}^m.$$

This quantity is independent of the choice of the covering and the partition of unity. Hence Λ defines a positive functional in the space $C_0(M)$ of continuous functions $f : M \to \mathbb{R}$ with compact support on M. By the Riesz representation Theorem (see Theorem 2.14 in [387]), there exists on M a unique positive Borel measure μ_g such that

$$\Lambda(f) = \int_M f \, d\mu_g.$$

In addition, the measure μ_g is regular and complete, and the measure of compact sets is finite. By regular we mean that the measure of a set E can be approximated by the one of open sets containing E or the one of compact sets contained in E and by complete that subsets of measure zero sets are measurable.

We shall often use the notation $dM = d\mu_g$ and write the integral of f with respect to the Riemannian measure μ_g as

$$\int_M f \, dM.$$

The spaces $L^p(M, g)$ are defined the usual way with respect to the Riemannian measure associated with the Riemannian metric g on M. We shall often omit the reference to the Riemannian metric and simply write $L^p(M)$. The Riemannian volume of a measurable set $E \subset M$ is denoted by $|E|$.

The Area Formula

The following particular case of the area formula for smooth maps between Riemannian manifolds follows easily from the definition of the Riemannian measure using partitions of unity.

Theorem 1.1 (Area Formula) *Let $F : M \to N$ be a diffeomorphism between Riemannian manifolds. Then we have*

$$\int_N f \, dN = \int_M (f \circ F) \,\mathrm{Jac}(F)\, dM \tag{1.1}$$

for any $f \in L^1(N)$.

1.2.2 The Coarea Formula

Now we briefly recall the coarea formula between Riemannian manifolds. Given a C^1 map $F : M \to N$ between Riemannian manifolds, let $\mathcal{C}(F)$ be the set of *critical points* $p \in M$ of F where dF_p is not a surjective map. The set of *regular points* $M \setminus \mathcal{C}(F)$ is composed of those points where dF_p is surjective. This set only exists when $m \geqslant n$ and it is an open subset of M. When p is a regular point of F, the set $F^{-1}(F(p))$ is locally a submanifold around p of dimension $m - n$.

Let $\mathcal{S}(F)$ be the set $F(\mathcal{C}(F))$ of singular values of F. Morse-Sard Theorem (see Theorem 1.3 in Chapter 3 of [230]) implies

Theorem 1.2 *Let $F : M \to N$ be a C^r map between Riemannian manifolds of dimension m, n such that $r > \max\{0, m - n\}$. Then $\mathcal{S}(F)$ has measure 0 in N.*

As for the coarea formula, we have

Theorem 1.3 (Coarea Formula) *Let $F : M \to N$ be a smooth map between Riemannian manifolds, with $m \geqslant n$. Let $h : M \to \mathbb{R}$ be a non-negative measurable function. Then*

$$\int_M h \operatorname{Jac}^{\perp}(F)\, dM = \int_N \left\{ \int_{F^{-1}(y)} h\, dF^{-1}(y) \right\} dN, \tag{1.2}$$

where $\operatorname{Jac}(h)^{\perp}$ is the normal Jacobian and the determinant of dF restricted to the subspace orthogonal to $\ker(dF)$.

If in addition $\mathcal{C}(F)$ has measure 0 in M, then the following formula also holds:

$$\int_M h\, dM = \int_N \left\{ \int_{F^{-1}(y)} \frac{h}{\operatorname{Jac}^{\perp}(F)}\, dF^{-1}(y) \right\} dN. \tag{1.3}$$

Proof Formula (1.2) is obtained from the Euclidean coarea formula taking local coordinates. Formula (1.3) as in the proof of Proposition 2.2 in [90]. □

1.2.3 Riemannian Volume of Submanifolds

The Riemannian measure can be induced in submanifolds of M. We briefly recall the notion of submanifold of a smooth manifold.

Definition 1.4 Given an m-dimensional manifold M and a non-negative integer $k \leqslant m$, a subset $S \subset M$ is a k-dimensional *submanifold* of M when, for every $p \in S$, there exists a coordinate system (U, φ) centered at p such that $U \cap S$ is, in coordinates in (U, φ), the set $x_{k+1} = \cdots = x_m = 0$.

We refer the reader to Chapter 1 in Warner's monograph [437] for basic properties of submanifolds with the warning that our notion of submanifold corresponds to Warner's notion of imbedding.

It is straightforward to check that a k-dimensional submanifold S becomes a k-dimensional manifold. We define a *curve* in a smooth manifold M as a 1-dimensional submanifold, a *surface* as a 2-dimensional submanifold, and a *hypersurface* as a submanifold of dimension $\dim(M) - 1$.

If (M, g) is a Riemannian manifold, then g restricted to tangent vectors to S makes S a Riemannian manifold. We denote by dS the Riemannian volume element in S. Given a measurable set $\Omega \subset S$, we denote its Riemannian measure by $A_S(\Omega)$, and we refer to it as the area of Ω (as a k-dimensional submanifold). We shall often omit the subscript S and simply write $A(\Omega)$. When $S = M$, we shall refer to the Riemannian volume of $\Omega \subset M$ simply as the volume of Ω, and we denote it by $|\Omega|$ or $V(\Omega)$.

The m-dimensional Riemannian volume of $\mathbb{S}^m$ will be denoted by $\mathbf{c}_m$ (see [102, p. 125 ff.]). The volume of a unit ball in $\mathbb{R}^m$ is denoted by ω_m. It follows from the coarea formula that

$$\mathbf{c}_{m-1} = m\omega_m. \tag{1.4}$$

If $B(r) \subset \mathbb{R}^m$ is any Euclidean ball of radius $r > 0$, then $A(\partial B(r)) = \mathbf{c}_{m-1} r^{m-1}$ and $|B(r)| = \omega_m r^m$. We define the constant $\mathbf{c}(m)$ by

$$\mathbf{c}(m) = \frac{\mathbf{c}_{m-1}}{\omega_m^{m/(m-1)}} = \frac{A(B(r))}{|B(r)|^{m/(m-1)}}, \tag{1.5}$$

which is called, for reasons that will be seen later, the *isoperimetric constant* of $\mathbb{R}^m$. Note that

$$\mathbf{c}(m) = m\omega_m^{1/m}.$$

1.2.4 The Divergence Theorem

We say that an open set Ω in a smooth manifold M has *smooth boundary* if, for every $p \in \partial\Omega$, there exists a coordinate system (U, φ) centered at p such that

$$U \cap \overline{\Omega} = \{p \in U : x_m(p) \geqslant 0\}. \tag{1.6}$$

This property implies $U \cap \Omega = \{p \in U : x_m(p) > 0\}$ and $U \cap \partial\Omega = \{p \in M : x_m(p) = 0\}$. The boundary of an open set Ω with smooth boundary is a hypersurface of M. Coordinate systems satisfying (1.6) are called *adapted to the boundary* $\partial\Omega$.

If M is a Riemannian manifold, there exists a unique vector field $N_{\partial\Omega} : \partial\Omega \to TM$ on $\partial\Omega$ such that $N_{\partial\Omega}$ has modulus one, is orthogonal to $T(\partial\Omega)$, and points *outside* Ω. This vector field is the *outer unit normal* to $\partial\Omega$.

We define now the divergence of a vector field on a submanifold. This notion is much more general than the one we required for the divergence theorem, but we shall need it when computing the first variation formula.

Definition 1.5 Given a k-dimensional submanifold S in M and a vector field X on M, we define the divergence of X in S as the function

$$(\operatorname{div}_S X)(p) = \sum_{i=1}^{k} \langle \nabla_{e_i} X, e_i \rangle, \tag{1.7}$$

where (e_i) is an orthonormal basis of T_pS.

It is immediate to check that the definition does not depend on the choice of the orthonormal basis on T_pS. Of course, it can be defined for smooth vector fields $X : S \to TM$ defined along the submanifold S. In case $S = M$, we denote div_M simply by div.

With these ingredients, we have the following:

Theorem 1.6 (Divergence Theorem) *Let $\Omega \subset M$ be an open subset with smooth boundary and X a smooth vector field on M. Assume that Ω is bounded or that X has compact support on M. Then we have*

$$\int_\Omega \operatorname{div} X \, dM = \int_{\partial\Omega} \langle X, \nu \rangle \, d(\partial\Omega), \tag{1.8}$$

where ν is the outer unit normal to $\partial\Omega$.

Theorem 1.6 is proven taken local coordinates systems (U, φ) so that either $U \subset \Omega$ or (U, φ) is adapted to the boundary. We extract a finite subcovering by cubes of the support of X (if X has compact support) or of $\overline{\Omega}$ (if $\overline{\Omega}$ is compact), prove (1.8) on each of these cubes, and use a partition of unity to get the global result. The proof is modelled on the proof of Stokes' Theorem (see §5 in Chapter V of Matsushima's monograph [278]). The divergence theorem can be directly obtained from Stokes' if M is an oriented manifold (see §V.7 in [278]).

Definition 1.7 Given a Riemannian manifold (M, g) and a function $f \in C^\infty(M)$, its *Laplacian* Δf is defined as

$$\Delta f = \operatorname{div}(\nabla f),$$

where ∇f is the *gradient* of f, defined as the only vector field satisfying $g(\nabla f, X) = X(f)$ for any smooth vector field $X \in \mathfrak{X}(M)$.

Is $S \subset M$ is a submanifold of M, then we denote by Δ_S and ∇_S the Laplacian and gradient on S.

1.2.5 The Doubling Property

A *metric measure space* (X, d, μ) is a metric space (X, d) together with a Borel regular measure μ on X. It is said that (X, d, μ) satisfies the *doubling property* in a subset $K \subset X$ if there exist constants $C \geqslant 1$ and $r_0 > 0$ such that

$$\mu(B(x, 2r)) \leqslant C\,\mu(B(x, r)) \tag{1.9}$$

for all $x \in K$ and $0 < r \leqslant r_0$, compare with [213] or page 76 in [224].

In a Riemannian manifold, we have the following doubling property

Theorem 1.8 (Doubling Property on Compact Subsets) *Let (M, g) be an m-dimensional Riemannian manifold and $K \subset M$ a compact subset. Then there exist constants $C_D \geqslant 1$ and $r_0 > 0$, only depending on K, such that for every $p \in K$ and $0 < r \leqslant r_0$ we have*

$$|B\,(p, 2r)\,| \leqslant C_D |B\,(p, r)\,|. \tag{1.10}$$

In particular, (1.10) *holds in the whole manifold if M is compact.*

Proof For every $p \in M$, there exist $\delta_p > 0$ and a totally normal neighborhood W_p of p, (i.e., such that $\exp_q$ is a diffeomorphism on $B(0, \delta_p) \subset T_qM$, and $W_p \subset \exp_q(B(0, \delta_p))$ for all $q \in W_p$) (see Theorem 3.7 in [141]). Then $(B(q, \delta_p), \exp_q^{-1})$ is a local chart on M.

As W_p is contained in a coordinate neighborhood, we can take an orthonormal frame on W_p. Let (g_{ij}^q) be the expression of the Riemann tensor in the normal coordinate system $(B(q, \delta_p), \exp_q^{-1})$ associated with this frame. The functions g_{ij}^q are differentiable functions of q.

Let V_p be a bounded open neighborhood of p such that $\overline{V}_p \subset W_p$. Then there are positive constants c_p, C_p such that the inequalities

$$c_p \leqslant \sqrt{\det(g_{ij}^q)} \leqslant C_p$$

hold on $\overline{B}(0, \delta_p/2)$ for all $q \in \overline{V}_p$. Integrating on $B(0, r) \subset T_qM$ for any $0 < r \leqslant \delta_p/2$, we have

$$c_p\,\mathcal{L}^m(B(0, r)) \leqslant |B(q, r)| \leqslant C_p\,\mathcal{L}^m(B(0, r)),$$

where $\mathcal{L}^m$ is the Lebesgue measure in T_qM. So we have

$$\frac{|B(q, 2r)|}{|B(q, r)|} \leqslant \frac{C_p\,\mathcal{L}^m(B(0, 2r))}{c_p\,\mathcal{L}^m(B(0, r))} = \frac{C_p}{c_p}2^m \tag{1.11}$$

for all $0 < r \leqslant \delta_p/4$ and $q \in \overline{V}_p$. Note that $2^m(C_p/c_p) \geqslant 1$.

If K is compact, for every $p \in K$, we take $\delta_p > 0$ and V_p as in the previous construction. Let $F \subset K$ be a finite subset so that $K \subset \bigcup_{p \in F} V_p$. Then inequality (1.10) follows from (1.11) taking $r_0 = \min\{\delta_p/4 : p \in F\}$ and $C_D = \min\{2^m C_p/c_p : p \in F\}$. □

1.3 Variation Formulas for the Riemannian Volume of Submanifolds

In this section, we compute the first and second variation of the Riemannian volume of a k-dimensional submanifold in a Riemannian manifold M. We first introduce the deformations we use to obtain these formulas.

1.3.1 Deformations

Definition 1.9 Given a manifold M, a one-parameter *family* of diffeomorphisms in M is a smooth map $\varphi : \mathbb{R} \times M \to M$ such that the maps $\varphi_t : M \to M$, defined by $\varphi_t(p) = \varphi(t, p)$ for all $t \in \mathbb{R}$, $p \in M$, satisfy

1. $\varphi_0 = \mathrm{Id}_M$
2. φ_t is a diffeomorphism of M for all $t \in \mathbb{R}$.

In addition, we say that the family of diffeomorphisms has compact support if there exists a compact subset $K \subset M$ such that $\varphi_t(p) = p$ for all $t \in \mathbb{R}$ and $p \in M \setminus K$.

We denote a one-parameter family of diffeomorphisms by $\{\varphi_t\}_{t\in\mathbb{R}}$. The *variational vector field* X or *velocity* associated with $\{\varphi_t\}_{t\in\mathbb{R}}$ is defined by

$$X_p = \left.\frac{d}{dt}\right|_{t=0} \varphi_t(p).$$

That is, X_p is the tangent vector to the curve $t \mapsto \varphi_t(p)$ at $t = 0$.

A special type of one-parameter families of diffeomorphisms is the one induced by a vector field on M.

Definition 1.10 Given a manifold M, a one-parameter *group* of diffeomorphisms in M is a smooth map $\varphi : \mathbb{R} \times M \to M$ such that the maps $\varphi_t : M \to M$, defined by $\varphi_t(p) = \varphi(t, p)$ for all $t \in \mathbb{R}$, $p \in M$, satisfy

1. $\varphi_0 = \mathrm{Id}_M$,
2. $\varphi_{s+t} = \varphi_s \circ \varphi_t$ for all $s, t \in \mathbb{R}$.

Moreover, we say that the one-parameter group has compact support if there exists a compact subset $K \subset M$ such that $\varphi_t(p) = p$ for all $t \in \mathbb{R}$ and $p \in M \setminus K$.

Conditions 1 and 2 in Definition 1.10 imply that φ_t is a diffeomorphism of M for all $t \in \mathbb{R}$ with inverse $\varphi_t^{-1} = \varphi_{-t}$. In particular, a one-parameter group of diffeomorphisms is a one-parameter family of diffeomorphisms.

A convenient way to produce one-parameter groups of diffeomorphisms is by using vector fields in M with compact support. Given $X \in \mathfrak{X}_0(M)$, we consider the *trajectories* or *integral curves* of X, the curves $\sigma_p : I \to M$ defined for all $p \in M$ by solving the ordinary differential equation $\sigma'(t) = X_{\sigma(t)}$ with initial condition $\sigma(0) = p$. When X has compact support, the curves σ_p are defined for all $t \in \mathbb{R}$, and the map $\varphi(t, p) = \sigma_p(t)$ is a one-parameter group of diffeomorphisms with compact support (see Chapter 1 in [438]). In this case, we say that φ is the *flow* associated with X.

1.3.2 The First Variation Formula

We derive here the first variation formula of the Riemannian volume for a submanifold of M. Our computations follow the ones by Simon in the Euclidean case (see §9 in Chapter 2 of [404]).

Theorem 1.11 (First Variation of the Riemannian Volume) *Let S be a k-dimensional submanifold of M and $\{\varphi_t\}_{t\in\mathbb{R}}$ a one-parameter family of diffeomorphisms of M with compact support. Let X be the associated variational vector field. Let $S' \subset S$ be a relatively compact subset of S. Then we have*

$$\left.\frac{d}{dt}\right|_{t=0} A(\varphi_t(S')) = \int_{S'} (\operatorname{div}_S X)\, dS. \tag{1.12}$$

Proof For t fixed, the restriction $\varphi_t|_S : S \to \varphi_t(S)$ is a diffeomorphism. By the area formula (1.1)

$$A(\varphi_t(S')) = \int_{S'} \operatorname{Jac}(\varphi_t|_S)\, dS. \tag{1.13}$$

Recall that the Jacobian of the map $\varphi_t|_S$ is defined, at any point $p \in S$, by

$$\operatorname{Jac}(\varphi_t|_S)(p) = \det(A(t, p))^{1/2},$$

where $A(t, p)$ is the square matrix of order k defined by

$$A(t, p) = \big(\langle (d\varphi_t)_p(e_i), (d\varphi_t)_p(e_j)\rangle\big)_{i,j}$$

for some orthonormal basis (e_i) of T_pS.

Differentiation in (1.13) under the integral sign provides

$$\frac{d}{dt}\bigg|_{t=0} A(\varphi_t(S')) = \int_{S'} \left(\frac{d}{dt}\bigg|_{t=0} \operatorname{Jac}(\varphi_t|_S)(p) \right) dS(p).$$

To compute the derivative of the Jacobian, we make use of Jacobi's formula: if $B(t)$ is a smooth curve in the space of non-singular matrices, we have

$$\frac{d}{dt} \det B = (\det B) \operatorname{trace}\left(B^{-1} \frac{dB}{dt} \right). \tag{1.14}$$

Using this formula with $B(t) = A(t, p)$, since $A(0, p)$ is the identity matrix, we get

$$\frac{d}{dt}\bigg|_{t=0} \operatorname{Jac}\varphi_t(p) = \frac{1}{2} \frac{d}{dt}\bigg|_{t=0} \det A(t,p) = \frac{1}{2} \operatorname{trace}\left(\frac{d}{dt}\bigg|_{t=0} A(t,p) \right) = \frac{1}{2} \sum_{i=1}^{k} a'_{ii}(0),$$

where $a_{ii}(t) = \langle (d\varphi_t)_p(e_i), (d\varphi_t)_p(e_i) \rangle$.

In order to compute $a'_{ii}(0)$, we consider the curve $\beta(t) = \varphi_t(p)$ and the smooth vector field $E_i(t) = (d\varphi_t)_p(e_i)$ along β. From Lemma 1.13, we obtain

$$\frac{D}{dt}\bigg|_{t=0} E_i(t) = \nabla_{e_i} X,$$

where D/dt is the standard covariant derivative along the curve β. Hence

$$\sum_{i=1}^{k} a'_{ii}(0) = \sum_{i=1}^{k} \frac{d}{dt}\bigg|_{t=0} \langle E_i(t), E_i(t) \rangle = \sum_{i=1}^{k} 2 \langle \nabla_{e_i} X, e_i \rangle = 2 \operatorname{div}_S X, \tag{1.15}$$

as desired. □

Remark 1.12 There is some flexibility in the hypotheses of Theorem 1.11. More precisely

1. For the proof of Theorem 1.11, we only need that φ be defined in $(-\varepsilon, \varepsilon) \times M$, for t in a small neighborhood of 0.
2. The hypothesis "X with compact support" can be omitted if we assume that S is compact. We simply modify X outside a bounded neighborhood of S to have compact support.

Lemma 1.13 *Let $\{\varphi_t\}_{t\in\mathbb{R}}$ be a one-parameter family of diffeormorphisms in M with velocity X. Let $p \in M$ and $e \in T_pM$. Define the smooth curve $\beta(t) = \varphi_t(p)$ and the smooth vector field $E(t) = (d\varphi_t)_p(e)$ along β. Then we have*

$$\left.\frac{D}{dt}\right|_{t=0} E(t) = \nabla_e X,$$

where D/dt is the covariant derivative along the curve β.

Proof We take a coordinate system (U, x) so that $p \in U$. By the continuity of $\varphi : \mathbb{R} \times M \to M$, there exists $\varepsilon > 0$ and an open neighborhood $V \subset U$ of p such that $\varphi_t(q) = \varphi(t, q)$ belongs to U for every $t \in (-\varepsilon, \varepsilon)$ and $q \in V$. Hence for all $t \in (-\varepsilon, \varepsilon)$, we have $\varphi_t(V) \subset U$.

In local coordinates, the vector field X takes the form

$$X = \sum_{i=1}^{m} f_i \frac{\partial}{\partial x_i}$$

for some functions $f_i \in C^\infty(U)$. Moreover, as $\varphi_t(U) \subset V$ for $t \in (-\varepsilon, \varepsilon)$, we can define $\varphi_i = x_i \circ \varphi$ for all $i = 1, \ldots, m$ and compute $(d\varphi_t)_p$ in the coordinate systems (V, x) and (U, x) as

$$(d\varphi_t)_p = \begin{pmatrix} \frac{\partial \varphi_1}{\partial x_1} & \cdots & \frac{\partial \varphi_1}{\partial x_m} \\ \vdots & \ddots & \vdots \\ \frac{\partial \varphi_m}{\partial x_1} & \cdots & \frac{\partial \varphi_m}{\partial x_m} \end{pmatrix} (t, p), \qquad t \in (-\varepsilon, \varepsilon). \tag{1.16}$$

Writing $e = \sum_{i=1}^m e_i (\frac{\partial}{\partial x_i})_p$, the coordinates of $E(t) = (d\varphi_t)_p(e)$ on the basis $\{(\partial/\partial x_i)_{\beta(t)}\}$ are given by

$$\left[\begin{pmatrix} \frac{\partial \varphi_1}{\partial x_1} & \cdots & \frac{\partial \varphi_1}{\partial x_m} \\ \vdots & \ddots & \vdots \\ \frac{\partial \varphi_m}{\partial x_1} & \cdots & \frac{\partial \varphi_m}{\partial x_m} \end{pmatrix} (t, p)\right] \begin{pmatrix} e_1 \\ \vdots \\ e_m \end{pmatrix}.$$

Hence

$$E(t) = \sum_{i=1}^{m} \left(\sum_{j=1}^{m} e_j \frac{\partial \varphi_i}{\partial x_j}(t, p) \right) \left(\frac{\partial}{\partial x_i} \right)_{\beta(t)} = \sum_{i=1}^{m} g_i(t) \left(\frac{\partial}{\partial x_i} \right)_{\beta(t)},$$

where

$$g_i(t) = \sum_{j=1}^{m} e_j \frac{\partial \varphi_i}{\partial x_j}(t, p).$$

So we have

$$\left.\frac{D}{dt}\right|_{t=0} E(t) = \sum_{i=1}^{m} g_i'(0)\left(\frac{\partial}{\partial x_i}\right)_p + \sum_{i=1}^{m} g_i(0)\nabla_{X_p}\frac{\partial}{\partial x_i}. \tag{1.17}$$

Letting $\beta_q(t) = (\varphi_1(t,q),\ldots,\varphi_m(t,q))$, we have $\beta_q'(0) = X_q$ for $t \in (-\varepsilon,\varepsilon)$, $q \in V$, and we get

$$\sum_{i=1}^{m} \frac{\partial\varphi_i}{\partial t}(0,q)\left(\frac{\partial}{\partial x_i}\right)_q = \sum_{i=1}^{m} f_i(q)\left(\frac{\partial}{\partial x_i}\right)_q.$$

Hence $(\partial\varphi_i/\partial t)(0,q) = f_i(q)$ for all $q \in V$ and

$$g_i'(0) = \sum_{j=1}^{m} e_j \frac{\partial^2\varphi_i}{\partial x_j \partial t}(0,p) = \sum_{j=i}^{m} e_j \frac{\partial f_i}{\partial x_j}(p) = e(f_i),$$

As $g_i(0) = e_i$, we get from (1.17) and the previous computation

$$\left.\frac{D}{dt}\right|_{t=0} E(t) = \sum_{i=1}^{m}\left(e(f_i)\left(\frac{\partial}{\partial x_i}\right)_p + e_i\nabla_{X_p}\frac{\partial}{\partial x_i}\right).$$

On the other hand

$$\nabla_e X = \sum_{i=1}^{m}\left(e(f_i)\left(\frac{\partial}{\partial x_i}\right)_p + f_i(p)\nabla_e\frac{\partial}{\partial x_i}\right).$$

As

$$\begin{aligned}\sum_{i=1}^{m} e_i\nabla_{X_p}\frac{\partial}{\partial x_i} &= \sum_{i,j=1}^{m} e_i f_j(p)\nabla_{(\partial/\partial x_j)_p}\frac{\partial}{\partial x_i}\\ &= \sum_{i,j=1}^{m} e_i f_j(p)\nabla_{(\partial/\partial x_i)_p}\frac{\partial}{\partial x_j} = \sum_{j=1}^{m} f_j(p)\nabla_e\frac{\partial}{\partial x_j},\end{aligned}$$

we get the desired result. □

Remark 1.14 Under the hypotheses of Lemma 1.13, we get

$$\frac{D}{dt}E(t) = \nabla_{E(t)}X_t,$$

where X_t is the vector field associated with the one-parameter family of diffeomorphisms $\{\psi_s\}_{s\in\mathbb{R}} = \{\varphi_{s+t} \circ \varphi_t^{-1}\}_{s\in\mathbb{R}}$. This is easily checked since $\psi_s(\varphi_t(p)) = \varphi_{s+t}(p)$ and

$$(d\psi_s)_{\varphi_t(p)}(E(t)) = (d\varphi_{s+t})_p(e) = E(s+t).$$

As $(D/dt)_{t=t_0}E(t) = (D/ds)_{s=0}E(s+t_0)$, the result follows.

Remark 1.15 Observe that the first variation formula evaluated at $t = 0$ is obtained in terms of the vector field X, the derivative at $t = 0$ of the one-parameter family $\{\varphi_t\}_{t\in\mathbb{R}}$. If we wish to compute the first derivative of the Riemannian area at any other $t_0 \in \mathbb{R}$, we can reduce ourselves to the previous case by considering the one-parameter family of diffeomorphisms $\{\varphi_{t+t_0} \circ \varphi_{t_0}^{-1}\}_{t\in\mathbb{R}}$, with associated vector field X_{t_0}, different from X in general. However, when $\{\varphi_t\}_{t\in\mathbb{R}}$ is a one-parameter *group*, then $\varphi_{t+t_0} \circ \varphi_{t_0}^{-1} = \varphi_t$ and $X_{t_0} = X$ for all $t_0 \in \mathbb{R}$. Examples where $X_{t_0} \neq X$ can be easily constructed taking a smooth function $f : \mathbb{R} \to \mathbb{R}^+$, and the diffeomorphisms $\varphi_t : \mathbb{R}^m \to \mathbb{R}^m$ defined as $\varphi_t(p) = f(t)p$ for all $p \in \mathbb{R}^m$ and $t \in \mathbb{R}$. Then

$$X_{t_0} = \frac{f'(t_0)}{f(t_0)} \sum_{i=1}^{m} x_i \frac{\partial}{\partial x_i}.$$

The Mean Curvature Vector

One advantage of the first variation formula (1.12) is that it is valid even for submanifolds with boundary. Its major disadvantage is that the geometry of the submanifold is not involved in the formula.

To write a formula equivalent to (1.12) when $k < \dim(M)$, but involving the geometry of S, we first give the following:

Definition 1.16 Given a k-dimensional submanifold S of a Riemannian manifold, with $k < \dim(M)$ and a local orthonormal basis $\nu_{k+1}, \ldots, \nu_m$ of the normal bundle to S, the *mean curvature vector* $\vec{H}_S$ of S is defined locally by

$$\vec{H}_S = \sum_{j=k+1}^{m} \left(\mathrm{div}_S\, \nu_j\right) \nu_j. \tag{1.18}$$

This expression is independent of the local orthonormal frame generating locally the normal bundle.

The first variation formula of a submanifold S is computed, on a bounded set $S' \subset S$ as $\int_{S'} (\mathrm{div}_S\, X)\, dS$, where X is the variational vector field. We decompose X over S as $X = X^\top + X^\perp$, where $X^\perp$ is the orthogonal projection of X to the tangent subbundle TS and $X^\top$ is the projection to the normal subbundle $TS^\perp$. Choosing a local orthonormal basis (ν_j), $j = 1, \dots, m-k$, of the normal subbundle $TS^\perp$, we have $X^\perp = \sum_{j=1}^{m-k} \langle X, \nu_j \rangle \nu_j$ and so

$$\operatorname{div}_S X^\perp = \sum_{j=1}^{m-k} \langle X, \nu_j \rangle \operatorname{div}_S \nu_j = \langle X, \sum_{j=1}^{m-k} (\operatorname{div}_S \nu_j) \nu_j \rangle.$$

Hence we obtain

$$\operatorname{div}_S X = \operatorname{div}_S X^\top + \operatorname{div}_S X^\perp = \operatorname{div}_S X^\top + \langle X, \vec{H}_S \rangle.$$

The first term can be computed using the divergence theorem in S. Hence we obtain from Theorem 1.11 the following:

Theorem 1.17 *Let S be a k-dimensional submanifold of M and $\{\varphi_t\}_{t\in\mathbb{R}}$ a one-parameter family of diffeomorphisms of M with compact support. Let X be the associated variational vector field. Let $S' \subset S$ be a relatively compact open subset of S with smooth boundary. Then we have*

$$\left.\frac{d}{dt}\right|_{t=0} A(\varphi_t(S')) = \int_{S'} \langle X, \vec{H}_S \rangle \, dS + \int_{\partial S'} \langle X, \eta \rangle \, d(\partial S'), \tag{1.19}$$

where η is the outer conormal to S' and $\vec{H}_S$ is the mean curvature vector of S. In particular, if $\operatorname{supp}(X) \cap \partial S' = \emptyset$ then

$$\left.\frac{d}{dt}\right|_{t=0} A(\varphi_t(S')) = \int_{S'} \langle X, \vec{H}_S \rangle \, dS. \tag{1.20}$$

In the case of a hypersurface with a globally defined unit normal vector, we can define its scalar mean curvature.

Definition 1.18 The *scalar mean curvature* of a hypersurface S with respect to a unit normal vector N is the function

$$H = \operatorname{div}_S N. \tag{1.21}$$

Given an orthonormal basis $e_1, \dots, e_{m-1}$ of T_pS, the scalar mean curvature $H(p)$ can be computed as

$$H(p) = \sum_{i=1}^{m-1} \langle \nabla_{e_i} N, e_i \rangle.$$

Observe that taking the unit normal $-N$ instead of N will change the sign of the scalar mean curvature. In any case, we have

$$\vec{H}_S = HN$$

for the mean curvature H with respect to the normal N. It is clear that, given $e \in T_pS$, the vector $\nabla_e N$ is orthogonal to N_p and hence tangent to S. The endomorphism of T_pS defined by $e \mapsto \nabla_e N$ is self-adjoint, and it is called the *Weingarten endomorphism* or the *shape operator* of S. Its eigenvalues $\kappa_1, \ldots, \kappa_{m-1}$ are called *principal curvatures* of S and the associated eigenvectors *principal directions*. With this notation, it is clear that

$$H(p) = \sum_{i=1}^{m-1} \kappa_i.$$

Given $e, v \in T_pS$, the bilinear form

$$\sigma(e, v) = \langle \nabla_e N, v \rangle \tag{1.22}$$

is the *second fundamental form* of S. The quantity

$$|\sigma|^2(p) = \sum_{i=1}^{m-1} \kappa_i^2 \tag{1.23}$$

is called the *norm of the second fundamental form* at $p \in S$. We trivially have

$$|\sigma|^2 = \frac{H^2}{m-1} + \frac{\sum_{i<j}(\kappa_i - \kappa_j)^2}{m-1} \geqslant \frac{H^2}{m-1}. \tag{1.24}$$

Equality holds in (1.24) if and only if $\kappa_1 = \kappa_2 = \cdots = \kappa_{m-1}$ (i.e., if the hypersurface S is *totally umbilical*).

Theorem 1.17 does not apply when $k = \dim M$ (i.e., when S is an open subset of M). In this case the variation of the volume is easily computed from (1.12) and the divergence theorem in M.

Proposition 1.19 *Let $\Omega \subset M$ be an open set with smooth boundary and $\{\varphi_t\}_{t\in\mathbb{R}}$ a one-parameter family of diffeomorphism with compact support and variational vector field X. Then*

$$\left.\frac{d}{dt}\right|_{t=0} |\varphi_t(\Omega)| = \int_{\partial\Omega} \langle X, N \rangle \, d(\partial\Omega), \tag{1.25}$$

where N is the outer unit normal to $\partial\Omega$.

1.3.3 The Second Variation Formula

We compute in this section the second derivative of the area functional. We recall (see Remark 1.15) that given a one-parameter family of diffeomorphisms $\{\varphi_t\}_{t\in\mathbb{R}}$ and $t_0 \in \mathbb{R}$, we can define the vector field X_{t_0} at $p \in M$ by means of the equality

$$(X_{t_0})_p = \frac{d}{dt}\bigg|_{t=t_0} (\varphi_{t+t_0} \circ \varphi_{t_0}^{-1})(p).$$

If $t_0 = 0$ then X_0 coincides with the vector field X associated with $\{\varphi_t\}_{t\in\mathbb{R}}$. If $\{\varphi_t\}_{t\in\mathbb{R}}$ is a one-parameter *group* of diffeomorphisms, then $\varphi_{t+t_0} \circ \varphi_{t_0}^{-1} = \varphi_t$ and so $X_{t_0} = X$ for all $t \in \mathbb{R}$.

Let us fix $t_0 \in \mathbb{R}$ and $p \in M$. We consider the curve $\beta(t) = \varphi_t(p)$. The vector $(X_{t_0})_{\varphi_{t_0}(p)}$ is the tangent vector to the curve $t \mapsto \varphi_{t+t_0}(p) = \beta(t+t_0)$ at $t = 0$. Hence

$$\beta'(t) = (X_t)_{\beta(t)}, \qquad \text{for all } t \in \mathbb{R}.$$

Given a one-parameter family of diffeomorphisms $\{\varphi_t\}_{t\in\mathbb{R}}$, we define the vector field X' by

$$(X')_p = \lim_{t\to 0} \frac{(X_t)_p - X_p}{t}. \tag{1.26}$$

Taking a coordinate system (U, x) such that $p \in U$, we can express the vector fields X_t in local coordinates

$$(X_t)_q = \sum_{i=1}^m f_i(t,q)\left(\frac{\partial}{\partial x_i}\right)_p, \qquad q \in U,$$

so that

$$(X')_q = \sum_{i=1}^m \frac{\partial f_i}{\partial t}(0,q)\left(\frac{\partial}{\partial x_i}\right)_q.$$

We shall need the following result in the proof of the second variation.

Lemma 1.20 *Let $B : I \subset \mathbb{R} \to GL(k,\mathbb{R})$ be a smooth map from an open interval $I \subset \mathbb{R}$ into the special linear group of invertible matrices of order k. Then we have*

$$(\det B)'' = (\det B)\, trace(B^{-1}B')^2 + (\det B) trace((B^{-1})'B' + B^{-1}B''). \tag{1.27}$$

If $B(0) = Id_k$ *then*

$$(\det B)''(0) = trace(B'(0))^2 + trace(-(B'(0))^2 + B''(0)). \tag{1.28}$$

Proof We differentiate Jacobi's identity (1.14) to obtain (1.27). To get (1.28) we simply take into account that $(B^{-1})' = -B^{-1}B'B^{-1}$. □

Theorem 1.21 (Second Variation of the Riemannian Volume) *Let S be a k-dimensional submanifold of M and $\{\varphi_t\}_{t\in\mathbb{R}}$ a one-parameter family of diffeomorphisms of M with compact support with associated vector field X. Let $S' \subset S$ be a bounded set in S. Then we have*

$$\begin{aligned}\frac{d^2}{dt^2}\bigg|_{t=0} A(\varphi_t(S')) = \int_{S'} \Big\{ &\operatorname{div}_S X' + \operatorname{div}_S Z + (\operatorname{div}_S X)^2 - \sum_{i=1}^{k} \langle R(e_i, X_p)X_p, e_i\rangle \\ &+ \sum_{i=1}^{k} |(D_{e_i}X)^{\perp}|^2 - \sum_{i,j=1}^{k} \langle D_{e_i}X, e_j\rangle\langle D_{e_j}X, e_i\rangle \Big\}\, dS,\end{aligned} \tag{1.29}$$

where $Z = \nabla_X X$ is the acceleration of X, $e_1, \ldots, e_k$ is an orthonormal basis of T_pS, R is the curvature operator in M, and X' the vector field defined in (1.26).

Proof We proceed as in the proof of the first variation formula expressing the area of $\varphi_t(S')$ as

$$A(\varphi_t(S')) = \int_{S'} \operatorname{Jac}(\varphi|_S)\, dS.$$

Differentiating under the integral sign yields

$$\frac{d^2}{dt^2}\bigg|_{t=0} A(\varphi_t(S')) = \int_{S'} \left(\frac{d^2}{dt^2}\bigg|_{t=0} \operatorname{Jac}(\varphi|_S) \right) dS.$$

We express $\operatorname{Jac}(\varphi_t|_S)(p) = \sqrt{\det A(t,p)}$, where $A(t,p)$ is the matrix

$$A(t,p) = \big(\langle (d\varphi_t)_p(e_i), (d\varphi_t)_p(e_j)\rangle\big)_{i,j=1,\ldots k},$$

with $\{e_1, \ldots, e_k\}$ any orthonormal basis of T_pS. From now on, we fix $p \in S$, and call $A(t) = A(t,p)$.

Since $A(0) = \mathrm{Id}_k$ we get

$$\frac{d^2}{dt^2}\bigg|_{t=0} \sqrt{\det A(t)} = -\frac{1}{4}\left(\frac{d}{dt}\bigg|_{t=0} \det A(t) \right)^2 + \frac{1}{2}\frac{d^2}{dt^2}\bigg|_{t=0} \det A(t). \tag{1.30}$$

Jacobi's identity (1.14) and Eq. (1.28) imply

$$\left.\frac{d}{dt}\right|_{t=0} \det A(t) = \operatorname{trace} A'(0),$$

$$\left.\frac{d^2}{dt^2}\right|_{t=0} \det A(t) = (\operatorname{trace} A'(0))^2 + \operatorname{trace}(-A'(0)^2 + A''(0)).$$

Both equations, together with (1.30), yield

$$\left.\frac{d^2}{dt^2}\right|_{t=0} \sqrt{\det A(t)} = \frac{1}{4}\,(\operatorname{trace} A'(0))^2 - \frac{1}{2}\operatorname{trace} A'(0)^2 + \frac{1}{2}\operatorname{trace} A''(0). \tag{1.31}$$

We compute now the terms in (1.31). The trace of $A'(0)$ was computed in (1.15) as

$$\frac{1}{2}\operatorname{trace} A'(0) = (\operatorname{div}_S X)(p). \tag{1.32}$$

To compute the trace of $A'(0)^2$, we first compute the coefficients $a'_{ij}(0)$ of $A'(0)$. We consider the vector fields $E_i(t) = (d\varphi_t)_p(e_i)$, defined along the curve $\beta(t) = \varphi_t(p)$. The coefficients of $A(t, p)$ are $\langle E_i(t), E_j(t)\rangle$ for $i, j = 1, \dots, k$. So we have

$$\begin{aligned} a'_{ij}(0) &= \left.\frac{d}{dt}\right|_{t=0} \langle E_i(t), E_j(t)\rangle \\ &= \langle \left.\frac{D}{dt}\right|_{t=0} E_i(t), E_j(0)\rangle + \langle E_i(0), \left.\frac{D}{dt}\right|_{t=0} E_j(t)\rangle = \langle \nabla_{e_i} X, e_j\rangle + \langle e_i, \nabla_{e_j} X\rangle. \end{aligned}$$

As the matrix $A'(0)$ is symmetric

$$\operatorname{trace} A'(0)^2 = \sum_{i,j=1}^{k} \big(\langle D_{e_i} X, e_j\rangle + \langle D_{e_j} X, e_i\rangle\big)^2. \tag{1.33}$$

We finally compute $\operatorname{trace} A''(0)$, for which we take into account that, for $i \in \{1, \dots, k\}$,

$$\begin{aligned} a'_{ii}(0) &= \left.\frac{d^2}{dt^2}\right|_{t=0} \langle E_i(t), E_i(t)\rangle = \left.\frac{d}{dt}\right|_{t=0} \Big(2\langle \frac{D}{dt} E_i(t), E_i(t)\rangle\Big) \\ &= 2\,\langle \left.\frac{D^2}{dt^2}\right|_{t=0} E_i(t), E_i(0)\rangle + 2\,\langle \left.\frac{D}{dt}\right|_{t=0} E_i(t), \left.\frac{D}{dt}\right|_{t=0} E_i(t)\rangle \\ &= 2\,\langle \left.\frac{D^2}{dt^2}\right|_{t=0} E_i(t), e_i\rangle + 2\,|\nabla_{e_i} X|^2. \end{aligned}$$

The second covariant derivative is computed from Lemma 1.22. From Eq. (1.35) we finally obtain

$$\frac{1}{2}\operatorname{trace} A''(0) = (\operatorname{div}_S \nabla_X X)(p) - \sum_{i=1}^{k} \langle R(e_i, X_p)X_p, e_i\rangle$$

$$+ (\operatorname{div}_S X')(p) + \sum_{i=1}^{k} |\nabla_{e_i} X|^2. \tag{1.34}$$

The proof is concluded by replacing in (1.31) the expressions obtained in (1.32), (1.33), and (1.34) and taking into account that

$$|(\nabla_{e_i} X)^{\perp}|^2 = |\nabla_{e_i} X|^2 - \sum_{j=1}^{k} \langle \nabla_{e_i} X, e_j\rangle^2$$

implies that

$$\sum_{i=1}^{k} |D_{e_i} X|^2 - \frac{1}{2} \sum_{i,j=1}^{k} \big(\langle D_{e_i} X, e_j\rangle + \langle e_i, D_{e_j} X\rangle\big)^2$$

is equal to

$$\sum_{i=1}^{k} |(D_{e_i} X)^{\perp}|^2 - \sum_{i,j=1}^{k} \langle e_i, D_{e_j} X\rangle \, \langle e_j, D_{e_i} X\rangle.$$

□

Lemma 1.22 *Let* $\{\varphi_t\}_{t\in\mathbb{R}}$ *be a one-parameter family of diffeomorphisms in* M *and* X *the associated variational vector field. Let* $p \in M$ *and* $e \in T_pM$*. Define the smooth curve* $\sigma(t) = \varphi_t(p)$ *and the smooth vector field* $E(t) = (d\varphi_t)_p(e)$ *along* σ*. Then we have*

$$\left.\frac{D^2}{dt^2}\right|_{t=0} E(t) = \nabla_e \nabla_X X + R(X_p, e)X_p + \nabla_e X', \tag{1.35}$$

where D/dt *is the covariant derivative along the curve* σ*,* R *is the Riemannian curvature operator on* M*, and* X' *is the pointwise derivative of* X_t *defined in* (1.26).

Proof We take local coordinates $(x_1, \dots, x_m)$ in an open neighborhood U of p. Reasoning as in the proof of Lemma 1.13, we get $\varphi_t(q) \in U$ for t small enough and q in a small neighborhood $V \subset U$. We have

$$(X_t)_q = \sum_{i=1}^{m} f_i(t, q)\left(\frac{\partial}{\partial x_i}\right)_q.$$

Letting $\varphi_i = x_i \circ \varphi$, we get

$$(d\varphi_t)_p = \begin{pmatrix} \frac{\partial \varphi_1}{\partial x_1} & \cdots & \frac{\partial \varphi_1}{\partial x_m} \\ \vdots & \ddots & \vdots \\ \frac{\partial \varphi_m}{\partial x_1} & \cdots & \frac{\partial \varphi_m}{\partial x_m} \end{pmatrix} (t, p).$$

We observe that $(\varphi_{t+t_0} \circ \varphi_{t_0}^{-1})(\varphi_{t_0}(p)) = \beta(t + t_0)$, so that

$$(X_{t_0})_{\varphi_{t_0}}(p) = \beta'(t_0).$$

We fix $t_0 \in \mathbb{R}$ and we consider the one-parameter family of diffeomorphisms $\psi_t = \varphi_{t+t_0} \circ \varphi_{t_0}^{-1}$, $t \in \mathbb{R}$. Let $q = \varphi_{t_0}(p)$, $v = (d\varphi_{t_0})_p(e)$ and $V(t) = (d\psi_t)_q(v)$. The vector field associated with $\{\psi_t\}_{t\in\mathbb{R}}$ is X_{t_0}. As $v = (d\varphi_{t_0})(e) = E(t_0)$ (Lemma 1.13), we get

$$\left.\frac{D}{dt}\right|_{t=0} V(t) = \nabla_v X_{t_0} = \nabla_{E(t_0)} X_{t_0},$$

where D/dt is the covariant derivative on the curve $t \mapsto \psi_t(q) = \varphi_{t+t_0}(p) = \beta(t + t_0)$. We also observe that

$$\begin{aligned} V(t) = (d\psi_t)_q(v) &= \left[(d\varphi_{t+t_0})_p \circ (d\varphi_{t_0}^{-1})_{\varphi_{t_0}(p)}\right]((d\varphi_{t_0})_p(e)) \\ &= (d\varphi_{t+t_0})_p(e) = E(t + t_0). \end{aligned}$$

Hence

$$\left.\frac{D}{dt}\right|_{t=0} E(t + t_0) = \left.\frac{D}{dt}\right|_{t=0} V(t),$$

where, in both cases, D/dt is the covariant derivative along $\beta(t)$, invariant by reparameterizations. We conclude that

$$\left.\frac{D}{dt}\right|_{t=t_0} E(t) = \nabla_{E(t_0)} X_{t_0}.$$

We express $E(t) = (d\varphi_t)_p(e) = \sum_{i=1}^m g_i(t)\left(\frac{\partial}{\partial x_i}\right)_{\beta(t)}$ in local coordinates to get

$$\frac{D^2}{dt^2}\bigg|_{t=0} E(t) = \frac{D}{dt}\bigg|_{t=0} \nabla_{E(t)} X_t = \frac{D}{dt}\bigg|_{t=0} \sum_{i=1}^m g_i(t) \nabla_{\left(\frac{\partial}{\partial x_i}\right)_{\beta(t)}} X_t$$
$$= \sum_{i=1}^m \left(g_i'(0) \nabla_{\left(\frac{\partial}{\partial x_i}\right)_{\beta(t)}} X + g_i(0) \frac{D}{dt}\bigg|_{t=0} \nabla_{\left(\frac{\partial}{\partial x_i}\right)_{\beta(t)}} X_t \right).$$

On the other hand

$$\frac{D}{dt}\bigg|_{t=0} \nabla_{\left(\frac{\partial}{\partial x_i}\right)_{\beta(t)}} X_t = \frac{D}{dt}\bigg|_{t=0} \left(\sum_{k=1}^m \frac{\partial f_k}{\partial x_i}(t, \beta(t)) \nabla_{\left(\frac{\partial}{\partial x_i}\right)_{\beta(t)}} \frac{\partial}{\partial x_k} \right.$$
$$\left. + f_k(t, \beta(t)) \nabla_{\left(\frac{\partial}{\partial x_i}\right)_{\beta(t)}} \frac{\partial}{\partial x_k} \right).$$

Moreover we have

$$\frac{d}{dt}\bigg|_{t=0} f_k(t, \beta(t)) = \frac{\partial f_k}{\partial t}(0, p) + \frac{d}{dt}\bigg|_{t=0} f_k(0, \beta(t)),$$
$$\frac{d}{dt}\bigg|_{t=0} \frac{\partial f_k}{\partial x_i}(t, \beta(t)) = \frac{\partial^2 f_k}{\partial x_i \partial t}(0, p) + \frac{d}{dt}\bigg|_{t=0} \frac{\partial f_k}{\partial x_i}(0, \beta(t)).$$

Using these formulas, we obtain

$$\frac{D}{dt}\bigg|_{t=0} \nabla_{\left(\frac{\partial}{\partial x_i}\right)_{\beta(t)}} X_t = \nabla_{\left(\frac{\partial}{\partial x_i}\right)_{\beta(t)}} X' + \frac{D}{dt}\bigg|_{t=0} \nabla_{\left(\frac{\partial}{\partial x_i}\right)_{\beta(t)}} X.$$

Henceforth

$$\frac{D^2}{dt^2}\bigg|_{t=0} E(t) = \sum_{i=1}^m \left(g_i'(0) \nabla_{\left(\frac{\partial}{\partial x_i}\right)_p} X + g_i(0) \frac{D}{dt}\bigg|_{t=0} \nabla_{\left(\frac{\partial}{\partial x_i}\right)_p} X \right) + \nabla_e X'$$
$$= \sum_{i=1}^m \left(e(f_i) \nabla_{\left(\frac{\partial}{\partial x_i}\right)_p} X + e_i \nabla_{X_p} \nabla_{\left(\frac{\partial}{\partial x_i}\right)} X \right) + \nabla_e X'.$$

As

$$\nabla_e(\nabla_X X) = \nabla_e \nabla_{\sum_{i=1}^m f_i(p)\left(\frac{\partial}{\partial x_i}\right)} X = \sum_{i=1}^m \left(e(f_i) \nabla_{\left(\frac{\partial}{\partial x_i}\right)_p} X + f_i(p) \nabla_e \nabla_{\frac{\partial}{\partial x_i}} X \right),$$

and

$$R(X_p, e)X_p = \sum_{i,j=1}^{m} f_i(p)e_j\, R\left(\left(\frac{\partial}{\partial x_i}\right)_p, \left(\frac{\partial}{\partial x_j}\right)_p\right)X_p$$

$$= \sum_{i,j=1}^{m} f_i(p)e_j\left(\nabla_{\left(\frac{\partial}{\partial x_i}\right)_p}\nabla_{\left(\frac{\partial}{\partial x_j}\right)}X - \nabla_{\left(\frac{\partial}{\partial x_j}\right)_p}\nabla_{\left(\frac{\partial}{\partial x_i}\right)}X\right)$$

$$= \sum_{j=1}^{m} e_j\nabla_{X_p}\nabla_{\left(\frac{\partial}{\partial x_j}\right)}X - \sum_{i=1}^{m} f_i(p)\nabla_e\nabla_{\left(\frac{\partial}{\partial x_i}\right)}X,$$

we get (1.35). □

We are interested in the particular form of the second variation (1.29) when S is is the boundary of a domain $\Omega \subset M$. In this case, we obtain the following result when the scalar mean curvature of S is constant.

Proposition 1.23 *Let $\Omega \subset M$ be a bounded set with smooth boundary $S = \partial\Omega$. Assume that S has constant scalar mean curvature H with respect to the outer unit normal N. Consider a one-parameter family of diffeomorphisms $\{\varphi_t\}_{t\in\mathbb{R}}$ with compact support and associated vector field X such that $X|_S = uN$. Then we have*

$$\left.\frac{d^2}{dt^2}\right|_{t=0} A(\varphi_t(S)) - H\left.\frac{d^2}{dt^2}\right|_{t=0} |\varphi_t(\Omega)| = \int_S \left\{|\nabla_S u|^2 - \left(|\sigma|^2 + \mathrm{Ric}(N, N)\right)u^2\right\} dS, \tag{1.36}$$

where ∇_S is the tangential gradient on S, $|\sigma|^2 = \sum_{i=1}^{k} \kappa_i^2$, the squared sum of the principal curvatures of S, and Ric *is the Ricci curvature of M.*

Proof We analyze the terms in the general second variation formula (1.29) under our hypotheses. We first observe that

$$\int_S \mathrm{div}_S\, W\, dS = \int_S H\,\langle W, N\rangle\, dS$$

for any vector field W. In particular

$$\int_S \mathrm{div}_S\, Z = \int_S H\langle Z, N\rangle\, dS, \qquad \int_S \mathrm{div}\, X'\, dS = \int_S H\langle X', N\rangle\, dS.$$

As for the other terms, we have

$$\nabla_{e_i} X = \nabla_{e_i}(uN) = e_i(u)\, N + u\nabla_{e_i} N,$$

so that

$$\mathrm{div}_S X = uH.$$

Choosing an orthonormal basis at a given point composed of principal vectors (e_i) with principal curvatures κ_i (so that $\nabla_{e_i} N = \kappa_i e_i$), we have

$$\sum_{i=1}^{m-1} |(\nabla_{e_i} X)^{\perp}|^2 = |\nabla_S u|^2,$$

where ∇_S is the gradient on S, and

$$\sum_{i,j=1}^{k} \langle \nabla_{e_i} X, e_j\rangle \langle \nabla_{e_j} X, e_i\rangle = \sum_{i=1}^{k} \langle \nabla_{e_i} X, e_i\rangle^2 = \sum_{i=1}^{k} \kappa_i^2 = |\sigma|^2.$$

On the other hand,

$$\sum_{i=1}^{m-1} R(e_i, uN, uN, e_i) = u^2 \mathrm{Ric}(N, N).$$

Summing up, we conclude that the second variation of perimeter is given by

$$\begin{aligned} \frac{d^2}{dt^2}\bigg|_{t=0} A(\varphi_t(S)) &= \int_S \big(|\nabla_S u|^2 - \big(\mathrm{Ric}(N,N) + |\sigma|^2\big)u^2\big)\, dS \\ &\quad + \int_S \big(H\langle X', N\rangle + H^2 u^2 + H\langle \nabla_X X, N\rangle\big)\, dS. \end{aligned} \tag{1.37}$$

Let us compute now the second derivative of the volume $|\varphi_t(\Omega)|$. We use an alternative method to the second variation formula. We first compute $d/dt\big|_{t=t_0} |\varphi_t(\Omega)|$ for arbitrary t_0. We consider the one-parameter family $\psi_t = \varphi_{t+t_0} \circ \varphi_{t_0}^{-1}$, whose associated vector field is X_{t_0}. Then $\psi_t(\varphi_{t_0}(\Omega)) = \varphi_{t+t_0}(\Omega)$, and we get

$$\frac{d}{dt}\bigg|_{t=t_0} |\varphi_t(\Omega)| = \frac{d}{dt}\bigg|_{t=0} |\varphi_{t+t_0}(\Omega)| = \frac{d}{dt}\bigg|_{t=t_0} |\psi_t(\Omega)| = \int_{S_{t_0}} \langle X_{t_0}, N_{t_0}\rangle\, dS_{t_0},$$

where N_{t_0} is the outer unit normal to Ω_{t_0}. Hence

$$\frac{d}{dt} |\varphi_t(\Omega)| = \int_{S_t} \langle X_t, N_t\rangle\, dS_t \quad \text{for all } t \in \mathbb{R}.$$

By the area formula

$$\int_{S_t} \langle X_t, N_t \rangle \, dS_t = \int_S \big(\langle X_t, N_t \rangle \circ \varphi_t \big) \operatorname{Jac}(\varphi_t|_S) \, dS$$

Hence

$$\begin{aligned}
\frac{d^2}{dt^2}\Big|_{t=0} |\varphi_t(\Omega)| &= \frac{d}{dt}\Big|_{t=0} \int_{S_t} \langle X_t, N_t \rangle \, dS_t \\
&= \frac{d}{dt}\Big|_{t=0} \int_S \big(\langle X_t, N_t \rangle \circ \varphi_t \big) \operatorname{Jac}(\varphi_t|_S) \, dS \\
&= \int_S \frac{d}{dt}\Big|_{t=0} \Big(\big(\langle X_t, N_t \rangle \circ \varphi_t \big) \operatorname{Jac}(\varphi_t|_S) \Big) \, dS.
\end{aligned} \tag{1.38}$$

We fix $p \in S$ and consider the curve $\beta(t) = \varphi_t(p)$. Let D/dt be the covariant derivative along β. Then we have

$$\frac{d}{dt}\Big|_{t=0} \langle X_t, N_t \rangle (\beta(t)) = \langle \frac{D}{dt}\Big|_{t=0} (X_t)_{\beta(t)}, N_p \rangle + \langle X_p, \frac{D}{dt}\Big|_{t=0} (N_t)_{\beta(t)} \rangle$$

As $|N_t| = 1$, we have $\langle D/dt|_{t=0} (N_t)_{\beta(t)}, N_p \rangle = 0$. Hence $\langle D/dt|_{t=0} (N_t)_{\beta(t)}, X_p \rangle = 0$ since X is proportional to N by assumption. To compute $D/dt|_{t=0} (X_t)_{\beta(t)}$, we take local coordinates in a neighborhood U of p, and we express

$$(X_t)_q = \sum_{i=1}^m f_i(t, q) \left(\frac{\partial}{\partial x_i} \right)_q, \quad q \in U.$$

Hence

$$\begin{aligned}
\frac{D}{dt}\Big|_{t=0} (X_t)_{\beta(t)} &= \sum_{i=1}^m \frac{\partial f_i}{\partial t}(0, p) \left(\frac{\partial}{\partial x_i} \right)_p + \sum_{i=1}^m \left(\frac{d}{dt}\Big|_{t=0} f_i(0, \beta(t)) \right) \left(\frac{\partial}{\partial x_i} \right)_p \\
&\quad + \sum_{i=1}^m f_i(0, p) \, \nabla_{\beta'(0)} \frac{\partial}{\partial x_i} \\
&= (X')_p + \frac{D}{dt}\Big|_{t=0} X_{\beta(t)} = (X')_p + \nabla_{X_p} X.
\end{aligned}$$

Applying the chain rule in (1.38) and the fact that $d/dt|_{t=0}\,\mathrm{Jac}(\varphi_t|_S) = \mathrm{div}_S\, X = uH$, we get

$$\frac{d^2}{dt^2}\bigg|_{t=0}|\varphi_t(\Omega)| = \int_S \big(\langle X' + \nabla_X X, N\rangle + u^2 N\big)\, dS \tag{1.39}$$

Substracting to (1.37) Eq. (1.39) multiplied by H, we get (1.46). □

We finally prove the following technical result. If a measurable set is a critical point of the volume with respect to a one-parameter family of diffeormorphisms, then the variation can be adjusted to preserve the volume for small positive time (see Theorem 3.7 in [4] for a similar result).

Lemma 1.24 *Let $E \subset M$ be a measurable set of finite volume and $\{\varphi_t\}_{t\in\mathbb{R}}$, $\{\xi_t\}_{t\in\mathbb{R}}$ one-parameter families of diffeomorphisms with compact support satisfying*

$$\frac{d}{dt}\bigg|_{t=0}|\varphi_t(E)| = 0, \qquad \frac{d}{dt}\bigg|_{t=0}|\xi_t(E)| \neq 0.$$

Let X be the velocity of $\{\varphi_t\}_{t\in\mathbb{R}}$. Then there exists another one-parameter family of diffeomorphisms $\{\psi_t\}_{\mathbb{R}}$ with compact support and velocity Y such that

1. *$Y = X$,*
2. *$|\psi_t(E)| = |E|$ for t small enough.*

Proof We consider the map $f : \mathbb{R}\times\mathbb{R} \to \mathbb{R}$ defined by

$$f(s,t) = |(\xi_s \circ \varphi_t)(E)|.$$

We have

$$\frac{\partial f}{\partial s}(0,0) = \frac{d}{ds}\bigg|_{s=0}|\xi_s(E)| \neq 0, \qquad \frac{\partial f}{\partial t}(0,0) = \frac{d}{dt}\bigg|_{t=0}|\varphi_t(E)| = 0.$$

Hence the implicit function theorem implies the existence of $\varepsilon > 0$ and a smooth function $s : (-\varepsilon, \varepsilon) \to \mathbb{R}$, such that $s(0) = 0$ and $f(s(t), t) = 0$. We also have $s'(0) = 0$ since $(\partial f/\partial t)(0,0) = 0$. We can extend $s(t)$ outside a small interval around 0 to a smooth function on $\mathbb{R}$ which is also denoted as $s(t)$. Then $\psi_t = \xi_{s(t)}\circ\varphi_t$ is a smooth one-parameter family of diffeomorphisms satisfying 2. To check that 1 holds, we observe that the velocity Y of $\{\psi_t\}_{t\in\mathbb{R}}$ is equal to $s'(0)Y + X = X$. □

1.3.4 Variation of the Scalar Mean Curvature of a Hypersurface

Given a hypersurface S with unit normal N and a one-parameter family of diffeomorphisms $\{\varphi_t\}_{t\in\mathbb{R}}$, we wish to compute the derivative of the scalar mean curvature of $\varphi_t(S)$. Of course we are assuming a smooth choice of unit normal along the deformation. This can be easily obtained taking a local frame $E_1, \ldots, E_{m-1}$ in S so that $E_1 \wedge \ldots \wedge E_{m-1}$ is positively proportional to the normal N of S. Letting $E_i(t)$ be the images of E_i under the map $d\varphi_t$, we take as normal to $S_t = \varphi_t(S)$ the unit vector field N_t positively proportional to $E_1(t) \wedge \ldots \wedge E_{m-1}(t)$. We first compute the variation of the unit normal N_t.

Lemma 1.25 *Let $\{\varphi_t\}_{t\in\mathbb{R}}$ be a one-parameter family of diffeomorphisms with velocity X. Let S be a hypersurface with unit normal N and let N_t a smooth choice of normals to $S_t = \varphi_t(S)$. For $p \in S$ fixed, let $\beta(t) = \varphi_t(p)$ and D/dt the covariant derivative along the curve β. Then we have*

$$\left.\frac{D}{dt}\right|_{t=0} N_t = \nabla_{X_p^\top} N - (\nabla_S u)(p), \tag{1.40}$$

where $u = \langle X, N\rangle$ and ∇_S is the gradient on S.

Proof The vector $(D/dt)|_{t=0}N_t$ is tangent to S since $|N_t| = 1$. Take $e \in T_pS$ and let $E(t) = (d\varphi_t)_p(e)$. By Lemma 1.13

$$\langle \left.\frac{D}{dt}\right|_{t=0} N_t, e\rangle = -\langle N_p, \left.\frac{D}{dt}\right|_{t=0} E(t)\rangle = -\langle N_p, \nabla_e X\rangle = -\langle N_p, \nabla_e X^\top\rangle - e(u).$$

We have $e(u) = \langle \nabla_S u(p), e\rangle$, and $-\langle N_p, \nabla_e X^\top\rangle = \sigma(e, X_p^\top) = \langle \nabla_{X_p^\top} N, e\rangle$, where σ is the second fundamental form of S with respect to N. These observations imply (1.40) □

For the variation of the mean curvature, we have the following result.

Lemma 1.26 *Let S be a hypersurface with unit normal N and mean curvature H with respect to N. Consider a one-parameter family of diffeomorphisms $\{\varphi_t\}_{t\in\mathbb{R}}$ with variational vector field X. Let H_t be the mean curvature of the hypersurface $\varphi_t(S)$ with respect to a smooth choice of unit normals N_t. Then*

$$\left.\frac{d}{dt}\right|_{t=0} H_t(\varphi_t(p)) = X_p^\top(H) - \big[\Delta_S u + (\mathrm{Ric}(N, N) + |\sigma|^2)\, u\big](p), \tag{1.41}$$

where $u = \langle X, N\rangle$, Δ_S is the Laplacian on S and $|\sigma|^2 = \sum_{i=1}^{m-1} \kappa_i^2$ is the squared norm of the second fundamental form σ of S.

In particular, if the tangent projection $X^\top$ of X onto S is equal to 0 or H is constant, then we have

$$\frac{d}{dt}\bigg|_{t=0} H_t(\varphi_t(p)) = -\big[\Delta_S u + (\mathrm{Ric}(N,N) + |\sigma|^2)\, u\big](p), \tag{1.42}$$

Proof We fix a point $p \in S$ and take an orthonormal basis $e_1, \ldots, e_{m-1}$ of T_pS consisting on principal vectors so that $e_1 \wedge \ldots \wedge e_{m-1} = N_p$. Let $\kappa_1, \ldots, \kappa_{m-1}$ be the associated principal curvatures. We consider the vector fields $E_i(t) = (d\varphi_t)_p(e_i)$ along the curve $\beta(t) = \varphi_t(p)$ and the unit vector field N_t orthogonal to $S_t = \varphi_t(S)$ coinciding with $\lambda\,(E_1(e) \wedge \ldots \wedge E_{m-1}(t))$ for some $\lambda > 0$ at $\beta(t)$. The mean curvature H_t of S_t with respect to the normal N_t at the point $\beta(t) = \varphi_t(p)$ can be computed as

$$H_t(\beta(t)) = \mathrm{trace}(A(t)^{-1}B(t)),$$

where $A(t), B(t)$ are the square matrices of order $m-1$ defined by $A(t) = \big(\langle E_i(t), E_j(t)\rangle\big)_{ij}$, $B(t) = \big(\langle \nabla_{E_i(t)} N_t, E_j(t)\rangle\big)_{ij}$. Since $A(0)$ is the identity matrix, we get

$$\frac{d}{dt}\bigg|_{t=0} H_t(\beta(t)) = \mathrm{trace}\left(-\left(\frac{d}{dt}\bigg|_{t=0} A(t)\right) B(0) + \frac{d}{dt}\bigg|_{t=0} B(t)\right).$$

Lemma 1.13 implies

$$\frac{d}{dt}\bigg|_{t=0} \langle E_i(t), E_j(t)\rangle = \langle \nabla_{e_i} X, e_j\rangle + \langle e_i, \nabla_{e_j} X\rangle.$$

Decomposing X as $X^\top + X^\perp$ and taking into account that $X^\perp = uN$, that $\langle \nabla_{e_i}(uN), e_j\rangle = \kappa_i \delta_{ij}$, and that $B(0)$ is the diagonal matrix with diagonal elements $\kappa_1, \ldots, \kappa_{m-1}$, we have

$$-\,\mathrm{trace}\left(\left(\frac{d}{dt}\bigg|_{t=0} A(t)\right) B(0)\right) = -2\,|\sigma|^2 u - \mathrm{trace}(CB(0)), \tag{1.43}$$

where C is the square matrix $\big(\langle \nabla_{e_i} X^\top, e_j\rangle + \langle e_i, \nabla_{e_j} X^\top\rangle\big)_{ij}$.

On the other hand

$$\frac{d}{dt}\bigg|_{t=0} b_{ii}(t) = \frac{d}{dt}\bigg|_{t=0} \langle \nabla_{E_i(t)} N_t, E_i(t)\rangle = \langle \frac{D}{dt}\bigg|_{t=0} \nabla_{E_i(t)} N_t, e_i\rangle + \langle \nabla_{e_i} N, \nabla_{e_i} X\rangle$$

again by Lemma 1.13. It is easy to check, along the same line as in the proof of Lemma 1.22, that

$$\frac{D}{dt}\bigg|_{t=0} \nabla_{E_i(t)} N_t = R(X, e_i)N + \nabla_{e_i}\nabla_X N_t.$$

From (1.40) and the decomposition $X = X^\top + X^\perp$, we get the following formula for the derivative of $B(t)$,

$$\operatorname{trace}\left(\frac{d}{dt}\bigg|_{t=0} B(t)\right) = -\operatorname{Ric}(N, N)\, u - \Delta_S u + |\sigma|^2 u + \operatorname{trace} D, \tag{1.44}$$

where D is the square matrix whose coefficient (i, i) is given by

$$\langle R(X^\top, e_i)N + \nabla_{e_i}\nabla_{X^\top} N_t, e_i\rangle + \langle \nabla_{e_i} N, \nabla_{e_i} X^\top\rangle.$$

From (1.43) and (1.44) we get

$$\frac{d}{dt}\bigg|_{t=0} H_t(\beta(t)) = -\big[\Delta_S u + (\operatorname{Ric}(N, N) + |\sigma|^2)\, u\big](p) + \operatorname{trace}\big(CB(0) + D\big). \tag{1.45}$$

This computation is valid for an arbitrary one-parameter family of diffeomorphisms with associated vector field X. The first summand in (1.45) corresponds to the normal part of X along the deformation and the second one to the tangent part. If we consider the vector field $X^\top$ and we extend it smoothly to a vector field Y in M with compact support, the flow $\{\psi_t\}_{t\in\mathbb{R}}$ associated with Y leaves S invariant, and we have $Y = X^\top$ on S. Formula (1.45) then implies

$$X_p^\top(H) = \frac{d}{dt}\bigg|_{t=0} H(\psi_t(p)) = \operatorname{trace}\big(-CB(0) + D\big).$$

This implies, together with (1.45), Eq. (1.41). □

Lemma 1.26 allows us to extend the result in Proposition 1.23 to general variations.

Theorem 1.27 *Let $E \subset M$ be a bounded set with smooth boundary $S = \partial E$. Assume that S has constant scalar mean curvature H with respect to the outer unit normal N. Consider a one-parameter family of diffeomorphisms $\{\varphi_t\}_{t\in\mathbb{R}}$ with compact support and associated vector field X such that $\langle X, N\rangle = u$. Then we have*

$$\frac{d^2}{dt^2}\bigg|_{t=0} A(\varphi_t(S)) - H\frac{d^2}{dt^2}\bigg|_{t=0} |\varphi_t(E)| = -\int_S \big\{\Delta_S u + \big(|\sigma|^2 + \operatorname{Ric}(N, N)\big)\, u\big\}\, u\, dS, \tag{1.46}$$

where Δ_S is the Laplacian on S, $|\sigma|^2 = \sum_{i=1}^{k} \kappa_i^2$, the squared sum of the principal curvatures of S, and Ric *is the Ricci curvature of M.*

Proof Let us compute first $(d/dt)_{t=t_0} A(\varphi_t(S))$ at some t_0. We consider the one-parameter family $\psi_t = \varphi_{t+t_0} \circ \varphi_{t_0}^{-1}$, whose associated vector field is X_{t_0}. Then $\psi_t \circ \varphi_{t_0} = \varphi_{t+t_0}$ and we get

$$\begin{aligned}\frac{d}{dt}\bigg|_{t=t_0} A(\varphi_t(S)) &= \frac{d}{dt}\bigg|_{t=0} A(\varphi_{t+t_0}(S)) = \frac{d}{dt}\bigg|_{t=0} A(\psi_t(S_{t_0})) \\ &= \int_{S_{t_0}} H_{t_0} \langle X_{t_0}, N_{t_0}\rangle \, dS_{t_0},\end{aligned}$$

where H_t is the mean curvature of $S_t = \varphi_t(S)$. We differentiate again and evaluate at $t = 0$ to obtain

$$\begin{aligned}\frac{d^2}{dt^2}\bigg|_{t=0} A(\varphi_t(S)) &= \frac{d}{dt}\bigg|_{t=0} \int_S \big[H_t \langle X_t, N_t\rangle) \circ \varphi_t\big] \operatorname{Jac}(\varphi_t) \, dS \\ &= \int_S \frac{d}{dt}\bigg|_{t=0} (H_t \circ \varphi_t) \, u \, dS + H \int_S \frac{d}{dt}\bigg|_{t=0} \\ &\quad \times \big[(\langle X_t, N_t\rangle \circ \varphi_t) \operatorname{Jac}(\varphi_t)\big] \, dS\end{aligned}$$

As H is constant on S, the derivative of the mean curvature $(d/dt)_{t=0}(H_t \circ \varphi_t)$ is computed from (1.41) and equals $-\Delta_S u - (\operatorname{Ric}(N, N) + |\sigma|^2) \, u$. The derivative $(d/dt)_{t=0} \int_{S_t} \langle X_t, N_t\rangle \, dS_t$ is computed from (1.38) and is equal to $(d^2/dt^2)_{t=0} |\varphi_t(E)|$. This way we get (1.46). □

1.3.5 Volume Comparison for Balls

Comparison results for the boundary area and the volume of geodesic balls in terms of the sectional or Ricci curvature are well-known (see [210], §III.4 in [102] and §7.1 in [345]). In this section, we briefly recall these results and indicate how to use the variation formulas of the previous sections to obtain such comparison results, starting from a comparison of the mean curvature of the spheres. We refer the reader to Chapter 5 in do Carmo [141] for the necessary background on Jacobi fields.

Let us consider some important functions first. For any $\kappa \in \mathbb{R}$ we define

$$\cos_\kappa(t) = \begin{cases} \cos \kappa^{1/2} t, & \kappa > 0, \\ 1, & \kappa = 0, \\ \cosh{(-\kappa)^{1/2}} t, & \kappa < 0, \end{cases} \tag{1.47}$$

and

$$\sin_\kappa(t) = \begin{cases} \frac{1}{\kappa^{1/2}} \sin \kappa^{1/2} t, & \kappa > 0, \\ t, & \kappa = 0, \\ \frac{1}{(-\kappa)^{1/2}} \sinh (-\kappa)^{1/2} t, & \kappa < 0. \end{cases} \tag{1.48}$$

Observe that $(\cos_\kappa)' = -\kappa \sin_\kappa$ for $\kappa \neq 0$ and $(\sin_\kappa)' = \cos_\kappa$ for any κ. In particular, $\cos_\kappa$ and $\sin_\kappa$ satisfy the differential equation

$$u'' + \kappa\, u = 0. \tag{1.49}$$

The solution to (1.49) with initial conditions $u(0) = 0$, $u'(0) = 1$ is $u(t) = \sin_\kappa(t)$. For future reference, we also define the functions

$$\tan_\kappa = \frac{\sin_\kappa}{\cos_\kappa}, \qquad \cot_\kappa = \frac{\cos_\kappa}{\sin_\kappa}. \tag{1.50}$$

Parameterization of Geodesic Spheres by the Exponential Map

Let $p \in M$ and $r_0 > 0$ such that the exponential map $\exp_p : B(0, r_0) \subset T_pM \to B(p, r)$ is a diffeomorphism. For any $0 < r < r_0$, we parameterize the unit sphere $S(p, r)$ by the map $F_r : S(0, 1) \to S(p, r)$ defined by $F_r(v) = \exp_p(rv)$, where $S(0, 1) \subset T_pM$ is the unit sphere centered at $0 \in T_pM$. The tangent space $T_vS(0, 1)$ is composed of those vectors in T_pM orthogonal to v. To compute $(dF_r)_v(w)$ when $w \perp v$, we take a curve $\alpha : I \to S(0, 1)$ with $\alpha(0) = v$ and $\alpha'(0) = w$, and we compute the tangent vector to the curve $s \mapsto \exp_p(r\alpha(s))$ at $s = 0$. The map $f(s, t) = \exp_p(t\alpha(s))$ is a variation by geodesics, so that $J(t) = (\partial f/\partial s)(0, t)$ is a *Jacobi field* along the geodesic $\gamma_v(t) = \exp_p(tv)$. This means that J satisfies the differential equation

$$J'' + R(J, \gamma_v')\, \gamma_v' = 0, \tag{1.51}$$

with initial conditions $J(0) = 0$, $J'(0) = w$. The prime applied to J denotes the covariant derivative along γ_v. By Corollary 2.5 in Chapter 5 in [141], we have

$$(dF_r)_v(w) = (d\exp_p)_{rv}(rw) = J(r).$$

The set of Jacobi fields satisfying the initial conditions $J(0) = 0$, $J'(0) = w$, with $|w| = 1$, $w \perp v$, evaluated at r, generates the tangent space $T_{\gamma_v(r)}S(p, r)$ at the point $\gamma_v(r)$. The second fundamental form of $S(p, r)$ at $\gamma_v(r)$ applied to the vectors $(J(r), J(r))$, yields

$$\frac{\langle J', J\rangle}{|J|^2}(r). \tag{1.52}$$

In case $M = \mathbb{M}^m_\kappa$, $m \geqslant 2$, has constant sectional curvatures, and we have (compare with §II.5 in [102])

Lemma 1.28 *Given a geodesic $\gamma : [0, a] \to \mathbb{M}^m_\kappa$ and a parallel vector field U orthogonal to γ', the vector fields*

$$\cos_\kappa(t)U_{\gamma(t)}, \quad \sin_\kappa(t)U_{\gamma(t)}$$

are Jacobi fields along γ.

If $J(t)$ is a Jacobi field along γ satisfying $J(0) = 0$ and $J'(0) \perp \gamma'(0)$, then $J(t) = \sin_\kappa(t)\, U_\gamma(t)$, where U_γ is the parallel vector along γ satisfying $U_{\gamma(0)} = J'(0)$. Moreover, for a Jacobi field J such that $J(0) = 0$, we have

$$\frac{\langle J', J\rangle}{|J|^2}(t) = \cot_\kappa(t).$$

Hence geodesic spheres in $\mathbb{M}^m_\kappa$ are totally umbilical. Moreover the mean curvature of the geodesic sphere of radius $r > 0$ in $\mathbb{M}^m_\kappa$ is

$$H^m_\kappa(r) = (m-1)\cot_\kappa(r).$$

Proof To prove that $\cos_\kappa(t)U_{\gamma(t)}$, $\sin_\kappa(t)U_{\gamma(t)}$ are Jacobi fields along γ, we just take covariant derivatives and use the expression of the curvature operator in spaces of constant sectional curvature (see Lemma 3.4 in [141]) that U is parallel and that the functions $\cos_\kappa$, $\sin_\kappa$ satisfy (1.49).

Let J be any Jacobi field satisfying $J(0) = 0$ and $J'(0) \perp \gamma'(0)$, and consider the Jacobi field $\tilde{J}(t) = \sin_\kappa(t)\, U_{\gamma(t)}$, where U is the parallel vector field aalong γ satisfying $U_{\gamma(0)} = J'(0)$. Since $\tilde{J}(0) = 0 = J(0)$ and $\tilde{J}'(0) = U_{\gamma(0)} \perp \gamma'(0) = J'(0)$, we get $J = \tilde{J}$ by the uniqueness of Jacobi fields with given initial conditions.

Finally, the expression for $\langle J', J\rangle/|J|(t)$ is a simple computation, which implies that $S(p, r)$ is totally umbilical and permits to compute the mean curvature of any geodesic sphere in $\mathbb{M}^m_\kappa$. □

Volume Comparison When the Sectional Curvatures Are Bounded Above

Let us assume that the sectional curvatures of M are bounded above by some constant δ. Using Jacobi equation (1.51) and the bound on the sectional curvatures, we get

$$|J|'' = \left(\frac{\langle J', J\rangle}{|J|}\right)' = \frac{\langle J'', J\rangle}{|J|} + \frac{|J'|^2|J|^2 - \langle J', J\rangle^2}{|J|^2} \geqslant -\delta|J|.$$

Observe that $|J|(0) = 0$ and that $|J|'(0) = \lim_{t\to 0}(\langle J', J\rangle/|J|)(t) = 1$.

We compare $|J|$ with the solution u_δ of the equation $u'' + \delta u = 0$ with initial conditions $u(0) = 0, u'(0) = 1$ using the classical Sturm comparison theorems, §XI.3 in [219] or Lemma 7 in [326], and we obtain

$$\frac{\langle J', J\rangle}{|J|^2}(t) \geqslant \frac{u'_\delta}{u_\delta}(t)$$

for all $0 < t \leqslant r_0$. Hence any principal curvature of the sphere $S(p, t)$ is bounded below by $(u'_\delta/u_\delta)(t)$, which is in turn the only principal curvature of the totally umbilical sphere of radius t in the space form $\mathbb{M}^m_\delta$ of constant sectional curvature δ. This implies

$$H_{S(p,t)} \geqslant H_\delta(t), \tag{1.53}$$

where we have denoted $H_\delta(t) = H^m_\delta(t)$. Calling $A_\delta(t)$ to the area of the sphere of radius t in $\mathbb{M}^m_\delta$, using the first variation formula for the deformation by outer parallels, we have

$$\left(\frac{A(S(p,t))}{A_\delta(t)}\right)' = \frac{\int_{S(p,t)} \left(H_{S(p,t)} - H_\delta(t)\right) dS(p,t)}{A_\delta(t)} \geqslant 0. \tag{1.54}$$

In particular we get

$$A(S(p,t)) \geqslant A_\delta(t) \tag{1.55}$$

for all $0 < t \leqslant r_0$. We can use the coarea formula to compute the volume of a ball as $|B(p,t)| = \int_0^t A(S(p,r)\, dr$ in any Riemannian manifold. Denoting by $V_\delta(t)$ the volume of the geodesic ball of radius t in $\mathbb{M}_\delta$ we have

$$|B(p,t)| \geqslant V_\delta(t) \tag{1.56}$$

for all $0 < t \leqslant r_0$.

So we have proven

Theorem 1.29 *Let (M, g) be a complete Riemannian manifold. Assume that the sectional curvatures of M are no larger than δ. Then*

1. $H_{S(p,t)} \geqslant H_\delta(t)$,
2. *$A(S(p,t))/A_\delta(t)$ is a monotone increasing function of t. In particular $A(p,t) \geqslant A_\delta(t)$.*
3. $|B(p,t)| \geqslant V_\delta(t)$,

for all $0 < t \leqslant r_0$, where r_0 is such as the ball $B(p, r_0)$ is contained in a normal neighborhood of p.

It is well-known that r_0 can be taken as $\min\{\operatorname{inj}(p), \pi/\delta^{1/2}\}$, where $\operatorname{inj}(p)$ is the injectivity radius of M at p and $\pi/\delta^{1/2}$ is taken as $+\infty$ if $\delta \leqslant 0$. In case equality holds for some $t > 0$ in any of the inequalities obtained in Theorem 1.29, it is known that $B(p,t)$ is isometric to the ball of radius t in $\mathbb{M}_\delta$. We refer to §III.4 in [102] for details of the analysis of the equality case.

Volume Comparison When the Ricci Curvature Is Bounded Below

Let us assume now that $\operatorname{Ric}(v,v) \geqslant (m-1)\,\delta$ for any unit vector v in M. This condition is usually expressed as $\operatorname{Ric} \geqslant (m-1)\,\delta$. We can apply formula (1.41) for the variation of the scalar mean curvature to the deformation of balls centered at the point p. This is a normal deformation with associated vector field X equal to the outer unit normal N_t to the ball of radius t. Hence the normal component of X is $u \equiv 1$. From the bound on the Ricci curvature, (1.24) and (1.41) we have

$$\frac{d}{dt} H_{S(p,t)} = -\operatorname{Ric}(N_t, N_t) - |\sigma_t|^2 \leqslant -(m-1)\,\delta - \frac{H^2_{S(p,t)}}{m-1}. \tag{1.57}$$

This inequality becomes an equality for balls in $\mathbb{M}^m_\delta$. Dividing by $m-1$ we have

$$\frac{d}{dt}\left(\frac{H_{S(p,t)}}{m-1}\right) + \left(\frac{H_{S(p,t)}}{m-1}\right)^2 \leqslant -\delta = \frac{d}{dt}\left(\frac{H_\delta(t)}{m-1}\right) + \left(\frac{H_\delta(t)}{m-1}\right)^2.$$

We recall the well-known relation

$$H_{S(p,t)} = \frac{m-1}{t} + O(t).$$

This can be easily checked since $|J(t)| = t + O(t^2)$ and $|J|'(t) = 1 + O(t)$ uniformly around p (for all $\xi \in T_pM$ with $|\xi| = 1$ and $v \perp \xi, |v| = 1$). By the Ricatti comparison estimate (see Corollary 6.4.2 in [345]), we have

$$H_{S(p,t)} \leqslant \frac{u'_\delta}{u_\delta}(t) = H^m_\delta(t). \tag{1.58}$$

This estimate is the opposite of (1.53). From this point on, we can reason as in the sectional curvature case to prove that

$$\left(\frac{A(S(p,t))}{A_\delta(t)}\right)' \leqslant 0,$$

that

$$A(S(p,t)) \leqslant A_\delta(t),$$

and that

$$|B(p,t)| \leqslant V_\delta(t).$$

So we have proven

Theorem 1.30 *Let (M, g) be a complete m-dimensional Riemannian manifold. Assume that* Ric $\geqslant (m-1)\,\delta$. *Then*

1. $H_{S(p,t)} \leqslant H_\delta^m(t)$,
2. $A(S(p,t))/A_\delta(t)$ *is a monotone decreasing function of t. In particular* $A(p,t) \leqslant A_\delta(t)$.
3. $|B(p,t)| \leqslant V_\delta(t)$,

for all $0 < t \leqslant r_0$, *where* r_0 *is such as the ball* $B(p, r_0)$ *is contained in a normal neighborhood of p.*

Theorem 1.30 is known to hold for any $r > 0$ (see Theorem III.4.4 in [102]). In addition, a Lemma by Gromov (see Lemma III.4.1) states that if f, g are positive integrable functions of a real variable and f/g is decreasing, then $\int_0^r f / \int_0^r g$ is also decreasing. This implies, using Theorem 1.29(2), that

Corollary 1.31 *Let (M, g) be a complete m-dimensional Riemannian manifold. Assume that* Ric $\geqslant (m-1)\,\delta$. *Then the function*

$$\frac{|B(p,t)|}{V_\delta(t)}$$

is decreasing for all $0 < t \leqslant r_0$, *where* r_0 *is such as the ball* $B(p, r_0)$ *contained in a normal neighborhood of p.*

Recall that a *normal neighborhood* U of p is one where the exponential map $\exp_p$ is a diffeomorphism when restricted to U.

As in the previous result, Corollary 1.31 is valid for all $t > 0$.

In Theorem 1.8 we proved the existence of a doubling constant in a compact subset of a Riemannian manifold for sufficiently small radius. An easy consequence of Corollary 1.31 is the existence of a doubling constant under a uniform lower bound on the Ricci curvature. In particular, this uniform lower bound would hold on any compact subset of a Riemannian manifold. Under the assumption Ric $\geqslant (m-1)\,\delta$, we would have

$$\frac{|B(p,t)|}{V_\delta(t)} \geqslant \frac{|B(p,2t)|}{V_\delta(2t)}$$

and so

$$\frac{|B(p,2t)|}{|B(p,t)|} \leqslant \frac{V_\delta(2t)}{V_\delta(t)}.$$

If $\delta \geqslant 0$, then we may assume $\delta = 0$, and the quotient $V_0(2t)/V_0(t)$ is bounded above by 2^m. If $\delta < 0$ then we have

$$\frac{V_\delta(2t)}{V_\delta(t)} = \frac{\int_0^{2t} \sinh(\sqrt{-\delta}s)^{m-1}ds}{\int_0^{t} \sinh(\sqrt{-\delta}s)^{m-1}ds}$$

and

$$\lim_{t\to 0} \frac{V_\delta(2t)}{V_\delta(t)} = \lim_{t\to 0} \left(\frac{\sinh(2\sqrt{-\delta}t)}{\sinh(\sqrt{-\delta}t)}\right)^{m-1} = 2^{m-1}.$$

Hence $V_\delta(2t)/V_\delta(t)$ is uniformly bounded when $t \in (0, r_0]$ for any $r_0 > 0$.

Remark 1.32 Consider a point $x \in M$ and the distance function d_x to x. The level sets of d_x are the geodesic spheres centered at x. If $z \neq x$ and d_x is differentiable near x, the gradient of d_x at z is the tangent vector to the unique unit-speed geodesic connecting x with z, which is normal to the geodesic sphere of radius $d_x(z)$ centered at x. The Laplacian of d_x at z is equal to

$$\sum_{i=1}^{m-1} \nabla^2 d_x(e_i, e_i) + \nabla^2 d_x(\nabla d_x, \nabla d_x)(z),$$

where $e_1, \ldots, e_{m-1}$ is an orthonormal basis of the geodesic sphere $S(x, d_x(z))$. Since $(\nabla d_x)_z$ is a unit normal to the sphere and $\nabla^2 d_x(\nabla d_x, \nabla d_x)(z) = 0$, we obtain

$$(\Delta d_x)(z) = H_{S(x,d_x(z))}(z).$$

Hence Theorems 1.29 and 1.30 provide comparison results for the Laplacian of the distance function. The reader is referred to Chapters 6 and 12 in Petersen [345] for a comprehensive approach to comparison results using the Laplacian of the distance function.

The following function is defined for future reference

Definition 1.33 Given a Riemannian manifold M, the function $b : (0, \infty) \to [0, \infty)$ is defined by

$$b(r) = \inf_{x\in M} |B(x, r)|, \quad r > 0. \tag{1.59}$$

Its main properties are the following:

Proposition 1.34 *Let M be a Riemannian manifold. Then*

1. *b is a non-decreasing function.*
2. *If M is compact, then b is continuous.*

The function $b(r)$ can be estimated using volume comparison Theorems 1.29 and 1.30.

1.3.6 Estimates of Normal Jacobians

We include here a brief discussion on normal geodesics and estimates of the normal Jacobian to a hypersurface as this technique plays an important role to obtain some isoperimetric inequalities. We do not intend to be exhaustive here. The reader is referred to the original paper by Heintze and Karcher [225] and to the discussions in §VI.3 in Chavel [102] and §31–§33 in Burago and Zalgaller [83] for complete details.

Let $S \subset M$ be a smooth hypersurface in a Riemannian manifold M with unit normal N. The normal exponential map $F : S \times [0, \infty) \to M$ is defined by

$$F(p, t) = \exp_p(tN_p),$$

where $\exp_p$ is the exponential map at p. We also define the map $F_t : S \to M$ by $F_t(q) = F(q, t)$ A *normal Jacobi field* J along the geodesic $\gamma_p(t) = \exp_p(tN_p)$ is a Jacobi field along γ_p that satisfies $J(0) \in T_pS$ and $J'(0) = W(J'(0))$, where W is the Weingarten endomorphism on S with respect to the normal N. Normal Jacobi fields are the ones associated with variations by geodesics starting from points in S with initial speed normal to S (see §31.3.3 in [83]).

The point $\gamma_p(t)$, $t > 0$, is a *focal* point if there exists a normal Jacobi field $J \not\equiv 0$ along γ_p such that $J(t) = 0$. There exists always some $\varepsilon > 0$ such that $\gamma_p(t)$ is not a focal point if $t \in [0, \varepsilon]$. When $\gamma_p(t)$ is not a focal point, $(dF_t)_p$ has maximal rank, and $F_t(S)$ is a hypersurface near $\gamma_p(t)$ whose tangent space is generated by normal Jacobi fields $J_1(t), \ldots, J_{m-1}(t)$ such that $J_1(0), \ldots, J_{m-1}(0)$ are linearly independent vectors that generate T_pS.

If $\gamma_p : [0, a] \to M$ is a geodesic normal to S without focal points, we wish to estimate the Jacobian $\mathrm{Jac}(T)$ of F_t at p. Given any basis $v_1, \ldots, v_{m-1}$ of T_pS and normal Jacobi fields $J_1, \ldots, J_{m-1}$ along γ_p with initial conditions $J_i(0) = v_i$, $J_i'(0) = W(v_i)$, $i = 1, \ldots, m-1$, the Jacobian of F_t at p is given by

$$\mathrm{Jac}(t) = \frac{\det^{1/2}\left(\langle J_i(t), J_j(t)\rangle\right)_{ij}}{\det\left(\langle J_i(0), J_j(0)\rangle\right)_{ij}}.$$

Given a complete simply connected m-dimensional manifold $\mathbb{M}^m_\kappa$ of constant sectional curvature κ, we denote by $\mathrm{Jac}^m_{\kappa,H}(t)$ the normal Jacobian of the geodesic sphere of constant mean curvature H in $\mathbb{M}^m_\kappa$. Since geodesic spheres are totally umbilical in $\mathbb{M}^m_\kappa$, normal Jacobi fields to a geodesic sphere of constant mean curvature H along inner geodesics are given by

$$\left(\cos_\kappa(t) - \frac{H}{m-1}\sin_\kappa(t)\right) U_{\gamma(t)}, \tag{1.60}$$

where U is a parallel and orthogonal vector field along the geodesic. Hence the normal Jacobian is given by

$$\mathrm{Jac}^m_{\kappa,H}(t) = \left(\cos_\kappa(t) - \frac{H}{m-1}\sin_\kappa(t)\right)^{m-1}. \tag{1.61}$$

We have the following result (see [225] and Theorem 33.3.9 in [83]).

Theorem 1.35 *Let M be an m-dimensional Riemannian manifold with* Ric $\geqslant (m-1)\kappa$. *Let $S \subset M$ be a hypersurface, $p \in S$, and H an upper bound on the mean curvature of S at p. Consider the normal geodesic $\gamma_p : [0, a] \to M$, and assume that there are no focal points of γ_p in $[0, a]$. Let* $\mathrm{Jac}(t)$ *be the normal Jacobian along γ_p. Then*

$$\mathrm{Jac}(t)\,\mathrm{Jac}^m_{\kappa,H}(t)^{-1} \tag{1.62}$$

is non-increasing in $[0, a]$. In particular $\mathrm{Jac}(t) \leqslant \mathrm{Jac}^m_{\kappa,H}(t)$.

Proof We have $(\mathrm{Jac}(t)(\mathrm{Jac}^m_{\kappa,H})^{-1})' \leqslant 0$ if and only if

$$(\log \mathrm{Jac})'(t) \leqslant (\log \mathrm{Jac}^m_{\kappa,H})'(t). \tag{1.63}$$

To compute the derivative of $\log \mathrm{Jac}$ (or $\log \mathrm{Jac}^m_{\kappa,H}$) at t_0, we choose Jacobi fields $J_1, \ldots, J_{m-1}$ so that the vectors $J_1(t_0), \ldots, J_{m-1}(t_0)$ are orthonormal. Then we have

$$\mathrm{Jac}'(t_0) = \frac{1}{\det\left(\langle J_i(0), J_j(0)\rangle\right)_{ij}} \sum_{i=1}^{m-1} \langle J_i'(t_0), J_i(t_0)\rangle.$$

and

$$(\log \mathrm{Jac})'(t_0) = \sum_{i=1}^{m-1} \langle J_i'(t_0), J_i(t_0)\rangle.$$

Using the Jacobi equation, we get

$$\begin{aligned}(\log \mathrm{Jac})'(t_0) &= \sum_{i=1}^{m-1} \langle J_i'(t_0), J_i(t_0)\rangle \\ &= \sum_{i=1}^{m-1} \langle J_i'(0), J_i(0)\rangle + \sum_{i=1}^{m-1} \int_0^{t_0} \big(|J_i'|^2 - \langle R(J_i, \gamma'), \gamma', J_i\rangle\big)(s)\, ds.\end{aligned}$$

This construction can be applied both to the normal geodesic γ_p on M and to a geodesic α on $\mathbb{M}_\kappa^m$ normal to a geodesic sphere of constant mean curvature H. Let us choose Jacobi fields $U_1, \dots, U_{m-1}$ along α so that $U_1(t_0), \dots, U_{m-1}(t_0)$ are orthonormal. From the expression (1.60), it turns out that

$$(\log \mathrm{Jac}_{\kappa,H}^m)'(t_0) = \frac{H}{m-1} \sum_{i=1}^{m-1} |U_i(0)|^2 + \sum_{i=1}^{m} \int_0^{t_0} \big(|U_i'|^2 - \kappa |U_i|^2\big)(s)\, ds.$$

We build now some vector fields along the geodesic γ_p. Let : $T_{\alpha(0)}\mathbb{M}_\kappa^m \to T_{\gamma_p(0)}M$ be a linear isometry taking $\alpha'(0)$ to $\gamma_p'(0)$. Let P_t, P_t^κ denote the parallel transport along γ_p and α, from 0 to t. We then define

$$Z_i(t) = (P_t \circ F \circ (P_t^\kappa)^{-1}) U_i(t).$$

As $|U_i| = |Z_i|$ is independent of i and $\mathrm{Ric} \geqslant (m-1)\kappa$, we have

$$\sum_{i=1}^{m-1} \langle R(Z_i, \gamma')\gamma', Z_i\rangle = \mathrm{Ric}(\gamma', \gamma')|Z_1|^2 \geqslant \sum_{i=1}^{m-1} \kappa |Z_i|^2$$

and so

$$\begin{aligned}(\log \mathrm{Jac}_{\kappa,H}^m)'(t_0) &\geqslant \frac{H}{m-1} \sum_{i=1}^{m-1} |Z_i(0)|^2 + \sum_{i=1}^{m-1} \int_0^{t_0} \big(|Z_i'|^2 - \langle R(Z_i, \gamma')\gamma', Z_i\rangle\big)(s)\, ds \\ &= \sum_{i=1}^{m-1} \langle Z_i'(0), Z_i(0)\rangle + \sum_{i=1}^{m-1} \int_0^{t_0} \big(|Z_i'|^2 - \langle R(Z_i, \gamma')\gamma', Z_i\rangle\big)(s)\, ds.\end{aligned}$$

Taking now normal Jacobi fields $J_1, \dots, J_{m-1}$ so that $J_i(t_0) = Z_i(t_0)$, a well-known minimizing property for Jacobi fields (see Theorem 32.1.1 in [83]), implies that the last term, the index form for the geodesic γ_p, is larger than or equal to

$$\sum_{i=1}^{m-1} \langle J_i'(0), J_i(0)\rangle + \sum_{i=1}^{m-1} \int_0^{t_0} \big(|J_i'|^2 - \langle R(J_i, \gamma'), \gamma', J_i\rangle\big)(s)\, ds = (\log \mathrm{Jac})'(t_0).$$

This implies (1.63). □

1.4 Sets of Finite Perimeter

The theory of sets of finite perimeter was extensively developed in Euclidean spaces and has been extended recently to more general spaces. Some basic results in Euclidean theory can be directly stated and proved in Riemannian manifolds without too much effort. The techniques in more general metric spaces can be, of course, applied to the metric structure of Riemannian manifolds.

1.4.1 Definitions

We recall in this section the notions of function of bounded variation and of set of finite perimeter in a Riemannian manifold (M, g).

Definition 1.36 Given an open subset $\Omega \subset M$, and a function $f \in L^1(\Omega)$, we define the *total variation* of f in Ω as

$$|Df|(\Omega) = \sup\left\{\int_\Omega f \operatorname{div} X\, dM : X \in \mathfrak{X}_0^1(\Omega),\ ||X|| \leqslant 1\right\}, \tag{1.64}$$

where $\mathfrak{X}_0^1(\Omega)$ is the space of vector fields of class C^1 in M with compact support inside Ω and $||X||$ is the supremum norm of X.

We say that a function $f \in L^1(\Omega)$ has *bounded variation* in Ω if $|Df|(\Omega) < \infty$. A function $f \in L^1(\Omega)$ is of *locally bounded variation* in Ω if, for every relatively compact open set $\Omega' \subset \Omega$, the function $f|_{\Omega'}$ has bounded variation in Ω'.

Definition 1.37 The *perimeter* of a measurable set $E \subset M$ *inside* an open set Ω is defined as

$$|\partial E|(\Omega) = P(E, \Omega) = \sup\left\{\int_E \operatorname{div} X\, dM : X \in \mathfrak{X}_0(\Omega),\ ||X|| \leqslant 1\right\}. \tag{1.65}$$

When $\Omega = M$, we write $P(E)$ instead of $P(E, M)$. A set E has *finite perimeter* in Ω if $P(E, \Omega) < \infty$. A measurable set $E \subset \Omega$ is of *locally finite perimeter* in Ω if, for every relatively compact open subset $\Omega' \subset \Omega$, we have $P(E, \Omega') < \infty$.

It is clear from the definitions that a measurable set $E \subset M$ has finite perimeter if and only if its characteristic function $\mathbf{1}_E$ is of bounded variation. The reader is referred to Giusti [184] or Maggi [275] for complete information on functions of locally bounded variation and sets of locally finite perimeter in Euclidean spaces. A good introduction to sets of finite perimeter in Riemannian manifolds can be found in Section 1 in [293].

When E is bounded and has C^1 boundary, the perimeter of E in Ω coincides with the Riemannian measure of $\partial E \cap \Omega$. This is obtained immediately from the divergence theorem.

An alternative definition of function of bounded variation was introduced by Miranda in [292, §3] (see also [17]). The total variation of a function $f \in L^1(\Omega)$ is defined as

$$|Df|(\Omega) = \inf\left\{ \liminf_{i\to\infty} \int_\Omega \operatorname{lip}(f_i)\, dM : \{f_i\}_{i\in\mathbb{N}} \in \operatorname{Lip}(\Omega),\ f_i \to f \text{ in } L^1_{loc}(\Omega) \right\}. \tag{1.66}$$

Here $\operatorname{lip}(f)$ is the pointwise Lipschitz constant of f, defined at a point p as

$$[\operatorname{lip}(f)](p) = \liminf_{r\to 0} \frac{1}{r}\left(\sup_{q\in \overline{B}(p,r)} d(f(p), f(q)) \right).$$

This notion of function of bounded variation coincides with the Euclidean one (see Proposition 1.1 in [292]) and with the one on Riemannian manifolds. It is important to remark that properties of functions of bounded variation and sets of finite perimeter defined this way are valid in Riemannian manifolds.

1.4.2 The Reduced Boundary

Let E be a set of locally finite perimeter in M. We consider a local chart (U, φ), where $U \subset M$ is a bounded open set, and an orthonormal frame $E_1, \ldots, E_m$ of vector fields globally defined on U. We define the functional $L : C^1_0(\varphi(U), \mathbb{R}^m) \to \mathbb{R}$ by

$$L(u) = \int_{U\cap E} \operatorname{div}\left(\sum_{i=1}^m (u_i \circ \varphi)\, E_i \right) dM,$$

for any function $u = (u_1, \ldots, u_m) \in C^1_0(\varphi(U), \mathbb{R}^m)$. Since E has locally finite perimeter in M and U is bounded there exists a constant $C(U) > 0$ such that

$$L(u) \leqslant C(U)||u||_{L_\infty(\varphi(U))} \tag{1.67}$$

for all functions $u \in C^1_0(\varphi(U), \mathbb{R}^m)$. Here

$$||u||_{L_\infty(\varphi(U))} = \sup_{\varphi(U)} \left(\sum_{i=1}^m u_i^2 \right)^{1/2}.$$

Approximating any continuous function $u \in C_0(\varphi(U), \mathbb{R}^m)$ by a sequence of functions in $C^1_0(\varphi(U), \mathbb{R}^m)$, we immediately see that L extends to a functional on $C_0(\varphi(U), \mathbb{R}^m)$ satisfying (1.67). We apply Riesz representation theorem (see

Theorem 1 in §1.8 in [149]) to ensure the existence of a Radon measure $\tilde{\mu}_{\varphi(U)}$ on $\varphi(U)$ and a $\tilde{\mu}_{\varphi(U)}$-measurable function $\tilde{\nu}_{\varphi(U)} : \varphi(U) \to \mathbb{R}^m$ with $|\tilde{\nu}_{\varphi(U)}| = 1$ for $\tilde{\mu}_{\varphi(U)}$-a.e. such that

$$L(u) = \int_{\varphi(U)} (u \cdot \tilde{\nu}_{\varphi(U)})\, d\tilde{\mu}_{\varphi(U)},$$

where the symbol $\cdot$ denotes the standard scalar product in $\mathbb{R}^m$. In addition, Riesz representation theorem implies that $\tilde{\mu}_{\varphi(U)}$ is defined on open sets $\varphi(V) \subset \varphi(U)$ by

$$\tilde{\mu}_{\varphi(U)}(\varphi(V)) = \sup\big\{L(u) : u \in C_0^1(\varphi(U), \mathbb{R}^m),\\ ||u||_{L_\infty(\varphi(U))} \leqslant 1, \operatorname{supp}(u) \subset \varphi(V)\big\}.$$

Then we may push forward the measure $\tilde{\mu}_{\varphi(U)}$ by the diffeomorphism φ^{-1} to get a Radon measure μ_U on U defined by

$$\mu_U(A) = \tilde{\mu}_{\varphi(U)}(\varphi(A)).$$

Moreover

$$\nu_U = \sum_{i=1}^{m} ((\tilde{\nu}_{\varphi(U)})_i \circ \varphi)\, E_i$$

is a μ_U-measurable vector field such that

$$L(u) = \int_U \langle \sum_{i=1}^{m} (u_i \circ \varphi) E_i, \nu_U\rangle\, d\mu_U.$$

Any C^1 vector field X with compact support on U can be expressed as $X = \sum_{i=1}^m (u_i \circ \varphi) E_i$, for $u \in C_0^1(\varphi(U), \mathbb{R}^m)$. So we have

$$\int_E \operatorname{div} X\, dM = \int_U \langle X, \nu_U\rangle\, d\mu_U. \tag{1.68}$$

for any C^1 vector field X in M with compact support in U. Moreover

$$\mu_U(V) = \sup\left\{\int_U \operatorname{div} X\, dM : ||X|| \leqslant 1, \operatorname{supp}(X) \subset V\right\} = P(E, V). \tag{1.69}$$

We conclude that μ_U coincides with the perimeter measure on M and is independent of the chosen frame $E_1, \ldots, E_m$ on U. Formula (1.68) implies that also ν_U is independent of the frame and of the local chart (U, φ). Hence there is a μ-measurable vector field ν on M which coincides with ν_U on every local chart

(U, φ). Moreover, for every vector field X with compact support of class C^1, we have

$$\int_E \operatorname{div} X \, dM = \int_M \langle X, \nu \rangle \, d\mu. \tag{1.70}$$

Let us define the *reduced boundary* $\partial^* E$ of a set $E \subset M$ of locally finite perimeter. Let ν be the measurable unit normal of E defined as above. We say that $x \in \partial^* E$ if

1. $P(E, B(x,r)) > 0$ for all $r > 0$.
2. The limit $\lim_{r \to 0} \nu_r(x)$, where $\nu_r(x)$ is defined by

$$\nu_r(x) = \frac{\int_{B(x,r)} \nu \, dP}{P(B(x,r))},$$

   exists and it is equal to $\nu(x)$.
3. $|\nu(x)| = 1$.

Note that we already know that $|\nu(x)| = 1$ for P-a.e. $x \in M$. The family of balls can be replaced by any other family $F_r \subset B(x,r)$ such that $P(F_r) \geqslant \alpha P(B(x,r))$ for some $\alpha > 0$ (see §7.9 in Rudin [387]).

Let us consider a local chart (U, φ) and a set $E \subset U$ of locally finite perimeter. For any $V \subset U$, we denote $\varphi(V)$ by $\tilde{V}$. We may consider the push forward $\tilde{P}$ of the Riemannian perimeter of E to $\varphi(U)$, defined by

$$\tilde{P}(\tilde{E}, \tilde{V}) = P(E, V).$$

We have the following result.

Lemma 1.38 *Let $E \subset M$ be a set of locally finite perimeter in a Riemannian manifold and (U, φ) a local chart. Let $\tilde{P}$ the push forward of the Riemannian perimeter of E to $\tilde{U} = \varphi(U)$ and let P_0 be the Euclidean perimeter of $\tilde{E} = \varphi(E \cap U)$ in $\varphi(U) \subset \mathbb{R}^m$. Then*

1. *$\tilde{P}$ and P_0 are absolutely continuous with respect to each other.*
2. *The reduced boundaries of $\tilde{E}$ in $\varphi(U)$ with respect to the perimeters $\tilde{P}$ and P_0 coincide for $\tilde{P}$, P_0-a.e. $x \in \tilde{U}$.*

Proof For any vector field $X = (f_1, \ldots, f_m)$ on $\tilde{U}$, the Riemannian divergence of the vector field $\sum_{i=1}^m (f_i \circ \varphi)\, \partial/\partial x_i$ on (M, g) at the point $x \in U$ is equal to

$$\left(\frac{1}{|G|^{1/2}} \sum_{i=1}^m \frac{\partial}{\partial x_i} (|G|^{1/2} f_i) \right) (\varphi(x)) = \left(\frac{1}{|G|^{1/2}} \operatorname{div}_0(|G|^{1/2} X) \right) (\varphi(x)),$$

where $x_1, \ldots, x_m$ are the coordinates in U, G is the matrix $g(\partial/\partial x_i, \partial/\partial x_j)$, $i, j = 1, \ldots, m$, and $|G|$ is the determinant of G (see p. 294 in Matsushima [278]). We have denoted by div_0 the standard divergence in $\mathbb{R}^m$. Then, for any open set $V \subset U$,

$$\int_V \mathrm{div}\left(\sum_{i=1}^{m}(f_i \circ \varphi)\frac{\partial}{\partial x_i}\right) dM = \int_{\tilde{V}} \sum_{i=1}^{m} \frac{\partial |G|^{1/2} f_i}{\partial x_i}\, d\mathcal{L}^m = \int_{\tilde{V}} \mathrm{div}_0(|G|^{1/2} X)\, d\mathcal{L}^m.$$

We take $\tilde{V} = \varphi(V) \subset \tilde{U}$ relatively compact in $\tilde{U}$ and $X = (f_i, \ldots, f_m)$ with compact support on $\tilde{V}$. The Euclidean norm of X is given by $(X \cdot X)^{1/2}$, where $\cdot$ is the standard scalar product in $\mathbb{R}^m$ and the Riemannian norm by $(X \cdot (GX))^{1/2}$. If $X \neq 0$ and $(X \cdot X)^{1/2} \leqslant 1$, then we have, for $\tilde{E} = \varphi(E \cap U)$

$$\begin{aligned}\int_{\tilde{E}} \mathrm{div}_0(X)\, d\mathcal{L}^m &= \left(\int_{\tilde{E}} \frac{\mathrm{div}_0\left(|G|^{1/2}\frac{X}{|G|^{1/2}}\right)}{\sup_{\tilde{V}}\left(\frac{X}{|G|^{1/2}} \cdot \frac{GX}{|G|^{1/2}}\right)^{1/2}}\, d\mathcal{L}^m\right) \sup_{\tilde{V}}\left(\frac{X}{|G|^{1/2}} \cdot \frac{GX}{|G|^{1/2}}\right) \\ &\leqslant \tilde{P}(\tilde{E}, \tilde{V}) \sup_{\tilde{V}} \frac{\lambda_{\max}}{|G|},\end{aligned}$$

where $\lambda_{\max}$ is the largest eigenvalue of the symmetric matrix G. Taking supremum over vector fields X with compact support in $\tilde{V}$ satisfying $(X \cdot X)^{1/2}$, we have

$$P_0(\tilde{E}, \tilde{V}) \leqslant \tilde{P}(\tilde{E}, \tilde{V}) \sup_{\tilde{V}} \frac{\lambda_{\max}}{|G|},$$

A similar reasoning implies

$$P_0(\tilde{E}, \tilde{V}) \geqslant \tilde{P}(\tilde{E}, \tilde{V}) \inf_{\tilde{V}} \frac{\lambda_{\min}}{|G|}.$$

So we get

$$\left(\inf_{\tilde{V}} \frac{\lambda_{\min}}{|G|}\right) \tilde{P}(\tilde{E}, \tilde{V}) \leqslant P_0(\tilde{E}, \tilde{V}) \leqslant \left(\sup_{\tilde{V}} \frac{\lambda_{\max}}{|G|}\right) \tilde{P}(\tilde{E}, \tilde{V}). \tag{1.71}$$

This implies that $\tilde{P}$ and P_0 are absolutely continuous with respect to each other. We also get that the Radon-Nikodym derivatives $dP_0/d\tilde{P}$ and $d\tilde{P}/dP_0$ are defined for $\tilde{P}$, P_0-a.e $x \in \tilde{U}$. This implies 1.

As for 2, we note first that (1.71) implies that $\tilde{P}(\tilde{E}, B(x,r)) > 0$ if and only if $P_0(\tilde{E}, B(x,r)) > 0$. We consider the measurable normals ν, ν_0. They satisfy $\nu \cdot (G\nu) = 1$ and $\nu_0 \cdot \nu_0 = 1$ for $\tilde{P}$, P_0-a.e. $x \in \tilde{U}$. Finally, for a vector field X with compact support in $\tilde{V}$

$$\begin{aligned}\int_{\tilde{V}} (X \cdot \nu_0)\, dP_0 &= \int_{\tilde{V}} \operatorname{div}_0(X)\, d\mathcal{L}^m \\ &= \int_{\tilde{V}} \operatorname{div}_0(|G|^{1/2} \tfrac{X}{|G|^{1/2}})\, d\mathcal{L}^m \\ &= \int_{\tilde{V}} (X \cdot \frac{G\tilde{\nu}}{|G|^{1/2}})\, d\tilde{P} \\ &= \int_{\tilde{V}} (X \cdot \frac{G\tilde{\nu}}{|G|^{1/2}}) \frac{d\tilde{P}}{dP_0}\, dP_0.\end{aligned}$$

So we get

$$\nu_0 = \frac{d\tilde{P}}{dP_0} \frac{G\tilde{\nu}}{|G|^{1/2}}$$

for $\tilde{P}$, P_0-a.e. $x \in \tilde{U}$. So we have, for $\tilde{P}$, P_0-a.e. $x \in \partial^* \tilde{E}$, the reduced boundary of $\tilde{E}$ with respect to $\tilde{P}$,

$$\lim_{r \to 0} \frac{\int_{B(x,r)} \nu_0\, dP_0}{\int_{B(x,r)} dP_0} = \lim_{r \to 0} \frac{\int_{B(x,r)} (G\tilde{\nu}/|G|^{1/2})\, d\tilde{P}}{\int_{B(x,r)} (dP_0/d\tilde{P})\, d\tilde{P}} = \left(\frac{G\tilde{\nu}}{|G|^{1/2}} \frac{d\tilde{P}}{dP_0} \right)(x) = \nu_0(x).$$

Hence $x \in \partial_0^* \tilde{E}$, the reduced boundary of $\tilde{E}$ with respect to P_0. The converse is similar. □

Using Lemma 1.38 we can translate most of the results for the boundary of a set of finite perimeter in Euclidean space to Riemannian manifolds (see Chapter 4 in Giusti [184]). In particular we have

Theorem 1.39 *Let $E \subset M$ be a set of locally finite perimeter. Then*

$$\partial^* E = \bigcup_{i=1}^{\infty} C_i \cup Z,$$

where $P(E, Z) = 0$, and each C_i is compact and is contained in the level set of a C^1 function with non-vanishing gradient (e.g., C_i is contained in a C^1 hypersurface).

Moreover, for every $B \subset \partial^* E$,

$$P(E, B) = \mathcal{H}^{m-1}(B),$$

and, for every open set $\Omega \subset M$,

$$P(E, \Omega) = \mathcal{H}^{m-1}(\partial^* E \cap \Omega).$$

Finally,

$$\overline{\partial^* E} = \partial E.$$

Theorem 1.40 *Let* $E \subset M$ *be a subset of locally finite perimeter in* M*. If the measurable unit normal* ν *is continuous, then* ∂E *is a* C^1 *hypersurface.*

1.4.3 Properties of Sets of Finite Perimeter

A natural topology in the space of measurable sets is given by the L^1-convergence of characteristic functions. We say that a sequence of measurable sets $\{E_i\}_{i\in\mathbb{N}}$ converges in $L^1(M)$ or *in measure* to a measurable set E when the characteristic functions $\mathbf{1}_{E_i}$ converge in $L^1(M)$ to $\mathbf{1}_E$. Since $|\mathbf{1}_{E_i} - \mathbf{1}_E|$ is the characteristic function of the symmetric difference $E_i \Delta E$, convergence in measure is equivalent to $\lim_{i\to\infty} |E_i \Delta E| = 0$. When the sequence $\{E_i\}_{i\in\mathbb{N}}$ L^1-converges to E, we simply write $E = \lim_{i\to\infty} E_i$ or $E_i \to E$. We say that a sequence of measurable sets $\{E_i\}_{i\in\mathbb{N}}$ converges in $L^1_{\text{loc}}(M)$ to a measurable set E if, for every bounded set $B \subset M$, the sequence $\{B \cap E_i\}_{i\in\mathbb{N}}$ converges in $L^1(M)$ to $B \cap E$. The space of functions of bounded variation in Ω is denoted by $BV(\Omega)$.

Two basic properties of sets of finite perimeter, which proofs can be found in the first chapter of Giusti [184], are the lower semicontinuity of the perimeter and the compactness of sets with uniformly bounded perimeter.

Proposition 1.41 (Lower Semicontinuity of Perimeter) *Let* $\Omega \subset M$ *be an open set in a Riemannian manifold. Let* $\{E_i\}_{i\in\mathbb{N}}$ *be a sequence of sets of finite perimeter in* Ω *converging in* $L^1_{loc}(\Omega)$ *to a measurable set* E*. Then*

$$P(E, \Omega) \leqslant \liminf_{i\to\infty} P(E_i, \Omega).$$

Theorem 1.42 (Compactness) *Let* $\Omega \subset M$ *be a bounded open set with Lipschitz boundary in a Riemannian manifold* M*. Let* $\{E_i\}_{i\in\mathbb{N}}$ *be a sequence of sets with uniformly bounded perimeters* $P(E_i, \Omega)$*. Then we can extract a subsequence converging in* $L^1(\Omega)$ *to a set of finite perimeter* $E \subset \Omega$*.*

Another interesting property is

Proposition 1.43 ([293, Proposition 1.4]) *For every* $u \in BV(M)$*, there exists a sequence* $\{u_i\}_{i\in\mathbb{N}} \subset C_0^\infty(M)$ *such that* u_i *converges to* u *in* $L^1(M)$ *and*

$$|Du|(M) = \lim_{i\to\infty} \int_M |\nabla u_i|\, dM.$$

The proof of Proposition 1.4 in [293] is modeled on the one of Theorem 1.17 in Giusti [184]. Another interesting result is the coarea formula for sets of finite perimeter (see Proposition 4.2 in [292] for a proof in metric measure spaces and Theorem 1.23 in [184] for a proof in Euclidean space).

Proposition 1.44 *Let $\Omega \subset M$ an open set and $u \in L^1_{loc}(\Omega)$. Letting $E_t = \{u > t\}$ we have*

$$\int_{\mathbb{R}} |\partial E_t|(\Omega)\, dt = |Du|(\Omega).$$

In case $u \in BV(\Omega)$, then E_t has finite perimeter in Ω for almost every $t \in \mathbb{R}$.

As a consequence of Propositions 1.43 and 1.44 we obtain, compare with Theorem 1.24 in [184].

Proposition 1.45 *Every bounded set E of finite perimeter can be approximated by a sequence of sets $\{E_i\}_{i\in\mathbb{N}}$ with C^∞ boundary such that*

1. *$E_i \to E$ in $L^1(M)$,*
2. *$P(E_i) \to P(E)$.*

1.4.4 A Uniform Relative Isoperimetric Inequality for Small Balls

Given a compact subset $K \subset M$, we can prove a uniform relative isoperimetric inequality in balls of small radius centered at points in K. More precisely we have:

Lemma 1.46 (Uniform Relative Isoperimetric Inequality for Balls) *Let M be an m-dimensional Riemannian manifold and $K \subset M$ a compact subset. There exist constants $C_I > 0$, $r_0 > 0$, only depending on K, so that, for any set $E \subset M$ with locally finite perimeter, we have*

$$P(E, B(x,r)) \geqslant C_I \min\left\{|E \cap B(x,r)|, |B(x,r) \setminus E|\right\}^{(m-1)/m}, \tag{1.72}$$

for any $x \in K$.

Proof Take $x \in K$. For $0 < r < \operatorname{inj}(x)$, the restriction of the exponential map to the ball in T_xM centered at the origin of radius $r > 0$ is a bilipschitz map to $B(x,r)$. Then we transfer the relative isoperimetric inequality in the Euclidean ball $B(0,r) \subset T_xM$ (Theorem 2 in §5.6.2 in [149]) to $B(x,r)$. The result follows by estimating uniformly the bilipschitz constant of the exponential map for points in K and small radius as in Theorem 1.8. □

1.4.5 Deformations of Sets of Finite Perimeter

In isoperimetric problems, it is essential to have ways of modifying the volume of a set while controlling the change of perimeter. We present in this subsection two ways of volume adjustment, one by adding or removing balls and a second one by using the flow associated with a given vector field.

Theorem 1.47 (Volume Adjustment Using Balls) *Let $E \subset M$ be a measurable set in a Riemannian manifold M of finite volume. Given $r > 0$, there exists some $x \in M$ such that*

$$|E \cap B(x,r)| \geqslant \frac{|E|}{2|M|}\, b(r). \tag{1.73}$$

Proof Recall that $b(r)$ was defined in (1.59) as $\inf_{y \in M} |B(y,r)|$. The proof follows from an application of Fubini-Tonelli's Theorem to the function $(x, y) \in M \times M \to (\mathbf{1}_E \mathbf{1}_{B(y,r)})(x)$ taking into account that $\mathbf{1}_{B(y,r)}(x) = \mathbf{1}_{B(x,r)}(y)$. We get

$$\int_E |B(y,r)|\, dM(y) = \int_M |E \cap B(y,r)|\, dM(y).$$

Hence $|E|\, b(r) \leqslant |M| \sup_{y \in M} |E \cap B(y,r)|$ and (1.73) is proven. □

Remark 1.48 Theorem 1.47 implies that, given E, we can find a point $x \in M$ so that the set $E \setminus B(x,r)$ has volume smaller than or equal to $|E|(1 - b(r)/(2|M|)$. By the continuity of the function $r \mapsto |E \setminus B(x,r)|$, removing from E a ball centered at x of suitable radius, we can subtract from E any volume smaller than $|E|\, b(r)/(2|M|)$.

A similar argument can be applied to the complementary set $E^c = M \setminus E$ to obtain a point $x \in M$ such that $|B(x,r) \setminus E| = |E^c \cap B(x,r)| \geqslant |E^c| b(r)/2|M|$. In this case, the volume of the set $E \cup B(x,r)$ is larger than or equal to $|E| + |E^c|\, b(r)/(2|M|)$. So we can add to E any volume between 0 and $|E^c||b(r)/(2|M|)$ by enlarging the radius of a ball of center x between 0 and r.

Remark 1.49 Given a set $E \subset M$ in a compact connected manifold with $0 < |E| < |M|$, one can obtain a point $x \in M$ so that

$$|E \cap B(x,r)| > 0 \text{ and } |E^c \cap B(x,r)| > 0 \text{ for any } r > 0 \tag{1.74}$$

by the following argument: consider the sets

$$E_0 = \{x \in M : \exists r > 0 \text{ with } |E \cap B(x,r)| = 0\},$$

and

$$E_1 = \{x \in M : \exists r > 0 \text{ with } |E \cap B(x,r)| = |B(x,r)|\},$$

Then E_0 and E_1 are open sets and $|E \cap E_0| = |E \setminus E_1| = 0$ (see Proposition 3.1 in [184]). The set $\tilde{E} = (E \cap E_1) \setminus E_0$ has the same volume as E and $E_1 \subset \operatorname{int}(\tilde{E})$, $E_0 \subset \operatorname{int}(\tilde{E}^c)$. By the connectedness of M and the condition on $|E|$, the boundary of $\tilde{E}$ is not empty. Since $\partial\tilde{E} \cap (E_0 \cup E_1) = \emptyset$, any point $x \in \partial\tilde{E}$ satisfies (1.74).

Theorem 1.50 (Volume Adjustment Using Vector Fields) *Let E be a set with finite volume and locally finite perimeter in an open set $\Omega \subset M$, and let $B \subset \Omega$ be an open set such that $P(E, B) > 0$. Then there exist two constants $C > 0$ and $\overline{m} > 0$ such that, for every $-\overline{m} < m < \overline{m}$, there exists a set $F \subset \Omega$ such that $F = E$ outside B, satisfying*

1. $|F| = |E| + m$,
2. $|P(F, \Omega) - P(E, \Omega)| \leqslant C|m|$.

Proof As $P(E, B) > 0$, there exists a vector field X with compact support in B such that $\int_E \operatorname{div} X \, d\mathcal{H}^n > 0$. The one-parameter group of diffeomorphisms $\{\varphi_t\}_{t\in\mathbb{R}}$ associated with X then satisfies $\frac{d}{dt}\big|_{t=0}|\varphi_t(E)| = \int_E \operatorname{div} X \, d\mathcal{H}^n > 0$. This implies the existence of $\overline{m} > 0$ and an open interval I around 0 such that the function $t \in I \mapsto (|\varphi_t(E)| - |E|) \in (-\overline{m}, \overline{m})$ is a C^1 diffeomorphism. This proves 1.

To prove (ii), we consider the reduced boundary $\partial^* E$ of E. Take $m \in (-\overline{m}, \overline{m})$ and $F = \varphi_{t_m}(E)$ so that $|F| = m$. Then $P(F, \Omega) - P(E, \Omega) = P(F, B) - P(E, B)$. By the area formula

$$|P(F, B) - P(E, B)| = \left| \int_{\partial^* E \cap B} (\operatorname{Jac}(\varphi_{t_m}) - 1) \, d\mathcal{H}^{n-1} \right| \leqslant C'|t_m| \, P(E, B),$$

where C' is a constant depending only on the vector field X. As $|t_m| \leqslant C''|m|$, where $C'' > 0$ is a constant only depending on X and E, we obtain 2. □

1.5 Other Notions of Boundary Area

1.5.1 Minkowski Content

The Minkowski content is a notion of boundary measure that can be defined simply in terms of volume and distance. Hence it may be considered in general metric spaces endowed with a measure. Given a bounded measurable set E, we consider the tubular neighborhood of E of radius $t > 0$ defined by

$$E_t = \{p \in M : d(p, E) \leqslant t\}.$$

Definition 1.51 The *lower* and *upper Minkowski contents* of a Borel set E are defined by

$$\operatorname{Mink}_-(E) = \liminf_{t \to 0} \frac{|E_t| - |E|}{t}, \quad \operatorname{Mink}_+(E) = \limsup_{t \to 0} \frac{|E_t| - |E|}{t}. \tag{1.75}$$

If the upper and lower Minkowski content coincide, their common value is called the *Minkowski content* of E. It is denoted by $\mathrm{Mink}(E)$.

When E is bounded with C^2 boundary, Steiner's formula implies that the Minkowski content of E is equal to the Riemannian measure of ∂E. Hence

$$P(E) = \mathrm{Mink}(E)$$

when E is bounded with C^2 boundary. In the general case, when E is merely a bounded measurable set, we have

$$P(E) \leqslant \mathrm{Mink}_-(E). \tag{1.76}$$

To prove this inequality, let X be a vector field in M with compact support and $||X|| \leqslant 1$, and let φ_t be its associated flow. Then $\varphi_t(E) \subset E_t$: if $q = \varphi_t(p)$, with $p \in E$, then the curve $s \in [0, t] \to \varphi_s(p)$ connects p and q with length

$$\int_0^t ||X_{\varphi_s(p)}||\, ds \leqslant \int_0^t ds = t.$$

Hence $d(q, E) \leqslant t$. This proves $\varphi_t(E) \subset E_t$. So we have $|\varphi_t(E)| \leqslant |E_t|$ for all t and

$$\int_M \operatorname{div} X\, dM = \frac{d}{dt}\bigg|_{t=0} |\varphi_t(E)| \leqslant \liminf_{t\to 0} \frac{|E_t| - |E|}{t} = \mathrm{Mink}_-(E).$$

Taking supremum over X, we finally get (1.76).

When considering variational problems, the perimeter is preferred to the Minkowski content because of the lack of semicontinuity properties of the latter. However, there is a closer link between perimeter and Minkowski content, as shown by Ambrosio, Di Marino, and Gigli [20]. Recall that we are assuming our Riemannian manifolds to be second-countable so that their Riemannian measures are σ-finite. Following [20] we define the relaxed Minkowski contents $\mathrm{Mink}_-^*(E)$ and $\mathrm{Mink}_+^*(E)$ of a Borel set E by

$$\mathrm{Mink}_-^*(E) = \inf\Big\{ \liminf_{i\to\infty} \mathrm{Mink}_-(E_i) : E_i \to E \text{ in measure}\Big\},$$

$$\mathrm{Mink}_+^*(E) = \inf\Big\{ \liminf_{i\to\infty} \mathrm{Mink}_+(E_i) : E_i \to E \text{ in measure}\Big\}.$$

Then it is clear that $\mathrm{Mink}_-^*(E) \leqslant \mathrm{Mink}_+^*(E)$. There holds also (see Lemma 2.1 in [20])

$$\mathrm{Mink}_-(E) \geqslant \mathrm{Mink}_+^*(E) \tag{1.77}$$

for any Borel set E, which implies $\text{Mink}^*_-(E) \geqslant \text{Mink}^*_+(E)$ and hence equality

$$\text{Mink}^*_-(E) = \text{Mink}^*_+(E).$$

To prove (1.77) we assume $\text{Mink}_-(E) < +\infty$. Hence $|E_r \setminus E| \to 0$ when $r \downarrow 0$. Since $(E_s)_t \subset E_{s+t}$ we have

$$\frac{|E_{s+t}| - |E_s|}{t} \geqslant \frac{|(E_s)_t| - |E_s|}{t}.$$

At a point of differentiability of the non-decreasing and left-continuous map $f(r) = |E_r|$, we have $f'(s) \geqslant \text{Mink}_+(E_s)$. Integrating with respect to s yields

$$|E_r| - |E| = f(r) - f(0_+) \geqslant \int_0^r f'(s)\, ds \geqslant \int_0^r \text{Mink}_+(E_s)\, ds$$

for all $r > 0$. Dividing by r and applying the mean value theorem, we get

$$\frac{|E_r| - |E|}{r} \geqslant \frac{1}{r} \int_0^r \text{Mink}_+(E_s)\, ds \geqslant \text{Mink}_+(E_{t(r)})$$

for some $0 < t(r) < r$. Taking inferior limits when $r \to 0$ yields (1.77).

The main result in Ambrosio, Di Marino, and Gigli [20] is the equality of the perimeter and the relaxed Minkowski content.

Theorem 1.52 ([20, Theorem 3.6]) *Let E be a Borel set with $|E| < +\infty$. Then*

$$P(E) = \text{Mink}^*_-(E). \tag{1.78}$$

Proof Let $E_i \to E$ in measure. By inequality (1.76), the lower semicontinuity of perimeter and the definition of Mink^*_- we get $P(E) \leqslant \text{Mink}^*_-(E)$. To prove the reverse inequality, we consider a sequence of Lipschitz functions $f_i : M \to [0, 1]$ with $\{f_i > 0\}$ bounded so that $f_i \to \mathbf{1}_E$ and $\int_M \text{lip}(f_i)\, dM \to P(E)$. Observe that

$$\int_0^\infty \text{Mink}_-(\{f_i \geqslant t\})\, dt \leqslant \int_M \text{lip}(f_i)\, dM$$

by (3.3) in [20]. Then, for every $i > 2$, we can find $t(i) \in (1/i, 1 - 1/i)$ such that

$$\left(1 - \frac{2}{i}\right) \text{Mink}_-(\{f_i \geqslant t(i)\}) \leqslant \int_M \text{lip}(f_i)\, dM.$$

Since $\{f_i \geqslant t(i)\}$ converge to E in measure, taking limits it follows that $\text{Mink}^*_-(E) \leqslant P(E)$. □

Remark 1.53 A well-known proof of the isoperimetric inequality in Euclidean space for the Minkowski content is obtained from the Brunn-Minkowski inequality (e.g., §8 in [83]). Given two compacts sets $E, B \subset \mathbb{R}^m$ we have

$$|E + B|^{1/m} \geqslant |E|^{1/m} + |B|^{1/m}, \tag{1.79}$$

where we have denoted by $E + B$ the Minkowski sum $\{e + b : e \in E, b \in B\}$. If we take B as the closed unit ball $\overline{B}(0, t)$ centered at 0 of radius $t > 0$ then $E + \overline{B}(0, t)$ is the closed tubular neighborhood E_t of E of radius t, and we obtain from (1.79) the inequality

$$\frac{|E_t|^{1/m} - |E|^{1/m}}{t} \geqslant \omega_m^{1/m}.$$

Taking lim inf when $t \to 0$ we get $\operatorname{Mink}_-(E) \geqslant m\omega_m^{1/m}\,|E|^{(m-1)/m}$.

Assume now that E has finite volume and perimeter, and let E_i converge to E in measure. Then we have

$$\liminf_{i\to\infty} \operatorname{Mink}_-(E_i) \geqslant m\omega_m^{1/m}\,|E|^{(m-1)/m}.$$

Taking infimum over any sequence converging in measure to E and using Theorem 1.52, we get

$$P(E) \geqslant m\omega_m^{1/m}|E|^{(m-1)/m}. \tag{1.80}$$

This is the classical isoperimetric inequality in $\mathbb{R}^m$ for the perimeter.

Remark 1.54 A *manifold with density* (M, g, Ψ) is a Riemannian manifold (M, g) together with a positive function Ψ on M. Then M is a metric measure space with distance d equal to the Riemannian distance and measure

$$V_\Psi(E) = \int_E \Psi\, dM$$

for any Borel set $E \subset M$. Given a bounded set $E \subset M$ with C^2 boundary S, its Minkowski content is

$$\frac{d}{dt}_{t=0} \int_{\varphi_t(E)} \Psi\, dM,$$

where φ_t is the flow associated with the restriction of the outer unit normal vector field X to the parallel hypersurfaces S_t at signed distance $|t| < \varepsilon$ from S. The first variation formula for the volume implies

$$\frac{d}{dt}_{t=0}\int_{\varphi_t(E)} \Psi\, dM = \int_E \big(X(\Psi) + \Psi\ \mathrm{div}_M X\big)\, dM$$
$$= \int_E \mathrm{div}_M(\Psi X)\, dM = \int_S \Psi\, dS,$$

since $X = N$. Hence the *weighted area* A_Ψ associated with the Minkowski content in (M, g, Ψ) is equal to

$$A_\Psi(S) = \int_S \Psi\, dS$$

for any smooth hypersurface $S \subset M$.

1.5.2 The Hausdorff Measure

The Hausdorff measure can be defined on any metric space (X, d). Let $A \subset X$. For any $s \in [0, \infty)$ and $\delta \in (0, \infty]$, we define

$$\mathcal{H}^s_\delta(A) = \inf\left\{c(s)\sum_{i=1}^{\infty}\big(\mathrm{diam}\, C_i\big)^s : A \subset \bigcup_{i=1}^{\infty} C_i,\ \mathrm{diam}\, C_i \leqslant \delta\right\}. \tag{1.81}$$

Here $c(s)$ is an arbitrary positive constant only depending on s. When s is a positive integer, a natural choice is $c(s) = 2^{-s}\omega_s$, where ω_s is the Lebesgue volume of the unit ball in $\mathbb{R}^s$. We always make this choice for integer s in Euclidean spaces or Riemannian manfolds.

When $\delta \leqslant \delta'$ we have $\mathcal{H}^s_\delta(A) \geqslant \mathcal{H}^s_{\delta'}(A)$. We define the s-dimensional Hausdorff measure by

$$\mathcal{H}^s(A) = \lim_{\delta\to 0} \mathcal{H}^s_\delta(A) = \sup_{\delta>0} \mathcal{H}^s_\delta(A). \tag{1.82}$$

By Carathéodory's criterion, $\mathcal{H}^s$ defines a measure on the Borel σ-algebra of X. The isodiametric inequality implies that the m-dimensional Hausdorff measure $\mathcal{H}^m$ coincides with Lebesgue measure $\mathcal{L}^m$ on $\mathbb{R}^m$ (see §2.1 in Evans and Gariepy [149]).

The Hausdorff dimension of a metric space is defined from the following result.

Lemma 1.55 *If X is a metric space, then there exists $d \in [0, \infty]$ such that $\mathcal{H}^s(X) = 0$ for all $s > d$ and $\mathcal{H}^s(X) = \infty$ for all $s < d$.*

Proof If $\mathcal{H}^s(X) = \infty$ for all $s \geqslant 0$, we let $d = \infty$. Otherwise we define $d = \inf\{s \geqslant 0 : \mathcal{H}^s(X) < \infty\}$. Obviously $\mathcal{H}^s(X) = \infty$ for all $s < d$. Now let

$s > t > d$ with $\mathcal{H}^t(X) < \infty$. Take $\delta > 0$ and consider a covering of X by a family of sets $\{C_i\}_{i\in\mathbb{N}}$ with $\operatorname{diam} C_i \leqslant \delta$ such that

$$c(t)\sum_{i=1}^{\infty}\big(\operatorname{diam} C_i\big)^t \leqslant \mathcal{H}^t_\delta(X) + 1 \leqslant \mathcal{H}^t(X) + 1.$$

Then

$$\begin{aligned}\mathcal{H}^s_\delta(X) &\leqslant c(s)\sum_{i=1}^{\infty}\big(\operatorname{diam} C_i\big)^s\\ &= \frac{c(s)}{c(t)}\,c(t)\sum_{i=1}^{\infty}\big(\operatorname{diam} C_i\big)^t\big(\operatorname{diam} C_i\big)^{s-t}\\ &\leqslant \frac{c(s)}{c(t)}\,\delta^{s-t}\big(\mathcal{H}^s(X)+1\big).\end{aligned}$$

Taking limits when $\delta \to 0$ we get $\mathcal{H}^s(A) = 0$. □

The *Hausdorff dimension* of X is then defined by

$$\dim_H(X) = \inf\big\{s \geqslant 0 : \mathcal{H}^s(X) < \infty\big\}.$$

The Riemannian volume μ_g of a Riemannian manifold (M, g) coincides with the m-dimensional Hausdorff measure $\mathcal{H}^m$ associated with the Riemannian distance. Fix a point $p \in M$ and take a normal chart (U, φ) centered at p so that the coordinates of the Riemann tensor are of the form $g_{ij} = \delta_{ij} + O(r)$. Then the Riemannian distance d on U and the Euclidean distance d_0 on $\varphi(U)$ satisfy the relation

$$d \approx d_0 \circ \varphi$$

on $B(p, r)$, where the symbol $\approx$ is used to express the existence of two functions $c, C = 1 + O(r)$ such that $c\,d \leqslant d_0 \circ \varphi \leqslant C\,d$. So, for any $A \subset B(p, r) \subset U$ we have

$$\mathcal{H}^m(A) \approx \mathcal{H}^m_0(\varphi(A)),$$

where $\mathcal{H}^m_0$ is the m-dimensional Hausdorff measure in $\mathbb{R}^m$. Since $\sqrt{\det(g_{ij})} = 1 + O(r)$ we get

$$\mu_g(A) \approx \mathcal{L}^m(\varphi(A)).$$

Finally, as $\mathcal{H}_0^m = \mathcal{L}^m$ we obtain

$$\mathcal{H}^m(A) \approx \mu_g(A).$$

This implies that $\mathcal{H}^m$ and μ_g are absolutely continuous with respect to each other and also that the Radon-Nikodym derivative of $\mathcal{H}^m$ with respect to μ_g is equal to 1. Hence $\mathcal{H}^m = \mu_g$.

We can also prove that the Riemmanian volume of a k-dimensional submanifold S of a Riemannian manifold M coincides with its k-dimensional Hausdorff measure. Let $p \in S$. We take a normal chart (U, φ) centered at p. Hence

$$\mathcal{H}^k(S \cap B(p,r)) \approx \mathcal{H}_0^k(\varphi(S) \cap B(0,r)).$$

Because of the area formula for Hausdorff measures, we have

$$A_0(\varphi(S) \cap B(0,r)) = \mathcal{H}_0^k(\varphi(S) \cap B(0,r)),$$

where A_0 is the k-dimensional Euclidean area of $\varphi(S) \cap B(0,r)$. Since

$$A_0(\varphi(S) \cap B(0,r)) \approx A(S \cap B(p,r)),$$

we finally get

$$\mathcal{H}^k(S \cap B(p,r)) \approx A(S \cap B(p,r)).$$

Note that the above argument also works for k-dimensional Lipschitz submanifolds defined by k-dimensional local Lipschitz parameterizations.

We refer the reader to the monographs by Federer [154, §2.10], Burago, Burago and Ivanov [81, §1.7.2], and Evans and Gariepy [149, Chapter 2] for further properties of the Hausdorff measure.

1.6 The Isoperimetric Profile

We introduce now the important notion of isoperimetric profile of a Riemannian manifold. This concept can be defined in any space with suitable notions of volume and perimeter as in metric measure spaces, where the measure plays the role of the volume and the Minkowski content the role of the perimeter.

Definition 1.56 Let (M, g) be a Riemannian manifold. The *isoperimetric profile* of M is the function I_M that assigns, to each $v \in (0, |M|)$, the value

$$I_M(v) = \inf\big\{P(E) : E \text{ measurable}\,, |E| = v\big\}. \tag{1.83}$$

The isoperimetric profile of M will be often denoted simply by I.

Definition 1.57 Let (M, g) be a Riemannian manifold. We say that a set $E \subset M$ is *isoperimetric* or that it is an *isoperimetric region* if

$$P(E) = I_M(|E|). \tag{1.84}$$

If $|E| = v$ then we say that E is an isoperimetric region of volume v.

The isoperimetric profile must be understood as an *optimal* isoperimetric inequality in M since, for any subset $F \subset M$ of volume $0 < |F| < |M|$,

$$P(F) \geqslant I_M(|F|),$$

with equality precisely for isoperimetric sets. Indeed, any function $f : (0, |M|) \to \mathbb{R}^+$ satisfying the inequality $f \leqslant I_M$ provides an isoperimetric inequality in M, namely, $P(F) \geqslant f(|F|)$, which it is not optimal in general.

As we will see in later chapters, existence of isoperimetric sets is not guaranteed in non-compact manifolds.

Example 1.58 The classical isoperimetric inequality in Euclidean space states that round balls are the unique isoperimetric sets in $\mathbb{R}^m$. Since the quantity $P(F)/|F|^{(m-1)/m}$ is invariant by Euclidean dilations we have, for any set $F \subset \mathbb{R}^m$ of finite perimeter and volume $|F|$ and any ball $B \subset \mathbb{R}^m$ of volume $|B| = |F|$, the inequality

$$\frac{P(F)}{|F|^{(m-1)/m}} \geqslant \frac{P(B)}{|B|^{(m-1)/m}} = \frac{P(B(0,1))}{|B(0,1)|^{(m-1)/m}} = m\omega_m^{1/m},$$

where $\omega_m = |B(0, 1)|$. Hence the isoperimetric profile of the m-dimensional Euclidean space is given by

$$I_{\mathbb{R}^m}(v) = m\omega_m^{1/m} v^{(m-1)/m}. \tag{1.85}$$

More examples of isoperimetric profiles will be given in the next chapters.

Remark 1.59 (Scaled Riemannian Metrics) Given a Riemannian manifold (M, g) and an scalar $\lambda \neq 0$, we consider the Riemannian metric $\lambda^2 g$ on M. Given a Borel set $E \subset M$ we get

$$|E|_{\lambda^2 g} = \lambda^m |E|_g, \qquad P_{\lambda^2 g}(E) = \lambda^{m-1} P_g(E).$$

This immediately implies

$$I_{(M,\lambda^2 g)}(v) = \lambda^{m-1} I_{(M,g)}\left(\frac{v}{\lambda^m}\right) \tag{1.86}$$

for every $0 < v < |M|_{\lambda^2 g} = \lambda^m |M|_g$ from the very definition of isoperimetric profile.

1.6.1 Regularity of Isoperimetric Sets

Regularity results for sets minimizing perimeter under a volume constraint were obtained by Morgan, who proved in Corollary 3.7 and 3.8 of [302] the following:

Theorem 1.60 *Let E be a measurable set of finite volume minimizing perimeter under a volume constraint in a smooth m-dimensional Riemannian manifold M. Then*

1. *If $m \leqslant 7$ then the boundary S of E is a smooth hypersurface.*
2. *If $m > 7$ then the boundary of E is the union of a smooth hypersurface S and a closed singular set S_0 of Hausdorff dimension at most $m-8$ (i.e., $\mathcal{H}^{m-8+\gamma}(S_0) = 0$ for all $\gamma > 0$).*

Morgan's proof is based on Bombieri's regularity results [69] for almost minimizing currents. The boundary of a set of finite perimeter is a rectifiable current (see [133], Theorem 14.3 in [404] or Chapter 4 in [184]). In local coordinates, the perimeter of a rectifiable set provides a parametric elliptic functional. Proposition 3.1 in [302] implies that the boundary of an isoperimetric set satisfies Bombieri's almost minimality condition.

When trying to get geometric information on the boundary of a set minimizing perimeter under a volume constraint, the following technical result proved by Sternberg and Zumbrum in Lemma 2.4 of [411] allows us to focus just on the regular part S of the boundary.

Lemma 1.61 *Let $E \subset M$ be a bounded minimizer of perimeter under a volume constraint in a smooth Riemannian manifold M. Let S be the regular part of the boundary of E and S_0 its singular part.*

Then for every $\varepsilon > 0$, there exist open sets $U' \Subset S \subset M$ with $S_0 \subset U'$, $U \subset M$ contained in an open tubular neighborhood in M of S_0 of radius ε, and a smooth function $\varphi_\varepsilon : M \to \mathbb{R}$ such that $0 \leqslant \varphi_\varepsilon \leqslant 1$,

$$\varphi_\varepsilon(x) = 0 \text{ in } U', \quad \varphi_\varepsilon(x) = 1 \text{ in } M \setminus U,$$

and

$$\int_S |\nabla_S \varphi_\varepsilon|^2 dS \leqslant C\varepsilon,$$

for some constant $C > 0$ depending on E but independent of ε.

Proof In the proof we shall use that $\mathcal{H}^{m-3}(S_0) = 0$. This holds since the Hausdorff dimension on S_0 is at most $m - 8$. Given any $\varepsilon > 0$ we may cover S_0 with a finite family of balls $B(p_i, r_i)$ such that

$$\sum_i r_i^{m-3} < \varepsilon.$$

For each i we take $\xi_i \in C_0^\infty(M)$ such that $0 \leqslant \xi_i \leqslant 1$,

$$\xi_i(q) = \begin{cases} 0, & q \in B(p_i, r_i), \\ 1, & q \in M \setminus B(p_i, 2r_i), \end{cases}$$

and

$$|\nabla \xi_i| \leqslant \frac{2}{r_i}.$$

We take $\varphi_\varepsilon = \prod_i \xi_i$. Then $0 \leqslant \varphi_\varepsilon \leqslant 1$, $\varphi_\varepsilon = 0$ on $U' = \bigcup_i B(p_i, r_i)$, and $\varphi_\varepsilon = 1$ on $M \setminus \bigcup B(p_i, 2r_i) = M \setminus U$. Moreover

$$\begin{aligned} \int_M |\nabla \varphi_\varepsilon|^2 dM &\leqslant \sum_i \int_M |\nabla \xi_i|^2 dM \\ &\leqslant \int_{B(p_i 2r_i)\setminus B(p_i, r_i)} |\nabla \xi_i|^2 dM \\ &\leqslant \frac{4}{r_i^2} C_0 (2r_i)^{m-1} \leqslant 2^{m+1} C_0 \sum_i r_i^{m-3} < C\varepsilon. \end{aligned}$$

□

1.6.2 Restricted and Free-Boundary Isoperimetric Profiles

Given an open set $\Omega \subset M$, we may consider the minimization problem

$$\inf\{P(E) : E \subset \overline{\Omega}, |E| = v\}. \tag{1.87}$$

We emphasize the fact that the perimeter here is the one in M. When Ω is bounded, compactness results for sets of finite perimeter imply that there is a solution E to (1.87) for any given volume. Such a set is called an *isoperimetric set* for the restricted isoperimetric problem (1.87) in Ω.

If Ω is a Euclidean domain with boundary $\partial\Omega$ of class C^1, then ∂E is of class C^1 near $\partial\Omega$ (see Theorem 3 in [185]). If Ω is a bounded convex set with C^2 boundary, it was shown in Theorem 3.6 in [412] that ∂E is of class $C^{1,1}$ near $\partial\Omega$. When Ω is a domain in a Riemannian manifold, the Proposition in page 418 of [444] implies that ∂E is $C^{1,1}$ near $\partial\Omega$ when $\partial\Omega$ is C^2 (see also Theorem 6.15 and Remark 6.16 in [295]).

We summarize this regularity result for future reference.

Theorem 1.62 *Let (M, g) be a complete Riemannian manifold and $\Omega \subset M$ be an open subset with C^2 boundary. Then for any $0 < v < |\Omega|$, there exists an isoperimetric region E of volume v in Ω. Moreover, the boundary of E satisfies the following properties:*

1. *∂E is of class $C^{1,1}$ near $\partial\Omega$.*
2. *$\partial E \cap \Omega$ in the union of a smooth hypersurface S with constant mean curvature H and a closed singular set S_0 of Hausdorff dimension at most $m - 8$.*

An interesting open problem is to prove the convexity of the isoperimetric regions for this problem inside convex domains of $\mathbb{R}^m$. Partial results have been obtained by Stredulinsky and Ziemer [412]; Rosales [384],; Alter, Caselles, and Chambole [16]; and Alter and Caselles [15].

A completely different problem is the following: given an open set $\Omega \subset M$, we consider the minimization problem

$$\inf\{P(E, \Omega) : E \subset \Omega, |E| = v\}. \tag{1.88}$$

We remark that for this problem only the boundary area of E inside Ω is taken into account.

In case Ω is bounded, compactness results for sets of finite perimeter imply the existence of a solution to (1.88) for any admissible volume. A set E realizing the minimum is called an *isoperimetric region*.

The boundary S of E is a "hypersurface" that separates the container Ω into two components with prescribed volume. Should it be regular, it would have constant mean curvature inside Ω and ∂S would meet $\partial\Omega$ in an orthogonal way. The boundary S is usually called a *free boundary*. Interior regularity of ∂E inside Ω is provided by the classical regularity result given in Theorem 1.60. Regularity results at $\partial\Omega$ in the Euclidean case have been provided in [203, 205, 206, 209].

1.7 Notes

Notes for Sect. 1.1 Riemannian Geometry became a classical subject in geometry during the twentieth century. The main results of the theory can be found in a large number of monographs. We shall mainly consider the ones by do Carmo [141], Warner [437], and Chavel [102], although some other exceptional ones are available, such as Kobayashi and Nomizu [252, 253], Petersen [345], and Bishop and Crittenden [60].

Notes for Sect. 1.2 The Riemannian volume is computed in local coordinates using the density

$$\sqrt{\det(g_{ij})}.$$

This allows us to define a positive functional on the set of continuous functions with compact support by means of a partition of unity and the use of Riesz representation theorem. This way, we immediately obtain regularity properties of the associated measure.

As observed in Note III.4 in Chavel [102], there is a way to define the Riemannian volume in an abstract way as the only functional that assigns volume 1 to the unit cube and is increasing under injective distance non-increasing maps (see Gromov [198] and pp. 193–195 in Burago, Burago, and Ivanov [81]).

The doubling constant obtained in Sect. 1.2.5 can also be obtained from comparison results for the volume of Riemannian balls when the Ricci curvature is bounded from below, as shown in Sect. 1.3.5.

Notes for Sect. 1.3 The first variation formula of the area for graphs in the three-dimensional Euclidean space was computed by Lagrange in 1762 [258]. The formalism we have followed appeared in Chapter 2 of Simon's treatise [404] for compactly supported variations associated with one-parameter groups of diffeomorphisms in Euclidean space. The use of one-parameter families instead of groups has some slight advantages, mainly because the composition of two one-parameter groups is merely a one-parameter family of diffeomorphisms. Variational formulas, specially the second one, have been used to characterize the second-order minima of perimeter under a volume constraint for sets with smooth boundary in several interesting spaces, such as the Euclidean one by Barbosa and do Carmo [38] and the sphere and hyperbolic space by Barbosa, do Carmo, and Eschenburg [39]. This approach is particularly useful in spaces where there are different types of isoperimetric sets, such as the three-dimensional real projective space; see Ritoré and Ros [365].

The doubling constant we have obtained from a lower bound on the Ricci curvature is not explicit. Under the hypothesis $\operatorname{Ric} \geqslant (m-1)\,\delta$, with $\delta < 0$, the more refined bound

$$|B(p, 2r)| \leqslant 2^m \exp\{2r(m-1)\sqrt{(-\delta)}\}\,|B(p, r)|,$$

valid for any $r > 0$, can be found in [107] and Section 10.1 in [213].

Notes for Sect. 1.4 Sets of finite perimeter were introduced by Caccioppoli in the decade of 1950. They have been extensively studied in Euclidean spaces by the Italian school of Geometric Measure Theory (see De Giorgi [134]). It is natural to use this theory in the study of isoperimetric problems. The definition and many properties of these sets can be easily extended to Riemannian manifolds, as we have outlined in Sect. 1.4. A classical monograph is Giusti's [184], and a more recent one, including the regularity theory for quasi-minimizers of the perimeter, is the one by Maggi [275]. The theory of functions of bounded variation and sets of finite perimeter has been extended to metric spaces in recent years. We refer the reader to [17, 21, 22, 292] for background on this topic.

Notes for Sect. 1.5 Minkowski content and Hausdorff measure also provide reasonable notions of boundary measure. Minkowski content was introduced by Minkowski [289, 290], starting from the observation that "volume is a more elementary concept than the length of a curve or the area of a surface." Perimeter is preferred over Minkowski content in variational problems, because of the lack of semi-continuity properties. See, however, the recent paper by Ambrosio, Di Marino, and Gigli [20], where it is proven that the perimeter and the relaxed Minkowski content coincide in metric measure spaces. In any case, the use of the Minkowski content is frequent in manifolds with density and more general metric measure spaces.

Manifolds with Density The m-dimensional Gauss space is the manifold with density with underlying Riemannian manifold $(\mathbb{R}^m, g_0)$ and density function $\Psi = (2\pi)^{-m/2}e^{-|x|^2/2}$. The function Ψ is known as the *Gaussian density*. The isoperimetric problem in Gauss space has been one of the leading problems of the theory. It was solved independently by Borell [75] in 1974 and Sudakov and Tirel'son [415] in 1975, who proved that half-spaces minimize perimeter under a volume constraint for this density. A new proof was given in 1983 by Ehrhard [147] using symmetrization. In 1997 Bobkov [65], after Talagrand [418], proved a functional version of this isoperimetric inequality, later extended to the sphere and used to prove isoperimetric estimates for the unit cube by Barthe and Maurey [44]. Barthe [42] proved an isoperimetric inequality for unconditional sets in Gauss space. Following [65], Bobkov and Houdré [66] considered "unimodal densities" with finite total measure on the real line. These authors explicitly computed the isoperimetric profile for such densities and found some of the isoperimetric solutions. Gromov [197, 199] studied manifolds with density as "metric measure spaces" and mentioned the natural generalization of mean curvature obtained by the first variation of weighted area. Bakry and Ledoux [33], Bayle [47], and Morgan [303] proved generalizations of the Levy-Gromov isoperimetric inequality and other geometric comparisons depending on a lower bound on the generalized Ricci curvature of the manifold. Isoperimetric comparisons results in manifolds with density were considered by Maurmann and Morgan [279]. Existence of isoperimetric sets in $\mathbb{R}^m$ with density under various hypotheses on the growth of the density were proven by Morgan and Pratelli [312] and E. Milman [285, 286] (see also De Philippis, Franzina, and Pratelli [135]). Rosales et al. [385] gave sufficient conditions for the existence of isoperimetric regions, described the stability of balls centered at 0 for radial densities, and stated in the *log-convex conjecture* by Brakke: that balls centered at the origin are isoperimetric sets for radial, log-convex densities, a result proven by Chambers [98]. For regularity of isoperimetric regions with density, see Sect. 3.10 in paper of Morgan [302], and see also Pratelli and Saracco [349]. Boundedness of isoperimetric regions was studied by Cinti and Pratelli [122] and Pratelli and Saracco [350]. Symmetrization techniques in manifolds with density developed by Ros [380] and Morgan, Howe, and Harman [309]. The Gaussian double bubble problem has been recently solved by Milman and Neeman [287].

For surveys on manifolds with density, the reader is referred to [47, 303, 308] and the references therein. Barthe [43] surveyed recent developments of the isoperimetric problem for product probability measures.

Notes for Sect. 1.6 As mentioned in the introduction, we will not treat isoperimetric inequalities related to mathematical physics. We refer the reader to Polya and Szego [347] and Payne [335] for background on such inequalities.

We are not considering neither quantitative isoperimetric inequalities. In this type of inequalities, the aim is to bound from below the *isoperimetric deficit*

$$\frac{P(E)}{P(B)} - 1,$$

where B is a Euclidean ball with $|E| = |B|$. Two-dimensional results were obtained by Bonnesen [70]. Pioneering work in Euclidean spaces was made by Fuglede [168, 169] and Hall [215]. Fusco, Maggi, and Pratelli [171] estimated optimally the isoperimetric deficit in $\mathbb{R}^m$ by a constant times the Fraenkel asymmetry. See also Figalli, Maggi, and Pratelli [160] for a mass transportation approach to the problem and Cicalese and Leonardi for a completely different approach [121]. The techniques by Cicalese and Leonardi have been extended to Riemannian manifolds by Chodosh, Engelstein, and Spolaor [112]. See the survey by Fusco [170].

In general, it is recognized that the regularity of the boundary of sets minimizing perimeter under a volume constraint follows from the Euclidean theory or, more generally, from the techniques employed in Euclidean theory as in Almgren [14] or Schoen and Simon [399]. The first proof of regularity in Euclidean space was given by Gonzalez, Massari, and Tamanini [186]. This was later extended by Tamanini to (Λ, r_0)-minimizers of the perimeter in [420] (see also [275]). Morgan's paper [302] treats the case of Riemannian manifolds as ambient spaces following ideas of Bombieri [69]. The result by Sternberg and Zumbrun [411] permits to focus on the geometry of the regular part S of Theorem 1.60.

The perimeters considered in Sect. 1.6.2 are particular cases of the functionals studied in capillary theory. We refer the reader to Finn [161], Wente [440, 441] and McCuan [281] for background on this subject.

Chapter 2
Isoperimetric Inequalities in Surfaces

In this chapter, we study isoperimetric inequalities in Riemannian surfaces. Firstly, we present Blaschke's variational proof of the Gauss-Bonnet theorem, a fundamental tool in the two-dimensional isoperimetric theory. Then we recall Hurwitz's simple proof of the isoperimetric inequality in the plane using Wirtinger's inequality and, afterward, Weil's proof of the validity of the planar isoperimetric inequality in Cartan-Hadamard planes (i.e., complete simply connected surfaces with non-positive sectional Gauss curvature).

In the next sections, we prove several isoperimetric inequalities depending on the Gauss curvature of the surface by deforming the regions while keeping controlled the perimeter and area along the deformation. We first consider the method of inner parallels. This is a classical deformation used to obtain general isoperimetric inequalities on surfaces. Then we continue with a result by Bandle using a deformation of a simply connected domain by level curves of a function u satisfying the differential inequality $\Delta_0 u + K_0 e^{2u} \geqslant 0$, where Δ_0 is the Laplacian of a flat metric conformal to the original Riemannian metric. Finally, we present the results of Benjamini and Cao using the curve shortening flow.

Then we present a completely different method: we use techniques of calculus of variations to characterize the curves minimizing length under an area constraint. This method is extremely versatile and allows us to obtain results on spheres that could not be achieved by using the deformation methods of the previous sections. We specialize this method to surfaces of revolution satisfying monotonicity conditions on the Gauss curvature. Of course, in order to apply this method to obtain isoperimetric inequalities, we need to prove the existence of minimizers. This is guaranteed in compact surfaces, but, unfortunately, not in non-compact surfaces. Indeed, we shall show that minimizers do not exist for any given area in complete surfaces of revolution with strictly increasing Gauss curvature.

We finish the chapter proving existence of minimizers in complete surfaces with non-negative Gauss curvature.

M. Ritoré, *Isoperimetric Inequalities in Riemannian Manifolds*,
Progress in Mathematics 348, https://doi.org/10.1007/978-3-031-37901-7_2

Along this chapter (M, g) is a complete Riemannian surface. The Gauss curvature of M will be denoted by K. If $\Omega \subset M$ is a region with smooth boundary $C = \partial\Omega$, we denote by h the geodesic curvature of C with respect to the outer unit normal N. This means that $h = \langle \nabla_T N, T\rangle$, where T is a unit tangent to C. With this choice, circles of radius $r > 0$ in the plane $\mathbb{R}^2$ have geodesic curvature $1/r$.

2.1 A Variational Proof of the Gauss-Bonnet Theorem

One of the main tools when using a deformation approach to obtain isoperimetric inequalities, either by inner parallels or by the curve shortening flow, is the Gauss-Bonnet theorem. As an application of the variational formulas for area and perimeter, we briefly recall Blaschke's variational proof of this result. For simplicity, when Ω is a bounded open set with smooth boundary, we shall denote by $\chi(\Omega)$ the Euler characteristic of $\overline{\Omega}$.

Theorem 2.1 (Gauss-Bonnet Theorem [62]) *Let M be a Riemannian surface and $\Omega \subset M$ an open bounded region with smooth boundary C and Euler characteristic $\chi(\Omega) = \chi(\overline{\Omega})$. Let K be the Gauss curvature of M and h the geodesic curvature of C with respect to the outer unit normal N. Then we have*

$$\int_\Omega K\,dM = 2\pi\chi(\Omega) - \int_C h\,dC. \tag{2.1}$$

Proof We consider first the case of a disk Ω with Euler characteristic $\chi(\Omega) = 1$.

Let U be any vector field with compact support in M, and let $u = \langle U, N\rangle$, where N is the outer unit normal to C. Consider the flow $\{\varphi_t\}_{t\in\mathbb{R}}$ associated with U, let $\Omega_t = \varphi_t(\Omega)$, $C_t = \varphi_t(C)$, and let h_t be the geodesic curvature of C_t with respect to the outer unit normal. Let us see that

$$\int_{\Omega_t} K dM + \int_{C_t} h_t\,dC_t$$

is a constant function of t.

The variation of the total Gauss curvature of Ω under the action of this flow is given by

$$\begin{aligned}\frac{d}{dt}\bigg|_{t=0} \int_{\Omega_t} K\,dM &= \int_\Omega \big(U(K) + K\operatorname{div} U\big)\,dM \\ &= \int_\Omega \operatorname{div}(KU)\,dM = \int_C Ku\,dC.\end{aligned}$$

On the other hand, the variation of the integral of the geodesic curvature is given by

$$\begin{aligned}\frac{d}{dt}\bigg|_{t=0}\int_{C_t} h_t\,dC_t &= \int_C \big(U(h_t) + h\ \mathrm{div}_C\, U\big)\,dC\\ &= \int_C \big(U^\top(h) + U^\perp(h_t) + h\ \mathrm{div}_C\, U^\top + h\ \mathrm{div}_C\, U^\perp\big)\,dC\\ &= \int_C \big(\mathrm{div}_C(hU^\top) + U^\perp(h_t) + h\ \mathrm{div}_C\, U^\perp\big)\,dC,\end{aligned}$$

where $U^\top$ and $U^\perp$ are the tangent and orthogonal projections of U to C. By the divergence theorem, $\int_C \mathrm{div}_C(hU^\top)\,dC = 0$. On the other hand, $\mathrm{div}_C\, U^\perp = hu$, and the derivative of the geodesic curvature can be computed as $-U^\perp(h_t) = u'' + (K + h^2)\,u$ by (1.41), where the prime indicates the derivative with respect to arc-length in C. So we have

$$\int_C U^\perp(h_t)\,dC = \int_C (K + h^2)\,u\,dC,$$

and we finally get

$$\frac{d}{dt}\bigg|_{t=0}\left(\int_{\Omega_t} K\,dM + \int_{C_t} h_t\,dC_t\right) = 0.$$

Hence the left side of (2.1) is invariant under the action of the flow φ_t associated with any vector field U.

Since $\chi(\Omega) = 1$, the surface with boundary $\overline{\Omega}$ is diffeomorphic to a closed disk by Theorem 3.7 in page 205 of Hirsch [230]. Let $f : \overline{\Omega} \to \mathbb{R}$ be a Morse function with a unique minimum $p \in \Omega$ so that $-\nabla f$ points inward on $C = \partial\Omega$. For any unit vector $v \in T_pM$ at p, we have $\nabla^2 f(v, v) > 0$. This implies the existence of $t_0 > 0$ such that $\langle(\nabla f)_{\gamma(t)}, \gamma'(t)\rangle > 0$ for any unit-speed geodesic γ with $\gamma(0) = p$ and any $t \in (0, t_0]$. Using a partition of unity, we obtain a vector field X on $\overline{\Omega}$ so that

- X vanishes only at p,
- $X = -\nabla f$ outside a neighborhood U of p,
- X coincides with the vector field $-\nabla d_p^2/2$, where d_p is the Riemannian distance to p, in a smaller neighborhood $V \subset U$ of p.

Then the flow associated with X contracts C to p in such a way that that the curves approaching p are geodesic circles. Hence the constant value of the left side of (2.1) is equal to 2π, which yields

$$\int_\Omega K\,dM = 2\pi - \int_C h\,dC \tag{2.2}$$

when Ω is a disk.

If Ω is not a disk, then we can express it as $\Omega = D_0 \setminus (\overline{D}_1 \cup \ldots \cup \overline{D}_k)$, where $D_0, D_1, \ldots, D_k$ are open disks with boundaries $C_0, C_1, \ldots, C_k$, respectively, such that $D_1, \ldots, D_k \subset D_0$. Observe that for the outer geodesic curvatures, we have the formula

$$\int_C h dC = \int_{C_0} h_0 dC_0 - \sum_{i=1}^{k} \int_{C_i} h_i dC_i.$$

Applying the Gauss-Bonnet theorem formula (2.2) for disks to each D_i we have

$$\begin{aligned}\int_\Omega K\,dM &= \int_{D_0} K\,dM - \sum_{i=1}^{k} \int_{D_i} K\,dM \\ &= 2\pi(1-k) + \int_{C_0} h_0 dC_0 - \sum_{i=1}^{k} \int_{C_i} h_i dC_i \\ &= 2\pi(1-k) + \int_C h dC,\end{aligned}$$

which implies (2.1) since $\chi(D_0 \setminus (\overline{D}_1 \cup \ldots \cup \overline{D}_k)) = 1 - k$. □

The Gauss-Bonnet theorem can be extended to regions with piecewise smooth boundary. In this case, the integral of the geodesic curvature must be replaced by a more general term involving the integral of the mean curvature of the regular part of C and also the angles at the corner points (see Theorem V.2.5 in Chavel's monograph [102] for details).

2.2 The Isoperimetric Inequality in Cartan-Hadamard Surfaces

In this section, we first recall the classical proof of the isoperimetric inequality in $\mathbb{R}^2$ due to Hurwitz [240] (see also [99]).

Theorem 2.2 *Let $\Omega \subset \mathbb{R}^2$ be a bounded open set with C^1 boundary C. Let L be the length of C and A the area of Ω. Then we have*

$$L^2 \geqslant 4\pi A. \tag{2.3}$$

Equality holds in (2.3) if and only if Ω is a disk.

Proof We consider first the case when Ω is simply connected. Let $\mathbf{x} : [0, L] \to C$ be a clockwise arc-length parameterization of C. We write $\mathbf{x}(t) = (x_1(t), x_2(t))$.

We translate Ω so that

$$\int_C \mathbf{x}(t)dt = 0.$$

Let $\dot{\mathbf{x}} = (\dot{x}_1, \dot{x}_2)$ be the tangent vector to C. As the parameterization is clockwise oriented, the outer unit normal N to C is $R(\dot{\mathbf{x}})$, where R is a $\pi/2$-degrees rotation in the plane. Applying the divergence theorem to the vector field $X = \sum_{i=1}^{2} x_i \frac{\partial}{\partial x_i}$ and Schwarz's inequality, we get

$$\begin{aligned} 2A = \int_\Omega \operatorname{div}(X)\, d\mathbb{R}^2 &= \int_C \langle X, R(\dot{\mathbf{x}})\rangle\, dC \leqslant \int_C |X||R(\dot{\mathbf{x}})|\, dC \\ &\leqslant \left(\int_C |X|\, dC\right)^{1/2} \left(|R(\dot{\mathbf{x}})|\, dC\right)^{1/2} \\ &= \left(\int_0^L (x_1^2 + x_2^2)(t)dt\right)^{1/2} L^{1/2}. \end{aligned}$$

By Wirtinger's inequality, we have

$$\int_0^L x_i^2(t)dt \leqslant \frac{L^2}{4\pi^2} \int_0^L \dot{x}_i^2(t)dt,$$

with equality if and only if $x_i(t)$ is a linear combination of $\cos(2\pi t/L)$ and $\sin(2\pi t/L)$. Hence

$$2A \leqslant \frac{L^2}{2\pi},$$

and (2.3) is obtained.

In case equality holds in (2.3), we have that (x_1, x_2) is proportional to $(-\dot{x}_2, \dot{x}_1)$. In particular, $|\mathbf{x}|$ is constant, and so $\mathbf{x}$ is an arc-length parameterization of the circle of radius $L/2\pi$.

If E is connected but not a disk, it can be written as $D_0 \setminus \bigcup_{i=1}^k \overline{D}_i$, where $D_0, D_1, \ldots, D_k$ are open disks such that $\overline{D}_i \subset D_0$, $i = 1, \ldots, k$. Then we have

$$L^2 > L_0^2 \geqslant 4\pi A_0 \geqslant 4\pi A,$$

so that in this case, the inequality is strict.

If E is not connected, let $(E_i)_i$ be the family of its connected components. Then we have

$$L^2 = \left(\sum_i L_i\right)^2 > \sum_i L_i^2 \geqslant 4\pi \sum_i A_i = 4\pi A. \qquad \square$$

Theorem 2.2 can be easily extended when Ω is a bounded set with C^1 boundary in a two-dimensional Cartan-Hadamard manifold, a complete simply connected two-dimensional manifold with non-positive Gauss curvature. The only relevant change is the choice of the vector field X. In this case, we fix a point $p \in M$ and consider the distance function d to p. Then we choose $X = \frac{1}{2}\nabla d^2$. The comparison results in Sect. 1.3.5 for the mean curvature of geodesic balls imply

$$\operatorname{div} X \geqslant 2.$$

An analysis of the equality yields that Ω is contained in a flat region of M. This result was first proven by André Weil [439] in 1926 with a different proof we briefly sketch in contemporary language.

Theorem 2.3 *Let M be a complete simply-connected surface with non-positive Gauss curvature. Let Ω be a bounded open set with C^1 boundary C. Let L be the length of C and A the area of Ω. Then*

$$L^2 \geqslant 4\pi A. \tag{2.4}$$

If equality holds in (2.4) then Ω is contained in a flat region of M.

Proof By the uniformization theorem for Riemann surfaces, M is conformally equivalent to the plane or the disk. We can choose global isothermal coordinates on M so that the Riemannian metric g of M can be written as $g = e^{2u} g_0$, where g_0 is the flat metric induced by the global coordinate system. The Gauss curvature of M is given by $-e^{-2u}\Delta_0 u$, where Δ_0 is the Laplacian for the metric g_0, and so $\Delta_0 u \geqslant 0$.

Assume that Ω is a disk. Let $v : \overline{\Omega} \to \mathbb{R}$ be the harmonic function in Ω with boundary value u on C. So $(u - v)$ satisfies

$$\begin{aligned} \Delta_0(u - v) &\geqslant 0 \text{ on } \Omega, \\ u - v &= 0 \text{ on } C. \end{aligned}$$

The maximum principle then implies $u \leqslant v$. If L', A' are the length of C and the area of Ω with respect to the flat metric $e^{2v} g_0$, we have

$$L = \int_C e^u \, ds_0 = \int_C e^v \, ds_0,$$

since $u = v$ on C. Here ds_0 is the length element for the metric g_0. As the metric $e^{2v} g_0$ is flat, the isoperimetric inequality $(L')^2 \geqslant 4\pi A'$ holds. So we get

$$L^2 = (L')^2 \geqslant 4\pi A' = 4\pi \int_\Omega e^{2v} \, dg_0 \geqslant \int_\Omega e^{2u} \, dg_0 = 4\pi A.$$

This proves (2.4).

In case equality holds in (2.4), we get $A = A'$, and so $u = v$. Hence the Riemannian metric of M restricted to Ω is flat.

In case Ω is connected but not a disk or has several components, we reason as in the final part of the proof of Theorem 2.2 to show that the strict inequality $L^2 > 4\pi A$ holds. □

Beckenbach and Radó (Theorem 2.4 in [49]) showed in 1933 that $K \leqslant 0$ is a sufficient and necessary condition for the validity of the Euclidean isoperimetric inequality in analytic disks. This property has been extended to metric spaces by Lytchak and Wenger [274].

2.3 The Method of Inner Parallels

The Gauss-Bonnet formula (2.1) allows us to express the integral $-\int_C h\, dC$, which is in fact the derivative of the length when we take inner parallels to the curve C, in terms of the Euler characteristic and of the total curvature of the enclosed region. Then the isoperimetric inequality (2.3) can be proven the following way: consider a disk Ω bounded by a Jordan curve C in a Riemannian surface with $K \leqslant K_0$ and the inner parallel regions $\Omega_t = \{p \in \Omega : d(p, C) < t\}$, which are bounded by the curves $C_t = \{p \in \Omega : d(p, C) = t\}$. At least while the sets C_t are smooth embedded curves (and we know that this happens when t is small enough if C is at least C^2), the derivatives of area and length are given, respectively, by

$$\frac{dA(\Omega_t)}{dt} = -L(C_t),$$
$$\frac{dL(C_t)}{dt} = -\int_{C_t} h_t\, dC_t = -2\pi + \int_{\Omega_t} K\, dM.$$

Hence we can take the area as a parameter for the deformation and, since $K \leqslant K_0$, we have

$$\frac{dL^2}{dA} = 2L\,\frac{dL}{dA} = 2\left(2\pi - \int_{\Omega_t} K\, dA\right) \geqslant 2\,(2\pi - K_0 A).$$

In case the deformation takes the regions Ω_t to some set of area zero, we merely integrate the above inequality between 0 and A to obtain

$$L^2 \geqslant 4\pi A - K_0 A^2.$$

Usually this deformation by inner parallels develops singularities, so it is necessary to estimate the evolution of length and area when crossing these singular times. In the analytic case, both $A(t)$ and $L(t)$ are continuous functions of t. This case was considered by Bol [68] and Fiala [157] (see also Chavel [102,

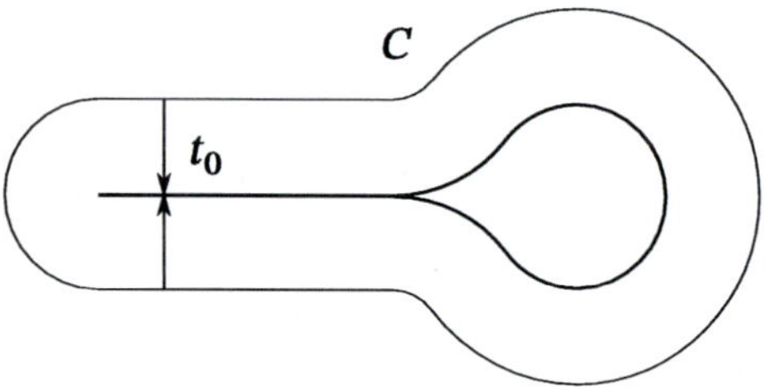

Fig. 2.1 A typical discontinuity of $L(t)$ at $t = t_0$

§V.5, pp. 255–263]). In the non-analytic case, the length function $L(t)$ may be discontinuous, as shown by Hartman [218], and the study of the evolution is much more involved (Fig. 2.1).

We briefly sketch how to deal with these problems in the general non-analytic case. Let $\Omega \subset M$ be a bounded region with smooth boundary, and let $C = \partial\Omega$ be its boundary. The inradius of Ω is defined by

$$r_m = \max\left\{d(p, C) : p \in \Omega\right\}.$$

For $r \in [0, r_m)$, we consider the sets

$$\Omega(r) = \{p \in \Omega : d(p, C) < r\},$$
$$C(r) = \{p \in \Omega : d(p, C) = r\},$$

and denote by $A(r)$ the area of $\Omega(r)$ and, when it has sense, the length of $C(r)$ by $L(r)$.

The sets $\Omega(r)$ exhibit some properties, which were proven by Hartmann [218]. We briefly recall the main results with precise references to [218]. The first one (see Lemma 5.2) is that for a closed set of non-exceptional points in $[0, r_m)$ of full measure, $C(r)$ is a piecewise smooth curve. Moreover, on every interval of non-exceptional values of r, we have

$$L'(r) \leqslant -\int_{\Omega(r)} K - \int_C h \tag{2.5}$$

(see Theorem 6.1). Although the function $L(r)$ may be discontinuous, there exists a positive "jump" function J such that $H = L + J$ is *absolutely continuous* (see Theorem 6.2). In addition (see Corollary 6.1)

$$L^2(b) - L^2(a) \leqslant \int_a^b 2L'(r)L(r)dr. \tag{2.6}$$

Moreover $A'(r) = L(r)$ for a.e. r and

$$A(r) = \int_0^r L(s)ds \tag{2.7}$$

(see Corollary 6.2).

Given a region $\Omega \subset M$ and $\lambda \in \mathbb{R}$, we consider the positive $(K-\lambda)^+ = \max\{K-\lambda, 0\}$ and negative $(K-\lambda)^- = -\min\{K-\lambda, 0\}$ parts of the function $K-\lambda$ and define the quantities

$$\omega_\lambda^+(\Omega) = \int_\Omega (K-\lambda)^+, \qquad \omega_\lambda^-(\Omega) = \int_\Omega (K-\lambda)^-,$$

so that

$$\int_\Omega K - \lambda|\Omega| = \int_\Omega (K-\lambda) = \omega_\lambda^+(\Omega) - \omega_\lambda^-(\Omega).$$

With this notation we have

Theorem 2.4 *Let (M, g) be a complete Riemannian surface and $\Omega \subset M$ be a region with smooth boundary $C = \partial\Omega$. Let L be the length of C and A the area of Ω. Then, for any $\lambda \in \mathbb{R}$, we have*

$$L^2 \geqslant [4\pi\chi(\Omega) - 2\omega_\lambda^+(\Omega)]A - \lambda A^2. \tag{2.8}$$

Proof We consider the deformation by inner parallels and apply the Gauss-Bonnet theorem to the region $\Omega(r)$. Taking (2.5) into account, we get

$$L'(r) \leqslant -\int_{\Omega(r)} K - \int_C h.$$

Applying Gauss-Bonnet theorem to the region Ω, we get $\int_\Omega K = 2\pi\chi(\Omega) - \int_C h$, and so

$$L'(r) \leqslant -2\pi\chi(\Omega) + \int_{\Omega\setminus\Omega(r)} K. \tag{2.9}$$

Since $\omega_\lambda^+(\Omega \setminus \Omega(r)) \leqslant \omega_\lambda^+(\Omega)$ we get

$$\int_{\Omega\setminus\Omega(r)} K \leqslant \lambda(A - A(r)) + \omega_\lambda^+(\Omega)$$

Multiplying both sides of (2.9) by $2L(r)$ and using the last estimate,

$$2L(r)L'(r) \leqslant 2L(r)\big[-2\pi\chi(\Omega) + \lambda(A - A(r)) + \omega_\lambda^+(\Omega)\big].$$

We use equality $A'(r) = L(r)$ on the right side, valid for a.e. r, to integrate this inequality from 0 to r_m. Using (2.5) we get

$$L(r_m)^2 - L(0)^2 \leqslant [-4\pi\chi(\Omega) + 2\omega_\lambda^+(\Omega)](A(r_m) - A(0)) \\ - \lambda(A(r_m)^2 - A(0)^2).$$

Now, using $A(0) = 0$, $A(r_m) = A$, $L(0) = L$, and $L(r_m) \geqslant 0$, we finally obtain

$$-L^2 \leqslant [-4\pi\chi(\Omega) + \omega_\lambda^+(\Omega)]A + \lambda A^2,$$

as desired. □

Inequality (2.8) can be specialized to different situations. If $K \leqslant K_0$, where $K_0 \in \mathbb{R}$, taking $\lambda = K_0$, we have $(K - \lambda)^+ = 0$, and so Eq. (2.8) is transformed into *Bol-Fiala's inequality*

$$L^2 \geqslant 4\pi\chi(\Omega)A - K_0A^2. \tag{2.10}$$

Taking $\lambda = 0$ we get from (2.8) the *Alexandrov-Fiala inequality*

$$L^2 \geqslant \left[4\pi\chi(\Omega) - \int_\Omega K^+\right]A. \tag{2.11}$$

Remark 2.5 If $D \subset M$ is a disk in a Riemannian surface with $K \leqslant K_0 \leqslant 0$, Bol-Fiala's inequality (2.10) implies $L(D)^2 \geqslant 4\pi A(D) - K_0A(D)^2$. If M is simply connected, this inequality also holds for any set with piecewise smooth boundary Ω: if Ω is connected, then it is a disk D with a finite numbers of disks removed, so that $L(\Omega) \geqslant L(D)$ and $A(\Omega) \leqslant A(D)$. Then, by the Bol-Fiala inequality for disks and the assumption $K_0 \leqslant 0$, we have

$$L(\Omega)^2 \geqslant L(D)^2 \geqslant 4\pi A(D) - K_0A(D)^2 \geqslant 4\pi A(\Omega) - K_0A(\Omega)^2.$$

If Ω is not connected, we apply this inequality to each connected component of E, making use of the inequality $(\sum_i a_i)^2 \geqslant \sum_i a_i^2$.

A further refinement of Bol-Fiala's inequality is obtained by considering, for any $a \in (0, |M|)$, the quantity

$$G(a) = \sup\left\{\int_\Omega K : |E| = a\right\}.$$

An isoperimetric inequality involving $G(a)$ was first obtained by Benjamini and Cao [50] using the curve shortening flow we will consider shortly. The proof we present now is a consequence of the method of inner parallels due to Pansu [332].

Theorem 2.6 *Let (M, g) be a Riemannian surface and $\Omega \subset M$ a bounded region with smooth boundary $C = \partial\Omega$. Let L be the length of C and A the area of Ω. Then we have*

$$L^2 \geqslant 2\pi\chi(\Omega)A - \int_0^A G(a)da. \tag{2.12}$$

Proof We use again the method of inner parallels. We consider the sets $\Omega(r)$ and $C(r)$ as in the proof of Theorem 2.4 and define $a(r)$ as the area of the region $\Omega \setminus \Omega(r)$. We observe that $a(r)$ is an strictly decreasing function and that, for a.e. r,

$$a'(r) = -L'(r)$$

Hence, the deformation can be parameterized by a instead of by r. Expressing L as a function of a, we have, for non-exceptional r,

$$\begin{aligned}\frac{dL^2}{da} = 2L\frac{dL}{da} &= 2L\frac{L'(r)}{A'(r)} \\ &= -2L'(r) \geqslant 2\pi\chi(\Omega) - \int_{\Omega\setminus\Omega(r)} K \\ &\geqslant 2\pi\chi(\Omega) - G(a).\end{aligned}$$

Here $L'(r)$ is estimated from inequality (2.9). Integrating this differential inequality in the interval $[0, A]$, inequality (2.12) is obtained. □

2.4 Bandle's Approach

The method we present in this section is also a deformation method, like the inner parallels one, so that the perimeter and the area along the deformation can be kept under control. It was introduced by Bandle (see [35] or §IV.2.6 in [36]) to obtain Bol-Fiala's inequality assuming that $K \leqslant K_0$ for a constant $K_0 > 0$. In this case, the deformation of a given region Ω is given by the level sets of a solution of a partial differential equation.

Assume that Ω is an *analytic* simply connected domain in the analytic Riemannian surface (M, g) and that there exists a flat metric g_0 in a neighborhood of Ω so that g and g_0 are conformal. Let $u : \overline{\Omega} \to \mathbb{R}$ be such that $g = e^{2u}g_0$. We denote by Δ_0 and ∇_0 the Laplacian and gradient of g_0 and by dg_0 and ds_0 the area and length elements associated with g_0. Because of the relation between the Gauss curvatures of conformal metrics (see, for instance, [245]), we have $\Delta_0 u = -Ke^{2u}$, and assuming $K \leqslant K_0$, we get

$$\Delta_0 u \geqslant -K_0 e^{2u}.$$

By suitably modifying g_0 and u, we can assume that $u = 0$ at $\partial\Omega$. This argument can be found in §2.6.1 in [36]: we consider the harmonic function h in Ω with the same boundary values as u. Then we replace u by the function $u - h$ and g_0 by the flat metric $e^{2h} g_0$. After this replacement, note that the length of $\partial\Omega$ for the metrics g and g_0 coincide.

Let $u_{\min}$ and $u_{\max}$ be the minimum and maximum values of u in $\overline{\Omega}$. For any $u_{\min} < t < u_{\max}$, we consider the set

$$\Omega(t) = \{p \in \Omega : u(p) \geqslant t\}.$$

When $t > 0$ the boundary of $\Omega(t)$ is the set of points $\{u = t\}$. However, when $t < 0$, we always have $\partial\Omega \subset \partial\Omega(t)$. Observe that $\Omega = \Omega(t_{\min})$, but $\Omega(0)$ is in general properly contained in Ω. Let us define

$$a(t) = \int_{\Omega(t)} dg_0, \qquad a = \int_{\Omega} dg_0.$$

These quantities are the areas of $\Omega(t)$ and Ω measured with respect to the metric g_0. Observe that $a(t)$ is a strictly increasing quantity by the analyticity of the function u, so that we can consider the inverse function $t(a)$. For later reference, we observe that equality

$$e^{2t(a)} = \frac{d}{da} \int_{\Omega(t(a))} e^{2u} dg_0. \tag{2.13}$$

holds for a.e. $t > 0$ ($a > 0$). This follows easily from

$$\frac{d}{da} \int_{\Omega(t(a))} e^{2u} dg_0 = \frac{dt}{da} \frac{d}{dt} \int_{\Omega(t)} e^{2u} dg_0$$

and equalities

$$\frac{da}{dt} = -\int_{\partial\Omega(t)} \frac{1}{|\nabla_0 u|} ds_0, \qquad \frac{d}{dt} \int_{\Omega(t)} e^{2u} dg_0 = -e^{2t} \int_{\partial\Omega(t)} \frac{1}{|\nabla_0 u|} ds_0.$$

We define the function

$$H(a) = K_0 \int_0^a e^{2t(\alpha)} d\alpha. \tag{2.14}$$

Note that integrating (2.13), we get

$$H(a) = K_0 \int_{\Omega(t(a))} e^{2u} dg_0 = K_0 \, |\Omega(t(a))|$$

whenever $t > 0$.

From (2.14) we obtain

$$H'(a) = K_0 e^{2t(a)}, \qquad H''(a) = K_0 e^{t(a)} \frac{dt}{da} = H'(a)\frac{dt}{da}$$

for a.e. a. We estimate da/dt using Schwarz's inequality to get

$$-\frac{da}{dt} = \int_{\partial\Omega(t)} \frac{1}{|\nabla_0 u|}\, ds_0 \geqslant \frac{L_0(\{u = t\})^2}{\int_{\partial\Omega(t)} |\nabla_0 u|\, ds_0}.$$

On the other hand

$$\int_{\partial\Omega(t)} |\nabla_0 u|\, ds_0 = -\int_{\Omega(t)} \Delta_0 u\, dg_0 \leqslant K_0 \int_{\Omega(t)} e^{2u} dg_0 = H(a(t)).$$

So we have, using the planar isoperimetric inequality $L_0(\partial\Omega(t))^2 \geqslant 4\pi a(t)$,

$$-\frac{da}{dt} \geqslant \frac{4\pi a}{H(a)}.$$

Hence we obtain

$$H''(a) = H'(a)\frac{dt}{da} \geqslant -H'(a)\frac{H(a)}{4\pi a}.$$

This is equivalent to $4\pi a H'' + H'H \geqslant 0$, and so

$$\Big(aH' - H + \frac{H^2}{4\pi}\Big)' \geqslant 0, \qquad \text{for } 0 < a < a(0).$$

Evaluating at $a = 0$ and $a = a(0)$, since $H(0) = 0$ ad $H'(a(0)) = K_0$, we have

$$a(0)K_0 - H(a(0)) + \frac{H^2(a(0))}{4\pi} \geqslant 0.$$

Dividing by K_0 and taking into account that $H(a(0))/K_0 = |\Omega(0)| = A_0$, we get

$$a(0) \geqslant A_0 - K_0 \frac{A_0^2}{4\pi} \tag{2.15}$$

On the other hand, where u is non-positive

$$a - a(0) = \int_{\Omega\setminus\Omega(0)} dg_0 \geqslant \int_{\Omega\setminus\Omega(0)} e^{2u} dg_0 = A - A_0. \tag{2.16}$$

Adding up Eqs. (2.15) and (2.16), we finally obtain

$$4\pi a \geqslant 4\pi A - K_0 A_0^2.$$

As $4\pi a \leqslant L^2$ by the isoperimetric inequality for the metric g_0 (since the length L of $\partial\Omega$ with respect to g coincides with the length with respect to g_0) and $A_0 \leqslant A$, we finally obtain

$$L^2 \geqslant 4\pi A - K_0 A^2, \tag{2.17}$$

as claimed.

2.5 Curve Shortening Flow and Isoperimetric Inequalities

When inner parallels are used to prove isoperimetric inequalities, we must deal with the singularities developed along the deformation. It would be desirable to replace this deformation by another one so that a good control of the derivatives of length and area is kept while avoiding formation of singularities. A natural candidate is the curve shortening flow on surfaces.

2.5.1 *Basic Results*

Given a Jordan curve $\alpha : \mathbb{S}^1 \to M$ in a Riemannian surface with inner unit normal n, we deform it by the curve shortening flow

$$\frac{\partial \alpha_t}{\partial t} = -h_t n_t, \tag{2.18}$$

where h_t denotes the geodesic curvature with respect to the unit normal vector field n_t to α_t. Note that $h_t n_t$ is the mean curvature vector of $\alpha_t(\mathbb{S}^1)$. A fundamental problem is to understand the evolution equation (2.18) for large values of t. Solutions of (2.18) are defined for $t \in [0, t_\infty)$, but not on a larger interval. The value t_∞ is called the *maximal time* for the solution $\alpha_t : \mathbb{S}^1 \times [0, t_\infty) \to M$. After several results by Angenent [23], Gage [173, 174], Gage-Hamilton [172], Abresch-Langer [1], and Grayson [189], the main result in this theory was proved by Grayson [190] (see Theorem 2.7).

The *convex hull* of a set $A \subset M$ is the smallest convex set containing A. We shall say that M is *convex at infinity* if the convex hull of every compact set is compact. Since an evolving curve cannot leave a locally convex region by the avoidance principle in Sect. 2.5.2, convexity at infinity is a hypothesis to ensure that the flow is confined to a compact region (Fig. 2.2).

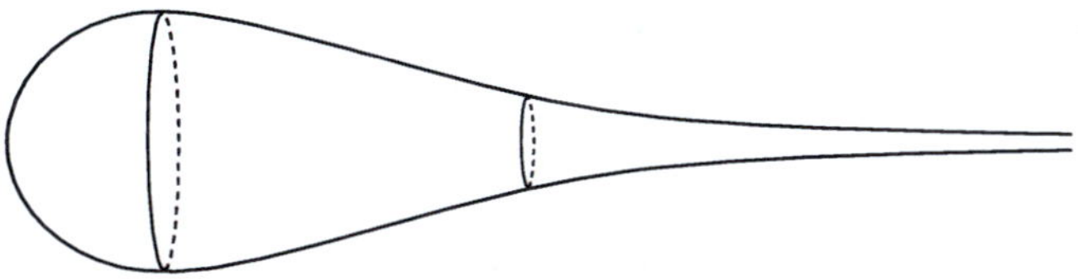

Fig. 2.2 A surface of revolution which is not convex at infinity

Theorem 2.7 ([190]) *Let M be a complete surface which is convex at infinity, α a Jordan curve in M, and α_t a solution of the curve shortening flow equation* (2.18). *Let t_∞ be the maximal time for this solution. Then*

1. *α_t remains embedded for all t.*
2. *If $t_\infty < +\infty$, then α_t converges to a point as $t \to t_\infty$.*
3. *If $t_\infty = +\infty$, then α_t converges to a closed geodesic as $t \to t_\infty$.*

Assume that the geodesic curvature flow is applied to a Jordan curve $C = \alpha(\mathbb{S}^1)$ enclosing a region Ω. Denote by C_t the Jordan curve $\alpha_t(C)$, and let Ω_t be the region enclosed by C_t. Let $A(t)$ be the area of the region Ω_t, and $L(t)$ be the length of C_t. The variation formula for the area and Gauss-Bonnet theorem implies

$$\frac{dA}{dt} = -\int_{C_t} h_t \, dC_t = -2\pi + \int_{\Omega_t} K \, dM \leqslant -2\pi + \int_M K^+ \, dM. \qquad (2.19)$$

Hence, geometric conditions must be imposed on M to guarantee $dA/dt < 0$. The derivative of length along the curve shortening flow is given by

$$\frac{dL}{dt} = -\int_{C_t} h_t^2 \, dC_t,$$

which is strictly negative whenever $h_t \not\equiv 0$. Assuming that the area is strictly decreasing along the flow, we have

$$\frac{dL^2}{dA} = 2L \frac{\int_{C_t} h_t^2 \, dC_t}{\int_{C_t} h_t \, dC_t} \geqslant 2 \int_{C_t} h_t \, dC_t, \qquad (2.20)$$

by the Schwarz inequality $\left(\int_{C_t} h_t \, dC_t\right)^2 \leqslant L(t) \int_{C_t} h_t^2 \, dC_t$. Hence we get from (2.20) the inequality

$$\frac{dL^2}{dA} \geqslant 2\left(2\pi - \int_{\Omega_t} K \, dA\right),$$

which is analogous to the formula obtained when taking inner parallels to the curve C.

2.5.2 The Avoidance Principle

A fundamental property of the curve shortening flow is the avoidance principle, which asserts that two disjoint curves remain disjoint under the action of the flow. The avoidance principle is based on the maximum principle for parabolic linear equations.

Theorem 2.8 ([351, Lemma 3]) *Let L be the differential operator*

$$L(u) \equiv a(x,t)\,\frac{\partial^2 u}{\partial x^2} + b(x,t)\,\frac{\partial u}{\partial x} - \frac{\partial u}{\partial t}, \tag{2.21}$$

where a and b are smooth and bounded and L is uniformly parabolic. Suppose that, in a domain E of the plane xt, u satisfies the inequality $L(u) \geqslant 0$. Assume that $u < M$ in the portion of E lying in the strip $t_0 < t < t_1$ for some fixed numbers t_0 and t_1. Then $u < M$ on the portion of the line $t = t_1$ contained in E.

Of course, if the function u is a solution of the linear parabolic equation

$$\frac{\partial u}{\partial t} = a\,\frac{\partial^2 u}{\partial x^2} + b\,\frac{\partial u}{\partial x} + cu,$$

with $cu \leqslant 0$, then u satisfies $L(u) \geqslant 0$.

Now we can prove the avoidance principle for the curve shortening flow.

Theorem 2.9 (Avoidance Principle [351, Prop. 1.6]) *Let $C_1, C_2 \subset M$ be disjoint Jordan curves in a complete Riemannian surface. Let $(C_1)_t$, $(C_2)_t$ solutions of the curve shortening flow defined in maximal time intervals $[0, T_1)$, $[0, T_2)$, respectively. Then $(C_1)_t$ and $(C_2)_t$ are disjoint for all $t < \min\{T_1, T_2\}$.*

Proof Assume $\bar{t} < \min\{T_1, T_2\}$ is the first contact time. Let $p \in (C_1)_{\bar{t}} \cap (C_2)_{\bar{t}}$, and let Γ be a curve tangent to both $(C_1)_{\bar{t}}$ and $(C_2)_{\bar{t}}$ at p. Consider Fermi coordinates (x, y), [102, p. 142], based on the curve $\Gamma = \Gamma(x)$ (Fig. 2.3).

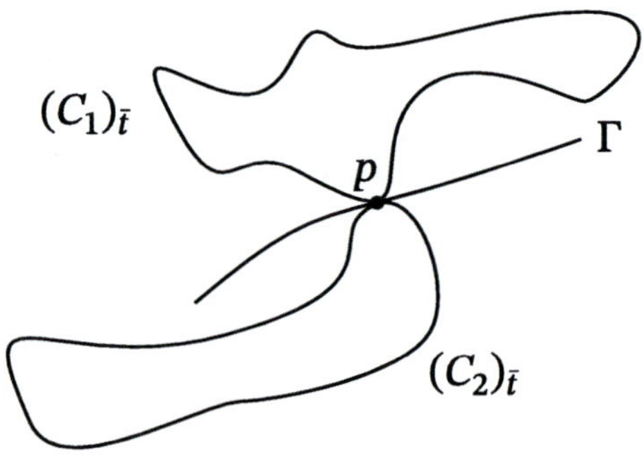

Fig. 2.3 The first contact time $\bar{t}$

Then the matrix of the Riemannian metric g in coordinates (x, y) is given by

$$\begin{pmatrix} g_{11} & 0 \\ 0 & 1 \end{pmatrix}.$$

The curves $(C_1)_{\bar{t}}$ and $(C_2)_{\bar{t}}$ are represented near $(\bar{x}, \bar{t})$, where $\bar{x}$ satisfies $\Gamma(\bar{x}) = p$, by the graphs of the functions $u_1(x, t)$, $u_2(x, t)$, respectively, and they satisfy the evolution equation

$$\frac{\partial u_i}{\partial t} = \frac{1}{(g_{11} + (u_i)_x^2)} (u_i)_{xx} + \tilde{\Phi}(x, t, u_i, (u_i)_x), \qquad i = 1, 2.$$

Assume that their difference $u = u_1 - u_2$ is strictly negative for $t < \bar{t}$. Then u satisfies a linear parabolic equation

$$\frac{\partial u}{\partial t} = a u_{xx} + b u_x + c u.$$

In order to apply the maximum principle for parabolic equations, we need $c \leqslant 0$, which does not hold in general. So instead of u, we consider the function $w(x, t) = e^{-\lambda t} u(x, t)$, and we have

$$\begin{aligned} \frac{\partial w}{\partial t} &= -\lambda w + e^{-\lambda t} \big(a u_{xx} + b u_x + c u\big) \\ &= a w_{xx} + b w_x + (c - \lambda)\, w, \end{aligned}$$

which satisfies $c - \lambda < 0$ for λ sufficiently large. From Theorem 2.8, we get a contradiction. □

2.5.3 *Applications of Curve Shortening Flow to Isoperimetric Inequalities*

Isoperimetric Inequalities in Planes

Using Grayson's Theorem, Benjamini and Cao [50] were able to prove the following isoperimetric result.

Theorem 2.10 ([50, Theorem 4]) *Let M be a complete plane of revolution about a given pole $o \in M$ so that the Gauss curvature is a non-increasing function of the distance to o. Let B_r be the geodesic ball of radius $r > 0$ centered at o, and assume that*

$$\int_{B_r} K^+ \, dM < 2\pi, \qquad \textit{for all } r > 0. \tag{2.22}$$

Then, given a relatively compact set $\Omega \Subset M$ with smooth boundary, we have

$$L(\partial\Omega) \geqslant L(\partial B_r), \tag{2.23}$$

where B_r is the geodesic ball centered at o with $A(B_r) = A(\Omega)$. Equality holds in (2.23) if and only if Ω is isometric to B_r.

This result can be applied to the paraboloid of revolution $z = x^2 + y^2$. Geodesic circles centered at o have constant geodesic curvature which is positive by Gauss-Bonnet theorem and (2.22), and so M is convex at infinity. Gauss-Bonnet theorem implies also that there are no closed geodesics in M, so the curve shortening flow deforms any Jordan curve to a point.

We shall give a simplified proof of Theorem 2.10 using the avoidance principle. We have also included the original proof by Benjamini and Cao because of its geometrical interest.

Proof of Theorem 2.10 Using the Avoidance Principle Prior to the proof we make the following observations

1. There are no closed geodesics in M by Gauss-Bonnet theorem.
2. Along the curve shortening flow, the area decreases by Eq. (2.19) and condition $\int_{B_r} K^+ \, dM < 2\pi$.
3. The curve shortening flow applied to circles centered at o provides a family of circles contracting to o.
4. M is convex at infinity since geodesic balls are convex.

We shall assume first that Ω is a disk. The curve shortening flow applied to $C = \partial\Omega$ collapses the curve to a point, since there are no closed geodesics in M. Hence the area can be taken as a parameter of the deformation, and we have

$$\begin{aligned}\frac{dL^2}{dA} = \frac{2L \int_{C_t} h_t^2 \, dC_t}{\int_{C_t} h_t \, dC_t} &\geqslant 2 \int_{C_t} h_t \, dC_t = 2\left(2\pi - \int_{\Omega_A} K \, dM\right) \\ &\geqslant 2\left(2\pi - \int_{B_{r(A)}} K \, dM\right) = \frac{dL^2}{dA}(B_{r(A)}),\end{aligned}$$

where Ω_A is the disk enclosed by some C_t of area $A \leqslant A(\Omega)$ and $B_{r(A)}$ is the ball centered at o of area A. In the first line of the inequalities, we have used the Schwarz inequality

$$\left(\int_{C_t} h_t \, dC_t\right)^2 \leqslant L \int_{C_t} h_t \, dC_t$$

and, in the second one, the fact that the ball centered at o of area a maximizes the total curvature among sets of area a. Integrating the differential inequality, we get

(2.23). In case equality holds we obtain that h_t is constant along the deformation and that $\Omega_A = B_{r(A)}$ for all $A \in (0, A(\Omega)]$.

Consider now the case that Ω is connected but the Euler characteristic $\chi(\Omega) \leqslant 0$. The set Ω is a disk D from which a finite number of disks have been removed. We trivially have

$$L(\partial\Omega) > L(\partial D), \qquad A(\Omega) < A(D).$$

Apply the curve shortening flow to D. Let t_0 be the instant for which $A(D_{t_0}) = A(\Omega)$. From the previous case, we get

$$L(\partial D_{t_0}) \geqslant L(\partial B_{r(A(D_{t_0}))}),$$

and we finally get

$$L(\partial\Omega) > L(\partial D) \geqslant L(\partial D_{t_0}) \geqslant L(\partial B_{r(A(\Omega))}).$$

Observe that in this case, the equality is never attained.

As a third case, assume that $\Omega = \Omega_1 \cup \ldots \cup \Omega_k$, where the sets Ω_i are disjoint disks. Let $T_1 \geqslant \ldots \geqslant T_k$ be the maximal times for the flows $(\partial\Omega_1)_t, \ldots, (\partial\Omega_k)_t$, and define

$$\Omega_t = \begin{cases} (\Omega_1)_t \cup \ldots \cup (\Omega_{k-1})_t \cup (\Omega_k)_t, & 0 \leqslant t \leqslant T_k, \\ (\Omega_1)_t \cup \ldots \cup (\Omega_{k-1})_t, & T_k \leqslant t \leqslant T_{k-1}, \\ \ldots & \ldots \\ (\Omega_1)_t, & T_2 \leqslant t \leqslant T_1. \end{cases}$$

Observe that, by the avoidance principle, the sets $(\Omega_1)_t, \ldots, (\Omega_k)_t$ are disjoint whenever they exist. Let $C_t = \partial\Omega_t$. Now we have on the interval $[T_i, T_{i-1}]$

$$\begin{aligned} \frac{dL^2}{dA} = \frac{2L \int_{C_t} h_t^2 \, dC_t}{\int_{C_t} h_t \, dC_t} \geqslant 2 \int_{C_t} h_t \, dC_t &= 2\left(2\pi\,(i-1) - \int_{\Omega_t} K \, dM\right) \\ &\geqslant 2\left(2\pi - \int_{\Omega_t} K \, dM\right), \end{aligned}$$

and the last inequality is strict if $i > 2$. Now we can reason as in the case of a single disk to conclude that

$$L(\partial\Omega) \geqslant L(\partial B_{r(A)}).$$

Moreover, the inequality is strict if Ω has more than one component.

Finally assume that $\Omega = \Omega_1 \cup \ldots \cup \Omega_k$, where the sets Ω_i are disjoint. Each Ω_i is a disk D_i from which a finite number of disks have been removed. The disks

D_i are either disjoint or nested. We say that D_i is *outermost* if it is not contained in another D_j. Let $\mathfrak{J}$ be the family of indexes corresponding to outermost disks, and

$$D = \bigcup_{i \in \mathfrak{J}} D_i.$$

Then we have

$$L(\partial\Omega) > L(\partial D), \qquad A(\Omega) < A(D),$$

and we may apply the curve shortening flow to the union of disks D as in the previous case to conclude that $L(\partial\Omega) > L(\partial B_{r(A)})$. □

Proof of Theorem 2.10 Following Benjamini and Cao Assume first that Ω is a region with $\chi(\Omega) \leqslant 0$. As M is a plane, Ω is obtained from a disk D deleting a finite number of disks with smooth boundary. Obviously $A(\Omega) < A(D)$ and $L(\partial\Omega) > L(\partial D)$. Apply the geodesic curvature flow to ∂D and stop at some curve C enclosing a region Ω^* with $A(\Omega^*) = A(\Omega)$. Obviously $L(\partial\Omega^*) \leqslant L(\partial D) < L(\partial\Omega)$. Hence Ω with $\chi(\Omega) \leqslant 0$ cannot be an isoperimetric region.

Assume now that Ω is a disk. One can use the curve shortening flow to obtain a family of domains Ω_a, for $a \in (0, A(\Omega))$. Let $r(a)$ be the radius of the geodesic ball centered at o of area a. Since ∂B_r has constant geodesic curvature for all r, we get that the geodesic curvature flow applied to a circle centered at o yields a family of contracting circles centered at o. By (2.20) and Gauss-Bonnet, we have

$$\begin{aligned} \frac{dL^2(\partial\Omega_a)}{da} &\geqslant 2\left\{2\pi - \int_{\Omega_a} K\,dM\right\} \\ &\geqslant 2\left\{2\pi - \int_{B_{r(a)}} K\,dM\right\} = \frac{dL^2(\partial B_{r(a)})}{da}, \end{aligned}$$

since K is non-increasing. As $\liminf_{a\to 0} L(\partial\Omega_a) \geqslant 0 = \lim_{a\to 0} L(\partial B_{r(a)})$, we get

$$L(\partial\Omega) = L(\partial\Omega_{A(\Omega)}) \geqslant L(\partial B_{r(A(\Omega))}).$$

In case $L(\partial\Omega) = L(\partial B_{r(A(\Omega))})$, we get

$$\int_{\Omega_a} K\,dM = \int_{B_{r(a)}} K\,dM,$$

for all $a \in (0, A(\Omega))$. Since K is non-increasing we easily conclude that Ω is isometric to $B_{r(A(\Omega))}$.

From now on, we assume that Ω has more than one component.

We treat first the special case $\Omega = \Omega_1 \cup \Omega_2$, with Ω_1, Ω_2 disjoint closed disks. Apply the curve shortening flow to each $\partial\Omega_1$ and $\partial\Omega_2$ to obtain families of disks

$(\Omega_1)_t$ and $(\Omega_2)_t$. Let t_1 and t_2 be the times for Ω_1 and Ω_2, respectively, to deform to points.

We assume first the case $\Omega_1 \cap (\Omega_2)_t = \emptyset$ for all $t \in [0, t_2)$. Consider the family

$$\Omega_t = \begin{cases} \Omega_1 \cup (\Omega_2)_t, & 0 \leqslant t \leqslant t_2, \\ (\Omega_1)_{t-t_2}, & t_2 \leqslant t \leqslant t_1 + t_2. \end{cases}$$

Let $A(t) = A(\Omega_t)$. Since $A'(t) < 0$ we consider the inverse function $t(a)$. Then

$$\begin{aligned} \frac{dL^2(\partial\Omega_{t(a)})}{da} &\geqslant 2\left(2\pi \,\#\{\text{components of } \Omega_{t(a)}\} - \int_{\Omega_{t(a)}} K\, dM\right) \\ &\geqslant 2\left(2\pi - \int_{\Omega_{t(a)}} K\, dM\right) \\ &= \frac{dL^2(B_{r(a)})}{da}. \end{aligned}$$

Observe that the inequality in the second line is strict for small t (for a close to $A(\Omega)$). Hence integrating we have

$$L(\partial\Omega) = L(\partial\Omega_1) + L(\partial\Omega_2) > L(B_{r(A(\Omega))}).$$

So in this case Ω cannot be an isoperimetric region.

As in the previous case, consider two disjoint disks Ω_1 and Ω_2, but now assume that there is $t_0 < t_2$ so that $\Omega_1 \cap (\Omega_2)_{t_0} \neq \emptyset$. We consider some $x_0 \in \partial\Omega_1 \cap (\Omega_2)_{t_0}$. Let α_1 and β be parametrizations of $\partial\Omega_1$ and $\partial(\Omega_2)_{t_0}$, respectively, so that $\alpha_1(0) = \beta(0) = x_0$ and $\alpha'(0) = \beta'(0)$. For s small enough, consider the geodesic segments $\sigma_1 = \overline{\alpha(s)\beta(s)}$, and $\sigma_2 = \overline{\alpha(-s)\beta(-s)}$. By the triangle inequality

$$L(\sigma_1) + L(\sigma_2) < L(\alpha|_{[-s,s]}) + L(\beta|_{[-s,s]}).$$

Hence the closed curve γ obtained from α, β, σ_1, σ_2, by removing the intervals $\alpha|_{[-s,s]}$, $\beta|_{[-s,s]}$ is a closed curve with length strictly less than $L(\partial\Omega_1)+L(\partial(\Omega_2)_{t_0})$ and area less than or equal to $A(\Omega_1) + A((\Omega_2)_{t_0})$. Round off the corners of γ to obtain a closed smooth curve enclosing a disk $\Omega^{(1)}$ with $L(\partial\Omega^{(1)}) < L(\partial\Omega_1) + L(\partial(\Omega_2)_{t_0})$, and $A(\Omega^{(1)}) > A(\Omega_1) + A((\Omega_2)_{t_0})$. Let $t_0 - \varepsilon$ be the time for which $A(\Omega^{(1)}) = A(\Omega_1) + A((\Omega_2)_{t_0-\varepsilon}$. We have $L(\partial\Omega^{(1)}) < L(\partial\Omega_1) + L(\partial(\Omega_2)_{t_0-\varepsilon})$. If $t^{(1)}$ is the time for $\Omega^{(1)}$ to collapse to a point, we define the family of domains

$$(\Omega)_t = \begin{cases} \Omega_1 \cup (\Omega_2)_t, & 0 \leqslant t \leqslant t_0 - \varepsilon, \\ (\Omega^{(1)})_{t-(t_0-\varepsilon)}, & t_0 - \varepsilon < t \leqslant t_0 - \varepsilon + t^{(1)}. \end{cases}$$

The area of $(\Omega)_t$ is decreasing and continuous and the length of $\partial(\Omega)_t$ is decreasing with a discontinuity at $t = t_0 - \varepsilon$. As in the previous case we get

$$L(\partial\Omega) > L(\partial B_{r(A(\Omega))}).$$

Assume now that $\Omega = \bigcup_{i=1}^k \Omega_i$, $k \geqslant 2$, with Ω_i disks. Apply the geodesic curvature flow to $\partial\Omega_k$ until either collapses to a point or touches for the first time t_0^k another connected component, which we may assume is Ω_{k-1}. Then replace $(\Omega_k)_{t_0^k} \cap \Omega_{k-1}$ by a smooth disk as in the previous paragraph, and apply the geodesic curvature flow to this new disk. By this procedure, we decrease the area continuously finally collapse Ω to a point, and we may conclude that

$$L(\partial\Omega) > L(\partial B_{r(A(\Omega))}),$$

as in the previous case.

Finally, assume that $\Omega = \bigcup_{i=1}^k \Omega_i$, $k \geqslant 2$, so that $\Omega_1, \ldots, \Omega_m$, with $1 \leqslant m \leqslant k$ are non-simply connected components of Ω. Each Ω_i, $i \leqslant m$, is a disk D_i from which some smooth disks have been removed, and

$$L(\partial\Omega_i) > L(\partial D_i), \qquad A(\Omega_i) < A(D_i), \qquad i \geqslant m.$$

Apply the deformation discussed in the previous paragraph to

$$\Omega' = \bigcup_{i=1}^{m} D_i \cup \bigcup_{i=m+1}^{k} \Omega_i$$

observing that some Ω_i, for $i \geqslant m$, perhaps could have disappeared in this process (as some Ω_i could be included in some D_j). This way we construct a deformation Ω'_t of continuous decreasing area and decreasing boundary length. Let t_0 so that $A(\Omega'_{t_0}) = A(\Omega)$. Then we have

$$L(\partial\Omega) > L(\Omega') \geqslant L((\Omega')_{t_0} \geqslant L(B_{r(A(\Omega))}),$$

and the proof is completed. □

The proof included here corresponds to Theorems 2, 4, and 5 in the paper by Benjamini and Cao [50]. They also obtained a Riemannian comparison theorem for isoperimetric profiles.

Theorem 2.11 ([50, Theorem 8]) *Let M be a rotationally symmetric surface with Riemannian metric $g = dr^2 + f(r)^2 d\theta^2$ such that $\int_M K^+ dM \leqslant 2\pi$ and $K(r)$ is a decreasing function of r. Suppose that N is a simply connected Riemannian surface* *which is convex at infinity and satisfies*

$$\sup_{A(\Omega)=A(B_r)} \int_\Omega K_N\, dN \leqslant \int_{B_r} K_M\, dM$$

for each r. Then

$$I_N(a) \geqslant I_M(a)$$

for each a. Moreover, if there exists a domain Ω with the same area as B_r and $L(\partial\Omega) = L(\partial B_r)$, then Ω must be isometric to B_r, where B_r is the geodesic disk of radius r centered at the origin in M.

Remark 2.12 Theorem 2.11 is a special result that compares the profile of M to the profile of a surface with non-constant Gauss curvature.

Proof In case $\int_{B_r} K\, dM < 2\pi$ for every $r > 0$, the proof of Theorem 2.10 applies without changes. So assume that $\int_B K\, dM = 2\pi$ for some rotationally symmetric ball B. Let $r_0 = \inf\{r : \int_{B_r} K\, dM = 2\pi\}$ and $r_1 = \sup\{r : \int_{B_r} K\, dM = 2\pi\}$. Since K is decreasing we have $K \equiv 0$ in $\overline{B}_{r_1} \setminus B_{r_0}$. On the other hand, for $r \in [r_0, r_1]$, we have

$$2\pi\, f'(r) = \int_{\partial B_r} h = 2\pi - \int_{B_r} K\, dM = 0.$$

Hence we conclude that $\overline{B}_{r_1} \setminus B_{r_0}$ is isometric to a flat cylinder that is foliated by the geodesics $\{\partial B_r\}$, $r \in [r_0, r_1]$.

Let $C \subset M$ be a simple closed geodesic and Ω the disk enclosed by C. By Gauss-Bonnet theorem

$$0 = \int_C h = 2\pi - \int_\Omega K\, dM$$

Since K is monotone, it follows from the definition of r_0 and r_1 that $C = \partial\Omega \subset \overline{B}_{r_1} \setminus B_{r_0}$. By the uniqueness of geodesics (or the maximum principle), we have that $C = \partial B_r$ for some $r \in [r_0, r_1]$

Let now $\Omega \subset N$. If $A(\Omega) < A(B_{r_0})$, then by the assumption on the curvature of N, we have $\int_\Omega K_N\, dN \leqslant \int_{B_{r_0-\varepsilon}} K\, dM < 2\pi$, where $A(B_{r_0-\varepsilon}) = A(\Omega)$. The proof of inequality $L(\partial\Omega) \geqslant L(\partial B_{r_0-\varepsilon})$ goes in the same way as in the proof of Theorem 2.10.

Assume now that $A(\Omega) \geqslant A(B_{r_0})$. To reproduce the arguments in the proof of Theorem 2.10, we have to take into account the case that the geodesic curvature flow takes a Jordan curve to a simple closed geodesic. So assume that $\Omega \subset N$ is a disk. Apply the curve shortening flow to $\partial\Omega$, and suppose that there is convergence to a simple closed geodesic C. Let D be the disk enclosed by C. For $\lambda \geqslant 0$, let C_λ

be the inner parallel to C at distance λ and D_λ the enclosed disk. Observe that

$$\frac{d}{d\lambda}A(D_\lambda) = -L(C_\lambda),$$

and

$$\frac{d}{d\lambda}L(C_\lambda) = -\int_{C_\lambda} h_\lambda = -2\pi + \int_{D_\lambda} K\,dM \leqslant 0.$$

Observe also that if $A(D_\lambda) < A(B_{r_0})$, then $\int_{D_\lambda} K_N\,dN < 2\pi$. Then there is λ_0 so that C_{λ_0} is an embedded curve in N with nonzero geodesic curvature. Geodesic curvature flow is then applied to C_{λ_0} to continue the evolution. In either case, there is a family $\{\Omega_a\}$, $a \in [0, A(\Omega)]$ so that

1. $A(\Omega_a) = a$, $A(\Omega_{A(\Omega)}) = \Omega$ and Ω_0 is a point.
2. $L(\partial\Omega_a)$ is non-decreasing.

Under these conditions, the proof of Theorem 2.10 can be used to prove the desired result. □

Isoperimetric Inequalities in Spheres

As remarked earlier, it is hard to obtain sharp isoperimetric inequalities in spheres with Gauss curvature decreasing from the poles by using the curve shortening flow. In this subsection, following Morgan and Johnson [311], we give a slight improvement of the Bol-Fiala inequality in spheres with Gauss curvature bounded above by a positive constant K_0. The inequality involves the classical Bol-Fiala inequality and the length of the shortest closed geodesic. We begin with a preliminary result

Lemma 2.13 *Let M be a smooth Riemannian surface with $K \leqslant 1$. Let C be a smooth Jordan curve of length L which flows under the curve shortening flow to a point and encloses area A. If $L^2 \leqslant 4\pi$ then*

$$A \leqslant 2\pi - (4\pi^2 - L^2)^{1/2}$$

Proof Whenever $dA/dt \neq 0$ we have, since $K \leqslant 1$,

$$\frac{dL^2}{dA} \geqslant 4\pi - 2A.$$

Since L is decreasing along the flow, we may integrate this differential inequality with respect to A to get $A^2 - 4\pi A + L^2 \geqslant 0$ along the deformation. The two roots of this polynomial in A are $2\pi \pm (4\pi^2 - L^2)^{1/2}$. Hence $A \geqslant 0$ if and only if either

$$A \leqslant 2\pi - (4\pi^2 - L^2)^{1/2}, \quad \text{or} \quad A \geqslant 2\pi + (4\pi^2 - L^2)^{1/2}.$$

As A approaches 0 when the flowing curves converge to the limit point, we find that the first equality is the one that really holds. □

The following Lemma will also be used in the proof of the main result of this section.

Lemma 2.14 *Let $f : [0, r] \to \mathbb{R}$ be a concave function. Assume that $a_1, \ldots, a_k, a_1 + \ldots + a_k \in [0, r]$. Then*

$$\sum_{i=1}^{k} \big(f(0) - f(a_i)\big) \leqslant f(0) - f\Big(\sum_{i=1}^{k} a_i\Big).$$

Proof It is simply a repeated application of the inequality $f(0) - f(a) \leqslant f(b) - f(a+b)$, which is obtained from

$$\frac{f(b) - f(0)}{b} \geqslant \frac{f(a+b) - f(a)}{b}.$$ □

Theorem 2.15 ([311, Thm. 5.3]) *Let M be a smooth Riemannian sphere with Gauss curvature $K \leqslant K_0$. Let L_0 be the infimum of the length of simple closed geodesics. Let A, L be the area and perimeter of a region Ω with smooth boundary C. Then*

$$L^2 \geqslant \min\big\{(2L_0)^2, 4\pi A - K_0 A^2\big\}.$$

Proof By scaling the Riemannian metric on M, we may assume that $K_0 = 1$. Let $C_1, \ldots, C_k$ denote the connected components of C. Let us assume that $L^2 < \max\{4\pi A - A^2\} = 4\pi$ and also that $L < 2L_0$. The last condition implies that, with the possible exception of C_1, all the components of C flow to points. By Lemma 2.13, each curve C_i, $i \geqslant 2$, bounds a disk D_i satisfying

$$A_i < 2\pi - (4\pi^2 - L_i^2)^{1/2},$$

where L_i is the length of ∂D_i and A_i the area of D_i.

From Bol-Fiala's inequality (2.10)

$$L_1^2 \geqslant 4\pi A_1 - A_1^2,$$

and we obtain $(2\pi - A_1)^2 \geqslant 4\pi^2 - L_1^2$. Hence, one of the following inequalities hold

$$A_1 \leqslant 2\pi - (4\pi^2 - L_1^2)^{1/2},$$

$$A_1 \geqslant 2\pi + (4\pi^2 - L_1^2)^{1/2}.$$

Assume A_1 satisfies $A_1 \leqslant 2\pi - (4\pi^2 - L_1^2)^{1/2}$. Then

$$A \leqslant \sum_{i=1}^{k} A_i \leqslant \sum_{i=1}^{k} 2\pi - (4\pi^2 - L_i^2)^{1/2} \leqslant 2\pi - (4\pi^2 - L^2)^{1/2},$$

by Lemma 2.14 since $x \mapsto (4\pi^2 - x^2)^{1/2}$ is concave. This implies the desired inequality $L^2 \geqslant 4\pi A - A^2$.

If C_1 satisfies $A_1 \geqslant 2\pi + (4\pi^2 - L_1^2)^{1/2}$, then $D_i \subset D_1$ for all $i \geqslant 2$ by the avoidance principle. Then

$$\begin{aligned} A \geqslant A_1 - \sum_{i=2}^{k} A_i &\geqslant \big(2\pi + (4\pi^2 - L_i^2)^{1/2}\big) - \big(2\pi - (4\pi^2 - Q^2)^{1/2}\big) \\ &\geqslant (4\pi^2)^{1/2} + (4\pi^2 - (L_1 + Q)^2)^{1/2} = 2\pi + (4\pi^2 - L^2)^{1/2}, \end{aligned}$$

by two more applications of Lemma 2.14. This implies again the inequality $L^2 \geqslant 4\pi A - A^2$. □

2.6 A Variational Approach

In this section, we present a completely different method to treat isoperimetric problems on surfaces. In the previous ones, we have considered deformation methods to collapse a given set while controlling the area and perimeter along the deformation. Here we consider classical methods of calculus of variations to classify the curves with constant geodesic curvature in surfaces of revolution, which are minimizers of the perimeter under a volume constraint. Such methods were already used by Schmidt [396] (see also Osserman [325, p. 1200]). They are powerful enough to provide a characterization of the isoperimetric sets in symmetric spheres of revolution with curvature decreasing from the poles to the equator. However, in order to apply this approach in non-compact surfaces, existence of isoperimetric sets must be proven first. While this is known in the compact case, existence may fail in the non-compact one. We prove that under monotonicity conditions on the Gauss curvature, such existence is guaranteed.

An interesting feature of this method is that singularities of certain types are allowed in the poles of our surfaces of revolution.

2.6.1 *Curves with Constant Geodesic Curvature in Surfaces of Revolution*

A *surface of revolution* is a complete Riemannian surface endowed with a one-parameter group of isometries so that the orbits are either points or circles. The former points are called *poles*, the latter curves *circles of revolution*.

We represent a surface of revolution without its poles as $\mathbb{S}^1 \times I$, where $\mathbb{S}^1$ is the unit circle and $I \subset \mathbb{R}$ is an open interval, endowed with the Riemannian metric

$$ds^2 = dt^2 + f(t)^2 d\theta^2 \tag{2.24}$$

for $\theta \in \mathbb{S}^1$ and $t \in I$. This Riemannian surface is a *warped product* (see Sect. 5.1.1).

On this surface, we define the *conformal* vector field $X = f(t)\,\partial_t$ (see [298]). From (2.24), it can be proven that $\nabla_u X = f'(t)\,u$ for any tangent vector u to M, where the prime denotes the derivative with respect to t and ∇ is the Riemannian connection. From this equality we get

$$\operatorname{div} X = 2f'(t), \tag{2.25}$$

where div is the divergence of the vector field X. The Gauss curvature of the metric depends only on t, and it is given by

$$K(t) = -\frac{f''(t)}{f(t)}.$$

The geodesic curvature of the circle $\mathbb{S}^1 \times \{t\}$, computed with respect to the normal ∂_t, is equal to $f^{-2}\langle D_{\partial_\theta}\partial_t, \partial_\theta\rangle$ and is given by

$$h(t) = \frac{f'(t)}{f(t)}.$$

The length of the closed curve $\mathbb{S}^1 \times \{t\}$ is given by $L(t) = 2\pi f(t)$.

A fundamental observation, to be used later, is that the function of t

$$(f')^2 - ff'' = (2\pi)^{-2} L^2(K + h^2),$$

has, up to a positive function, the same derivative with respect to t that the Gauss curvature $K(t)$. Hence $K(t)$ and $L^2(K + h^2)(t)$ are simultaneously increasing or decreasing, and they have the same critical points. If the metric extends smoothly to the pole $t = 0$, as a consequence of the Gauss-Bonnet theorem, we get

$$\lim_{t\to 0^+} [(f')^2 - ff''](t) = 1.$$

Let us compute now the equations satisfied by the curves with prescribed geodesic curvature. Consider a curve $\gamma(s) = (\theta(s), t(s))$ parameterized by arc-length s. Assume that the surface is oriented by the form $d\theta \wedge dt$, and let σ be the oriented angle $\angle(d\gamma/ds, \partial_\theta)$. Then the tangent vector $d\gamma/ds$ is given by

$$\frac{d\gamma}{ds} = \left(\frac{d\theta}{ds}, \frac{dt}{ds}\right) = \cos\sigma \frac{\partial_\theta}{f(t)} + \sin\sigma\, \partial_t.$$

We consider the unit normal vector field N to γ given by

$$N = -\sin\sigma \frac{\partial_\theta}{f(t)} + \cos\sigma\, \partial_t.$$

From the expression of the Riemannian metric (2.24) and Koszul formulas (e.g., (9) in the proof of Theorem 3.6 in do Carmo [141]), we get

$$\nabla_{d\gamma/ds} N = \left\{-\frac{d\sigma}{ds} + \frac{f'(t)}{f(t)} \cos\sigma\right\} \frac{d\gamma}{ds}.$$

So the equations for a unit-speed curve $\gamma(s) = (\theta(s), t(s))$ with geodesic curvature h are

$$\begin{aligned} \frac{d\theta}{ds} &= \frac{\cos\sigma}{f(t)}, \\ \frac{dt}{ds} &= \sin\sigma, \\ \frac{d\sigma}{ds} &= -h + \frac{f'(t)}{f(t)} \cos\sigma, \end{aligned} \tag{2.26}$$

Moreover, if h is constant then, for c in the closure of the domain of definition of f, the energy function

$$E = f(t)\cos\sigma - h \int_c^t f(\xi)\, d\xi. \tag{2.27}$$

is constant on any solution of (2.26).

The function (2.27) is usually called a *first integral* of the system (2.26). First integrals are usually obtained from Noether's Theorem [181] (see also [255] for the case of constant mean curvature surfaces). For $h = 0$ the expression $f(t)\cos\sigma =$ constant is nothing but Clairaut relation for geodesics in surfaces of revolution. From the uniqueness of solutions to (2.26) with respect to the initial conditions, we easily obtain these symmetry properties:

1. If $(dt/ds)(s_0) = 0$, then γ is symmetric with respect to the geodesic $\theta = \theta(s_0)$.
2. The curve γ can be translated along the θ-axis.

Now we describe the qualitative behavior of solutions of (2.26) with h constant

Proposition 2.16 *Let $\gamma = (\theta, t)$ be a solution to (2.26) with constant geodesic curvature h. Assume that s_0 is a maximum of t satisfying $\cos\sigma(s_0) = 1$ and that $s_1 > s_0$ is the critical point of t closest to s_0. Then*

1. *If $\cos\sigma(s_1) = 1$ then γ is a periodic θ-graph. The curve γ yields a closed embedded curve in $\mathbb{S}^1 \times I$ if and only if the θ-distance between two consecutive maxima or minima of t is equal to $2\pi/k$, for some $k \in \mathbb{N}$.*
2. *If $\cos\sigma(s_1) = -1$ then there exists $s \in (s_0, s_1)$ such that γ is vertical at $\gamma(s)$. The curve γ yields a closed embedded curve if and only if $\theta(s_0) = \theta(s_1)$.*

Proof The function t is strictly decreasing for $s > s_0$ close enough to s_0. As there are no critical points of t in (s_0, s_1), we conclude that t is decreasing in this interval. Choosing $c < t(s_1)$ in (2.27), we observe that $f(t)\cos\sigma$ is a monotone function of t.

In case 1, $\cos\sigma(s_1) = 1$, and so $f(t)\cos\sigma > 0$ for all $s \in [s_0, s_1]$. This implies that $d\theta/ds > 0$, and so the curve is a graph over θ. Reflecting with respect to the lines $\theta =$ constant corresponding to the extrema of $\theta(t)$, we see that γ is periodic in the coordinate θ. It is then closed and embedded trivially if its minimum θ-period is exactly $2\pi/k$, for $k \in \mathbb{N}$.

In case 2, $\cos\sigma(s_1) = -1$, and as $f(t)\cos\sigma$ is an increasing function of t, there exists $s \in (s_0, s_1)$ with tangent vector $-\partial_t$. Reflecting with respect to the lines $\theta =$ constant corresponding to the extrema of t it follows that γ yields an embedded curve if and only if $\theta(s_0) = \theta(s_1)$. □

The curves described in Proposition 2.16 have the same behavior as the ones obtained by Delaunay [137] as the generating curves of surfaces of revolution with non-zero constant mean curvature in $\mathbb{R}^3$. By this analogy, the curves in Proposition 2.16, which are graphs over θ, will be referred to as *unduloid type curves* or *unduloids* and the ones whose tangent vector is vertical somewhere as *nodoid type curves* or *nodoids*.

These curves are depicted in Fig. 2.4.

The following equation will play a key role in the characterization of curves with constant geodesic curvature. It is obtained from a direct computation using Eqs. (2.26). Let γ be a solution of (2.26) and $u = \sin\sigma$. When $d\theta/ds \neq 0$ we have

$$\frac{d^2u}{d\theta^2} + [(f')^2 - ff'']\,u = 0. \tag{2.28}$$

In the following two results, we describe the behavior of closed curves with constant geodesic curvature in a surface of revolution.

Lemma 2.17 *Consider a rotationally symmetric surface with metric $ds^2 = dt^2 + f(t)^2\,d\theta^2$ and a pole at $t = 0$. Let $C \subset M$ be a curve with constant geodesic curvature h so that $t|_C$ achieves local maxima and minima.*

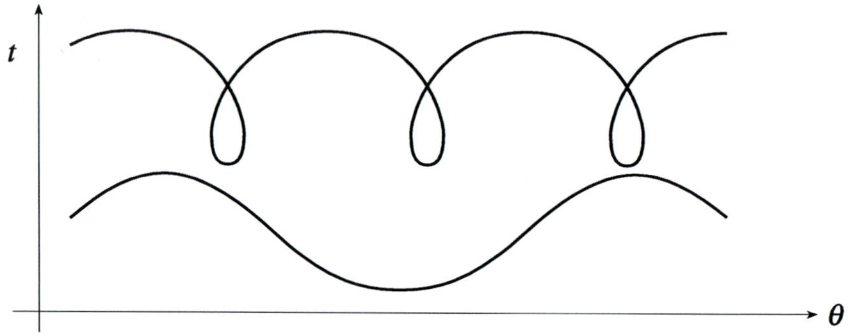

Fig. 2.4 Nodoid and unduloid type solutions to Eqs. (2.26)

Then C is a nodoid, an unduloid, a geodesic circle around the pole, or a curve approaching the pole. In the last case, the curve C is a graph over θ with exactly one maximum for the t-coordinate and meets the line $t = 0$ orthogonally.

Proof Parameterize C by a solution $(\theta(s), t(s))$ to (2.26) so that $t(0) = T$ is a maximum of the t-coordinate and $\cos\sigma(0) = 1$. Let E be the energy of the parameterized curve.

If C is not a geodesic circle around the pole, then $t|_C$ has a strict maximum at $s = 0$. If the minimum of $t|_C$ is positive, then C is either an unduloid or a nodoid by Proposition 2.16.

If C approaches the pole (when the minimum of $t|_C$ is 0), then $E = 0$. By the first integral (2.27), we have $f(T) - h\int_0^T f(\xi)\,d\xi = 0$, and so $h > 0$. Moreover

$$\cos\sigma = \frac{h\int_0^t f(\xi)\,d\xi}{f(t)},$$

and so $\cos\sigma > 0$ and C is a graph over θ. When $t \to 0$ the above fraction goes to 0. Hence C meets $t = 0$ orthogonally. □

Lemma 2.18 *Consider a rotationally symmetric surface with metric $ds^2 = dt^2 + f(t)^2\,d\theta^2$. Then*

1. *There are no closed embedded unduloids in regions where the function $(f')^2 - ff''$ is less than or equal to 1 but not identically 1. If $(f')^2 - ff'' \geqslant 1$ but not identically 1 in the region where the unduloid is defined, then the half-period of the unduloid is smaller than π.*
2. *There are no closed embedded nodoids in regions where the Gauss curvature is decreasing or increasing and not constant.*
3. *There are no closed embedded curves touching a pole inside regions where the Gauss curvature is increasing or decreasing but not constant.*

Proof Take an arc-length parameterized curve $C = (\theta, t)$ with constant geodesic curvature.

(i) If C is an unduloid, we may assume that there is a minimum of $t|_C$ at $\theta = 0$. This is easily obtained by translating the curve in the θ direction. Assume that the first maximum of $t|_C$ in $\{\theta > 0\}$ is achieved at θ_0. The curve C yields a closed embedded curve in M if and only if $\theta_0 = \pi/k$, for some $k \in \mathbb{N}$. The function $\sin\sigma(\theta)$ is a positive solution of (2.28) on the interval $(0, \theta_0)$ which vanishes at $\theta = 0$. Since $(f')^2 - ff'' \leqslant 1$ and $\not\equiv 1$, we can compare it with $\sin\theta$, the positive solution to $u'' + u = 0$ in $(0, \pi)$ which vanishes at $\theta = 0$. Using the classical Sturm comparison results (see Chapter IX.3 in [219, 326], or Lemma 2.3 in [360]), we obtain $\theta_0 > \pi$. If $(f')^2 - ff'' \geqslant 1$ but not identically 1, then Sturm's Comparison results prove that $\theta_0 < \pi$. This proves 1.

(ii) Assume that C is a nodoid. Translate it in the direction of θ until a point satisfying $\sin\sigma = 1$ lies over $\theta = 0$ and $C \subset \{\theta \geqslant 0\}$. We can suppose that $\sigma = 0$ at this point. Let θ_0 and $\theta_1 > 0$ be the projection over θ of the closest points with $\sigma = -\pi/2$ and $\sigma = \pi/2$, respectively. We obtain a closed embedded curve if and only if $\theta_0 = \theta_1$. The pieces of C corresponding to the σ-intervals $[-\pi/2, 0)$, $(0, \pi/2]$ are graphs over θ. The function $\sin\sigma$, restricted to each piece, gives two positive solutions to (2.28) over $[0, \theta_0]$, $[0, \theta_1]$, respectively. Observe that $\sin\sigma|_{\theta=0} = 1$ and that $(d\sin\sigma)/d\theta|_{\theta=0} = -hf$. We can compare them as in the previous case using Sturm comparison results. If $K(t)$ is decreasing and not constant, then $(f')^2 - ff''$ is also, and so $\theta_0 < \theta_1$. If $K(t)$ is increasing and not constant, then $\theta_0 > \theta_1$. This proves 2.

(iii) Assume now that C approaches the pole. Then the energy of the curve given by (2.27) is equal to 0. We know that C is a graph over θ with one maximum for the t-coordinate and that C is symmetric with respect to this maximum. Translate C until it meets $t = 0$ at $\theta = 0$ and the maximum of $t|_C$ lies over $\theta_0 > 0$. Then C is smooth at the pole if and only if $\theta_0 = \pi/2$. We compare $\sin\sigma$ with $\sin\theta$, using Sturm comparison and the hypotheses on $(f')^2 - ff''$. If $K(t)$ is decreasing and not constant, then $(f')^2 - ff'' \leqslant 1$ and not identically 1 over C. It follows that $\theta_0 > \pi/2$. If $K(t)$ is increasing and not constant we have $\theta_0 < \pi/2$. In any case 3 follows. □

We recall that a curve C enclosing a set E is *stable* if it has constant geodesic curvature and the second derivative of length for variations keeping constant the enclosed area is non-negative. Such a curve is always two-sided. Analytically a two-sided curve C is stable if and only if

$$Q(u, u) = -\int_C u\left(\frac{d^2u}{ds^2} + (K + h^2)\,u\right) ds \geqslant 0, \tag{2.29}$$

for all functions $u : C \to \mathbb{R}$ such that $\int_C u\,ds = 0$ (see Sect. 1.3). In the above formula, K is the Gauss curvature of M, h is the geodesic curvature of C, d/ds is the derivative with respect to arc-length on C, and ds is the Riemannian measure on C. So d^2/ds^2 is the one-dimensional Laplacian. The left side of (2.29) is the quadratic form, often called the *index form*, associated with the Sturm-Liouville operator

$$J(u) = \frac{d^2u}{ds^2} + (K + h^2)\,u, \tag{2.30}$$

which will be referred to as the *Jacobi operator* or the *second variation operator*. Associated with each connected component C' of the curve, there is an increasing sequence of eigenvalues $\{\lambda_i(C')\}_{i\in\mathbb{N}}$. We refer to the reader to Chavel's book [100] for standard properties of eigenvalues. A *Jacobi field u* is a solution to the equation $J(u) = 0$. From (2.26), one can prove that

$$u = -\langle \partial_\theta, N\rangle = f(t)\sin\sigma$$

is a Jacobi field over any solution γ to (2.26). Note that ∂_θ is a Killing field, with associated flow composed of isometries, so that its normal component u must satisfy the equation $J(u) = 0$ because of Lemma 1.26.

If a connected curve is stable, then at most the first eigenvalue is negative. If C_1 and C_2 are connected curves with the same constant geodesic curvature h and the first eigenvalues $\lambda_1(C_1)$ and $\lambda_1(C_2)$ are non-positive and at least one of them is negative, then $C_1 \cup C_2$ is an unstable curve. This is proven by taking a linear combination of the first eigenfunctions of C_1 and C_2 with mean zero.

The notion of stability is related to the isoperimetric problem since the boundary of an isoperimetric region is a stable curve. Of course, each component of a stable one is also stable.

The following result characterizes the stability of the closed curves $\mathbb{S}^1 \times \{t\}$ in the rotationally symmetric surfaces we introduced at the beginning of the section.

Lemma 2.19 *For t fixed, the curve $\mathbb{S}^1 \times \{t\}$ is stable if and only if*

$$[(f')^2 - ff''](t) \leqslant 1, \tag{2.31}$$

or, equivalently, $L^2(K + h^2)(t) \leqslant 4\pi^2$.

Proof The curve $\mathbb{S}^1 \times \{t\}$ is isometric to the circle of radius $f(t)$ and the function $K + h^2$ equals $-f''/f + (f'/f)^2$, which is a constant function over $\mathbb{S}^1 \times \{t\}$. The stability of the curve is then equivalent (see [39]) to that the first non-zero eigenvalue $\lambda_2 = 1/f^2(t)$ of the Laplacian of the curve is greater than or equal to $[-f''/f + (f'/f)^2](t)$, which implies (2.31). □

We now write down the condition for the boundary of an annulus to be stable.

Lemma 2.20 *The boundary of the annulus $\Omega = \mathbb{S}^1 \times [t_1, t_2]$ is stable if and only if each connected component $\partial\Omega_1 = \{t = t_1\}$, $\partial\Omega_2 = \{t = t_2\}$ is stable itself and*

$$\frac{K + h^2}{L}(t_1) + \frac{K + h^2}{L}(t_2) \leqslant 0. \tag{2.32}$$

Moreover if $f(t) = f(-t)$ for all t and $t_1 = -t_2$, then the annulus $\mathbb{S}^1 \times [-t_2, t_2]$ is stable if and only if

$$(K + h^2)(t_2) \leqslant 0. \tag{2.33}$$

Proof If $\partial\Omega$ is stable, then $\partial\Omega_i$ is also stable for $i = 1, 2$. Take the function

$$u = \begin{cases} -L(t_2), & \partial\Omega_1, \\ L(t_1), & \partial\Omega_2, \end{cases}$$

which has mean zero when integrated over $\partial\Omega$. Inserting this function in the index form, we obtain $Q(u, u) \geqslant 0$ by stability. Inequality (2.32) then follows immediately.

Suppose now that each connected component $\partial\Omega_i$, $i = 1, 2$, is stable and that inequality (2.32) holds. Let $u : \partial\Omega \to \mathbb{R}$ be a mean zero function. Call u_i, $i = 1, 2$, to the restriction of u to $\partial\Omega_i$. Then $u_i = c_i + v_i$, where c_i is a constant and the integral of v_i over $\partial\Omega_i$ is zero. We have

$$Q(u_i, u_i) = Q(c_i, c_i) + Q(v_i, v_i) + 2\, Q(c_i, v_i).$$

Note that $Q(c_i, v_i) = 0$ since c_i is constant and v_i has mean zero. Moreover $Q(v_i, v_i) \geqslant 0$ since v_i has mean zero and $\partial\Omega_i$ is stable. Hence

$$\begin{aligned} Q(u, u) &= Q(u_1, u_1) + Q(u_2, u_2) \geqslant Q(c_1, c_1) + Q(c_2, c_2) \\ &= -(K + h^2)(t_1)\, c_1^2 L(t_1) - (K + h^2)(t_2)\, c_2^2 L(t_2). \end{aligned}$$

Since u has mean zero, we have $c_1 L(t_1) = c_2 L(t_2)$, and we conclude

$$Q(u, u) \geqslant \left(-\frac{K + h^2}{L}(t_1) - \frac{K + h^2}{L}(t_2) \right) c_1^2 L(t_1)^2 \geqslant 0.$$

The last inequality by (2.32) holds. So we have proved that $\partial\Omega$ is stable.

Suppose now that $f(t) = f(-t)$ and that $t_1 = -t_2$. Observe that $K(t) = K(-t)$, that $h(t)^2 = h(-t)^2$, and that $L(t) = L(-t)$. So in this case inequalities (2.32) and (2.33) are equivalent. Moreover inequality (2.33) and Lemma 2.19 show that each component of $\partial\Omega$ is stable. From these observations, the assertion on $\mathbb{S}^1 \times [-t_2, t_2]$ follows easily. □

2.6.2 *Planes of Revolution with Monotone Gauss Curvature*

We prove two main results in this section using the properties of constant geodesic curvature curves already obtained. In Theorem 2.23, we show that isoperimetric regions exist on a plane of revolution with Gauss curvature non-increasing from the pole and in Theorem 2.24 that they are bounded by geodesic circles around the pole. Uniqueness holds unless there is a region with constant Gauss curvature around the pole, and in this case, every geodesic disk in this region is an isoperimetric set for its area. Our second main result is Theorem 2.27, where we prove that on planes of revolution with strictly increasing Gauss curvature, there are no isoperimetric sets for any given area, yet the isoperimetric profile is explicitly computable.

To prove both results, we need to analyze the behavior of minimizing sequences for the isoperimetric problem. We say that $\{E_i\}_{i\in\mathbb{N}}$ is a *minimizing sequence* for area $0 < A < A(M)$ if $A(E_i) = A$ for all $i \in \mathbb{N}$ and $I_M(A) = \lim_{i\to\infty} A(E_i)$.

Lemma 2.21 *Let M be a Riemannian surface, $A \in (0, A(M))$, and let $\{E_i\}_{i\in\mathbb{N}}$ be a minimizing sequence for area A of sets with smooth boundary.*

Then there is a possibly empty set $E \subset M$ with $A(E) \leqslant A$ and a non-relabeled subsequence such that $E_i = E_i^c \cup E_i^d$, where E_i^c and E_i^d are union of connected components of E_i, and

1. *E_i^c converges to E in $L^1_{loc}(M)$.*
2. *E_i^c is relatively compact for each i, and the sequence $\{E_i^d\}_{i\in\mathbb{N}}$ diverges,*
3. *The sequences $P(E_i^c)$ and $P(E_i^d)$ converge to L_c and L_d, respectively, and $L_c + L_d = I_M(A)$.*
4. *E is an isoperimetric region for its area. In particular, ∂E is smooth and has constant geodesic curvature, and it is stable.*

We prove a more general result valid for any dimension in Theorem 4.21. However we present here a more elementary proof in the two-dimensional case to make this chapter self-contained.

Proof of Lemma 2.21 We fix $p \in M$ and consider the distance r to p. Take a sequence $\{r_i\}_{i\in\mathbb{N}}$ of positive numbers diverging to infinity. As $P(E_j, B(p, r_i))$ is uniformly bounded in j for i fixed, we may assume, passing to a subsequence, the existence of a set E of locally finite perimeter in M such that

$$\int_{B(p,r_i)} |\mathbf{1}_{E_i} - \mathbf{1}_E|\, dM \leqslant \frac{1}{i}. \tag{2.34}$$

Take as E_i^c the union of the connected components of E_i intersecting the ball $B(p, r_i)$. Formula (2.34) implies that $\{E_i^c\}_{i\in\mathbb{N}}$ converges to E in $L^1_{\text{loc}}(M)$. Take $E_i^d = E_i \setminus E_i^c$.

The set E_i^c is relatively compact since $P(E_i^c) \leqslant C$ and $E_i^c \cap B(p, r_i) \neq \emptyset$ imply $E_i^c \subset B(p, r_i + C)$. The sequence $\{E_i^d\}_{i\in\mathbb{N}}$ diverges by construction since $E_i^d \cap B(p, r_i) = \emptyset$.

Passing again to a subsequence, we may assume that the limits

$$L_c = \lim_{i\to\infty} P(E_i^c), \quad L_d = \lim_{i\to\infty} P(E_i^d)$$

exist. Moreover

$$I_M(A) = \lim_{i\to\infty} P(E_i) = \lim_{i\to\infty} \left[P(E_i^c) + P(E_i^d)\right] = L_c + L_d.$$

Finally, let us prove that E is an isoperimetric region for its area. This would imply that its boundary has constant geodesic curvature and is stable. Assume $A(E) > 0$ and that there exists $F \subset M$ with $A(E) = A(F)$ such that $P(E) > P(F) > 0$. As F has finite area, we can find $s_i < r_i$ such that $\lim_{i\to\infty} A(F \setminus B(p, s_i)) = 0$ and $\lim_{i\to\infty} P(F \cap B(p, r_i)) \leqslant P(F)$. This is done using the coarea formula to find $s_i < r_i$ such that

$$\mathcal{H}^1(F \cap \partial B(p, s_i)) \leqslant \frac{A(F)}{r_i}.$$

Using Theorem 1.50 we can make small adjustments to get a set G_i with $G_i \subset B(p, r_i)$, $A(G_i) + A(E_i^d) = A$ and $\lim_{i\to\infty} P(G_i) \leqslant P(F)$. This way the sequence $G_i \cup E_i^d$ would be a sequence of sets of area A with strictly smaller limit of perimeter than the minimizing sequence E_i. This contradiction proves that E is an isoperietric region for its volume. □

The following result allows us to compare the perimeter of two disks with the same area

Lemma 2.22 *Let* $\{D_a^1\}$, $\{D_a^2\}$, $a \in (0, a_0)$, *be smooth families of disks such that for all* a

1. $A(D_a^1) = A(D_a^2) = a$,
2. ∂D_a^1 *and* ∂D_a^2 *have constant geodesic curvature (not necessarily the same),*
3. $\int_{D_a^1} K\, dM \leqslant \int_{D_a^2} K\, dM$.

Then $L(\partial D_a^1) \geqslant L(\partial D_a^2)$ *for all* a. *If inequality in 3 is strict for some subset in* $(0, a_0)$ *with positive measure, then* $L(\partial D_a^1) > L(\partial D_a^2)$.

Proof Fix a and $i = 1, 2$, and let φ be the normal component of the variational field associated with the deformation ∂D_a^i. Then we have

$$\frac{dL(\partial D_a^i)^2}{da} = 2\, L(\partial D_a^i)\, \frac{\int_{\partial D_a^i} h(\partial D_a^i)\varphi}{\int_{\partial D_a^i} \varphi} = 2 \int_{\partial D_a^i} h(\partial D_a^i),$$

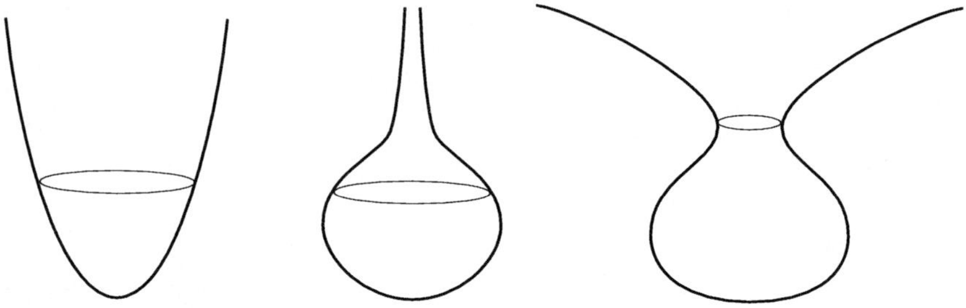

Fig. 2.5 Examples of planes with decreasing curvature

since $h(\partial D_a^i)$ is constant. By Gauss-Bonnet theorem

$$\frac{dL(\partial D_a^i)^2}{da} = 2\left(2\pi - \int_{D_a^i} K\, dM\right).$$

Comparing the derivatives using 3 and integrating from 0 to a the Lemma follows. □

Planes with Non-increasing Curvature

We consider here the cases of planes of revolution with non-increasing Gauss curvature as a function of the distance to the pole. Let us describe first the behavior of the function f. If we represent the plane minus the pole as $\mathbb{S}^1 \times (0, +\infty)$ and the Riemannian metric by $dt^2 + f(t)^2 d\theta^2$, then the function $h(t) = f'(t)/f(t)$ is the geodesic curvature of the circles of revolution with respect to the normal ∂_t, pointing outside the geodesic ball centered at the pole of the same radius. We know that h satisfies the equation $h' = -(K + h^2)$ and that the function $L^2(K + h^2)$ is decreasing since K is decreasing. As $K + h^2 > 0$ for t small enough, we get two values $0 < t_m \leqslant t_M \leqslant +\infty$ such that $h' < 0$ on $(0, t_m)$, $h' = 0$ on $[t_m, t_M]$ and $h' > 0$ for $t > t_M$. Then f is initially increasing, then eventually constant on some interval, decreases up to some eventual minimum, and then increases again. Some possibilities for the behavior of f are depicted in Fig. 2.5.

Now we prove existence of isoperimetric regions in such planes.

Theorem 2.23 *Let M be a rotationally symmetric plane with a metric with Gauss curvature non-increasing from the pole. Then isoperimetric regions exist on M.*

Proof By Lemma 2.21 we only have to show that there is no loss of area at infinity. This clearly cannot happen when the area of M is finite. So we assume that $A(M) = \infty$. Take $A > 0$ and a minimizing sequence $\{E_i\}_{i\in\mathbb{N}}$ for area A. Using Lemma 2.21 we decompose this sequence as $E_i = E_i^c \cup E_i^d$, where E_i^c converges to a set E enclosing area $A_c \leqslant A$. The curve ∂E is smooth, and it has constant geodesic curvature h and is stable. Let $K_\infty \leqslant 0$ be the limit of the curvature at the end of M. We distinguish two cases.

Case 1. $K_\infty < 0$.

In this case, K is strictly negative near infinity. As the area of M is infinite, the injectivity radius of M goes to infinity when $t \to \infty$.

Let $L_d = \lim_{i\to\infty} L(\partial E_i^d)$ and $A_d = \lim_{i\to\infty} A(E_i^d)$ be the limit length and area of the divergent sequence E_i^d. By Remark 2.5, Bol-Fiala's inequality for disks is valid for E_i^d, so that passing to the limit we have

$$L_d^2 \geqslant 4\pi A_d - K_\infty A_d^2. \tag{2.35}$$

If $K_\infty = -\infty$ then A_d should be 0 since otherwise we would have $L_d = +\infty$, which is not possible since L_d is finite. In this case, there is no loss of area. So let us assume from now on that K_∞ is finite. Let us see that in fact equality holds in inequality (2.35). Consider a disk D_∞ in a plane $M(K_\infty)$ of constant curvature K_∞ such that $A(\Omega_\infty) = A_d$. Its perimeter equals $4\pi A_d - K_\infty A_d^2$. If inequality (2.35) were strict then $L(\partial D_\infty) < L_d$. Approximating D_∞ by geodesic disks in M centered in a divergent sequence of points with the same area as E_i^d, we obtain an improved minimizing sequence. This contradiction shows that $L(\partial D_\infty) = L_d$.

Let us see now that A_c, A_d cannot be positive at the same time. Denote by h_∞ the geodesic curvature of D_∞. We first claim that

$$h = h_\infty$$

holds. Otherwise, using the first variation formula, we could find deformations E_t, $(D_\infty)_t$ of E in M and of D_∞ in $M(K_\infty)$ such that $A(E_t)+A((D_\infty)_t) = A_c+A_d = A$ and the sum of the boundary lengths of E_t and of $(D_\infty)_t$ would be less than $L_c + L_d$ for small t. Fixing one of such t and approximating the domain $(D_\infty)_t$ by a sequence of geodesic disks D_i in M with area $A - A(E_t)$ (this can be done since the injectivity radius of M goes to infinity), we obtain a minimizing sequence for area A with limit perimeter smaller than the one of $\{E_i\}_{i\in\mathbb{N}}$. This contradiction shows that $h_\infty = h$.

A similar argument implies that if we make a variation of $E \cup D_\infty$ by parallels keeping the area enclosed constant, then the second derivative of perimeter is non-negative. Hence if we consider the mean zero function

$$u = \begin{cases} L_d, & \partial E, \\ -L_c, & \partial D_\infty, \end{cases}$$

on $\partial E \cup \partial D_\infty$, then the index form (2.29) applied to u should give a non-negative value but

$$Q(u,u) = -\int_{\partial E} (K+h^2)\, L_d^2 - \int_{\partial D_\infty} (K_\infty + h^2)\, L_c^2$$
$$\leqslant -\int_{\partial E} (K_\infty+h^2)\, L_d^2 - \int_{\partial D_\infty} (K_\infty + h^2)\, L_c^2,$$

since $K \geqslant K_\infty$. As $K_\infty + h^2$ is positive for geodesic circles in $M(K_\infty)$, the last quantity is negative. This gives us a contradiction that shows that either $A_d = 0$ or $A_c = 0$. If $A_d = 0$ the proof is complete. If $A_c = 0$ then a geodesic disk centered at the pole is better than a disk at infinity since the curvature of the former is larger than the one of the latter.

Case 2. $K_\infty = 0$.

In this case we cannot reason as in the previous case since it is not guaranteed that the injectivity radius of M is infinite. Hence, we cannot always approximate a disk in $M(K_\infty) = \mathbb{R}^2$ by embedded disks in M.

Since the curvature of M is non-negative, $\int_M K \leqslant 2\pi$ by Cohn-Vossen inequality. As $f'' < 0$, there follows $-1 < \int_0^t f''(\xi)\, d\xi < 0$, and so we have $0 \leqslant f' < 1$.

We consider the conformal field $X = f(t)\, \partial_t$ and the associated one-parameter group φ_t of diffeomorphisms. As $f' \geqslant 0$ Eq. (2.25) and the first variation formula for the area imply that the function $t \mapsto A(\varphi_t(E))$ is increasing. Let us see that φ_t also increases the length of curves. Consider an embedded curve $C \subset M$. Let v be a unit vector tangent to C at p. Let Z be a vector field around p that coincides with $d\varphi_s(v)$ for s small and such that $[X, Z] = 0$. Then

$$\frac{d}{ds}\bigg|_{s=0} |d\varphi_s(v)| = \langle D_{X_p} Z, v\rangle = \langle D_{Z_p} X, v\rangle + \langle [X,Z](p), v\rangle = f'|v| = f' \geqslant 0.$$

Integrating on s it follows that $|\varphi_s(v)| \geqslant |v|$ when $s > 0$ and, using the area formula, we deduce that $L(\varphi_s(C)) \geqslant L(C)$ when $s > 0$.

We can assume that each set E_i in the minimizing sequence is a union of disks. To prove this claim observe that each connected component of E_i is a disk D_m with a finite number of holes removed and that there are at most a countable number of components. Since two different disks in this family are either disjoint or one of them lies inside the other, the set $\bigcup_m D_m$ is a union of disks. We claim that $\bigcup_m D_m$ has finite area: otherwise there is a subsequence D_{m_k} such that ∂D_{m_k} diverges. The Alexandrov-Fiala inequality (2.11) implies that a divergent sequence of disjoint disks has area bounded above by the perimeter of the union of the disks, which is bounded by the one of E_i. So we may assume that $D_{m_k} \subset D_{m_{k+1}}$ for all k. But then $L(\partial D_{m_k})$ is uniformly bounded below by some positive constant, and we get a contradiction since $\sum_k L(\partial D_{m_k}) \leqslant L(\partial E_i)$, which proves the claim. Hence $\bigcup_m D_m$ has finite area larger than the one of E_i. On the other hand, the perimeter of $\bigcup_m D_m$ is smaller than or equal to the one of E_i. Applying the family of diffeomorphisms $\{\varphi_t\}_{t<0}$ associated with the conformal field X, we decrease the area and the perimeter of $\bigcup_m D_m$ to obtain a family of disks enclosing area A and with perimeter less than or equal to the one of E_i. We denote this minimizing sequence in the same way and apply Lemma 2.21 to get a limit set E.

The isoperimetric inequality for disk type domains in surfaces, the divergence of the sequence $\{E_i^d\}$, and the fact that the positive Gauss curvature goes uniformly to zero outside of compact subsets imply

$$L_d^2 \geqslant 4\pi A_d.$$

Consider now the geodesic circle C centered at the pole enclosing area A. Let L be the length of ∂C. We have

$$(L_c + L_d)^2 \geqslant L_c^2 + L_d^2 \geqslant 4\pi(\pi f^2(A_c) + A_d),$$

where $f(A)$ denotes the composition $f(t(A))$, for $t(A)$ the inverse function of the area of the geodesic circles of the radius t around the pole. We claim that the following inequality holds:

$$\pi f^2(A_c) + A_d > \pi f^2(A), \quad \text{whenever} \quad A = A_c + A_d, \quad A_d > 0.$$

Indeed this inequality becomes an equality if $A_d = 0$. Fixing A_c and thinking about the two terms in the above inequality as functions of A_d, we easily see that the derivative of $\pi f^2(A_c + A_d)$ with respect to A_d equals f' (the derivative of f with respect to t), which is smaller than one. This proves the strict inequality and the claim. Hence

$$(L_c + L_d)^2 \geqslant L^2,$$

and equality implies that $L_d = 0$ and so $A_d = 0$. We conclude that strict inequality holds in the above displayed inequality. So if $A_d > 0$ the geodesic circle C encloses a better solution. This is clearly not possible. So $A_d = 0$. □

Now we describe the isoperimetric regions in M.

Theorem 2.24 *Let M be a rotationally symmetric plane with a metric with Gauss curvature non-increasing from the pole. Then the isoperimetric regions in M are bounded by circles of revolution or geodesic circles in regions of constant Gauss curvature. Moreover, an isoperimetric region bounded by circles of revolution can always be chosen and has no more than two boundary components.*

Proof Existence of isoperimetric regions is guaranteed by Theorem 2.23. Isoperimetric regions are bounded by smooth curves with constant geodesic curvature. We recall that $(f')^2 + ff'' \leqslant 1$ and that the region $(f')^2 - ff'' = 1$ is a disk of radius t_0 around the pole with constant Gauss curvature.

Let C be a connected component of the boundary of an isoperimetric region. According to Lemma 2.17, it is an unduloid, a nodoid, a geodesic circle, or a curve approaching the pole. From Lemma 2.18(1), a closed unduloid must be contained in the region $(f')^2 - ff'' = 1$, which is a region with constant Gauss curvature in M. Hence, it must be a geodesic circle bounding a geodesic disk. If C is a nodoid, then C must be contained in a region of constant Gauss curvature by Lemma 2.18(2). If C approaches the pole, then Lemma 2.18(3) implies that C must be contained in a disk centered at the pole with constant Gauss curvature. We conclude that C is either a circle of revolution or a geodesic circle in a region with constant Gauss curvature.

If C is not a circle of revolution, then it is a geodesic circle in a region with constant Gauss curvature, and so $(K + h^2)|_C$ is strictly positive. A standard stability argument as in the proof of Case 1 in Theorem 2.23 shows that there cannot be more than one of such components. A similar stability argument also shows that there cannot be other connected components of the isoperimetric region in the disk centered at the pole of radius t_1, where $t_1 = \inf_C t$, since $K + h^2 \geqslant (K + h^2)|_C > 0$ on this disk. Using Lemma 2.22, we replace C by the circle of revolution centered at the pole enclosing the same area as C. This way we can select an isoperimetric region enclosed by circles of revolution. Such a region cannot have more than two components by Lemma 2.25. □

Lemma 2.25 *Let E be an isoperimetric region bounded by circles of revolution in a plane with non-increasing Gauss curvature. Then E is bounded by at most two circles of revolution, and so E is a disk or an annulus.*

Proof Assume E is bounded by circles of revolution of radii $t_1 < t_2 < \ldots < t_k$. We know that the geodesic curvatures of these circles (with respect to the outer unit normal) must be constant. This implies that $h(t_i)$ must satisfy the conditions

$$h(t_1) = -h(t_2) = h(t_3) = \ldots$$

Recall there exists at most one interval $[t_m, t_M]$, with $0 < t_m < t_M \leqslant \infty$ such that $h' < 0$ on $(0, t_m)$, $h' = 0$ on $[t_m, t_M]$, and $h' > 0$ on (t_M, ∞).

In case $h(t_1) < 0$ this implies that there exists at most another value $t_2 > t_1$ such that $h(t_2) = -h(t_1)$. Hence E is a disk or an annulus.

If $h(t_i) = 0$ for all i then at least $t_1, \ldots, t_{k-1}$ lie in the region $\{t : h'(t) \leqslant 0\}$. Hence $h \equiv 0$ in the interval $[t_1, t_{k-1}]$. So f is constant on $[t_1, t_{k-1}]$, and we can merge annular components of E in $[t_1, t_{k-1}]$ by translation in the direction of t until at most three boundary curves of E remain. In this process, we have preserved the area and reduced perimeter.

If $h(t_1) > 0$, then the monotonicity properties of h imply that there are at most three components of E. Hence $E = D(t_1) \cup A(t_2, t_3)$, where $D(t_1)$ is the closed geodesic ball of radius t_1 centered at the pole and $A(t_2, t_3) = D(t_3) \setminus \operatorname{int} D(t_2)$.

So it remains to discuss the case when E has three boundary components and $h(t_1) \geqslant 0$. We observe first that $C(t_1) = \partial B(t_1)$ is contained in the region $h' \leqslant 0$ (since otherwise E would not have three components). If $C(t_2)$ also lies in the region $\{h' \leqslant 0\}$, then at least one of the circles must be in the region $h' < 0$: if neither is then $K + h^2 \equiv 0$ in $[t_1, t_2]$. Hence h is constant in this interval, and as $h(t_2) = -h(t_1)$, we have $h(t) \equiv 0$ in $[t_1, t_2]$. We conclude that f is constant and $[t_1, t_2]$ is a cylindrical region. For $t > t_2$, we have $h' > 0$, and so $h(t_3)$ could not be zero. This shows that at least $C(t_1)$ lies in the region $\{h' < 0\} = \{K + h^2 > 0\}$. Since the configuration $C(t_1) \cup C(t_2)$ is stable, Lemma 2.20 implies that $C(t_2)$, and so $C(t_3)$, must be contained in the region $h' > 0$.

The following claim (see Lemma 3.2(C) in [310]) then allows us to discard this configuration.

Claim Let $A(t_1, t_2)$ be a closed annulus with $h(t_2) \leqslant -h(t_1) < 0$. Then $A(t_1, t_2)$ has more perimeter than the disk $D(T)$ of the same area as $A(t_1, t_2)$.

Once we prove the claim we have that $A(T, t_3)$ has the same area as $D(t_1) \cup A(t_1, t_2)$, but less perimeter. This way we discard the three boundary components case by direct comparison.

To prove the claim we consider all annuli with $h(t_2) \leqslant -h(t_1) < 0$ with less perimeter than the disk of the same area, and we take the one with the largest $h(t_1)$ ($C(t_1)$ closest to the pole). Let $D(T)$ be the disk with the same area as $A(t_1, t_2)$. Then $L(T) > L(t_1)$ since otherwise the perimeter of $D(T)$ would be smaller than the one of $A(t_1, t_2)$. We have

$$h(T) > -h(t_1). \tag{2.36}$$

If $h(T) \geqslant 0$, this is obvious. If $h(T) < 0$, we take T^* in the region $f' > 0$ so that $L(T^*) = L(T)$. Recall that the function f initially increases and then decreases in an interval to which T belongs. As $L(T^*) > L(t_1)$ we have $h(T^*) > h(t_1)$. Formula

$$\frac{1}{2}\frac{dh^2}{dL} = -(K + h^2)$$

and the monotonicity of the Gauss curvature imply $h^2(T^*) \geqslant h^2(T)$ and so (2.36) holds.

Now we reduce t_1 slightly. Because of (2.36), if we slightly increase T, the perimeter of the new annulus would be still smaller than the one of the new disk. Since $h(t_1)$ is larger, in the new annulus, $-h(t_1) < h(t_2) < 0$. We continuously reduce t_1 and t_2, keeping the enclosed area constant. The perimeter decreases whenever $-h(t_1) < h(t_2)$. As $h(t_1)$ increases as t_1 decreases, equality $-h(t_1) = h(t_2)$ can never hold by the choice of the original annulus. Hence, the process continues until a disk is obtained with less perimeter than the annulus. □

Planes with Non-decreasing Curvature

We assume in this section that the Gauss curvature $K(t)$ is an non-decreasing function of t. We know that the function $(f')^2 - ff'' = (2\pi)^{-2}L^2(K + h^2)(t)$

is also non-decreasing. If M is regular at the pole, then $(f')^2 - ff''$ approaches 1 when t goes to zero and so

$$(f')^2 - ff'' \geqslant 1, \qquad t > 0.$$

Note that $(f')^2 - ff''$ equals 1 precisely on some disk D of constant curvature around the pole. Hence the geodesic circles centered at the pole and contained in D are stable, and the ones in the complementary region $M \setminus D$ are unstable by Lemma 2.19.

Since $K(t)$ is increasing and M is complete, there exists the limit $K_\infty = \lim_{t\to+\infty} K(t)$, and it is non-positive. So $K(t) \leqslant 0$ everywhere and M is a Cartan-Hadamard manifold. As $f'(0) = 1$ and $f''(t) \geqslant 0$, we have $f'(t) \geqslant 1$ for all t, and $f(t)$ is a strictly increasing function. The injectivity radius goes to $+\infty$ when we approach the end.

The result characterizing the stable embedded curves in such a plane is

Theorem 2.26 *Let M be a rotationally symmetric plane with non-decreasing curvature. Then there are no stable curves embedded in M, except possibly the boundaries of geodesic disks with constant curvature. If the Gauss curvature is strictly increasing, then there are no stable embedded curves in M.*

Proof By Lemma 2.17, an embedded connected curve with constant geodesic curvature in M must be a nodoid, an unduloid, a curve touching the pole, or a circle of revolution. We have $((f')^2 - ff'')(t) \geqslant 1$, and equality holds at some geodesic disk of radius t_0 around the pole with constant Gauss curvature. Of course t_0 can be either 0 or $+\infty$. Any curve with constant geodesic curvature contained in $t \leqslant t_0$ is the boundary of a geodesic disk with constant Gauss curvature.

Curves approaching the pole and intersecting the region $\{t > t_0\}$ cannot exist by Lemma 2.18(3).

Assume that C is an unduloid touching the region $\{t > t_0\}$. By Lemma 2.18(1) the θ-distance between two consecutive maxima or minima of $t|_C$ is less than 2π. Hence we need at least two pieces between adjacent maxima or minima of the t-coordinate to obtain a closed embedded curve. In this case the restriction of the Jacobi field $u = -\langle N, \partial_\theta \rangle = f(t) \sin\sigma$ to the curve has at least four nodal domains. Courant's nodal domain [100] then shows that the Jacobi operator has at least three negative eigenvalues, and so the curve is unstable.

Nodoids touching the region $\{t > t_0\}$ exist if they are contained inside a region with constant Gauss curvature and they are the boundaries of geodesic disks with constant Gauss curvature by Lemma 2.17.

Circles of revolution are stable if and only if they are contained in $t \leqslant t_0$ by Lemma 2.19. We conclude that the only connected stable embedded curves are the boundaries of geodesic disks with constant curvature in M. Since these curves have negative first eigenvalue for the Jacobi operator J, we conclude that a stable curve has to be connected.

When the Gauss curvature is strictly increasing, the only embedded curves with constant geodesic curvature are circles of revolution, which are not stable. □

Now we characterize the isoperimetric regions in the following result.

Theorem 2.27 *Let M be a rotationally symmetric plane of non-decreasing curvature as a function of the distance to the pole. Let $K_\infty = \sup_M K \leqslant 0$.*

Then the only isoperimetric regions in M are geodesic disks in the set $\{p \in M : K(p) = K_\infty\}$. If this region is empty then isoperimetric regions do not exist on M.

In any case, the isoperimetric profile of M is given by

$$I_M(a)^2 = 4\pi a - K_\infty a^2. \tag{2.37}$$

for all $a > 0$.

Proof Fix some area $A \in (0, A(M))$, and let $\{E_i\}_{i\in\mathbb{N}}$ be a minimizing sequence for area A. By Lemma 2.21 we have $E_i = E_i^c \cup E_i^d$, where E_i^c is convergent to some region E and E_i^d is divergent. Moreover ∂E is stable, and so it is a geodesic circle in a region of constant curvature by Theorem 2.26. From Bol-Fiala's inequality (2.10), we conclude that E is a geodesic disk with constant Gauss curvature K_0, and we have

$$L(\partial E)^2 = 4\pi A(E) - K_0 A(E)^2.$$

On the other hand, for the divergent part E_i^d of the sequence, we have

$$L_d^2 = 4\pi A_d - K_\infty A_d^2.$$

It is clear that the perimeter of a disk in $M(K_\infty)$ enclosing area A is smaller than or equal to $L(\partial E) + L_d$ (equality holds only if $L(\partial E) = 0$ or $L_d = 0$). If the region where $K = K_\infty$ is not empty, then a geodesic disk of area A inside this region is an isoperimetric region.

Since $K \leqslant K_\infty$ we get from Bol-Fiala's inequality (2.10) that $I_M(a) \geqslant I_{M_\infty}(a) = 4\pi a - K_\infty a^2$ for all $a > 0$. On the other hand, as M is a Cartan-Hadamard surface, we can take a divergent sequence of embedded geodesic disks D_i of area a. For such a family, it is clear that $\lim_{i\to\infty} L(D_i) = L_\infty(D_\infty)$, where $D_\infty \subset M_\infty$ is a geodesic disk of area a. Hence we have

$$I_{M_\infty}(a) = \lim_{i\to\infty} L(\partial D_i) \geqslant I_M(a).$$

This inequality, together with $I_M \geqslant I_{M_\infty}$, implies (2.37). □

2.6.3 Spheres of Revolution with Monotone Gauss Curvature

We consider now spheres of revolution with an equatorial symmetry and Gauss curvature K monotone as a function of the distance to the equator. Thus K either decreases or increases from the poles to the equator. In both cases we are able to solve the isoperimetric problem, but the solutions are completely different.

Let M be a rotationally symmetric sphere. Removing its two poles, the resulting Riemannian surface is isometric to $\mathbb{S}^1 \times (0, t_0)$ with Riemannian metric $ds^2 = dt^2 + f(t)^2 d\theta^2$. The point corresponding to $t = 0$ will be called the *south pole* and the one corresponding to $t = t_0$ the *north pole*. The curves $\mathbb{S}^1 \times \{t\}$ will be called *parallels* and the ones with θ-coordinate constant *meridians*. A *complete meridian* in the union of the meridians $\theta = \theta_0$ and $\theta = \theta_0 + \pi$ for some θ_0. Complete meridians are closed geodesics of M. The *equator* is the curve $\mathbb{S}^1 \times \{t_0/2\}$. We shall impose the existence of an equatorial symmetry with respect to $t = t_0/2$, that is

$$f(t) = f(t_0 - t).$$

Since M is a regular Riemannian surface, the function $f(t)$ and their derivatives up to third order extend to $t = 0$ and to $t = t_0$, and we have $f(0) = f(t_0) = 0$, $f''(0) = f''(t_0) = 0$, $f'(0) = 1$, $f'(t_0) = -1$, $f'''(0) = -K_s$, $f'''(t_0) = K_n$, where $K_s = K_n$ are the values of the Gauss curvature at $t = 0$ and $t = t_0$, respectively.

We remark that there are no rotationally symmetric metrics with monotone curvature in a sphere. This follows from an old argument by Kazdan and Warner [245], integrating the derivative of $(f')^2 - ff''$ between 0 and t_0 or by the arguments in Howards, Hutchings, and Morgan [310].

Spheres with Curvature Non-decreasing from the Equator

In this subsection, we assume that the Gauss curvature $K(t)$ is non-increasing for $t < t_0/2$ and so, by the equatorial symmetry, non-decreasing for $t > t_0/2$.

The function $(f')^2 - ff''$ takes the value 1 at $t = 0$, $t = t_0$, it decreases from $t = 0$ up to $t = t_0/2$, and it increases from $t = t_0/2$ up to $t = t_0$. So

$$(f')^2 - ff'' \leqslant 1$$

on $(0, t_0)$. Equality holds on regions with constant Gauss curvature around both poles. Lemma 2.19 shows that any circle of revolution is stable.

The geometric behavior of M is described in the next Lemma. See Fig. 2.6.

Lemma 2.28 *The Gauss curvature $K(t)$ is positive around the poles and either non-negative everywhere or negative on a symmetric annulus around the equator.*

1. *If $K(t) \geqslant 0$ everywhere, then $f'(t) \geqslant 0$ for $t \in (0, t_0/2]$ and $f'(t) = 0$ precisely at some closed interval containing $t_0/2$.*

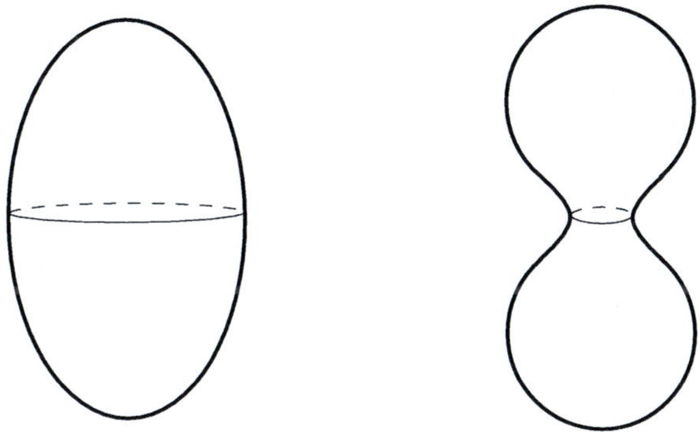

Fig. 2.6 A convex sphere and a sphere with points of negative curvature

2. *If $K(t)$ changes its sign, then there exists $t_1 \in (0, t_0/2)$ such that $f'(t) > 0$ for $t \in (0, t_1)$ and $f'(t) < 0$ for $t \in (t_1, t_0/2)$.*

Proof The Gauss curvature is positive somewhere by Gauss-Bonnet theorem. It achieves its maxima at the poles and decreases up to the equator, so it is either non-negative and eventually vanishes around the equator or it becomes negative around the equator.

1 and 2 follow since $f'(0) = 1$, $f'(t_0/2) = 0$, $f'(t_0) = -1$ and $(f')' + Kf = 0$. □

In order to obtain a classification of isoperimetric regions, we present first a classification of stable curves with constant geodesic curvature. The main difference with the planar case is the existence of a family of embedded nodoids, which are, nevertheless, unstable. We prove this in the following result.

Lemma 2.29 *Consider the surface $\mathbb{S}^1 \times (0, t_0)$ with the Riemannian metric $ds^2 = dt^2 + f(t)^2 d\theta^2$. Assume that $f(t - t_0) = f(t)$ and that $K(t)$ is monotone for $t \in (t_0/2, t_0)$. Let $C \subset \mathbb{S}^1 \times (0, t_0)$ be a nodoid type curve which is not contained inside a region with constant Gauss curvature. Then we have*

1. *The curve C is embedded if and only if it is symmetric with respect to $t = t_0/2$.*
2. *If $K(t)$ is increasing for $t \in (t_0/2, t_0)$ and C is embedded then C is unstable.*

Proof First we prove 1. Obviously if C is symmetric with respect to $t = t_0/2$ then it is embedded. Let $T = \max t|_C$, $S = \min t|_C$. We parameterize C by a solution (θ, t) to (2.26) with initial conditions $(\theta(0), t(0), \sigma(0)) = (0, T, 0)$. Let E be the energy defined in (2.27) and h the geodesic curvature of the solution. As C is a nodoid, we have $Eh < 0$ ($Eh \geqslant 0$ implies that C is a graph over θ by (2.27)). Let us assume that $E < 0$ and $h > 0$. The other possibility $E > 0$, $h < 0$ is treated in a similar way.

Since C is embedded it cannot be contained in a region where the curvature is strictly monotone by Lemma 2.18. So $S < t_0/2 < T$. Moreover

$$f(T) - h\int_0^T f(\xi)\,d\xi = E, \qquad -f(S) - h\int_0^S f(\xi)\,d\xi = E.$$

If $E = -h\int_0^{t_0/2} f(\xi)\,d\xi$, then C meets the equator orthogonally, and it is symmetric with respect to $t = t_0/2$. So let us assume $E \neq -h\int_0^{t_0/2} f(\xi)\,d\xi$.

If we reflect C with respect to $t = t_0/2$ we obtain a new curve $\widetilde{C}$ with the same geodesic curvature h that can be parameterized by a solution $\widetilde{\theta}, \widetilde{t}, \widetilde{\sigma}$ to (2.26) with initial conditions $(0, t_0 - S, 0)$. The energy $\widetilde{E}$ of this solution is given by

$$\widetilde{E} = f(t_0 - S) - h\int_0^{t_0 - S} f(\xi)\,d\xi,$$

and a computation using the symmetry $f(t_0 - t) = f(t)$ yields

$$\left(E + h\int_0^{t_0/2} f(\xi)\,d\xi\right) + \left(\widetilde{E} + h\int_0^{t_0/2} f(\xi)\,d\xi\right) = 0.$$

So replacing C by $\widetilde{C}$ if necessary, we may assume

$$E > -h\int_0^{t_0/2} f(\xi)\,d\xi, \tag{2.38}$$

From (2.38) and (2.27), we have $\cos\sigma > 0$ when $t \geqslant t_0/2$, and we obtain that $C \cap \{t > t_0/2\}$ is a graph over θ.

Claim For $t \in (t_0 - T, t_0/2)$, we have $-f(t) - h\int_0^t f(\xi)\,d\xi < E$ and $S < t_0 - T$.

In order to prove the claim fix $t \in (t_0 - T, t_0/2)$. Using the symmetry $f(t - t_0) = f(t)$

$$\begin{aligned}- f(t) - h\int_0^t f(\xi)\,d\xi &= -f(t_0 - t) + h\int_0^{t_0 - t} f(\xi)\,d\xi - 2h\int_0^{t_0/2} f(\xi)\,d\xi = \\ &< -f(t_0 - t)\sin\sigma + h\int_0^{t_0 - t} f(\xi)\,d\xi \\ &\quad - 2h\int_0^{t_0/2} f(\xi)\,d\xi = \\ &= E - 2E - 2h\int_0^{t_0/2} f(\xi)\,d\xi < E.\end{aligned}$$

Fig. 2.7 Reflection of C about $t = t_0/2$

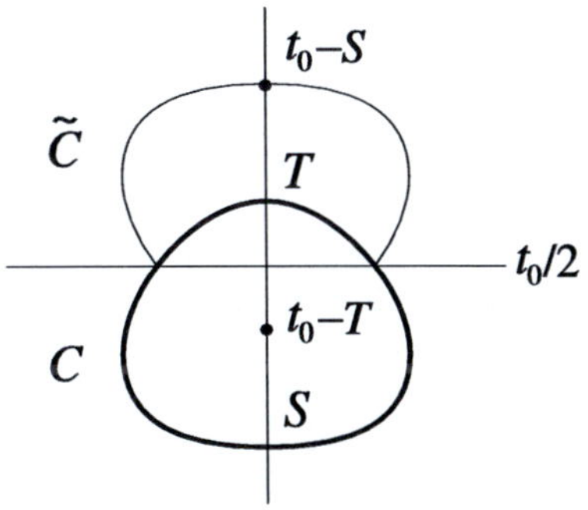

In the first line, we have used the equality $\int_0^{t_0-t} f(\xi)d\xi = \int_t^{t_0} f(\xi)d\xi$. Inequality $S < t_0 - T$ follows trivially and proves the claim.

Let us see that C lies below $\widetilde{C}$ in the region $t > t_0/2$ (see Fig. 2.7). If both curves meet at some point, we have $t = \widetilde{t}$ and

$$f(t)(\cos\sigma - \cos\widetilde{\sigma}) = E - \widetilde{E}.$$

Evaluating at $t_0/2$ we conclude, as $\cos\sigma > 0$ and $\cos\widetilde{\sigma} < 0$, that $E - \widetilde{E} > 0$. So $\sin\sigma > \sin\widetilde{\sigma}$ at any point where both curves cross themselves. Initially $\widetilde{t} > t$. Let $\theta_0 < 0$ be the smaller value where $\widetilde{t} = t$. At this value, both $\widetilde{t}$ and t are graphs over θ, so we have $(dt/d\theta)(\theta_0) \geqslant (d\widetilde{t}/d\theta)(\theta_0)$, which implies $\tan\sigma(\theta_0) \geqslant \tan\widetilde{\sigma}(\theta_0)$, and so $\cos\sigma(\theta_0) \leqslant \cos\widetilde{\sigma}(\theta_0)$, which contradicts the above.

Get two points (θ_0, t_1) and (θ_0, t_2) in C with the same θ_0 and $t_1 < t_2$. Assume that $K(t)$ is increasing in $(0, t_0/2)$. If $t_2 < t_0/2$, then $((f')^2 - ff'')(t_1) \leqslant ((f')^2 - ff'')(t_2)$. If $t_2 > t_0/2$, then the property proved in the previous paragraph shows that $t_0 - t_2 > t_1$. As $t_0 - t_2 < t_0/2$ and the function $(f')^2 - ff''$ is symmetric with respect to $t = t_0/2$, we conclude that $((f')^2 - ff'')(t_1) \leqslant ((f')^2 - ff'')(t_2)$. Moreover there are at least two such points where the inequality is strict since C is not contained in a region with constant curvature. If $K(t)$ is decreasing in the interval $(0, t_0/2)$, reverse inequalities are obtained. Hence, we can apply the same argument as in the proof of Lemma 2.18(2) to show that C cannot be closed.

To see 2, we consider the test function on C defined by

$$u(s) = (f'(t) - hf(t)\cos\sigma)(s) = \frac{d^2}{ds^2}\int_0^{t(s)} f(\xi)\,d\xi,$$

which has mean zero when integrated over C. Since C is symmetric by 1, we see that u vanishes over the equator. At any point where $\cos\sigma = \pm 1$, we have $dt/ds = 0$

and

$$\frac{d^2t}{ds^2} = -h\cos\sigma + \frac{f'(t)}{f(t)} = \frac{u}{f(t)}.$$

Hence $u < 0$ at the maxima of $t|_C$ and $u > 0$ at the minima of $t|_C$.

A direct computation using Eqs. (2.26) shows that

$$J(u) = -K'(t)\, f(t)\, \sin^2\sigma, \tag{2.39}$$

and so $J(u) < 0$ in the north hemisphere and $J(u) > 0$ in the south hemisphere. This implies that the first eigenvalue for the Dirichlet problem of the Jacobi operator is negative in both $C \cap \{t \geqslant t_0/2\}$ and $C \cap \{t \leqslant t_0/2\}$. We conclude that C is unstable. □

Lemma 2.29 now easily implies the classification of closed embedded stable curves with constant geodesic curvature.

Theorem 2.30 *Let M be a rotationally symmetric sphere with an equatorial symmetry which Gauss curvature is a non-increasing function of the distance from the poles. Then the only connected closed embedded curves in M with constant geodesic curvature are parallels, meridians, nodoids symmetric with respect to the equator which are graphs over the equator, and the boundaries of geodesic disks with constant Gauss curvature.*

Moreover, the only stable ones are the parallels and the boundaries of geodesic disks with constant Gauss curvature.

Proof Let $C \subset M$ be a closed embedded curve with constant geodesic curvature h.

If C meets the south pole, then we parameterize C by a solution of (2.26) with $E = 0$. If, in addition, C touches the north pole, then $f(t_0) = 0$ and (2.27) shows that $h = 0$. Again by (2.27) we have $\cos\sigma = 0$ and C is a meridian.

So let us assume that C does not touch both poles. Eventually after an equatorial reflection, we may assume that C stays away from the north pole. We can also assume that C is not contained in a region with constant Gauss curvature since in this case C would be a geodesic circle.

As $(f')^2 - ff'' \leqslant 1$ and $\not\equiv 1$, Lemma 2.17 implies that C is neither an unduloid nor a curve approaching the pole.

If C is a nodoid, then we conclude by Lemma 2.17 and by next Lemma 2.29 that C is either the boundary of a geodesic disk with constant Gauss curvature or a unstable nodoid symmetric with respect to the equator.

Finally circles of revolution are stable by Lemma 2.19. □

An analysis of the candidates using Theorem 2.30 gives the classification of isoperimetric regions in M (Fig. 2.8).

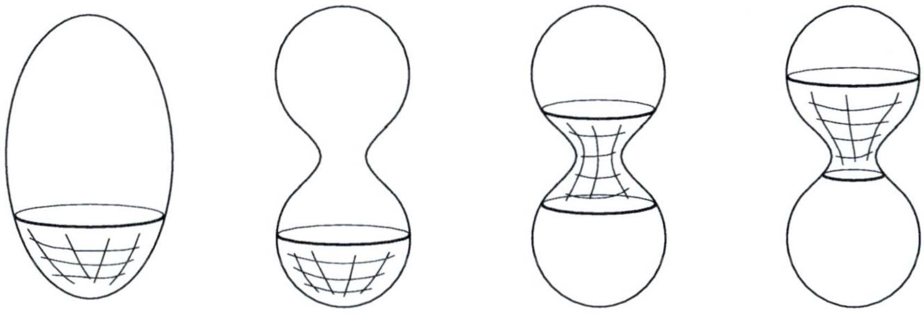

Fig. 2.8 Isoperimetric regions in spheres with curvature increasing from the equator

Theorem 2.31 *Let M be a rotationally symmetric sphere with an equatorial symmetry and Gauss curvature non-increasing as a function of the distance from the poles. Then the isoperimetric regions in M are*

1. *geodesic disks centered at the poles and possibly geodesic disks in regions of constant Gauss curvature about the poles and their complements,*
2. *annuli symmetric with respect to the equator which are contained in the region $K + h^2 \leqslant 0$ and their complements,*
3. *non-symmetric annuli containing the least length meridian, which have one boundary component in $K + h^2 < 0$ and the other one in $K + h^2 > 0$, and their complements.*

If $K > 0$ then 2 and 3 cannot happen. If $K \geqslant 0$ then 3 cannot hold.

Proof Let $E \subset M$ be an isoperimetric region. Then ∂E is a curve with constant geodesic curvature h. The components of ∂E are either circles of revolution or the boundaries of geodesic disks in a region of constant Gauss curvature by Theorem 2.30.

Suppose that one component of ∂E is the boundary of a geodesic disk D with constant curvature. Replacing E by $M \setminus E$ if necessary, we may assume $D \subset E$. Note that in this case $h(D) > 0$ and that $\lambda_1(\partial D) < 0$. Let us see that we can replace E by another set bounded by circles of revolution and with no larger perimeter.

If $\partial E = \partial D$ then $E = D$. Lemma 2.22 shows that a geodesic disk D^* centered at a pole with the same area as D satisfies $L(\partial D^*) \leqslant L(\partial D)$. Moreover, if $L(\partial D^*) = L(\partial D)$, then both D and D^* are contained in a region with constant Gauss curvature around a pole.

If ∂E is not connected, then any other boundary component C' must have $\lambda_1(C') > 0$, and so it must be a circle of revolution inside $K + h^2 < 0$. Inside this region, the geodesic curvature of circles of revolution is monotone ($h'(t) = -(K + h^2)(t)$), and there are exactly two circles with geodesic curvature $\pm h$. Then E is the union of D with either a symmetric annulus inside $K + h^2 < 0$ or the union of D with a geodesic disk around a pole. In both cases we can replace D by a geodesic disk D^* about a pole with the same area as that of D to obtain a new region with perimeter less than or equal to $L(\partial E)$. Equality holds if and only if D and D^* are contained in a region with constant curvature around a pole. So we may assume that E is bounded by circles of revolution.

So $\partial\Omega = \bigcup_{i=1}^n \mathbb{S}^1 \times \{t_i\}$ with $\{t_i\}$ increasing and $h(t_i) = -h(t_{i+1})$, for $i = 1, \ldots, n-1$. If $h \neq 0$ then Lemma 2.28 shows that there are at most four circles of revolution in $\partial\Omega$. If $h = 0$ then we could have a large number of geodesics of revolution contained in a flat region, but if we have more than two, then we can join the annuli they bound in the flat region to obtain a region with least perimeter.

If two components of ∂E are in $K + h^2 > 0$, then each one has negative first eigenvalue for the Jacobi operator, and so ∂E is unstable. Hence, at most, one component of ∂E is contained in $K + h^2 > 0$. This implies that if $K > 0$, then there is at most one boundary component. Moreover, if there are four circles in ∂E, then two of them are inside $K + h^2 > 0$, as the region where $f' > 0$ (resp. $f' < 0$) around the south (resp. north) pole is a subset of $K + h^2 > 0$. So ∂E can have at most three boundary components.

If ∂E is connected, then E is a geodesic disk around a pole or complement.

When ∂E has two components, then E is an annulus or complement. If the annulus is symmetric with respect to the equator, then it is contained in the region $K + h^2 \leqslant 0$ by the stability condition (2.33). If the annulus is non-symmetric, then one component must lie in $K + h^2 > 0$ (if both components are inside $K + h^2 \leqslant 0$ then the annulus is symmetric) and the other one in $K + h^2 < 0$.

Finally assume that ∂E has three boundary components. Then $h > 0$, and E is the union of a symmetric annulus contained in the region $K + h^2 \leqslant 0$ and a geodesic disk around a pole or its complement. There are annuli with the same area as E and less perimeter than E by Theorem 3.1 in [310]. □

Remark 2.32 If we allow $0 < f'(0) < 1$, all the results in this subsection hold. The metric so obtained is singular at the poles.

Remark 2.33 Assume that f is C^1 and piecewise C^2. Then the second variation of length formula is valid at least for a variation by curves meeting the parallels where $K(t)$ is discontinuous at a finite number of points. This is enough to discard the constant geodesic curvature curves, which are symmetric with respect to the equator.

Remark 2.34 As in [310], one can show that sometimes annuli have smaller perimeter than disks of the same area by attaching a long narrow symmetric hyperbolic annulus to two curvature 1 spheres in a C^1 way. The metric of this surface is C^1 and piecewise C^2.

Remark 2.35 Let us see that non-symmetric annuli can be isoperimetric regions. Attach a hyperbolic annulus to two disks with curvature 1 in a C^1 way. This can be done so that the hyperbolic annulus has strictly less perimeter than a disk of curvature 1 of the same area. A slightly larger symmetric annulus has still less perimeter than the disk of the same area. We deform the annulus by parallels keeping the area enclosed constant until we get a geodesic disk about the pole. The perimeter along this deformation initially decreases since the starting annulus is unstable. As the perimeter of the initial annulus is less than the one of the final disk, there is a stable annulus in the deformation by Lemma 2.20, which cannot be symmetric since it has larger area than the largest symmetric stable annulus.

Spheres with Curvature Non-increasing from the Equator

We assume now that $K(t)$ is a decreasing function of the distance from the equator $t = t_0/2$. We assume for simplicity that the sphere is convex. Gauss-Bonnet theorem then implies that the Gauss curvature is strictly positive around the equator. The monotonicity assumption on the Gauss curvature implies

$$(f')^2 - ff'' \geqslant 1$$

on $(0, t_0)$. Equality holds at some region with constant Gauss curvature around the poles.

We follow the scheme in the previous sections to treat the isoperimetric problem in these spheres of revolution. First we compute the closed embedded curves with constant mean curvature that are stable and then the isoperimetric regions, whose boundaries are the former curves.

Let us prove first the existence of a family $\{C_T\}$, for $T \in (t_0/2, t_0)$, of nodoids foliating the hemisphere $|\theta| \leqslant \pi/2$ minus the point $(0, t_0/2)$. Given $T \in (t_0/2, t_0)$, consider the solution $C_T = (\theta_E, t_E, \sigma_T)$ of Eqs. (2.26) with geodesic curvature

$$h_T = f(T)\left(\int_{t_0/2}^{T} f(\xi)d\xi\right)^{-1} > 0$$

and initial conditions $(0, T, 0)$ at $s = 0$. The value of the energy is $E_T = -h \int_0^{t_0/2} f(\xi)d\xi$. The first integral (2.27) then implies that C_T is a graph on the hemispheres $t \neq t_0/2$ and meets the equator orthogonally. This proves the existence of C_T. The convexity of M implies $\lambda_1(C_T) < 0$ for all T.

As for the stable curves, we have the following classification result.

Theorem 2.36 *Let M be a rotationally symmetric convex sphere with an equatorial symmetry and Gauss curvature increasing as a function of the distance to the poles. Then the only stable embedded curves in M can be nodoids symmetric with respect to the equator, complete meridians or geodesic circles in regions of constant Gauss curvature. In particular, any of these curves is connected.*

Proof Let C be a *connected* embedded curve with constant geodesic curvature. Assume C is not the boundary of a constant curvature disk. Then it cannot be a curve touching one pole by Lemma 2.18(3), neither a stable unduloid by the same argument as in the proof of Theorem 2.26 since we would need at least two copies between consecutive maxima to close the curve. If C is a nodoid, then it is

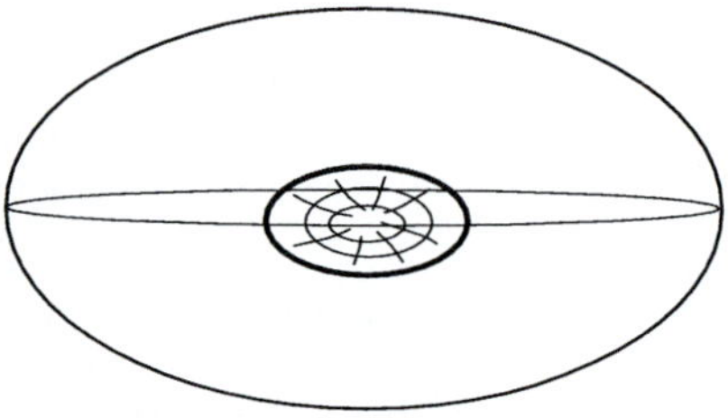

Fig. 2.9 Isoperimetric regions in convex spheres with curvature decreasing from the equator

symmetric by Lemma 2.29(1). Circles of revolution not contained in a region of constant Gauss curvature around the poles are not stable by Lemma 2.6.3.

As all these types of curves have negative first eigenvalue for the Jacobi operator, we conclude that an embedded stable closed curve must be connected, and the Theorem follows. □

Once we have classified the isoperimetric boundaries in Theorem 2.36, a classification of isoperimetric regions follows (Fig. 2.9).

Theorem 2.37 *Let M be a rotationally symmetric convex sphere with an equatorial symmetry and whose Gauss curvature is a non-decreasing function of the distance to the poles. Then isoperimetric regions in M are disks enclosed by nodoids symmetric with respect to the equator, geodesic disks inside a region with constant curvature around the equator, and their complements.*

Proof Let E be an isoperimetric region enclosing area A. Its boundary ∂E is a stable curve with constant mean curvature. By Theorem 2.36, the curve ∂E is connected, and it is either a geodesic disk in a region with constant Gauss curvature, a complete meridian, or nodoid C_T. The E is one of the regions delimited by such a curve. □

2.6.4 Surfaces with Singularities

Let $I \subset \mathbb{R}$ be an interval of the form $(0, a)$, with $a \in \mathbb{R} \cup \{+\infty\}$. Consider the Riemannian metric $ds^2 = dt^2 + f(t)^2 d\theta^2$, for some smooth function $f : I \to \mathbb{R}$. Assume that $\lim_{t\to 0} f(t) = 0$ so that the surface can be completed near $t = 0$ by adding a point. A simple computation yields, for $r \in (0, a)$,

$$\int_{D_r} K + \int_{S_r} h(r) = 2\pi \lim_{t\to 0+} f'(0), \tag{2.40}$$

where $D_r = \{(s, \theta) : 0 < s < r\}$, $S_r = \{(s, \theta) : s = r\}$, K is the Gauss curvature, and $h(r)$ the geodesic curvature of the circle S_r. Gauss-Bonnet theorem implies that if the completion of the surface is regular, then $f'(0) = 1$.

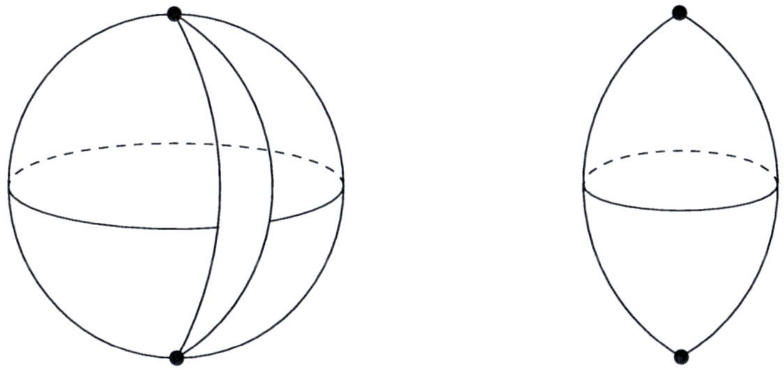

Fig. 2.10 Construction of a sphere with constant Gauss curvature and two conical singularities

We say that the completed surface has a *conical singularity* at $t = 0$ if

$$0 < \lim_{t \to 0+} f'(t) < 1.$$

For instance, we can consider any of the surfaces of revolution considered in the previous section, removing a wedge $2\pi - \alpha \leqslant \theta \leqslant 2\pi$ and pasting the resulting surface along the boundary geodesics or considering a quotient orbifold by the finite group generated by the rotation of angle $2\pi/n$, with $n \in \mathbb{N}$. In both cases, a surface with one (in the case of a plane) or two (in the case of a sphere) conical singularities is obtained (Fig. 2.10).

In these singular surfaces, area and length are defined as in the smooth case. Out of the singular points, the boundaries of isoperimetric sets are smooth curves with constant geodesic curvature. At singular points (Lemma 2.3 in [124]), the lateral unit tangent vectors must make an angle larger than or equal to π.

Using the techniques in the previous sections, we can prove the following.

Theorem 2.38 *Let M be a symmetric sphere of revolution with two conical singularities and positive Gauss curvature non-increasing from the poles. Then the only isoperimetric sets are geodesic disks centered at the poles.*

Proof Since M is a compact manifold, isoperimetric sets exist for any $0 < A < A(M)$. The boundary of an isoperimetric set is smooth out of the singularities at the poles.

Since $(f')^2 - ff'' < 1$, the proof of Lemma 2.18(3) implies that that curves touching the pole do it at an angle smaller than π. Hence none of these curves can be part of the boundary of an isoperimetric set by Lemma 2.3 in [124]. This implies that the boundary of the isoperimetric set is contained in the regular part of M and must be connected by a stability argument since $K \geqslant 0$.

Again by condition $(f')^2 - ff'' < 1$ (Lemma 2.18(1)), there are no closed non-constant unduloid type solution to the geodesic curvature equation. For nodoids, we can apply Lemma 2.29 to conclude that it must be contained in a region with

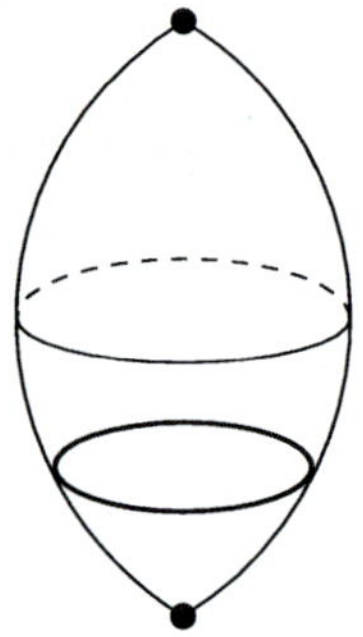

Fig. 2.11 The only isoperimetric sets in a sphere of revolution with constant Gauss curvature and two conical singularities are geodesic balls about the poles

constant Gauss curvature K_0. In the latter case, the boundary length would satisfy

$$L^2 = 4\pi A - K_0 A^2.$$

Let us see that the disk of area A centered at a conical singularity satisfies $L^2 < 4\pi A - K_0 A^2$. For such a disk we have $K \geqslant K_0$. We fix the singularity at $t = 0$, and we let D_a be the disk centered at this point of area a. Let S_a be its boundary, L_a its perimeter, and h_a the geodesic curvature of its boundary with respect to the outer normal. By the Gauss-Bonnet formula with singularities (2.40), we have

$$\frac{dL_a^2}{da} = 2\int_{S_a} h_a = 2\left[2\pi \lim_{t\to 0+} f'(0) - \int_{D_a} K\right] < 4\pi - 2K_0 a.$$

Integrating from 0 to A, we have $L_A^2 < 4\pi A - K_0 A^2$. Hence, the disk centered at the singularity is the isoperimetric set for area A. □

This result holds in particular when M has constant positive curvature. This singular surface is obtained by cutting a wedge bounded by two geodesics from the north to the south pole and pasting the two boundaries (see [427, 428]) (Fig. 2.11).

2.7 Existence of Isoperimetric Regions in Complete Surfaces with Non-negative Gauss Curvature

In a non-compact surface, existence of isoperimetric regions is a difficult question. As was shown in Theorem 2.27, on planes of revolution with strictly increasing Gauss curvature, isoperimetric regions do not exist for any given value of area. However, under the assumption of non-negative Gauss curvature, it is possible to prove existence. Our main result is as follows.

Theorem 2.39 *Let M be a complete Riemannian plane with non-negative Gauss curvature. Then isoperimetric regions exist on M for any given area. Each one is a disk bounded by a closed embedded curve with constant geodesic curvature.*

The proof of Theorem 2.39 follows by analyzing the behavior of minimizing sequences in M and using Alexandrov-Fiala's inequality (2.11)

$$L^2 \geqslant \left[4\pi - \int_E K \right] A$$

for convex surfaces.

Let us recall that for a complete non-compact surface with non-negative Gauss curvature, Cohn-Vossen's inequality (see [123] and Corollary 2 in [443]) implies

$$C(M) = \int_M K dA \leqslant 2\pi \chi(M).$$

Thus, we immediately obtain Cohn-Vossen's classification: M is either a plane, a flat cylinder, or a flat Möbius band. In the flat cases, closed curves with constant geodesic curvature are geodesic circles. The existence of a sufficiently large number of isometries so that the quotient of M by this group is a compact set implies existence of isoperimetric regions for any given area. This way, we obtain the following classification result.

Theorem 2.40

1. *Let M be a complete flat cylinder. Then isoperimetric regions exist on M for any value of A. Moreover, there exists $A_0 > 0$ such that the geodesic disks for $A \leqslant A_0$ and annuli bounded by two equidistant geodesics if $A \geqslant A_0$.*
2. *Let M be a complete flat Möbius strip. Then isoperimetric regions exist on M for any value of A. Moreover, there exists $A_0 > 0$ such that the geodesic disks for $A \leqslant A_0$ and Möbius bands bounded by a closed geodesic if $A \geqslant A_0$.*

We don't give the proof of existence here since it is easily obtained from the general methods of Chap. 4. Once we have existence, the characterization of isoperimetric sets is rather straightforward.

In the proof of Theorem 2.39, we use a Lipschitz contraction as a fundamental tool. There are two ways of producing such a deformation: either embedding M in $\mathbb{R}^3$ and considering the negative gradient of the height function or using the well-known Sharafutdinov retraction in manifolds with non-negative curvature. The proof of the following result, due to V. A. Sharafutdinov [435], can be found in [453]. The *soul* of a complete manifold with non-negative sectional curvature was defined by Cheeger and Gromoll [106] and is a compact totally geodesic submanifold without boundary. When M is a surface, it is either a point or a geodesic. If it is a geodesic, Cheeger-Gromoll splitting theorem [105] implies that M is flat.

Lemma 2.41 *Let M be a complete unbounded non-flat Riemannian surface with non-negative sectional curvature. Then there exists a distance non-increasing strong deformation retract from M onto a point of M.*

2.7.1 Proof of the Existence Result

Theorem 2.39 will be proven in this subsection.

We fix some positive area A and take a minimizing sequence of sets $\{E_i\}_{i\in\mathbb{N}}$ with smooth boundary such that $A(E_i) = A$ for all $i \in \mathbb{N}$ and $P(E_i) \to I(A)$, where I is the isoperimetric profile of M. Let $L = I(A)$. We may assume that each E_i has a finite number of components. This can be done discarding an infinite number of connected components of E_i with negligible area and correcting the loss of area enlarging sightly the remaining components.

We assume that the Gauss curvature of M is not identically zero, since the flat case is well-known (see Theorem 2.2).

The proof will be given into several steps.

Step 1 We can replace E_i by a finite union of disks.

Proof of Step 1 The set E_i is the union D_i of a finite collection of disks from which a finite number of disks have been removed. The set D_i has finite area and perimeter and $A(D_i) \geqslant A(E_i)$ and $P(D_i) \leqslant P(E_i)$. Let $\{\varphi_t\}_{t>0}$ be the Lipschitz deformation in Lemma 2.41. Since φ_t contracts M to a point we can find $t_i > 0$ so that $A(\varphi_{t_i}(D_i)) = A$. As φ_{t_i} is a contracting map

$$P(\varphi_{t_i}(D_i)) \leqslant P(D_i) \leqslant P(E_i).$$

So $\{\varphi_{t_i}(D_i)\}_{i\in\mathbb{N}}$ is also a minimizing sequence for area A which elements are finite unions of disks. □

By Lemma 2.21 we have the decomposition $E_i = E_i^c \cup E_i^d$ and the existence of a set $E \subset M$ such that a non-relabeled subsequence E_i^c converges to E in $L^1_{\text{loc}}(M)$. Denote by L_c and A_c the perimeter and area of E. Let $L_d = L - L_c$, $A_d = A - A_c$. Since E_i^d is a divergent sequence and $C(M)$ is finite, the total curvature of E_i^d converges to 0 when $i \to \infty$. Using the Alexandrov-Fiala inequality (2.11) for the sets E_i^d with Lipschitz boundary (we can approximate them by sets with smooth boundary) and taking limits, we have

$$L_d^2 \geqslant 4\pi A_d. \tag{2.41}$$

Step 2 The set E is either empty or a disk.

Proof of Step 2 The boundary of E has constant mean curvature h, and the stability inequality (2.29) holds for all smooth function in ∂E with mean zero. If ∂E is not connected, we insert a locally constant nowhere vanishing test function with mean zero in (2.29). As $K + h^2 \geqslant 0$ this implies $K + h^2 \equiv 0$ on ∂E. Hence ∂E is composed of a finite number of closed geodesics. If C is one of such geodesics and encloses a disk D, Gauss-Bonnet theorem implies

$$0 = \int_C h = 2\pi - \int_D K$$

and so D has total curvature 2π. Hence $M \setminus D$ is flat. So ∂E is contained in a flat portion of M.

Consider now two geodesics C_1 and C_2 in ∂E enclosing disks D_1 and D_2 in M. We can always assume that $D_2 \subset D_1$ and so $D_1 \setminus D_2$ is flat. Then $D_1 \setminus D_2$ is isometric to a cylinder and C_1, C_2 is isometric to the product $[a, b] \times C_1$. We conclude that ∂E is contained in a cylindrical region of M and that all geodesics are equidistant. By translating the components inside the cylindrical region, we can join them to obtain just one annulus or a disk with the same area and less perimeter. If we get an annulus, we can fill in to get a disk with less perimeter and more area, and using the contracting maps φ_t, we can reduce the area until it is equal to $A(E)$. □

Step 3 A subsequence of the minimizing sequence E_i converges to E.

Proof We consider the sets $E(r) = \{p \in M : d(p, E) \leqslant r\}$, for $r > 0$. If E is empty, we take $E(r)$ a geodesic disk of radius r centered at a point of positive curvature. By results of Hartman we have already used in this chapter, $\partial E(r)$ is a piecewise smooth curve for a full measure set of non-exceptional $r \in (0, \infty)$. Moreover, $L(r) = L(\partial E(r))$, defined for non-exceptional r, can be extended so as to be a function of locally bounded variation, and we have

$$L(r)^2 - L_c^2 \leqslant \int_0^r 2LL'(s)ds \tag{2.42}$$

and, for non-exceptional s,

$$L'(s) \leqslant \int_{\partial E} \int_{\partial E} h - \int_{E(r)\setminus E} K.$$

Using Gauss-Bonnet and $K \geqslant 0$ the above inequality is transformed into

$$L'(s) \leqslant 2\pi - \int_{E(r)} K \leqslant 2\pi. \tag{2.43}$$

So we conclude from (2.42) and (2.43)

$$L(r)^2 - L_c^2 \leqslant 4\pi \int_0^r L(s)ds = 4\pi(A(E(r)) - A_c).$$

Letting $A(E(r)) \uparrow A$ we see that there exists $r_0 > 0$ such that $E' = E(r_0)$ is a finite perimeter set of area A.

$$P(E')^2 \leqslant L_c^2 + 4\pi(A - A_c)$$

Using $A_c + A_d = A$ and (2.41) we get

$$P(E')^2 \leqslant L_c^2 + L_d^2 \leqslant L^2 \tag{2.44}$$

so that E' has area A and perimeter less than or equal to L. As $L = I(A)$ we conclude $P(E') = A$. So equality holds in the last inequality of (2.43) and the total curvature of $E(r)$ is 0. This shows that $A_c \neq 0$ since E has been chosen in case $A_c = 0$ to contain points of positive curvature. From the equality in (2.44), we get $L_d = 0$, and by (2.41), $A_d = 0$. So E keeps all the area of the minimizing sequence, and it is an isoperimetric region of area A. Of course, it is an isoperimetric region, and it is a disk.

Hence, the proof of Theorem 2.39 is completed. □

2.7.2 *Consequences of the Existence Result*

Once we have proven existence of isoperimetric regions, further properties of the isoperimetric profile can be obtained.

Corollary 2.42 *Let M be a complete plane with non-negative curvature. Then the isoperimetric profile of M is a positive concave function.*

Proof The isoperimetric profile I of M is positive by Theorem 2.39 and the Alexandrov-Fiala inequality (2.11).

To prove the concavity of I, we fix some $A_0 > 0$ and take an isoperimetric region E_0 of area A_0. Let $L_0 = P(E_0) = I(A_0)$. We consider a variation using a vector field X with normal component $\langle X, N\rangle = 1$, where N is the outer unit normal to E_0. The first variation of area allows us to build a variation of E_0 parameterized by area near A_0. If $f(A)$ is the perimeter of the set in the deformation with area A, then the variational formulas for area and perimeter yield

$$f''(A) = -\frac{1}{L_0^2}\left(\int_{\partial E_0} (K + h^2)\, dA\right) \leqslant 0$$

(see Sect. 3.4). Observe that $I \leqslant f$ and that $I(A_0) = f(A_0)$. Thus I admits a support convex function. The elementary Lemma 3.11 then implies that I itself is convex. □

Corollary 2.42 implies that the profile I possesses all regularity properties of concave (or convex) functions. In particular, left and right derivatives I'_-, I'_+ exist for every $A > 0$, I and I' are differentiable almost everywhere, and I is absolutely continuous in $(0, +\infty)$. Since $I > 0$ by Corollary 2.42, the profile I is a non-decreasing function (see [376]).

Now we relate the geodesic curvature to the differential properties of the profile.

Corollary 2.43 *Let M be a complete plane with non-negative curvature. Then the geodesic curvature h of any isoperimetric region of area A satisfies*

$$I'_+(A) \leqslant h \leqslant I'_-(A). \tag{2.45}$$

In particular, when A is a regular value of I, all isoperimetric regions of area A have geodesic curvature $I'(A)$.

Moreover, for every non-regular value A of I, there are two different isoperimetric regions of area A and geodesic curvatures $I'_-(A)$, $I'_+(A)$.

Proof We only consider the non-flat case.

Let E be an isoperimetric region of area A and geodesic curvature h. We build a deformation of E by sets E_a of area a, for a close to A. Let $f(a) = (E_a)$. The first variation formulas of area and perimeter imply

$$f'(A) = h.$$

As $I \leqslant f$ for a close to A, and $I(A) = f(A)$, we get (2.45).

Assume now that A is not a regular value of I. Consider a sequence $\{A_i\}_{i \in \mathbb{N}}$ of regular values of I decreasing to A. For each i, take an isoperimetric region E_i of area A_i whose boundary has geodesic curvature h_i. By the concavity of I

$$0 \leqslant h_i = I'(A_i) \leqslant I'_+(A) < +\infty. \tag{2.46}$$

The sequence $\{E_i\}_{i \in \mathbb{N}}$ cannot diverge. If this were the case, we would have

$$I(A)^2 = \lim_{i \to \infty} I(A_i)^2 \geqslant 4\pi A,$$

but a geodesic disk of area A centered at point of positive curvature would have less perimeter as in the proof of Theorem 2.39.

As $\{\partial E_i\}_{i \in \mathbb{N}}$ has an accumulation point and $\{h_i\}_{i \in \mathbb{N}}$ is bounded, standard results on ordinary differential equations imply that we can extract a convergent subsequence of ∂E_i to a curve C with constant mean curvature which encloses an isoperimetric region of area A. Since

$$h = \lim_{i \to \infty} h_i = \lim_{i \to \infty} I'(A_i) \geqslant I'_+(A)$$

and from (2.46) we get $h \leqslant I_+(A)$, we obtain $h = I'_+(A)$.

Finally, an isoperimetric region of area A and geodesic curvature $I'_-(A)$ is obtained as a limit of isoperimetric sets E_i of area $A_i < A$. □

Regularity results similar to the ones in Theorem 2.43 can be found in Section 6 of Hsiang's paper [237].

Another consequence of the existence result of isoperimetric sets is the following pair of properties of the isoperimetric profile for large values of the area.

Corollary 2.44 *Let M be a complete plane with non-negative curvature.*

1. *If there is a closed geodesic of length ℓ_0, then $C(M) = 2\pi$, the isoperimetric profile of M satisfies $I \leqslant \ell_0$, and there exists $A_0 > 0$ such that $I(A) = \ell_0$ for all $A \geqslant A_0$.*
2. *If $C(M) < 2\pi$, then $I(A) \to \infty$ when $A \to \infty$.*

Proof 1. Let C be a closed geodesic enclosing a region Ω. By Gauss-Bonnet theorem, $C(\Omega) = 2\pi$, and so $M \setminus \Omega$ is flat and isometric to the product $(0, +\infty) \times C$. Hence, for a large enough area, we have $I(A) \leqslant \ell_0$, and since the isoperimetric profile is non-decreasing, we get $I \leqslant \ell_0$. If E is an isoperimetric region of large area, then $\partial E \subset M \setminus \Omega$ since $P(E)$ is uniformly bounded by ℓ_0. A usual calibration argument implies that $P(E) \geqslant \ell_0$ with equality if ∂E is a geodesic in $M \setminus \Omega$ equidistant from C. We conclude that ∂E is a geodesic in $M \setminus \Omega$ equidistant from C.

2. As I is an increasing function, let $I_\infty = \lim_{A \to \infty} I(A)$. For every $A > 0$, choose an isoperimetric region E_A of area A with geodesic curvature h_A. Taking a sequence of regular values of I when $A \to \infty$, we get $h_A \downarrow h_\infty$. If $h_\infty = 0$ then Gauss-Bonnet implies

$$h_A I(A) = 2\pi - C(E_A) \geqslant 2\pi - C(M) > c_0 > 0.$$

and so $I_\infty = \infty$. If $h_\infty > 0$ then $h_A \geqslant h_\infty$ for almost every A, and the absolute continuity of I implies that $I_\infty = \infty$. □

2.8 Notes

Notes for Sect. 2.2 While there are many proofs of the planar isoperimetric inequality, we have chosen to include Hurwitz's [240] because of its simplicity. Its extension to Cartan-Hadamard surfaces was done by Chavel [99], although the original result in Cartan-Hadamard manifolds is due to A. Weil [439] following Carleman's proof of the isoperimetric inequality for disks on minimal surfaces [95] and, independently, by Beckenbach and Radó [49]. Beckanbach and Radó results have been extended to metric measure spaces by Lytchak and Wenger [274]. See also Osserman's survey [325].

During the decade of 1920, Bonnesen proved several lower estimates of the *isoperimetric deficit* $L^2 - 4\pi A$, where L is the boundary length of a set of area

A. In particular, he proved

$$L^2 - 4\pi A \geqslant \pi^2 (R - r)^2,$$

where r, R are the inradius and circumradius of the set (see [70, 71] and Osserman [326]).

Proofs of the planar isoperimetric inequality using methods of integral geometry or geometric probability were given by Santaló (pp. 37–38 in [391]; see also Blåsjö, §11 in [64], and [389]). Using similar methods, he also gave a proof of a Bonnesen type isoperimetric inequality in the hyperbolic plane. More precisely he proved the inequality

$$L^2 - 4\pi A - A^2 \geqslant \frac{1}{4} \left(\coth(R/2) - \coth(r/2)\right),$$

where L is the boundary length of a set of area A and r, R are the inradius and circumradius of the set (see [390]).

The isoperimetric inequality in the sphere was proven by Bernstein [57] for regions contained in a hemisphere and by Radó [352] in the general case.

Hales [214] gave a proof of the planar isoperimetric inequality by polygonal approximation and a passage to the limit using the same techniques as in the original proof of the Jordan Curve theorem.

Notes for Sect. 2.3 The method of equidistants goes back to early theory of convex sets. The first applications of this method to isoperimetric inequalities in surfaces with variable curvature were made by Bol [68] and Fiala [157, 158] for analytic surfaces. The extension to smooth surfaces was made by Hartman [218] (see also [403]). The main difficulty is to describe the behavior of equidistants for almost all values. We refer the reader to Chapter 1 in Burago and Zalgaller's monograph [83].

Notes for Sect. 2.4 Bandle's method is a nice application of the use of level sets of a solution of a partial differential equation to deform a domain in order to get an isoperimetric inequality. The reader is referred to the very interesting monograph by Bandle [36], where many applications can be found. Huber [239] gave a proof of the isoperimetric inequality based on inequalities for differences of subharmonic functions.

Notes for Sect. 2.5 The paper by Benjamini and Cao [50] was a major breakthrough at the time it was published. In addition to the results exposed in Sect. 2.5, the authors also gave some applications of their results to get bounds for the Laplacian on regions in rotationally symmetric surfaces with non-increasing curvature [50, Thm. 10] and comparison results for eigenvalues [50, Corollary B]. They also tried to apply their methods to rotationally symmetric spheres, but unfortunately, as stated in [310, p. 4904], their proof is wrong since equation (4.6) in [50] is wrong.

The methods employed in [50] were also used by Topping [425] to get similar results. Shortly after Benjamini and Cao's paper, Pansu [332] showed that the classical deformation by inner parallels can be used to obtain the results in Benjamini and Cao's paper. We have included Pansu's proof as Theorem 2.6 in Sect. 2.3. Topping also recovered these results by polyhedral approximation [426]. Howards, Hutchings, and Morgan [310] solved the isoperimetric problem in rotationally symmetric spheres in some particular cases.

Let us also mention that a generalization of the Banchoff-Pohl inequality using the curve shortening flow was given by Süssmann [416].

Notes for Sect. 2.6 The variational methods used in this section finally permitted to characterize the isoperimetric regions in symmetric convex spheres of revolution. This problem was finally solved in [360] from a classification of simple closed curves with constant geodesic curvature, an approach used by Schmidt [396] in the early 1940s. Using these techniques, Cañete [92] classified the stable regions in tori of revolution, and Cañete and the author [86] completed the classification of isoperimetric regions in quadrics of revolution. In the paper [92], a very interesting analysis of the stability of unduloids is made.

More results for some other different topological types such as projective planes and cylinders are also obtained in [360] under monotonicity conditions on the Gauss curvature. As mentioned in the introduction of [360], this paper originated from the desire to understand to what extent the classical Jellet's theorem for star-shaped constant mean curvature hypersurfaces was valid in the two-dimensional case (see [298]).

As mentioned at the end of the section, many of the results remain valid if we assume the correct metric singularity at the poles. For instance, if we consider a plane of revolution with a single pole satisfying

$$0 < [(f')^2 - ff''](0) < 1$$

most results of Sect. 2.6.2 for planes of decreasing Gauss curvature are valid.

A quite systematic study of isoperimetric curves in surfaces with singularities can be found in [124]. Morgan [306] and Petean [340] considered isoperimetric sets in orbifolds. Petean gave some applications of his results to obtain lower bounds on the Yamabe invariant. See also Petean [340], Petean and Ruiz [341, 342], and Henry and Petean [228].

On hyperbolic surfaces, Morgan and Adams [5] obtained a classification of isoperimetric boundaries in hyperbolic surfaces. They are

- a circle,
- horocycles around cusps,
- two neighboring curves at constant distance from a geodesic,
- geodesics or single neighboring curves.

Pansu obtained in [332] some interesting results on the isoperimetric profile of compact surfaces. In Lemma 10 he considered a smooth function L, defined on a neighborhood of 0, such that $L(0) = 0$, $L'(0) = 4\pi$ and $L'''(0) = 0$. Then f defined by

$$L\left(2\pi \int_0^t f(s)ds\right) = 2\pi f(t)$$

generates a surface of revolution with decreasing Gauss curvature. Two disks of revolution obtained this way from two functions L_1, L_2 collated along an equator produce a surface of revolution whose isoperimetric profile crosses itself infinitely many times (see Théorème 2). Moreover, if the surface is analytic, then the isoperimetric profile is semi-analytic (see Théorème 3). In Théorème 1, he proved that isoperimetric regions with small area in a compact surface are disks close to the maxima of curvature. Ye [451] showed that punctured neighborhoods of non-degenerate critical points of the Gauss curvature are foliated by closed embedded curves with constant geodesic curvature.

Notes for Sect. 2.7 The first author to notice that a sequence of disks escaping to infinity in a convex surface must satisfy the planar isoperimetric inequality in the limit seems to be Fiala [157]. A fundamental tool to prove this property is Alexandrov-Fiala inequality (2.11) for surfaces of non-negative curvature. This is the main idea used in the proof of existence of isoperimetric regions for any area in a complete plane with non-negative curvature. The result is in striking contrast with Theorem 2.37, which provides many examples of complete planes with non-positive curvature for which isoperimetric regions do not exist for any value of area.

Chapter 3
The Isoperimetric Profile of Compact Manifolds

In this chapter, the basic properties of the isoperimetric profile I_M of a compact Riemannian manifold M, and of their isoperimetric sets, are described. We recall that the isoperimetric profile is the function that assigns to any volume the infimum of the perimeter of sets of this volume and should be thought of as the best possible isoperimetric inequality in M. Isoperimetric regions are the ones with the smallest possible perimeter for a given volume.

In the first part, we introduce the concept of normalized isoperimetric profile and prove some basic properties: existence of isoperimetric sets for any given volume, positivity of the isoperimetric profile, and symmetry with respect to $|M|/2$. We also show that $I_M(v)$ approaches 0 when v is close to 0 or to the total volume $|M|$. Then continuity and local Hölder continuity of the isoperimetric profile are proven, and the behavior of I_M for small volumes is described.

Afterward, we focus on regularity properties of the profile and that I_M^α, for $1 \leqslant \alpha \leqslant m/(m-1)$, satisfies a differential inequality in a weak sense. One of the main consequences of this differential inequality is that I^α, when $1 \leqslant \alpha \leqslant m/(m-1)$, is locally the sum of a concave function and a smooth one, and hence, it enjoys the same regularity properties as concave functions. In particular, it is differentiable except on a countable set and twice differentiable almost everywhere, it has lateral derivatives everywhere, and it is an absolutely continuous function. The differential inequality is then used to provide a description of the isoperimetric sets in round spheres and also for comparison purposes: following Bayle, we give a proof of the classical Lévy-Gromov isoperimetric inequality. Under the hypothesis that the Ricci curvature of the manifold is non-negative, we obtain the concavity of I^α, strict concavity indeed when $1 \leqslant \alpha < m/(m-1)$.

We conclude the chapter by looking at the continuity properties of the isoperimetric profile under Lipschitz convergence and provide some density estimates for isoperimetric sets following Leonardi and Rigot. From these estimates, some basic regularity properties of isoperimetric sets such as Ahlfors regularity of the

M. Ritoré, *Isoperimetric Inequalities in Riemannian Manifolds*,
Progress in Mathematics 348, https://doi.org/10.1007/978-3-031-37901-7_3

boundary are obtained. These techniques also improve the convergence in measure of isoperimetric sets to Hausdorff convergence.

3.1 The Normalized Isoperimetric Profile

The isoperimetric profile of the Riemannian manifold M, defined in Sect. 1.6, is denoted by I_M. Sometimes, it is convenient to normalize it so that it is defined in a fixed interval (e.g., when comparing the profiles of two manifolds or looking at convergence properties of a sequence of isoperimetric profiles).

Definition 3.1 Let M be a compact Riemannian manifold. Its *normalized isoperimetric profile* $h_M : (0, 1) \to \mathbb{R}^+$ is the function

$$h_M(\lambda) = \frac{I_M(\lambda|M|)}{|M|}. \tag{3.1}$$

As for the isoperimetric profile, the normalized isoperimetric profile provides an optimal isoperimetric inequality on M. If $F \subset M$ is a measurable set, then inequality

$$P(F) \geqslant |M|\, h_M\left(\frac{|F|}{|M|}\right) \tag{3.2}$$

is satisfied. Equality holds just for isoperimetric regions.

Since the isoperimetric profile I_M and the normalized isoperimetric profile h_M just differ in composition with affine maps, they share most of their relevant properties.

Definition 3.2 The *Cheeger constant* of a Riemannian manifold M is defined by

$$\mathrm{Ch}(M) = \inf \frac{P(E)}{\min\left\{|E|, |M \setminus E|\right\}}.$$

A measurable set $E \subset M$ satisfying

$$\mathrm{Ch}(M) = \frac{P(E)}{\min\left\{|E|, |M \setminus E|\right\}}$$

is called a *Cheeger set*.

That the Cheeger constant is finite follows by taking a finite perimeter set E with $|E|, |M \setminus E| > 0$. If $E \subset M$ is a Cheeger set, then also $M \setminus E$ is a Cheeger set. The definition of the Cheeger constant immediately implies the inequality

$$P(E) \geqslant \mathrm{Ch}(E) \min\left\{|E|, |M \setminus E|\right\}$$

for any measurable set $E \subset M$. A Cheeger set $E \subset M$ is an isoperimetric set since for any measurable set F with $|F| = |E|$, we also have $\min\{|E|, |M \setminus E|\} = \min\{|F|, |M \setminus F|\}$, and so

$$P(F) \geqslant \mathrm{Ch}(M) \min\{|F|, |M \setminus F|\} = \mathrm{Ch}(M) \min\{|E|, |M \setminus E|\} = P(E).$$

3.1.1 Basic Properties

Let us establish now the existence of isoperimetric regions in compact Riemannian manifolds. This follows directly from the compactness result in Theorem 1.42 and the lower semicontinuity of perimeter in Proposition 1.41.

Theorem 3.3 *Let M be a compact Riemannian manifold, $v \in (0, |M|)$. Then isoperimetric sets of volume v exist on M.*

Proof Let $\{E_i\}_{i\in\mathbb{N}}$ be a minimizing sequence of measurable subsets in M of volume v such that $\lim_{i\to\infty} P(E_i) = I_M(v)$. Since $P(E_i)$ is uniformly bounded, we can extract a subsequence converging in $L^{(}M)$ to a measurable set $E \subset M$ of volume v by Theorem 1.42. The lower semicontinuity of the perimeter (Proposition 1.41) implies

$$I_M(v) \leqslant P(E) \leqslant \liminf_{i\to\infty} P(E_i) = I_M(v).$$

Hence, $E \subset M$ is an isoperimetric set of volume v. □

Let us show now that the isoperimetric profile is a positive symmetric function that extends continuously to $v = 0$ and $v = |M|$.

Lemma 3.4 *The isoperimetric profile of a compact Riemannian manifold is a positive function in the interval $(0, |M|)$, symmetric with respect to $|M|/2$, and extends continuously to 0 at the endpoints of the interval $(0, |M|)$.*

Similar properties hold for the normalized isoperimetric profile: it is positive, symmetric with respect to $1/2$, and extends continuously to 0 at $\lambda = 0, 1$.

Proof Let $0 < v < |M|$ and let $E \subset M$ be an isoperimetric set of volume v. A standard argument in measure theory (see Proposition 3.1 in Giusti's monograph [184]) implies the existence of $x \in M$ and $r > 0$, smaller than the injectivity radius of M, so that $|E \cap B(x, r)|, |B(x, r) \setminus E| > 0$. We can then apply the relative isoperimetric inequality (1.72) in $B(x, r)$ to conclude that

$$P(E) \geqslant P(E, B(x, r)) = C_I \min\{|E \cap B(x, r)|, |B(x, r) \setminus E|\}^{(m-1)/m} > 0.$$

The symmetry of the profile follows because if $E \subset M$ is an isoperimetric region of volume v, then $M \setminus E$ is an isoperimetric region of volume $|M| - v$ with the same perimeter as E.

To prove that I_M extends continuously as 0 at the endpoints of the interval $(0, |M|)$, it is sufficient to do it at $v = 0$ by the symmetry property. We simply fix a point $x \in M$ and consider the function $r \mapsto |B(x, r)|$, which is continuous, increasing, and approaches 0 when $r \to 0^+$. For $r > 0$ small enough the balls $B(x, r)$ have compact smooth boundary, and so they have finite perimeter $P(B(x, r))$ converging to 0 when $r \to 0^+$. Hence, $P(B(x, r)) \geqslant I(|B(x, r)|)$ and, letting $r \to 0^+$, we get the result. □

3.2 Continuity of the Isoperimetric Profile

We show here that the isoperimetric profile function is locally Hölder continuous with exponent $(m - 1)/m$. Although we shall obtain later stronger regularity properties, the continuity of the isoperimetric profile is a requirement for the improvement of regularity.

Theorem 3.5 *The isoperimetric profile of a compact Riemannian manifold is a continuous function. Moreover, it is locally Hölder continuous with exponent $(m - 1)/m$ in the interval $(0, |M|)$.*

Proof The proof follows the arguments by Gallot (see Lemme 6.2 in [176]).

Let $r_0 > 0$ be smaller than the injectivity radius of M, and let $C_m, C'_m > 0$ such that

$$P(B(x, s)) \leqslant C_m s^{m-1}, \quad |B(x, s)| \geqslant C'_m s^m, \quad \text{for all } x \in M,\ 0 < s < r_0.$$

The existence of C_m, C'_m is guaranteed by the compactness of M. Observe that this implies

$$P(B(x, s)) \leqslant \frac{C_m}{(C'_m)^{(m-1)/m}} |B(x, s)|^{(m-1)/m} \tag{3.3}$$

for all $x \in M$ and $0 < s < r_0$.

Fix some $v_0 > 0$ and take $0 < v < v_0$ close enough to v_0, so that we can choose $0 < r < r_0$ satisfying

$$\frac{b(r)}{2|M|} v_0 = v_0 - v, \tag{3.4}$$

where $b(r) = \inf_{x \in M} |B(x, r)|$, defined in (1.59), is continuous and satisfies $\lim_{r \to 0} b(r) = 0$.

Fix $\varepsilon > 0$ and take a finite perimeter set $E \subset M$ of volume v_0 such that $P(E) \leqslant I_M(v_0) + \varepsilon$. Then

$$\int_M |B(x, r) \cap E| \, dM(x) = \int_E |B(x, r)| \, dM(x). \tag{3.5}$$

This formula is obtained by applying Fubini-Tonelli's Theorem to the function $(x, y) \in M \times M \mapsto (\mathbf{1}_{B(x,r)}\mathbf{1}_E)(y) = \mathbf{1}_{B(x,r)\cap E}(y)$. Equation (3.5) implies the existence of $x \in M$ such that

$$|B(x,r) \cap E| \geqslant \frac{b(r)}{2|M|}|E| = v - v_0$$

by the choice of r. From the continuity of the non-decreasing function $t \mapsto |B(x,t) \cap E|$, we can find $s \in (0, r]$ such that $|B(x,s) \cap E| = v_0 - v$. Hence, $|E \setminus B(x,s)| = v$, and we get

$$I_M(v) \leqslant P(E\setminus B(x,s)) \leqslant P(E)+P(B(x,s)) \leqslant I_M(v_0)+\varepsilon+C\left(\frac{v_0-v}{v_0}\right)^{(m-1)/m},$$

for some constant $C > 0$ depending on C_m, C'_m and M. Here we have used (3.3), inequality $|B(x,s)| \leqslant |B(x,r)| \leqslant C'b(r)$, and (3.4). As $\varepsilon > 0$ is arbitrary, we get

$$I_M(v) \leqslant I_M(v_0) + C\left(\frac{v_0-v}{v_0}\right)^{(m-1)/m}$$

for $0 < v < v_0$ close enough to v_0.

Considering the volumes $|M| - v_0$, $|M| - v$, the symmetry of the isoperimetric profile implies a similar inequality for $|M| > v > v_0$. Hence, we get

$$|I_M(v) - I_M(v_0)| \leqslant C\left|\frac{v_0-v}{v_0}\right|^{(m-1)/m},$$

for v close enough to v_0 so that $b(r)/(2|M|)v_0 = |v_0 - v|$ is satisfied for some $0 < r < r_0$. □

Remark 3.6 A proof showing merely continuity (not Hölder continuity) follows using standard properties of the perimeter and the deformation Theorem 1.50.

Take a sequence $\{v_i\}_{i\in\mathbb{N}}$ of volumes satisfying $0 < v_i < |M|$ converging to $0 < v < |M|$. For each i, take an isoperimetric set E_i of volume v_i. As the perimeters $P(E_i)$ are uniformly bounded by $\sup_{(0,|M|)} I_M$, we can take a convergent subsequence to some set $E \subset M$ of finite perimeter and volume v. By the lower semicontinuity of perimeter, we get

$$I_M(v) \leqslant P(E) \leqslant \liminf_{i\to\infty} P(E_i) = \liminf_{i\to\infty} I_M(v_i).$$

This shows that I_M is lower semicontinuous.

To prove the upper semicontinuity of I_M, we take an isoperimetric set $E \subset M$ of volume v. By the deformation Theorem 1.50 we can find, for large i, sets E_i of volume v_i and finite perimeter so that

$$I_M(v_i) \leqslant P(E_i) \leqslant P(E) + C|v - v_i| \leqslant I_M(v) + C|v - v_i|.$$

Taking lim sup when $i \to \infty$ we obtain the upper semicontinuity of perimeter.

As a general rule, convergence of isoperimetric sets implies the lower semicontinuity of the perimeter, while existence of isoperimetric sets implies upper semicontinuity.

3.3 An Asymptotic Expansion for Small Volumes

In this subsection, we show that that the isoperimetric profile of a compact manifold is asymptotically, for small volumes, the one of the Euclidean space of the same dimension. We closely follow the proof by Berard and Meyer (see Theorem in Appendix C of [52]).

Theorem 3.7 *Let (M, g) be a compact Riemannian manifold. For every $\varepsilon > 0$, there exists a positive constant $v_0 = v_0(M, g, \varepsilon)$ such that any measurable set $E \subset M$ of volume $0 < v \leqslant v_0$ satisfies*

$$P(E) \geqslant (1 - \varepsilon)\, \mathbf{c}(m)|E|^{(m-1)/m}, \tag{3.6}$$

where $\mathbf{c}(m) = m\omega_m^{1/m}$ is the isoperimetric constant in $\mathbb{R}^m$.

Proof Let $\rho > 0$ so that for any set F contained in a geodesic ball in M of radius 2ρ, we have

$$P(F) \geqslant (1 - \varepsilon/2)\, \mathbf{c}(m)|F|^{(m-1)/m}. \tag{3.7}$$

The existence of ρ follows because the Riemannian metric on M is uniformly and asymptotically Euclidean.

Let $x_1, \ldots, x_\ell$ be a maximal family of points in M so that the balls $B(x_i, \rho/2)$ are disjoint. By the coarea formula, for each i, we have

$$|E \cap B(x_i, 2\rho)| \geqslant \int_\rho^{2\rho} A(\partial B(x_i, r) \cap E)\, dr,$$

and so there exists $\rho(i) \in [\rho, 2\rho]$ such that

$$A(\partial B(x_i, \rho(i)) \cap E) \leqslant \frac{|E \cap B(x, 2\rho)|}{\rho} \leqslant \frac{|E|}{\rho}.$$

Let $\mathcal{B}$ be the set of connected components of $M \setminus \bigcup_{i=1}^{\ell} \partial B(x_i, \rho(i))$ and E' the disjoint union of the sets $E \cap F$, where $F \in \mathcal{B}$. Then we have

$$P(E') \leqslant P(E) + 2\ell \frac{|E|}{\rho}$$

and since each component of E' is contained in a ball of radius 2ρ, we obtain, using the concavity of the function $s \mapsto s^{(m-1)/m}$ and (3.7),

$$\begin{aligned}(1-\varepsilon/2)\,\mathbf{c}(m)|E|^{(m-1)/m} &= (1-\varepsilon/2)\,\mathbf{c}(m)\Big(\sum_{F\in\mathcal{B}} |E \cap F|\Big)^{(m-1)/m}\\ &\leqslant (1-\varepsilon/2)\,\mathbf{c}(m)\sum_{F\in\mathcal{B}} |E\cap F|^{(m-1)/m}\\ &\leqslant P(E') \leqslant P(E) + 2\ell\frac{|E|}{\rho}.\end{aligned}$$

Hence,

$$\frac{P(E)}{|E|^{(m-1)/m}} \geqslant (1-\varepsilon/2)\,\mathbf{c}(m) - 2\ell\frac{|E|^{1/m}}{\rho}.$$

So it is enough to take

$$|E| \leqslant \left(\frac{\varepsilon\mathbf{c}(m)\rho}{4\ell}\right)^m$$

to obtain (3.6). □

Corollary 3.8 *Let M be a compact Riemannian manifold. Then*

$$\lim_{v\to 0} \frac{I_M(v)}{v^{(m-1)/m}} = \mathbf{c}(m), \tag{3.8}$$

and

$$\lim_{\lambda\to 0} \frac{h_M(\lambda)}{\lambda^{(m-1)/m}} = \mathbf{c}(m)|M|^{-1/m}. \tag{3.9}$$

Proof Let $\{v_i\}_{i\in\mathbb{N}}$ be a sequence of positive numbers converging to 0, and let $\{E_i\}_{i\in\mathbb{N}}$ be a sequence of isoperimetric sets of volume v_i. Theorem 3.7 implies, for any $\varepsilon > 0$,

$$\liminf_{v_i\to 0} \frac{I_M(v_i)}{v_i^{(m-1)/m}} \geqslant (1-\varepsilon)\,\mathbf{c}(m).$$

Hence,

$$\liminf_{v_i \to 0} \frac{I_M(v_i)}{v_i^{(m-1)/m}} \geqslant \mathbf{c}(m).$$

On the other hand, taking small balls B_i of volume v_i centered at a fixed point of M, we have

$$\limsup_{v_i \to 0} \frac{I_M(v_i)}{v_i^{(m-1)/m}} \leqslant \limsup_{i \to \infty} \frac{P(B_i)}{|B_i|^{(m-1)/m}} = \mathbf{c}(m).$$

This implies (3.8).

Equation (3.9) is obtained by writing

$$\frac{h_M(\lambda)}{\lambda^{(m-1)/m}} = \frac{I_M(\lambda|M|)}{(\lambda|M|)^{(m-1)/m}} |M|^{-1/m}$$

and taking limits when $\lambda \to 0$. □

Remark 3.9 The symmetry $I_M(v) = I_M(|M| - v)$ of the isoperimetric profile implies

$$\frac{I_M(|M| - v)}{(|M| - v)^{(m-1)/m}} = \frac{I_M(v)}{(|M| - v)^{(m-1)/m}}.$$

So we have

$$\lim_{v \to |M|} \frac{I_M(v)}{(|M| - v)^{(m-1)/m}} = \mathbf{c}(m).$$

Analogously, since $h_M(\lambda) = h_M(1 - \lambda)$, we obtain

$$\lim_{\lambda \to 1} \frac{h_M(\lambda)}{(1 - \lambda)^{(m-1)/m}} = \mathbf{c}(m)|M|^{-1/m}.$$

Remark 3.10 Theorem 3.7 implies the existence of a Cheeger set in M with volume $0 < v < |M|$ and that the Cheeger constant $\mathrm{Ch}(M)$ is positive. Let $\{E_i\}_{i \in \mathbb{N}}$ be a sequence of measurable sets such that

$$\frac{P(E_i)}{\min\big\{|E_i|, |M \setminus E_i|\big\}} \to \mathrm{Ch}(M) < +\infty.$$

Then $P(E_i)$ is uniformly bounded above, and we can extract a non-relabelled subsequence of $\{E_i\}_{i \in \mathbb{N}}$ converging in $L^1(M)$ to a set E of volume $v \in [0, |M|]$.

If $|E_i| \to 0$ then (3.8) implies

$$c(m) = \lim_{i\to\infty} \frac{P(E_i)}{|E_i|^{(m-1)/m}} = \lim_{i\to\infty} \frac{P(E_i)}{|E_i|} |E_i|^{1/m} = \mathrm{Ch}(M)\cdot 0 = 0,$$

which is clearly not possible. In a similar way, $|E_i| \to |M|$ gives a contradiction. Hence, $0 < v < |M|$. By the lower semicontinuity of perimeter

$$\mathrm{Ch}(M) \leqslant \frac{P(E)}{\min\{|E|, |M \setminus E|\}} \leqslant \liminf_{i\to\infty} \frac{P(E_i)}{\min\{|E_i|, |M \setminus E_i|\}} \leqslant \mathrm{Ch}(M).$$

Hence, E is a Cheeger set. We also know that it is an isoperimetric set so that

$$\mathrm{Ch}(M) = \frac{I_M(|E|)}{\min\{|E|, |M \setminus E|\}} > 0.$$

3.4 Differentiability of the Isoperimetric Profile

In this section, we prove that the functions I_M^α, with $1 \leqslant \alpha \leqslant m/(m-1)$, of the isoperimetric profile I_M of a compact manifold M are locally the sum of a concave function and a smooth one. Hence, the isoperimetric profile shares the regularity properties of concave functions such as existence of side derivatives everywhere, first derivatives except on a countable set, and second derivatives almost everywhere. We prove this result by using deformations of the regular part of an isoperimetric boundary, so that it is necessary to use the full regularity theory for isoperimetric minimizers, to derive a differential inequality satisfied in a weak sense by the isoperimetric profile. We refer the reader to Bavard and Pansu [45], Proposition 3.3 in Morgan and Johnson [311], Bayle [46], and Bayle and Rosales [48], for similar proofs.

We first prove the following characterization of concave functions. A function $f : J \subset \mathbb{R} \to \mathbb{R}$ defined on an interval J is *strictly concave* if

$$f((1-\lambda)x + \lambda y) > (1-\lambda) f(x) + \lambda f(y)$$

for all $x, y \in J$, $x \neq y$, and $0 < \lambda < 1$.

Lemma 3.11 *Let $f : J \to \mathbb{R}$ be a continuous function defined on an open interval $J \subset \mathbb{R}$. Assume that for all $x \in J$, there exists a sequence of smooth functions $\{f_{x,i}\}_{i\in\mathbb{N}}$, each one defined on a neighborhood of x, such that $f \leqslant f_{x,i}$, $f(x) = f_{x,i}(x)$, and* $\limsup_{i\to\infty} f''_{x,i}(x) \leqslant 0$. *Then f is a concave function.*

If the stronger condition $\limsup_{i\to 0} f''_{x,i}(x) < 0$ *is satisfied at every point $x \in J$, then f is strictly concave.*

Proof If f is not concave, then there exists $\delta > 0$ such that the function $f_\delta(x) = f(x) - \delta x^2$ is not concave. To see this, represent the subgraph of f as the closure of the union of the increasing family (when $\rho \to 0$) of the subgraphs of f_ρ. If the subgraph of f is not a convex set, then the subgraph of f_ρ is not convex for ρ small enough.

As f_δ is not concave, there exist two points $x_1 < x_2$ on I such that the function $L(x) - f_\delta(x)$ has a positive maximum $x_0 \in (x_1, x_2)$. Here $L(x)$ is the linear function containing $(x_1, f_\delta(x_1))$ and $(x_2, f_\delta(x_2))$ in its graph. Then all the smooth functions $L(x) - f_{x_0,i}(x) + \delta x^2$ have a maximum at x_0. Hence, $f''_{x_0,i}(x_0) \geqslant 2\delta > 0$ for all $i \in \mathbb{N}$, contradicting the hypothesis $\limsup_{i\to\infty} f''_{x_0,i}(x) \leqslant 0$.

Assume finally that condition $\limsup_{i\to 0} f''_{x,i}(x) < 0$ is satisfied. If f is not strictly concave then, with the same notation as in the previous case, there exist $x_1 < x_2$ and $x_0 \in (x_1, x_2)$ such that $L - f$ has a maximum at x_0. Then all the smooth functions $L - f_{x_0,i}$ have a maximum at x_0, and so $f''_{x_0,i}(x_0) \geqslant 0$. This is a contradiction to the hypothesis $\limsup_{i\to\infty} f''_{x_0,i}(x_0) < 0$. □

Let us now give sense to the inequality $f'' \leqslant C$ for non-smooth functions.

Definition 3.12 Let $f : (a, b) \to \mathbb{R}$ be a continuous function and $x_0 \in (a, b)$. Let $C \in \mathbb{R}$. We say that $f''(x_0) \leqslant C$ *in a weak sense* if there exists a sequence of smooth functions $\{f_i\}_{i\in\mathbb{N}}$, each one defined on an open interval containing x_0, so that

1. $f \leqslant f_i$ in the common domain of definition of f anf f_i,
2. $f(x_0) = f_i(x_0)$ for all i,
3. $\limsup_{i\to\infty} f_i''(x_0) \leqslant C$.

We say that $f''(x_0) = C$ in a weak sense if there exists a sequence of smooth functions $\{f_i\}_{i\in\mathbb{N}}$ satisfying 1,2, and $\limsup_{i\to\infty} f_i''(x_0) = C$.

3.4.1 A Differential Inequality for the Isoperimetric Profile

In this section, we prove that the isoperimetric profile of a compact manifold raised to some powers satisfies a differential inequality in a weak sense.

Theorem 3.13 *Let M be an m-dimensional compact Riemannian manifold satisfying* $\mathrm{Ric} \geqslant (m-1)\delta$. *Let I_M be its isoperimetric profile and* $1 \leqslant \alpha \leqslant m/(m-1)$. *Then I_M^α satisfies in a weak sense the differential inequality*

$$(I_M^\alpha)'' \leqslant -\alpha(m-1)\,\delta\,(I_M^\alpha)^{(\alpha-2)/\alpha}, \tag{3.10}$$

in the interval $(0, |M|)$.

Equality holds in (3.10) *for some* $v_0 \in (0, |M|)$ *if* $\alpha = m/(m-1)$, *and every isoperimetric set of volume* v_0 *has totally umbilical boundary and satisfies*

$\operatorname{Ric}(N, N) = (m-1)\delta$, where N is the unit normal to the regular part of the boundary.

Moreover, I^α is locally the sum of a concave function and a smooth function.

Proof We fix some $0 < v_0 < |M|$ and take an isoperimetric region $E \subset M$ of volume v_0. Let S be the regular part of its boundary and S_0 the singular set. By Lemma 1.61, there exists a sequence of smooth functions $\{f_i\}_{i\in\mathbb{N}}$ with compact support in S satisfying

1. $0 \leqslant f_i \leqslant 1$ for all i,
2. the sequence $\{f_i\}_{i\in\mathbb{N}}$ converges pointwise to the constant function 1 on S.
3. $\lim_{i\to\infty} \int_S |\nabla f_i|^2 dS = 0$.

For any i, we take a vector field X_i with compact support on M so that $X_i = f_i N$ on S, where N is the outer unit normal to E on S. Let $\{\varphi_t^i\}_{t\in\mathbb{R}}$ the associated flow. We have

$$\left.\frac{d}{dt}\right|_{t=0} |\varphi_t^i(E)| = \int_S f_i \, dS > 0,$$

and so we can take the volume as a parameter of the deformation $\varphi_t^i(E)$ for v close to v_0 and write $A_i(v) = P(\varphi_{t(v)}^i(E))$. The function A_i is smooth and satisfies $A_i^\alpha(v) \geqslant I_M^\alpha(v)$, with equality at v_0. Let us compute the second derivative of A_i^α with respect to v at $v = v_0$. Taking derivatives of A_i with respect to the volume, we get

$$\frac{d}{dv} A_i^\alpha = \alpha A_i^{\alpha-1} \frac{dA_i/dt}{dv/dt},$$

and

$$\frac{d^2}{dv^2} A_i^\alpha = \alpha A_i^{\alpha-1} \Bigg((\alpha - 1) \frac{1}{A_i} \left(\frac{dA_i/dt}{dv/dt} \right)^2 + \frac{1}{(dv/dt)^2} \left(\frac{d^2 A_i}{dt^2} - \frac{dA_i/dt}{dv/dt} \frac{d^2 v}{dt^2} \right) \Bigg).$$

Note that $A_i(v_0) = A(S)$ and that $(dA/dt)/(dv/dt)_{t=0}$ is equal to the constant mean curvature H of S. Letting $' = d/dt$, the expression $A'' - HV''$ equals the second variation operator (1.46). Evaluating at $v = v_0$

$$\frac{d^2}{dv^2} A_i^\alpha(v_0) = \alpha A(S)^{\alpha-1} \Bigg((\alpha - 1) \frac{H^2}{A(S)} + \frac{1}{\left(\int_S f_i dS\right)^2} \left(\int_S (|\nabla f_i|^2 - (\operatorname{Ric}(N, N) + |\sigma|^2) f_i^2 \, dS \right) \Bigg),$$

Taking lim sup when $i \to \infty$ we get

$$\limsup_{i\to\infty} \frac{d^2}{dv^2} A_i^\alpha(v_0) = -\alpha A(S)^{\alpha-3} \int_S \left(\mathrm{Ric}(N, N) + \left(|\sigma|^2 - (\alpha-1)H^2\right)\right) dS$$
$$\leqslant -\alpha A(S)^{\alpha-2}(m-1)\delta = -\alpha(m-1)\delta \,(I_M(v_0)^\alpha)^{(\alpha-2)/\alpha},$$

from $\lim_{i\to\infty} \int_S f_i \, dS = A(S)$ and inequality

$$|\sigma|^2 - (\alpha-1)H^2 \geqslant |\sigma|^2 - \frac{H^2}{m-1} \geqslant 0,$$

which follows from $\alpha \leqslant m/(m-1)$ and (1.24). Thus (3.10) is satisfied in a weak sense. In case equality holds in (3.10), we must have $\alpha = m/(m-1)$, $|\sigma|^2 = H^2/(m-1)$ (and so S is totally umbilical), and $\mathrm{Ric}(N, N) = (m-1)\delta$.

To prove that I_M^α is the sum of a concave function and a smooth function, we take $0 < v_1 < v_2 < |M|$. The boundedness of the isoperimetric profile in $[v_1, v_2]$ implies the existence of a constant $C > 0$ such that

$$-\alpha I_M(v)^{\alpha-2}(m-1)\delta \leqslant C, \tag{3.11}$$

for all $v_1 < v < v_2$. This implies that $f(v) = I_M(v)^\alpha - Cv^2$ satisfies $f''(v) \leqslant 0$ in a weak sense for all $v \in (v_1, v_2)$. By Lemma 3.11, f is a concave function on (v_1, v_2). □

Remark 3.14 Given $1 \leqslant \alpha \leqslant m/(m-1)$, we have $h_M^\alpha(\lambda) = |M|^{-\alpha} I_M^\alpha(\lambda|M|)$, and so

$$(h_M^\alpha)'' \leqslant |M|^{2-\alpha} (I_M^\alpha)''$$

in a weak sense. By (3.10) we get

$$(h_M^\alpha)'' \leqslant -\alpha(m-1)\delta \,(h_m^\alpha)^{(\alpha-2)/\alpha}. \tag{3.12}$$

In particular, if $\alpha = m/(m-1)$ then

$$(h^{m/(m-1)})'' \leqslant -m\delta (h^{m/(m-1)})^{-(m-2)/m}, \tag{3.13}$$

Equality holds in (3.13) for some $\lambda_0 \in (0, 1)$ if every isoperimetric set in M of volume $\lambda_0|M|$ has totally umbilical boundary and $\mathrm{Ric}(N, N) = (m-1)\delta$, where N is the outer unit normal to the regular part of the isoperimetric boundary.

3.4.2 The Isoperimetric Profile of the Sphere

One of the main consequences of Theorem 3.13 is the characterization of the isoperimetric sets in the sphere $\mathbb{S}^m(\delta)$ of constant sectional curvature $\delta > 0$.

Theorem 3.15 *The isoperimetric sets in the sphere $\mathbb{S}^m(\delta)$ are the geodesic balls.*

Proof Let $f = I_{\mathbb{S}^m(\delta)}^{m/(m-1)}$. By Eq. (3.10), f satisfies the differential inequality

$$f'' \leqslant \Lambda(f), \tag{3.14}$$

where

$$\Lambda(x) = -m\delta\, x^{-(m-2)/m}.$$

Let now I_0 be the function that assigns to each volume $v \in (0, |\mathbb{S}^m(\delta)|)$ the perimeter of the geodesic ball B_v of volume v. Define $f_0 = I_0^{m/(m-1)}$. Let H_v be the mean curvature of B_v computed with respect to the outer normal. The first variation formulas for perimeter and volume (applied to a family of concentric geodesic spheres in $\mathbb{S}^m(\delta)$) and the formula for the derivative of the mean curvature (1.41) imply

$$\frac{dI_0}{dv} = H_v, \quad \frac{d^2 I_0}{dv^2} = -\frac{1}{I_0}\left((m-1)\,\delta + \frac{H_v^2}{m-1}\right).$$

A straightforward computation then implies

$$\frac{d^2 I_0^{m/(m-1)}}{dv^2} = -m\delta I_0^{-(m-2)/(m-1)}$$

and so

$$f_0'' = \Lambda(f_0). \tag{3.15}$$

Since $I_M \leqslant I_0$ we have

$$f - f_0 \leqslant 0.$$

From (3.14) and (3.15), we get

$$(f - f_0)'' \leqslant \Lambda(f) - \Lambda(f_0) \leqslant 0$$

in a weak sense since Λ is an increasing function. Hence, $f - f_0$ is a non-positive concave function on the interval $(0, |\mathbb{S}^m(\delta)|)$ vanishing at the endpoints. This implies $f = f_0$. In particular, geodesic spheres are isoperimetric sets, and since equality $f'' = \Lambda(f)$ holds, Theorem 3.13 implies that any isoperimetric boundary is a geodesic sphere. □

3.4.3 Consequences of the Differential Inequality

We prove now some analytical, geometrical, and topological corollaries of the differential inequality (3.10).

Corollary 3.16 *Let M be a compact Riemannian manifold with strictly positive Ricci curvature. Then, for $1 \leqslant \alpha \leqslant m/(m-1)$, the function I_M^α is strictly concave. In particular, the isoperimetric profile is a strictly concave function.*

Proof We take $\delta > 0$ such that $\operatorname{Ric} \geqslant (m-1)\delta$. Then (3.10) implies $(I_M^\alpha)'' < 0$. The last part of Lemma 3.11 implies that I_M^α is strictly concave. □

The strict concavity of the isoperimetric profile is an important property that implies the connectedness of isoperimetric regions in M.

Theorem 3.17 *Let M be a compact Riemannian manifold, and assume that I_M is strictly concave. Then isoperimetric regions in M are connected. In particular, the hypothesis is satisfied when M has strictly positive Ricci curvature.*

Proof Assume that an isoperimetric region $E \subset M$ of volume $v = v_1 + v_2$ has two components E_1 and E_2 of volumes $v_1 > 0$ and $v_2 > 0$, respectively. Then we have

$$I_M(v_1) + I_M(v_2) \leqslant P(E_1) + P(E_2) = P(E) = I_M(v).$$

But this is a contradiction since for the strictly concave function I_M satisfying $I(0) = 0$, assuming $v_1 \leqslant v_2$, we should have

$$\frac{I_M(v_1) - I_M(0)}{v_1 - 0} > \frac{I_M(v_1 + v_2) - I_M(v_2)}{(v_1 + v_2) - v_2},$$

which implies $I_M(v) = I_M(v_1 + v_2) < I_M(v_1) + I_M(v_2)$. □

Remark 3.18 The notion of *indecomposable sets* of finite perimeter is related to the topological notion of connectedness. A set E is indecomposable if it cannot be expressed as disjoint union of two sets E_1, E_2 of positive measure such that $P(E) = P(E_1) + P(E_2)$. In our case, since we are only working with isoperimetric sets, which enjoy high regularity properties, we do not use this notion. The interested reader is referred to Ambrosio et al. [19].

Lemma 3.19 *Let M be a compact Riemannian manifold and $I = I_M$ its isoperimetric profile. Then the left and right derivatives I'_-, I'_+ exist everywhere. Moreover, for $0 < v < |M|$ we have*

$$I'_-(v) \geqslant I'_+(v). \tag{3.16}$$

In particular, the set of points where $I'_+(v) < I'_-(v)$ is at most countable. We also have

$$\lim_{w\downarrow v} I'_+(w) = I'_+(v) \qquad \lim_{w\uparrow v} I'_+(w) = I'_-(v)$$

$$\lim_{w\downarrow v} I'_-(w) = I'_+(v) \qquad \lim_{w\uparrow v} I'_-(w) = I'_-(v)$$

Finally, if $v, w \in (0, |M|)$, then we have

$$I(w) - I(v) = \int_v^w I'_+(\xi)\, d\xi = \int_v^w I'_-(\xi)\, d\xi. \tag{3.17}$$

Proof All claims follow from similar properties of concave (convex) functions since I is locally the sum of a concave and a smooth function. Equation (3.16) and the limit properties follow from Theorem 24.1 in Rockafellar [376] and Eq. (3.17) from Corollary 24.2.1 in [376]. □

Theorem 3.20 *Let M be a compact Riemannian manifold and let $I = I_M$ be its isoperimetric profile.*

1. *The constant mean curvature H of the boundary of an isoperimetric set of volume v_0 satisfies*

$$I'_+(v_0) \leqslant H \leqslant I'_-(v_0).$$

   In particular, if I is regular at v_0, then the boundary of any isoperimetric set of volume v_0 has mean curvature $H = I'(v_0)$.
2. *If I is not regular at v_0, then there exist two isoperimetric sets in M of volume v_0 which boundaries have mean curvatures $I'_+(v_0) \neq I'_-(v_0)$.*
3. *For any regular value ξ of I, let $H(\xi)$ be the mean curvature of the boundary of any isoperimetric set of volume ξ. Then*

$$I(w) - I(v) = \int_v^w H(\xi)\, d\xi \tag{3.18}$$

   for all $0 < v < w < |M|$.

Proof For the first assertion, we take an isoperimetric region $E \subset M$ of volume v_0. Let S, S_0 be the regular and singular parts of the boundary of E. We take the deformation $\{E_t\}_{t\in\mathbb{R}}$ associated to any vector field X with compact support in M such that $\operatorname{supp}(X) \cap S_0 = \emptyset$ and $\int_E \operatorname{div} X \, dM \neq 0$. Then we can express this deformation taking the volume as a parameter to obtain a function $A(v)$ whose derivative at $v = v_0$ is the mean curvature H of S. Since $I(v) \leqslant A(v)$, $I(v_0) = A(v_0)$, we get $I'_+(v_0) \leqslant A'(v_0) \leqslant I'_-(v_0)$. This implies the claim.

To prove the second assertion, we take a sequence $\{v_i\}_{i\in\mathbb{N}}$ of regular values of I so that v_i decreases to v_0. We take a sequence of isoperimetric regions $\{E_i\}_{i\in\mathbb{N}}$ of volumes v_i and boundary mean curvature $H_i \in \mathbb{R}$. Then E_i converges in the L^1 topology to a set $E \subset M$ of volume v that is isoperimetric and has boundary mean curvature equal to $H \in \mathbb{R}$. Near a regular point at the boundary of E, we have $C^{1,\alpha}$ convergence of the regular part of the boundaries of E_i by Allard's regularity theorem for rectifiable varifolds (e.g., Chapter 5 in Simon [404]). Higher-order convergence is obtained from elliptic theory, and so we have $H = \lim_{i\to\infty} H_i$. Hence,

$$H = \lim_{i\to\infty} H_i = \lim_{i\to\infty} I'(v_i) = I'_+(v_0).$$

Analogously, we can construct an isoperimetric region with boundary mean curvature $I'_-(v_0)$.

Finally, (3.18) follows from (3.17), taking into account that $I'_+ = I'_- = H$ for regular values of I. □

3.5 Lévy-Gromov's Isoperimetric Inequality

We present in this section the Lévy-Gromov isoperimetric inequality comparing the normalized profiles of an m-dimensional compact Riemannian manifold with Ricci curvature satisfying $\operatorname{Ric} \geqslant (m-1)\,\delta$ with the one of the sphere $\mathbb{S}^m(\delta)$ with sectional curvatures equal to δ. We give a proof of this result using the differential inequality (3.13). For the most part of the proof, we follow Bayle [47].

Theorem 3.21 (Lévy-Gromov Isoperimetric Inequality) *Let M be a compact Riemannian manifold satisfying*

$$\operatorname{Ric} \geqslant (m-1)\,\delta,$$

for some $\delta > 0$. Then we have

$$h_M(\lambda) \geqslant h_{\mathbb{S}^m(\delta)}(\lambda), \tag{3.19}$$

for any $\lambda \in (0, 1)$. If equality holds for some $\lambda_0 \in (0, 1)$, then M is isometric to $\mathbb{S}^m(\delta)$.

3.5.1 A Proof Using the Differential Inequality

Proof Let $f = h_M^{m/(m-1)}$ and $f_0 = h_{\mathbb{S}^m(\delta)}^{m/(m-1)}$. Then $f'' \leqslant \Lambda(f)$ in a weak sense, where

$$\Lambda(x) = -m\delta x^{-(m-2)/m}$$

by Remark 3.14. We also have $f_0'' = \Lambda(f_0)$ by the proof of Theorem 3.15.

Assume that there exists $\lambda_0 \in (0, 1)$ so that $f(\lambda_0) < f_0(\lambda_0)$. Take a maximal interval $J \subset [0, 1]$ containing λ_0 so that $f < f_0$ in the interior of J and $f - f_0$ vanishes at the endpoints of J. Then

$$(f - f_0)'' \leqslant \Lambda(f) - \Lambda(f_0) < 0$$

in the interior of J by the strict monotonicity of Λ. But this not possible since $f - f_0$ would be a non-positive concave function in J vanishing at the end points of the interval. This shows that $f \geqslant f_0$ and so $h_M \geqslant h_{\mathbb{S}^m(\delta)}$.

Let us now discuss the equality case. Assume the existence of $\lambda_0 \in (0, 1)$ such that $f(\lambda_0) = f_0(\lambda_0)$. By the symmetry of the normalized isoperimetric profiles with respect to $\lambda = \frac{1}{2}$, we may assume that $\lambda_0 \in (0, \frac{1}{2}]$. Let us distinguish two cases:

Case 1. If $\lambda_0 < \frac{1}{2}$, we take any $\varepsilon > 0$ so that $\lambda_0 + \varepsilon < \frac{1}{2}$ and define

$$f_0^\varepsilon(\lambda) = f_0(\lambda + \varepsilon) - f_0(\lambda_0 + \varepsilon) + f_0(\lambda_0).$$

The function f_0^ε is obtained by translating the graph of f_0 to the left at distance ε and then translating it down so that $f_0^\varepsilon(\lambda_0) = f_0(\lambda_0) = f(\lambda_0)$. We have the following properties:

- $f_0^\varepsilon(\lambda_0) = f(\lambda_0)$,
- $f_0^\varepsilon > f$ in $(\lambda_0 - \delta, \lambda_0)$ for some small $\delta > 0$,
- $f_0^\varepsilon(0) > 0 = f(0)$,
- $(f_0^\varepsilon)'' \geqslant \Lambda(f_0^\varepsilon)$,

The first property follows from the definition of f_0^ε. For the second one, we observe that $f'_-(\lambda_0) = f_0'(\lambda_0) > f_0'(\lambda_0 + \varepsilon) = (f_0^\varepsilon)'(\lambda_0)$. For the third one, the strict concavity of f_0 implies $f_0(\lambda_0 + \varepsilon) - f_0(\lambda_0) < f_0(\varepsilon)$ since $f_0(0) = 0$. For the last property, we compute

$$(f_0^\varepsilon)''(\lambda) = f_0''(\lambda + \varepsilon) = \Lambda(f_0(\lambda + \varepsilon)) \geqslant \Lambda(f_0^\varepsilon(\lambda)),$$

the last inequality since $f_0(\lambda+\varepsilon) > f_0^\varepsilon(\lambda)$, an inequality equivalent to $f_0(\lambda_0+\varepsilon) > f_0(\lambda_0)$.

We claim that $f_0^\varepsilon \geqslant f$ in the interval $(0, \lambda_0]$. Otherwise, there exists some $\mu_0 \in (0, \lambda_0)$ such that $f - f_0^\varepsilon < 0$ in (μ_0, λ_0) and $(f_0^\varepsilon - f)$ vanishes at the end points of the interval. But $(f - f_0^\varepsilon)'' \leqslant \Lambda(f) - \Lambda(f_0^\varepsilon) \leqslant 0$, a contradiction.

So we have $f_0^\varepsilon \geqslant f$ in the interval $(0, \lambda_0]$ for all $\varepsilon > 0$ such that $\lambda_0 + \varepsilon < \frac{1}{2}$. Letting $\varepsilon \to 0$, we get $f_0 \geqslant f$, and since $f \geqslant f_0$, we obtain $f = f_0$ in the interval $[0, \lambda_0]$. By the asymptotic expansion for h, h_0 at $\lambda = 0$, we get $|M| = |\mathbb{S}_\delta|$. We conclude that M is isometric to $\mathbb{S}^m(\delta)$ by the rigidity part of Bishop's volume comparison theorem (see Theorem III.4.4 in [102]).

Case 2. Assume $\lambda_0 = \frac{1}{2}$. Consider, for $\varepsilon > 0$ small enough, the function

$$f_0^\varepsilon(\lambda) = f_0(\lambda + \varepsilon) + f(\tfrac{1}{2} - \varepsilon) - f_0(\tfrac{1}{2}).$$

The following properties can be obtained using arguments similar to the ones in Case 1:

- $f_0^\varepsilon(\frac{1}{2} - \varepsilon) = f(\frac{1}{2} - \varepsilon)$.
- $f_0^\varepsilon > f$ in $(\frac{1}{2} - \varepsilon - \delta, \frac{1}{2} - \varepsilon)$ for some $\delta > 0$,
- $f_0^\varepsilon(0) > 0 = f(0)$,
- $(f_0^\varepsilon)'' \geqslant \Lambda(f_0^\varepsilon)$

Now we reason as the first case to conclude that $f_0^\varepsilon \geqslant f$ in the interval $(0, \frac{1}{2} - \varepsilon]$. Letting $\varepsilon \to 0$, we get $f_0 \geqslant f$ in $(0, \frac{1}{2}]$, and so $f_0 = f$ in $[0, \frac{1}{2}]$. We conclude the proof as in the previous case. □

3.5.2 The Original Proof

We give in this section a sketch of the original proof of inequality (3.19) by Gromov (see §2.2 in Appendix I in [288], Appendix C in [198] or §VI.6 in [102]). Inequality (3.19) is equivalent to

$$\frac{P(F)}{|M|} \geqslant \frac{P(B)}{|\mathbb{S}^m(\delta)|}, \tag{3.20}$$

where $B \subset \mathbb{S}^m(\delta)$ is the geodesic ball of volume

$$|B| = |\mathbb{S}^m(\delta)| \, \frac{|F|}{|M|}$$

and $F \subset M$ is any measurable set.

Consider a compact Riemannian manifold M satisfying $\operatorname{Ric} \geqslant (m-1)\,\delta$, with $\delta > 0$. It is enough to prove (3.20) for isoperimetric sets. Let $E \subset M$ be an isoperimetric region of volume $\lambda|M|$, $\lambda \in (0, 1)$. Let $B \subset \mathbb{S}^m(\delta)$ be the geodesic ball of volume $\lambda|\mathbb{S}^m(\delta)|$, and let $r > 0$ be the radius of B.

We know that the boundary of E is composed of a smooth hypersurface S and a singular set $S_0$with Hausdorff dimension at most $m-8$. Let H be the mean curvature of S and Ω a geodesic sphere of constant mean curvature H in the sphere $\mathbb{S}^m(\delta)$

with Ricci curvature $(m-1)\,\delta$. Every $q \in E$ can be connected to ∂E by a geodesic $\gamma : [0, d(p, S)] \to M$ minimizing the distance to $M \setminus E$. The point $p = \gamma(0)$ is a *regular* point of S since there is a ball inside E centered at p touching ∂E at q. Hence, blowing up E at p produces an area-minimizing set E_∞ in Euclidean space containing a halfspace. This implies that the limit set E_∞ is indeed a halfspace and p is a regular point at the boundary of E. The geodesic γ is then orthogonal to S.

Since S has constant mean curvature H, the Jacobian $\operatorname{Jac}_p(t)$ of the normal map to S satisfies $\operatorname{Jac}_p(t) \leqslant \operatorname{Jac}_{\delta,H}(t)$, where $\operatorname{Jac}_{\delta,H}(t)$ is the Jacobian of the geodesic sphere Ω of mean curvature H in $\mathbb{S}^m(\delta)$ by the Heintze-Karcher inequality (see Theorem 1.35). Hence,

$$|E| \leqslant \int_S \left\{ \int_0^{c(p)} \operatorname{Jac}_p(t)\,dt \right\} dS \leqslant \int_S \left\{ \int_0^{t^+} \operatorname{Jac}^m_{\delta,H}(t)\,dt \right\} dS = A(S)\,\frac{|\Omega|}{A(\partial\Omega)},$$

where $c(p)$ is the cut distance, which is smaller than or equal to the first zero t^+ of the function $\operatorname{Jac}^m_{\delta,H}(t)$. Note that $t^+ = \cot_\delta^{-1}(H/(m-1))$ and that Ω is a geodesic disc of radius t^+. So we have

$$\frac{A(\partial\Omega)}{|\Omega|} \leqslant \frac{A(S)}{|E|}.$$

For geodesic balls $B(x,r) \subset \mathbb{S}^m(\delta)$, the function $r \mapsto P(B(x,r))/|B(x,r)|$ is decreasing. Hence, if the radius r of B satisfies

$$r \geqslant t^+$$

then

$$\frac{A(\partial B)}{|B|} \leqslant \frac{A(\partial\Omega)}{|\Omega|} \leqslant \frac{A(S)}{|E|}.$$

As $|E|/|M| = |B|/|\mathbb{S}^m(\delta)|$, inequality (3.20) follows.

In case

$$r \leqslant t^+,$$

we consider the isoperimetric set $M \setminus E$ with mean curvature $-H/(m-1)$. As before, an application of the Heintze-Karcher inequality implies

$$\frac{A(\partial\Omega')}{|\Omega'|} \leqslant \frac{A(S)}{|M \setminus E|},$$

where Ω' is the geodesic ball of radius $t^- = \cot_\delta^{-1}(-H/(m-1)) = \pi/\delta^{1/2} - t^+$. Then $\pi/\delta^{1/2} - r \geqslant \pi/\delta^{1/2} - t^+ = t^-$, and so

$$\frac{A(\partial B')}{|B'|} \leqslant \frac{A(\partial\Omega)}{|\Omega|} \leqslant \frac{A(S)}{|M \setminus E|},$$

where $B' \subset \mathbb{S}^m(\delta)$ is the geodesic ball of radius $\pi/\delta^{1/2} - r$ complementary to B. Since

$$|B'| = |\mathbb{S}^m(\delta)| - |B| = |\mathbb{S}^m(\delta)| \left(1 - \frac{|E|}{|M|}\right) = |\mathbb{S}^m(\delta)| \frac{|M \setminus E|}{|M|},$$

we obtain (3.20).

3.6 Continuity Under Lipschitz Convergence

In this section, we consider a sequence of compact Riemannian manifolds $\{M_i\}_{i\in\mathbb{N}}$ converging in some topology to a limit compact Riemannian manifold M. We ask ourselves whether the sequence of associated normalized isoperimetric profiles $\{h_i\}_{i\in\mathbb{N}}$converges to $h = h_M$. It turns out that Lipschitz convergence of the underlying metric spaces is enough to ensure the convergence of the profiles. In some restricted classes of Riemannian manifolds, Lipschitz convergence follows from the weaker Gromov-Hausdorff convergence.

We recall some basic definitions (see Chapter 3 in Gromov [198]). A map $f : (X, d) \to (X', d')$ between metric spaces is called *Lipschitz* is there exists a constant $C > 0$ such that

$$d'(f(x), f(y)) \leqslant C\, d(x, y),$$

for any $x, y \in X$. A map is *bilipschitz* if it is bijective and both f, f^{-1} are Lipschitz.

Definition 3.22 The *dilatation* $\mathrm{dil}(f)$ of a Lipschitz map $f : (X, d) \to (X', d')$ between two metric spaces is defined as

$$\mathrm{dil}(f) = \sup_{x \neq y} \frac{d'(f(x), f(y))}{d(x, y)}.$$

The *Lipschitz distance* between two metric spaces is the infimum of the quantities

$$|\log \mathrm{dil}(f)| + |\log \mathrm{dil}(f^{-1})|,$$

where f varies in the set of bilipschitz homeomorphisms between the metric spaces.

The dilatation of a Lipschitz map between two metric spaces is the infimum of the Lipschitz constants of f. It is denoted sometimes by $\mathrm{Lip}(f)$. If the Lipschitz distance between two metric spaces (X, d), (X', d') is smaller than ε, then there exists a bilipschitz map $f : X \to X'$ between them so that

$$e^{-\varepsilon} d(x, y) \leqslant d'(f(x), f(y)) \leqslant e^{\varepsilon} d(x, y).$$

On Riemannian manifolds, we consider the structure of metric space induced by the Riemannian distance. A sequence of Riemannian manifolds $\{M_i\}_{i\in\mathbb{N}}$ converge in Lipschitz distance to a Riemannian manifold M if there is a sequence of bilipschitz maps $f_i : M_i \to M$ so that

$$\frac{1}{C_i} d_{M_i}(x, y) \leqslant d_M(f_i(x), f_i(y)) \leqslant C_i\, d_{M_i}(x, y), \tag{3.21}$$

where $\lim_{i\to\infty} C_i = 1$.

We have the following result (see [454]).

Theorem 3.23 *Let $\{M_i\}_{i\in\mathbb{N}}$ be a sequence of compact Riemannian manifolds converging in Lipschitz distance to a limit compact Riemannian manifold M. Then the sequence of normalized isoperimetric profiles $\{h_i\}_{i\in\mathbb{N}}$ converges pointwise to the normalized isoperimetric profile h of M in* $(0, 1)$.

Proof Recall that for a Lipschitz map f between m-dimensional Riemannian manifolds with Lipschitz constant no larger than $C > 0$, we have $|f(E)| \leqslant C^m|E|$ and $P(f(E)) \leqslant C^{m-1}P(E)$. Volume and perimeter in each case correspond to the Riemannian manifold containing the set.

Take a sequence of bilipschitz maps $f_i : M_i \to M$ so that (3.21) is satisfied with $\lim_{i\to\infty} C_i = 1$. For any measurable set $F_i \subset M_i$, we have

$$\frac{1}{C_i^m} |F_i| \leqslant |f_i(F_i)| \leqslant C_i^m|F_i|, \quad \frac{1}{C_i^{m-1}} P(F_i) \leqslant P(f_i(F_i)) \leqslant C_i^m(F_i).$$

Fix some $\lambda_0 \in (0, 1)$, and take a sequence of isoperimetric sets $E_i \subset M_i$ of volumes $\lambda_0|M_i|$. From (3.22) we get $\lim_{i\to\infty} |M_i| = |M|$ and also $\lim_{i\to\infty} |f_i(E_i)| = \lambda_0|M|$. Then we have

$$|M|\, h\left(\frac{|f_i(E_i)|}{|M|}\right) \leqslant P(f_i(E_i)) \leqslant C_i^{m-1} P(E_i) = C_i^{m-1}|M_i|h_i(\lambda_0).$$

We take lim inf when $i \to \infty$ and, since $C_i \to 1$, $|M_i| \to |M|$, $|f_i(E_i)| \to \lambda_0|M|$ and h is continuous, we get

$$h(\lambda_0) \leqslant \liminf_{i\to\infty} h_i(\lambda_0). \tag{3.22}$$

Let now $E \subset M$ be an isoperimetric region of volume $\lambda_0|M|$. Using Theorem 1.50, we can find $\varepsilon > 0$ and a deformation of E by sets E_t of volumes $|E_t| = |E| + t$ such that $|P(E_t) - P(E)| \leqslant C|t|$, for some constant $C > 0$. For large $i \in \mathbb{N}$, we can find $t(i) \to 0$ when $i \to \infty$ so that $|f_i(E_{t(i)})| = \lambda_0|M_i|$, and we have

$$\begin{aligned}|M_i|h_i(\lambda_0) &\leqslant P(f_i(E_{t(i)})) \leqslant C_i^{m-1}P(E_{t(i)})\\ &\leqslant C_i^{m-1}(P(E) + C\,|t(i)|) = C_i^{m-1}(|M|h(\lambda_0) + C\,|t(i)|).\end{aligned}$$

Taking lim sup when $i \to \infty$, since $C_i \to 1$, $|M_i| \to |M|$ and $t(i) \to 0$, we get

$$\limsup_{i\to\infty} h_i(\lambda_0) \leqslant h(\lambda_0). \tag{3.23}$$

Inequalities (3.22) and (3.23) then imply the pointwise convergence of $\{h_i\}_{i\in\mathbb{N}}$ to h. □

Remark 3.24 If, in addition, all manifolds in the statement of Theorem 3.23 have non-negative Ricci curvature, the normalized isoperimetric profiles are concave. Then we get uniform convergence on compact subsets of the interval $(0, 1)$.

A weaker notion of convergence in metric spaces is given by the Gromov-Hausdorff distance.

Definition 3.25 The Gromov-Hausdorff distance between two metric spaces X, Y is defined as the infimum of the quantities

$$d_H^Z(f(X), g(Y)),$$

where $f : X \to Z$, $g : Y \to Z$ are isometric embeddings and d_H^Z is the usual Hausdorff distance between sets of Z.

It is well-known that Lipschitz convergence implies Gromov-Hausdorff convergence (see Example 7.4.3 in [81]). If we consider the class $\mathfrak{M}(n, d, \Lambda, v_0)$ of compact n-dimensional Riemannian manifolds with $\operatorname{diam}(M) \leqslant d$, $|M| \geqslant v_0$ and sectional curvatures $|K_M| \leqslant \Lambda^2$, it is known that Gromov-Hausdorff convergence implies Lipschitz convergence (see §D in Chapter 8 of Gromov [198] and also [343] and [192]).

3.7 Density Estimates for Isoperimetric Regions

We show in this section some elementary regularity properties for isoperimetric regions. In particular, we prove that any isoperimetric region E is equivalent, up to a set of measure 0, to an open set E_1 of the manifold and that the boundary ∂E is

Ahlfors-regular in the sense that there exist constants $C, r_0 > 0$ such that

$$C^{-1}r^{m-1} \leqslant P(E, B(x,r)) \leqslant Cr^{m-1}$$

for all $0 < r < r_0$. Let us prove first the upper Ahlfors estimate (see Lemma 5.1 in [266]).

Lemma 3.26 *Let (M, g) be a compact Riemannian manifold and let $E \subset M$ be an isoperimetric set. There exists $r_0 > 0$ and a constant $C > 0$ such that*

$$P(E, B(x,r)) \leqslant Cr^{m-1}, \tag{3.24}$$

for all $0 < r < r_0$. The radius r_0 depends on a local Hölder constant of the isoperimetric profile near $|E|$ and on the geometry of M, and C only depends on $r_0 > 0$ and the geometry of M.

Proof We know that the isoperimetric profile I_M of M satisfies a local Hölder condition by Theorem 3.5. Let $C' > 0$ be a constant such that

$$\left|I_M(v) - I_M(|E|)\right| \leqslant C'\left(v - |E|\right)^{(m-1)/m}$$

whenever $|v - |E|| < \varepsilon$. We choose $r_0 > 0$ and a constant $C'' > 0$ so that $|B(z, r_0)| \leqslant C''r^m < \varepsilon$ for all $z \in M$ and $C''' > 0$ so that $P(B(z,r)) \leqslant C'''r^{m-1}$ for all $z \in M$ and $0 < r < r_0$. The constants C'', C''' only depend on the geometry of M (for instance, on a lower bound on the Ricci curvature of M).

Now we fix $x \in M$, and we consider the set $F_r = E \cup B(x,r)$, for $0 < r < r_0$. Then

$$\begin{aligned} P(F_r) &\leqslant P(E, M \setminus \overline{B}(x,r)) + P(E \cup B(x,r), \partial B(x,r)) \\ &\leqslant P(E) - P(E, B(x,r)) + C'''r^{m-1}, \end{aligned}$$

for a.e. r by Ambrosio's localization Lemma 3.5 in [17]. So we have

$$I_M(|F|) - I_M(|E|) + P(E, B(x,r)) \leqslant C'''r^{m-1}. \tag{3.25}$$

By the Hölder continuity of the isoperimetric profile

$$\begin{aligned} I_M(|F|) - I_M(|E|) &\geqslant -C'\left(|F| - |E|\right)^{(m-1)/m} \\ &= -C'|B(x,r) \setminus E|^{(m-1)/m} \geqslant -C'|B(x,r)|^{(m-1)/m} \\ &\geqslant -C'C''r^{m-1}. \end{aligned}$$

This inequality, together with (3.25), implies (3.24). □

Now we prove the lower Ahlfors estimate. Given a measurable set $E \subset M$, we consider the function

$$h(x,r) = r^{-m} \min\big\{|E \cap B(x,r)|, |B(x,r) \setminus E|\big\}.$$

Let us prove first the following result (see Lemma 5.2 in [266] for a similar property for sub-Riemannian Carnot groups).

Lemma 3.27 *Let $E \subset M$ be an isoperimetric set. Then there exist $r_0 > 0$ and $\varepsilon > 0$, only depending on the geometry of M and on $|E|$, such that for all $x \in M$ and $0 < r < r_0$, if $h(x,r) \leqslant \varepsilon$, then $|B(x,r/2) \cap E| = 0$, or $|E \setminus B(x,r/2)| = 0$.*

Proof Assume first that $h(x,r) = r^{-m}|B(x,r)\setminus E|$. We consider the set $E \cup B(x,r)$ and let $\mu(t) = |B(x,t) \setminus E|$.

The perimeter of $E \cup B(x,r)$ can be estimated, a.e. $t > 0$, by

$$P(E \cup B(x,t)) \leqslant P(E) - P(E, B(x,t)) + \mu'(t). \tag{3.26}$$

We have

$$P(E, B(x,t)) \geqslant C_I\, |B(x,t) \setminus E|^{(m-1)/m} = C_I\, \mu(t)^{(m-1)/m},$$

where $C_I > 0$ is the constant appearing in the local isoperimetric inequality (1.72). Here $0 < t < r_0$, where r_0 is the radius in the local isoperimetric inequality, which only depends on the geometry of the compact manifold M. Hence, from (3.26), we obtain

$$I_M(v + \mu(t)) - I_M(v) \leqslant -C_I\, \mu(t)^{(m-1)/m} + \mu'(t)$$

for a.e. $0 < t < r_0$.

Since the isoperimetric profile is locally a Lipschitz function by Theorem 3.13, there exists a positive constant $C' > 0$ such that

$$I_M(v) - I_M(|E|) \geqslant -C'|v - |E||$$

whenever $|v - |E|| < \rho$. We reduce r_0 if necessary so that $|B(z, r_0)| < \rho$ for all $z \in M$. So we have for $0 < t < r_0$

$$I_M(|E| + \mu(t)) - I_M(|E|) \geqslant -C'\mu(t).$$

Hence,

$$\mu'(t) \geqslant C_I \mu(t)^{(m-1)/m} - C'\mu(t) \geqslant \frac{C_I}{2}\, \mu(t)^{(m-1)/m}$$

for a.e. $t > 0$ whenever $\mu(t)^{1/m} \leqslant C'' = C_I/2C'$. Suppose that $\mu(t) > 0$ for all $t \in [r/2, r]$. Integrating the inequality $\mu'/C''\mu^{(m-1)/m} \geqslant 1$ between $r/2$ and $r < r_0$, we get

$$\frac{r}{2} \leqslant C'' \left(\mu(r)^{1/m} - \mu(r/2)^{1/m}\right) \leqslant C''\mu(r)^{1/m} \leqslant C''\varepsilon^{1/m} r.$$

This is clearly not possible when ε is small enough ($\varepsilon^{1/m} < 1/(2C'')$). Therefore, $\mu(t) = 0$ for all $t \in [r/2, 2]$, and so $|E \setminus B(x, r/2)| = 0$.

If $h(x, r) = r^{-m}|E \cap B(x, r)|$, we consider the isoperimetric region $M \setminus E$ and we reduce this second case to the previous one. □

Definition 3.28 Given an isoperimetric set $E \subset M$, we define:

$$E_1 = \{x \in M : \text{there exists } r > 0 \text{ such that } |B(x, r) \setminus E| = 0\}.$$

$$E_0 = \{x \in M : \text{there exists } r > 0 \text{ such that } |E \cap B(x, r)| = 0\},$$

$$S = \{x \in M : h(x, r) > \varepsilon \text{ for small } r\},$$

where $\varepsilon > 0$ is given in Lemma 3.27.

Then we have the following.

Theorem 3.29 *Let $E \subset M$ be an isoperimetric region. Then*

1. *E_1, E_0, S form a partition of M.*
2. *E_1 and E_0 are open.*
3. *S is Ahlfors-regular.*
4. *E_1 is the set of points of density 1 of E, and E_0 the set of points of density 0 of E. Hence, E_1 and E are equivalent and also E_0 and $M \setminus E$.*

Proof Property 1 follows from Lemma 3.27 and property 2 from the definitions of E_1 and E_0.

To prove 3, we take a point $x \in S$ and $0 < r \leqslant r_0$. By the definition of S and the relative isoperimetric inequality, we have

$$P(E, B(x, r)) \geqslant C(r^m h(x, r))^{(m-1)/m} \geqslant C\varepsilon^{(m-1)/m} r^{m-1},$$

thus proving the lower Ahlfors regularity. The upper one was obtained in Lemma 3.24.

To prove 4, assume that x is a point of density 1 of E. Then $h(x, r) = r^{-m}|B(x, r) \setminus E|$ if r is small and $\lim_{r \to 0} h(x, r) = 0$. Hence, $x \in E_1$ by Lemma 3.27. The other inclusion follows from the definition of E_1. Hence, E_1 and E are equivalent. Similar arguments replacing E by $M \setminus E$ show that E_0 is the set of points of density 1 of $M \setminus E$ and then $M \setminus E$ and E_0 are equivalent. □

We say that an open set $\Omega \subset M$ satisfies *condition* B if there exist $C > 0$ and $r_0 > 0$ such that for any ball B centered at a point in $\partial\Omega$ with radius $0 < r \leqslant r_0$, there exist two balls of radius Cr such that $B_1 \subset \Omega \cap B$, $B_2 \subset B \setminus \overline{\Omega}$. Condition B ensures that the boundary of Ω separates well Ω from $M \setminus \Omega$. Euclidean sets satisfying the Ahlfors property and condition B have strong rectifiability properties (see [132] and [131]).

Theorem 3.30 *Let $E \subset M$ an isoperimetric region in M. Then E_1 and E_0 satisfy condition* B.

Proof Take $x \in S$ and $0 < r \leqslant r_0$ small enough. We consider the set

$$Z = \{z \in B(x, r/2) : d(z, S) \leqslant sr/2\}$$

for some $0 < s < 1$ to be fixed later.

Let us estimate $|Z|$ from above. We take a maximal family A of points in $S \cap B(x, r)$ at distance at least $sr/2$. Then the balls $B(y, sr/4)$, for $y \in A$, are disjoint and satisfy $\bigcup_{y \in A} B(y, sr/4) \subset B(x, 2r)$. By the Ahlfors regularity of S

$$\begin{aligned} C^{-1}(sr)^{m-1}\text{card}(A) &\leqslant \sum_{y \in A} P(E, B(y, sr/4)) \\ &\leqslant P(E, B(x, 2r)) \leqslant Cr^{m-1}. \end{aligned}$$

Hence, $\text{card}(A) \leqslant Cs^{1-m}$. Since the family of balls $\{B(y, sr) : y \in A\}$ cover $S \cap B(x, r)$ we have

$$|Z| \leqslant C \,\text{card}(A)(sr)^m \leqslant Csr^m.$$

Since $x \in S$ we have

$$|E_1 \cap B(x, r)| \geqslant \varepsilon r^m, \quad |E_0 \cap B(x, r)| \geqslant \varepsilon r^m.$$

Then, for s small enough, we can find $z_1 \in E_1 \cap B(x, r) \setminus Z$ and $z_0 \in E_0 \cap B(x, r) \setminus Z$ so that

$$B(z_1, sr/2) \subset E_1 \cap B(x, r), \quad B(z_0, rs/2) \subset E_0 \cap B(x, r).$$

□

Let us finish this section proving an interesting consequence of Theorem 3.29.

Proposition 3.31 *Let $\{E_i\}_{i \in \mathbb{N}}$ be a sequence of isoperimetric regions in M converging in $L^1(M)$ to an isoperimetric region E. Assume that E_i coincides with the set of Lebesgue points of E_i for all $i \in \mathbb{N}$ and that the same property holds for E. Then $\{E_i\}_{i \in \mathbb{N}}$ also converges to E in Hausdorff distance.*

Proof For any $s > 0$, we have to check that $E_i \subset E_s$ and that $E \subset (E_i)_s$ for i large enough.

If $E_i \subset E_s$ does not hold for i large enough, replacing $\{E_i\}_{i\in\mathbb{N}}$ by a subsequence if necessary, we may assume the existence of a sequence of points $x_i \in E_i$ converging to some $x \in M$ and such that $d(x_i, E) \geqslant s$ for all i. Hence, $d(x, E) \geqslant s$, and as $B(x, s) \subset M \setminus E$, we have $|E \cap B(x, s)| = 0$. Since $B(x_i, s/2) \subset B(x, s)$ for i large enough and E_i converges to E in the L^1 topology, we get

$$\limsup_{i\to\infty} |E_i \cap B(x_i, s/2)| \leqslant \lim_{i\to\infty} |E_i \cap B(x, s)| = 0.$$

Lemma 3.27 then implies $|E_i \cap B(x_i, s/4)| = 0$, a contradiction to the fact that $x_i \in (E_i)_1$.

If $E \subset (E_i)_s$ does not hold for i large enough, eventually replacing $\{E_i\}_{i\in\mathbb{N}}$ by a subsequence, we can find a sequence of points $x_i \in E$, converging to some $x \in M$, such that $d(x_i, E_i) \geqslant s$. Hence, $|E_i \cap B(x_i, s)| = 0$ and so $|E_i \cap B(x, s/2)| = 0$ for i large. This implies $|E \cap B(x, s/2)| = 0$ by the L^1 convergence of E_i to E. As $B(x_i, s/2) \subset B(x, s)$ for i large enough we have $|E \cap B(x_i, s/4)| = 0$ for such i, a contradiction to the fact that $x_i \in E = E_1$. □

3.8 Notes

Notes for Sect. 3.2 The notion of isoperimetric profile appeared in Bérard [53] and Gallot [176] under the name of *isoperimetric function*. It already appers under the name of isoperimetric profile in Bérard [54] and Bavard and Pansu [45]. Basic properties of the profile for compact manifolds, including continuity, were proved by Gallot, including the proof of the local Hölder continuity of the profile.

Notes for Sect. 3.3 The asymptotic expansion of the isoperimetric profile near $v = 0$ was obtained by Berard and Meyer [52]. Further asymptoyic properties of the isoperimetric profile will be considered in Chap. 7. This aymptotic expansion is no longer valid in singular spaces such as cones.

The Cheeger constant was introduced by Cheeger in [103], where he obtained the inequality

$$\lambda_1(M) \geqslant \frac{\mathrm{Ch}(M)}{4}$$

for the smallest positive eigenvalue of the Laplacian $\lambda_1(M)$. An upper bound for $\lambda_1(M)$ in terms of $\mathrm{Ch}(M)$ was later obtained by Buser [85]. The Cheeger constant and the Cheeger sets have been intensively studied in domains of the Euclidean space. It is known that inside an open bounded convex set of $\mathbb{R}^m$, the Cheeger set is unique, and it is a convex set (see Alter and Caselles [15]).

Notes for Sect. 3.4 A differential inequality for the isoperimetric profile was obtained by Bavard and Pansu in [45], under the assumption of boundary regularity of an isoperimetric region. On surfaces they obtained the inequality $(I^2)'' \leqslant -2 \inf K$, where K is the Gauss curvature of the surface. They indicate that this differential inequality was obtained independently by Gallot (p. 480 in [176]). The differential inequalities by Bavard and Pansu were later extended by Bayle [46] to include the power $m/(m-1)$ of the profile. Simultaneously, Morgan and Johnson [311] obtained related differential inequalities. The argument of using functions with arbitrarily small capacity supported in the regular part of the boundary of an isoperimetric regions was used by Sternberg and Zumbrum [411]. Is it worth noting that the isoperimetric profile of the sphere can be obtained from the differential inequality, as indicated in Theorem 3.15. Regularity and concavity properties of the profile also follow from the differential properties. The Hölder continuity property has been extended to a class of non-compact Riemannian manifold by Muñoz Flores and Nardulli [316].

Grimaldi, Nardulli, and Pansu [195] have proven that the isoperimetric profile of a compact analytic manifold is real analytic in a neighborhood of 0 relying on results by Bierstone and Milman [58], and that the isoperimetric profile is semianalytic in dimension less than 8 (no singularities on isoperimetric boundaries) relying on results by Tamm [421]. Pansu [332] proved that in analytic surfaces, the isoperimetric profile consists of a finite number of smooth graphs and provided a counteraxamples in the smooth category. Grimaldi, Nardulli, and Pansu [194] proved that an analytic manifold where all isoperimetric boundaries are smooth, the scalar curvature achieves a unique non-degenerate maximum and the isoperimetric profile is smooth is homeomorphic to a sphere.

Notes for Sect. 3.5 Lévy proved [268] that the isoperimetric profile of a convex hypersurface $M \subset \mathbb{R}^{m+1}$ with principal curvatures larger than or equal to one was bounded below by the isoperimetric profile of the sphere $\mathbb{S}^m$. Gromov extended this result to manifolds with a positive lower bound on the Ricci curvature in Appendix C_+ of [198] using Heintze-Karcher inequalities [225]. The proof we have followed here is taken from Bayle's Thesis [47] and only uses the differential inequality for $h^{m/(m-1)}$. Gromov's original proof, following Lévy's ideas, using the Heintze inequality appeared in Appendix I in [288]. See also Appendix C in Gromov [198] and §VI.6 in Chavel [102] for a more detailed proof.

The Lévy-Gromov inequality was extended by Bérard, Besson, and Gallot [51] replacing a lower bound on the Ricci curvature by a lower bound of $\big(\inf_M \operatorname{Ric}\big) \operatorname{diam}(M)^2$.

Regularity properties of the isoperimetric profile, as well as geometric and topological restrictions for isoperimetric sets, follow from the differential inequality.

The Lévy-Gromov inequality has been extended to manifolds with density by E. Milman [286] and to metric measure spaces satisfying a curvature-dimension condition by Cavalleti and Mondino [97]. See also Klartag [246] for related ideas. For the extension of the notion of Ricci curvature to metric measure spaces, see Sturm [413, 414] and Lott and Villani [273].

Notes for Sect. 3.6 Gromov-Hausdorff distance was introduced by Gromov in §6 of [196] (see also Chapter 3 of [198]), as a generalization of the classical Hausdorff distance for subsets of a metric space. It was also proved in [198] that this Lipschitz convergence implies generalized Hausdorff convergence. Precompactness properties for this distance were also proved by Gromov (see also Peters [343] and Greene and Wu [192]). Bayle studied in Chapter 4 of his PhD thesis [47] the continuity of the isoperimetric profile with respect to Gromov-Hausdorff convergence. A detailed discussion of convergence of Riemannian manifolds is given in Petersen's monograph [345]. In Chap. 9, we shall see convergence properties of the normalized isoperimetric profile of convex bodies in Euclidean spaces with respect to classical Hausdorff convergence. Some of the conjectures in Bayle's thesis have been recently solved by Antonelli et al. (see §7 in [28]) by using new techniques from RCD theory.

Notes for Sect. 3.7 Most of the results in this section are based on the approach to isoperimetric sets in Carnot groups by Leonardi and Rigot [266]. Note that the regularity of perimeter-minimizing sets under a volume constraint is still an open problem in sub-Riemannian geometry.

Convergence in Hausdorff distance of isoperimetric sets in non-collapsed spaces with lower Ricci curvature bounds have been obtained by Antonelli et al. in §5 of [28].

Chapter 4
The Isoperimetric Profile of Non-compact Manifolds

In this chapter, we consider *complete non-compact* Riemannian manifolds. From the isoperimetric point of view, several differences with the compact case immediately arise, like the lack of continuity of the isoperimetric profile and the phenomenon of non-existence of isoperimetric regions.

The first example of a Riemannian manifold with a discontinuous isoperimetric profile was provided by Nardulli and Pansu [319] by collating copies of compact quotients of three-dimensional Riemannian Heisenberg groups. Afterward, a two-dimensional example was built by Papasoglu and Swenson [333] by patching up standard two-dimensional spheres arranged as the vertices of an expanding graph. However, despite the existence of manifolds with discontinuous isoperimetric profile, there are geometric conditions under which continuity follows. According to the results in [364], the existence of a convex Lipschitz continuous exhaustion function implies that the isoperimetric profile is a continuous and non-decreasing function. Examples of such manifolds are the simply connected manifolds with strictly positive sectional curvatures and the Cartan–Hadamard manifolds (i.e., complete simply connected manifolds with non-positive sectional curvatures). Of particular interest are the results in Sect. 4.2.3, where it is shown that the isoperimetric profile of a complete manifold can be approximated by the ones of an exhaustion of M. In particular, in the definition of isoperimetric profile, we can restrict ourselves to the class of *bounded* sets of finite perimeter.

Concerning existence of isoperimetric sets, we have already seen, in Theorem 2.27 of Chap. 2, examples of planes of revolution with decreasing Gauss curvature where isoperimetric sets do not exist for any given area. Hence, additional conditions on the manifold are necessary to ensure existence of isoperimetric sets. We also present in this chapter an existence result when the action of the isometry group $\operatorname{Isom}(M)$ of M is *cocompact* (i.e., when $M/\operatorname{Isom}(M)$ is compact), whose validity was first indicated by Morgan [307] following ideas by Almgren [14]. In the case of two-dimensional surfaces with non-negative Gauss curvature, existence of isoperimetric sets for any given area was established in Sect. 2.7. As a consequence,

M. Ritoré, *Isoperimetric Inequalities in Riemannian Manifolds*,
Progress in Mathematics 348, https://doi.org/10.1007/978-3-031-37901-7_4

we show that geodesic balls are the only isoperimetric sets in Euclidean and hyperbolic spaces. We also provide a generalized existence result for manifolds with bounded Lipschitz geometry, extended results first obtained by Nardulli in [318].

4.1 A Manifold with Discontinuous Isoperimetric Profile

The results in this section are due to Nardulli and Pansu [319], who provided an example of a Riemannian manifold whose isoperimetric profile is not continuous.

The strategy of their proof consists on building a sequence of compact Riemannian manifolds $\{M_i\}_{i\in\mathbb{N}}$ of volumes $1+\tau_i$, with $\tau_i \to 0$, such that I_{M_i} is uniformly bounded below by a positive constant in the interval $[\tau_i, 1]$ and so that the isoperimetric profile of the disjoint union $M' = \amalg_i M_i$ satisfies $I_{M'}(1) > 0$, but there is a sequence $v_i > 1$, $v_i \to 1$, such that $\liminf_{i\to\infty} I_{M'}(v_i) = 0$. If we wish to obtain a connected example, then we join M_i to M_{i+1} by means of a tube of small volume. The manifolds M_i are obtained as quotients of the Riemannian Heisenberg group $\mathbb{H}^1$.

4.1.1 Geometry of the First Heisenberg Group $\mathbb{H}^1$

We briefly recall the definition of the Heisenberg group $\mathbb{H}^1$ and refer the reader to [368] for further details on the geometry of $\mathbb{H}^1$. The underlying space of $\mathbb{H}^1$ is the Euclidean space $\mathbb{R}^3$ with coordinates (x, y, t). The standard product $*$ in $\mathbb{H}^1$ is defined by

$$(x, y, t) * (x', y', t') = (x + x', y + y', t + t' + x'y - xy').$$

A basis of left-invariant vector fields is given by

$$X = \frac{\partial}{\partial x} + y\frac{\partial}{\partial t}, \quad Y = \frac{\partial}{\partial y} - x\frac{\partial}{\partial t}, \quad T = \frac{\partial}{\partial t}.$$

The vector fields X and Y generate a non-integrable two-dimensional distribution $\mathcal{H}$ making $\mathbb{H}^1$ a contact manifold. Vector fields U such that $U_p \in \mathcal{H}_p$ for all $p \in \mathbb{H}^1$ are called *horizontal*.

We consider on $\mathbb{H}^1$ the left-invariant Riemannian metric g on $\mathbb{H}^1$ making X, Y, T an orthonormal frame. Since left-translations are affine mappings with Jacobian 1, Haar's measure coincides, up to a constant, with the Lebesgue measure in $\mathbb{R}^3$ and also with the Riemannian volume element induced by g.

For any $\varepsilon > 0$, we may also consider on $\mathbb{H}^1$ the Riemannnian metric g_ε so that X, Y, T are orthogonal vectors, with

$$g_\varepsilon(X, X) = g_\varepsilon(Y, Y) = 1, \quad g_\varepsilon(T, T) = \frac{1}{\varepsilon^2}.$$

When ε approaches 0, the Riemannian distances d_ε converge to the Carnot-Carathéódory distance in $\mathbb{H}^1$, defined by:

$$d_{cc}(p, q) = \inf\left\{L(\gamma) : \gamma \in C^1([0, 1], \mathbb{H}^1), \gamma(0) = p, \gamma(1) = q, \gamma'(t) \in \mathcal{H}_{\gamma(t)}\ \forall t\right\},$$

where $L(\gamma)$ is the length of the horizontal curve γ, measured with respect to any of the Riemannian metrics g_ε. The metric g (or any of the g_ε) restricted to $\mathcal{H}$ gives a structure of *sub-Riemannian manifold*to $\mathbb{H}^1$ (see [296]).

We consider the subgroup Γ of left-translations by integer coordinate points. The discrete group Γ acts freely and properly on $\mathbb{H}^1$, and the quotient $N = \mathbb{H}^1/\Gamma$ is a Riemannian manifold. A fundamental region for N in $\mathbb{R}^3$ is any Euclidean cube with sides of length 1, and so N has volume 1.

4.1.2 The Isoperimetric Profile of Some Quotients of $\mathbb{H}^1$

In $\mathbb{H}^1$ we consider the family of anisotropic dilations δ_λ, $\lambda \geqslant 0$, defined by

$$\delta_\lambda((x, y, t)) = (\lambda x, \lambda y, \lambda^2 t).$$

For λ fixed, we let Γ_λ be the group of left-translations by the vectors $(\lambda n_1, \lambda n_2, \lambda^2 n_3)$, where n_i are integer numbers.

Definition 4.1 We define $N_{\lambda,\varepsilon}$ as the manifold $\mathbb{H}^1/\Gamma_\lambda$ endowed with the Riemannian metric g_ε.

Associated with each metric g_ε, we have a perimeter P_ε. In addition we can also define a *sub-Riemannian perimeter* P_{sr} as

$$P_{sr}(E) = \sup\left\{\int_E \operatorname{div}(U)\, dv_g : U \in \mathfrak{X}_0^1(\mathbb{H}^1) \cap \mathcal{H}, ||U||_\infty \leqslant 1\right\},$$

where dv_g is the Riemannian volume element on $(\mathbb{H}^1, g)$. The only difference with the definition of the Riemannian perimeter associated with the metric g is that the supremum is taken here only over *horizontal* vector fields. If $S \subset \mathbb{H}^1$ is a surface of class C^1 bounding a relatively compact region $E \subset \mathbb{H}^1$, then its sub-Riemannian perimeter equals the sub-Riemannian area

$$P_{sr}(E) = \int_S |N_h| \, dS,$$

where N is a unit normal to S for the metric g and N_h is the orthogonal projection to the horizontal distribution $\mathcal{H}$.

The sub-Riemannian perimeter satisfies

$$P_{sr}(\delta_\lambda(E)) = \lambda^3 P_{sr}(E).$$

Observe also that the measure $|\cdot|$ associated with g satisfies

$$|\delta_\lambda(E)| = \lambda^4 |E|.$$

Hence the sub-Riemannian isoperimetric profile I_{sr}, defined as the infimum of the sub-Riemannian perimeter of sets with a given volume, satisfies $I_{sr}(v) = cv^{3/4}$ for some constant $c \geqslant 0$, which is indeed strictly positive as shown by Garofalo and Nhieu [179]. So we have

Theorem 4.2 ([179, 266, 330, 331]) *The sub-Riemannian isoperimetric profile of* $\mathbb{H}^1$ *satisfies*

$$I_{sr}(v) = cv^{3/4}, \tag{4.1}$$

for some constant $c > 0$.

The compact manifolds $N_{\lambda,\varepsilon}$ in Definition 4.1 are the building blocks of the construction. We begin by analyzing the isoperimetric profile of $N_{\lambda,\varepsilon}$.

Proposition 4.3 ([319, Corollary 3]) *The isoperimetric profile of* $N_{\lambda,\varepsilon}$ *satisfies*

$$I_{N_{\lambda,\varepsilon}}(v) \geqslant \frac{c}{\varepsilon^{1/4}} \min\left\{v, \frac{\lambda^4}{\varepsilon} - v\right\}^{3/4}, \tag{4.2}$$

for some constant $c > 0$ *independent of* λ *and* ε.

Proof Let us prove first that, for any measurable set $E \subset N = N_{\lambda,\varepsilon}$, we have

$$P_{sr}(E) \geqslant c \min\left\{v, 1 - v\right\}^{3/4}, \tag{4.3}$$

for some positive constant c.

To prove (4.3) we lift E to $\mathbb{H}^1$ to get a set $\widetilde{E}$ and consider the intersection $E' = \widetilde{E} \cap [-\frac{1}{2}, \frac{1}{2}]^3$. Since $[-\frac{1}{2}, \frac{1}{2}]^3$ is a fundamental region of N, the sets E and E' have the same volume. It is not difficult to check that

$$P_{sr}(E') \leqslant P_{sr}(E) + P_{sr}(E' \cap \{\text{faces of } [-\tfrac{1}{2}, \tfrac{1}{2}]^3\}).$$

A simple application of the Fubini's theorem for the Lebesgue measure (equal to Haar's measure up to a constant) implies that after left-translations, we can assure that the Euclidean area of the intersection of E' with the faces of the cube is smaller than or equal to $|E'| = |E|$. The sub-Riemannian area of a domain in a vertical plane is exactly the Euclidean area, while the sub-Riemannian area of a set U in a horizontal plane $\{t = \text{constant}\}$ can be computed as

$$\int_U r\,dxdy,$$

where $r = \sqrt{x^2 + y^2}$. Hence the sub-Riemannian perimeter of the intersection of E' with a horizontal face of the cube is smaller than or equal to $(1/2)^{1/2}$ times the Euclidean area. We conclude

$$P_{sr}(E) \geqslant P_{sr}(E') - 4v - \tfrac{1}{2^{1/2}}(2v) \geqslant P_{sr}(E') - 6v.$$

We can now apply the isoperimetric inequality (4.1) to E' to get

$$P_{sr}(E) \geqslant cv^{3/4} - 6v = v^{3/4}(c - 6v^{1/4}) \geqslant \tfrac{c}{2}v^{3/4}$$

whenever $v \leqslant (c/12)^4$. This implies

$$(I_N)_{sr}(v) \geqslant \frac{c}{2}v^{3/4}, \quad \text{whenever } v \leqslant \left(\frac{c}{12}\right)^4.$$

We can now reason as in the compact Riemannian case (see the proof of Lemma 3.4, to prove that the isoperimetric profile $(I_N)_{sr}$ is positive; the local Poincaré and isoperimetric inequalities proven by Garofalo and Nhieu [179] in this context are necessary). This way, we get the existence of a constant $c > 0$, different from the previous one, so that (4.3) holds.

Next we observe that the behavior of the sub-Riemannian perimeter and volume under the anisotropic dilations δ_λ, together with inequality (4.3), imply

$$(I_{N_\lambda})_{sr}(v) \geqslant c \min\left\{v, \lambda^4 - v\right\}^{3/4}.$$

Finally we use the inequalities $\varepsilon P_\varepsilon \geqslant P_{sr}$ and $|\cdot|_\varepsilon = \frac{1}{\varepsilon}|\cdot|$ to obtain (4.2). □

Definition 4.4 We define the compact manifolds $M_i = N_{\lambda_i,\varepsilon_i}$ where $\tau_i = 1/i$, $\lambda_i = \tau_i^{3/4}(1 + \tau_i)^{1/4}$, and $\varepsilon_i = \tau_i^3$.

From its definition, it is clear that the Riemannian volume of M_i is equal to $\lambda_i^4/\varepsilon_i = 1 + \tau_i$. From (4.2) we get

$$I_{M_i}(v) \geqslant \frac{c}{\tau_i^{3/4}} \min\{v, (1+\tau_i) - v\}^{3/4}.$$

In particular, $I_{M_i}(1) \geqslant c > 0$ for all i.

4.1.3 *Proof of the Existence Result*

Theorem 4.5 *There exists a complete connected Riemannian manifold M such that its isoperimetric profile I_M is not continuous.*

Proof The manifold M will be built from the compact Riemannian manifolds M_i defined in 4.4.

We consider first the disjoint union $M' = \amalg_i M_i$. Let us prove that

$$I_{M'}(v) \geqslant \frac{c}{8}, \qquad \text{whenever } v \in [\frac{1}{16}, 1], \tag{4.4}$$

where $c > 0$ is the constant in inequality (4.2).

Let $\Omega \subset M'$ be a region with $|\Omega| = v \in [\frac{1}{16}, 1]$. We write $\Omega = \amalg_i \Omega_i$, where $\Omega_i \subset M_i$, $|\Omega_i| = v_i$, and $\sum_{i=1}^{\infty} v_i = v$. If $v_i \geqslant \frac{1}{2}|M_i| = \frac{1}{2}(1+\tau_i)$ for some $i \in \mathbb{N}$, then (4.2) implies

$$A(\partial\Omega) \geqslant A(\partial\Omega_i) \geqslant \frac{c}{\tau_i^{3/4}}(1+\tau_i - v_i)^{3/4} \geqslant c \geqslant \frac{c}{16}.$$

In case $v_i \leqslant \frac{1}{2}|M_i|$ for all i, we get from (4.2)

$$A(\partial\Omega_i) \geqslant \frac{c}{\tau_i^{3/4}} v_i^{3/4} \geqslant c v_i^{3/4},$$

and so

$$A(\partial\Omega) = \sum_{i=1}^{\infty} A(\partial\Omega_i) \geqslant c \sum_{i=1}^{\infty} v_i^{3/4} \geqslant c\left(\sum_{i=1}^{\infty} v_i\right)^{3/4} = c v^{3/4} \geqslant \frac{c}{16^{3/4}} = \frac{c}{8},$$

which implies the claim.

Now we proceed to build the connected example M. Starting from the sequence of compact manifolds $\{M_i\}_{i\in\mathbb{N}}$, we pick a sequence of positive volumes $\{w_i\}_{i\in\mathbb{N}}$ and areas $\{a_i\}_{i\in\mathbb{N}}$ with $a_1 = 0$, $a_i > 0$ for $i \geqslant 2$, and such that

$$\sum_i w_i < \frac{1}{2}, \qquad \sum_i a_i < \frac{c}{16}.$$

For every $i \in \mathbb{N}$, we consider two small disjoint balls $B_{i,1}$, $B_{i,2}$ inside M_i (we take $B_{1,1} = \emptyset$) such that the total boundary area is smaller than or equal to a_i and the total volume is smaller than τ_i. Let $M_i' = M_i \setminus \operatorname{int}(B_{i,1} \cup B_{i,2})$. Consider cylinders T_i homeomorphic to $\mathbb{S}^2 \times [0, 1]$ endowed with a Riemannian metric chosen so that the volume of T_i is less than or equal to w_i and the connected sum $M_i'' = M_i' \# T_i$, where the gluing is done along $\partial B_{i,2}$ and $\mathbb{S}^2 \times \{0\}$. We now consider the Riemannian connected sum

$$M = M_1'' \# M_2'' \# \ldots \# M_i'' \# M_{i+1}'' \# \ldots$$

where M_i'' and M_{i+1}'' are attached along $\mathbb{S}^2 \times \{1\} \subset T_i$ and $\partial B_{i+1,1}$. The Riemannian metric in $M \setminus \bigcup_i (B_{i,1} \cup B_{i,2})$ coincides with the one of $M' = \amalg_i M_i$.

Let us prove that there exists a sequence $\alpha_i > 0$ with $\alpha_i \to 0$ such that $\lim_{i\to\infty} I_M(1 + \alpha_i) = 0$. For that we consider, for every $i \in \mathbb{N}$, the region M_i', which has volume

$$|M_i'| = |M_i| - |B_{i,1} \cup B_{i,2}| = 1 + \tau_i - |B_{i,1} \cup B_{i,2}| = 1 + \alpha_i,$$

with $\alpha_i > 0$ and $\alpha_i \to 0$ (recall that $|B_{i,1} \cup B_{i,2}| < \tau_i$). Hence, we have

$$I_M(1 + \alpha_i) \leqslant A(\partial M_i') = A(\partial B_{i,1}) + A(\partial B_{i,2}) \to 0.$$

Finally, let us show that $I_M(1) > 0$. Consider a region $\Omega \subset M$ with $|\Omega| = 1$. Let $\Omega' = \bigcup_i \Omega_i'$, where $\Omega_i' = \Omega \cap M_i'$. Then

$$1 \geqslant |\Omega'| \geqslant 1 - \sum_i w_i \geqslant \frac{1}{2}.$$

By (4.4) we have

$$A(\partial\Omega') \geqslant \frac{c}{8} > 0.$$

Since

$$A(\partial\Omega) \geqslant A(\partial\Omega') - \sum_i a_i \geqslant \frac{c}{8} - \frac{c}{16}.$$

This shows that $I_M(1) \geqslant c/16 > 0$. □

4.2 Continuity of the Isoperimetric Profile Under Sectional Curvature Conditions

In the previous section, we gave an example of a non-compact Riemannian manifold whose isoperimetric profile is discontinuous. In this one we give geometric conditions ensuring the continuity of the isoperimetric profile.

4.2.1 *Geometry of Manifolds with a Convex Exhaustion Function*

We assume that our complete non-compact Riemannian manifold M possesses a strictly convex Lipschitz continuous exhaustion function $f \in C^\infty(M)$. These manifolds were studied in depth by Greene and Wu [191], who derived interesting topological and geometric properties from the existence of such a function (e.g., they are diffeomorphic to the Euclidean space of the same dimension; see Theorem 3 in [191]). Complete non-compact manifolds with strictly positive sectional curvatures possess such a C^∞ convex function. This follows from the existence of a continuous strictly convex function obtained by Cheeger and Gromoll [106] and an approximation result by C^∞ functions by Greene and Wu (see Theorem 2 in [191]). In Cartan–Hadamard manifolds, the squared distance function is known to be a C^∞ strictly convex exhaustion function (see [34]), although it is not (globally) Lipschitz. Composing with a certain real function a C^∞ strictly convex Lipschitz continuous exhaustion function is obtained. Details are given in the proof of Theorem 4.17.

The main result in this section is Theorem 4.16, where we prove the continuity of the isoperimetric profile I_M assuming the existence of a C^∞ strictly convex Lipschitz continuous exhaustion function. One of the main ingredients of the proof consists on approximating I_M of M by the profiles of the compact sublevel sets of f (see Sect. 4.2.3).

We start with some definitions.

Definition 4.6 A continuous function $f : M \to \mathbb{R}$ is *convex* (resp. *strictly convex*) if $f \circ \gamma$ is convex (resp. strictly convex) for any geodesic $\gamma : I \to M$.

It follows that a smooth function $f \in C^\infty(M)$ is strictly convex if and only if $(f \circ \gamma)'' > 0$ on I for any geodesic $\gamma : I \to M$ or, equivalently, if the Hessian $\nabla^2 f$ is definite positive.

We recall that a function $f : M \to \mathbb{R}$ is *L-Lipschitz* for some $L > 0$ if $|f(p) - f(q)| \leqslant L\, d(p, q)$ for any pair of points $p, q \in M$.

Definition 4.7 A continuous function $f : M \to \mathbb{R}$ is an *exhaustion function* if, for any $r > \inf f$, the set $C_r = \{p \in M : f(p) \leqslant r\}$ is a compact subset of M.

In the sequel we assume the existence of a strictly convex L-Lipschitz exhaustion function $f \in C^\infty(M)$. The following properties are well-known:

1. f has a unique minimum x_0, which is the only critical point of f.
2. The sets $\partial C_r = \{p \in M : f(p) = r\}$ are strictly convex hypersurfaces whenever $r > f(x_0)$. In particular, their mean curvatures are strictly positive.
3. If $f(x_0) = 0$, then $B(x_0, L^{-1}r) \subset \mathrm{int}(C_r)$, and so $\overline{B}(x_0, L^{-1}r) \subset C_r$.
4. If $f(x_0) = 0$, then there exists a positive constant K such that $C_r \subset \overline{B}(x_0, K^{-1}r + 1)$ for all $r \geqslant 1$.
5. M is diffeomorphic to $\mathbb{R}^n$. In particular, M is unbounded.
6. The volume of M is infinite.

We remark that we can always normalize the function f, by adding a constant, so that $f(x_0) = 0$.

Existence of a minimum of f follows from the compacity of the sub-level sets of f which are compact sets. The uniqueness of the minimum is obtained from the arguments at the beginning of the proof of Theorem 3(a) in the paper by Greene and Wu [191]: we assume the existence of two mimima x_0, x_1 of M, which can be connected by a length-minimizing geodesic $\gamma : [0, 1] \to M$. Since $f \circ \gamma$ is a convex function, it must be constant on [0, 1], but as $\nabla^2 f$ is definite positive, we conclude that $x_0 = x_1$.

Property 2 is well-known since the second fundamental form σ of ∂C_r at $p \in \partial C_r$ with respect to the outer normal $\nabla f / |\nabla f|$ is given by

$$\sigma_p(u, v) = \langle \nabla_u \left(\frac{\nabla f}{|\nabla f|} \right), v \rangle = \frac{1}{|\nabla f_p|} (\nabla^2 f)_p(u, v), \qquad u, v \in T_p \partial C_r.$$

Hence, $\sigma_p(u, u) > 0$ for any $u \in T_p \partial C_r$, $u \neq 0$.

Property 3 is obtained from the inequality $f(x) \leqslant f(x_0) + L d(x_0, x)$.

Property 4 is a consequence of the arguments used to prove Theorem 5 in [191]. We sketch the short proof: choose $t_0 \in (0, 1]$ such that $\overline{B}(x_0, t_0) \subset C_1 \subset C_r$. Let $K = \inf\{f(\exp_{x_0}(t_0 v)) : v \in T_{x_0}M, |v| = 1\} > 0$. If $x \in C_r \setminus \overline{B}(x_0, t_0)$, using Hopf-Rinow's theorem, we can connect x_0 to x by a unit-speed length-minimizing geodesic $\gamma : [0, d(x_0, x)] \to M$. By the convexity of $(f \circ \gamma)$, since $f(x_0) = f(\gamma(0)) = 0$ and $0 < t_0 \leqslant 1$,

$$\frac{(f \circ \gamma)(d(x, x_0)) - (f \circ \gamma)(t_0)}{d(x, x_0) - t_0} \geqslant \frac{(f \circ \gamma)(t_0)}{t_0} \geqslant (f \circ \gamma)(t_0)$$

As $f \geqslant 0$ and $f(\gamma(t_0)) \geqslant K$, for $x \in C_r$, we have

$$\begin{aligned} r \geqslant f(x) = (f \circ \gamma)(d(x_0, x)) &\geqslant (f \circ \gamma)(t_0) + (f \circ \gamma)(t_0)(d(x_0, x) - t_0) \\ &\geqslant K(d(x_0, x) - t_0). \end{aligned}$$

Thus, $d(x_0, x) \leqslant K^{-1}r + t_0 < K^{-1}r + 1$. If $x \in \overline{B}(x_0, t_0)$, the same inequality holds and proves that $C_r \subset \overline{B}(x_0, K^{-1}r + 1)$.

Property 5 is proven in Theorem 3(a) in [191].

Let us finally see that the volume of M is infinite. Since f is L-Lipschitz, there holds $|\nabla f| \leqslant L$. By the coarea formula

$$|C_r| = \int_0^r \left\{ \int_{\partial C_t} \frac{1}{|\nabla f|} \, d(\partial C_t) \right\} dt \geqslant \frac{1}{L} \int_0^r A(\partial C_t) \, dt. \tag{4.5}$$

The mean curvature H_t of ∂C_t is computed as

$$H_t(p) = \sum_{i=1}^{m-1} \langle \nabla_{e_i} \left(\frac{\nabla f}{|\nabla f|} \right), e_i \rangle = \sum_{i=1}^{m-1} \frac{\nabla^2 f(e_i, e_i)}{|(\nabla f)_p|} \geqslant 0,$$

where $e_1, \dots, e_{m-1}$ is an orthonormal basis of $T_p(\partial C_t)$ for $p \in \partial C_t$. On the other hand, the derivative of the area ∂C_t with respect to t is given by

$$\frac{d}{dt} A(\partial C_t) = \int_{\partial C_t} H_t \, \langle \frac{\nabla f}{|\nabla f|}, \frac{\nabla f}{|\nabla f|^2} \rangle \, d(\partial C_t) = \int_{\partial C_t} \frac{H_t}{|\nabla f|} \, d(\partial C_t) \geqslant 0.$$

Hence, we obtain from (4.5), fixing $r_0 > 0$,

$$|C_r| \geqslant \frac{1}{L} A(\partial C_{r_0}) \, (r - r_0).$$

Taking limits when $r \to \infty$, we get $|M| = \infty$.

4.2.2 A Volume Estimate

The isoperimetric profiles $I_r : (0, |C_r|) \to \mathbb{R}^+$ of the sublevel sets of f are defined by

$$I_r(v) = \inf \{ P(E) : E \subset C_r \text{ measurable}, |E| = v \}.$$

eq The compactness of C_r and the lower semicontinuity of perimeter imply the existence of isoperimetric regions in C_r for all $v \in (0, |C_r|)$, as well as the continuity of the isoperimetric profile of C_r. From the definitions of I_r and I_M, we have $I_M \leqslant I_r \leqslant I_s$ for all $r \geqslant s > f(x_0)$. This problem is the *restricted isoperimetric problem* considered in Sect. 1.6.2.

Given $\delta \in \mathbb{R}$, we denote by $V_{\delta,m}(r)$ the volume of the geodesic ball in the n-dimensional complete simply connected manifold with constant sectional curvatures equal to δ. When $\delta = 0$, $V_{0,m}(r) = \omega_m r^m$, where ω_m is the m-dimensional volume of the unit ball in $\mathbb{R}^m$. In case $\delta > 0$, the radius r will be taken smaller than $\pi/\delta^{1/2}$. In what follows, we shall take $\pi/\delta^{1/2} = +\infty$ when $\delta \leqslant 0$. The injectivity radius of $x_0 \in M$ will be denoted by $\operatorname{inj}(x_0)$. If $K \subset M$, the injectivity radius of K will be defined by $\operatorname{inj}(K) = \inf_{x\in K} \operatorname{inj}(x)$. When K is relatively compact, $\operatorname{inj}(K) > 0$.

The following result is a non-compact version of Gallot's result (see Lemme 6.2 in [176] or Theorem 3.5). It will play a crucial role in the sequel.

Lemma 4.8 *Let M be an m-dimensional complete non-compact Riemannian manifold, $E \subset M$ a measurable set of finite volume, $B \subset M$ a bounded measurable set such that $|B| - |E| > 0$, and δ the supremum of the sectional curvatures of M in B. Fix $r_0 > 0$ and choose $D \supset B$ bounded and measurable such that $d(B, \partial D) > r_0$. For any $0 < r < \min\{r_0, \operatorname{inj}(B), \pi/\delta^{1/2}\}$ define*

$$\Lambda(r) = \frac{|B| - |E|}{|D|} V_{\delta,m}(r). \tag{4.6}$$

Then there exists $x \in D$ such that

$$|B(x,r) \setminus E| \geqslant \Lambda(r) > 0.$$

Proof Given two measurable sets $D, F \subset M$ of finite volume, Fubini-Tonelli's theorem applied to the function $(x,z) \in D \times M \mapsto \mathbf{1}_{F\cap B(x,r)}(z)$ yields

$$\int_D |F \cap B(x,r)|\, dM(x) = \int_F |B(z,r) \cap D|\, dM(z).$$

For $F = M \setminus E$, this formula reads

$$\int_D |B(x,r) \setminus E|\, dM(x) = \int_{M\setminus E} |B(z,r) \cap D|\, dM(z).$$

Since $r \leqslant r_0$, we have $B(z,r) \cap D = B(z,r)$ for any $z \in B$, and we get the bound

$$\begin{aligned}\int_{M\setminus E} |B(z,r) \cap D|\, dM(z) &\geqslant \int_{B\setminus E} |B(z,r) \cap D|\, dM(z)\\ &= \int_{B\setminus E} |B(z,r)|\, dM(z) \geqslant |B \setminus E|\, V_{\delta,m}(r)\\ &\geqslant \big(|B| - |E|\big)\, V_{\delta,m}(r),\end{aligned}$$

where inequality $|B(z,r)| \geqslant V_{\delta,m}(r)$ follows from Günther-Bishop's volume comparison theorem [102, Thm. III.4.2] or Theorem 1.29. On the other hand,

$$\int_D |B(x,r) \setminus E|\, dM(x) \leqslant |D| \sup_{x\in D} |B(x,r) \setminus E|.$$

This way we obtain

$$\sup_{x\in D} |B(x,r)\setminus E| \geqslant \frac{|B|-|E|}{|D|} V_{\delta,m}(r),$$

and the result follows. □

4.2.3 Approximation of the Isoperimetric Profile

The following proof follows the lines of Lemma 3.1 in [372] with the modifications imposed by the geometry of M.

Lemma 4.9 *Let M be an m-dimensional complete Riemannian manifold possessing a Lipschitz continuous exhaustion function. Then*

$$I_M(v) = \inf_{r>\inf f} I_r(v),$$

for every $v \in (0, |M|)$.

Proof From the definition of I_r, it follows that $I_s \geqslant I_r \geqslant I_M$ in the interval $(0, |C_s|)$ for $r \geqslant s$. Hence, $I_M \leqslant \inf_{r>\inf f} I_r$. From now on, we assume f is normalized so that $f(x_0) = 0$ at the unique minimum x_0 of f.

To prove the opposite inequality $I_M \geqslant \inf_{r>\inf f} I_r$, we follow the arguments in [367]. Fix $0 < v < |M|$, and let $\{E_i\}_{i\in\mathbb{N}} \subset M$ be a sequence of sets of finite perimeter satisfying $|E_i| = v$ and $\lim_{i\to\infty} P(E_i) = I_M(v)$.

Since $|E_i| = v < |M|$, there exists $R_i > 0$ such that $|E_i \setminus C_{R_i}| < \frac{1}{i}$. We now define a sequence of real numbers $\{r_i\}_{i\in\mathbb{N}}$ by taking $r_1 = R_1$ and $r_{i+1} = \max\{r_i, R_{i+1}\} + i$. Then $\{r_i\}_{i\in\mathbb{N}}$ satisfies

$$r_{i+1} - r_i \geqslant i, \qquad |E_i \setminus C_{r_i}| < \frac{1}{i}.$$

In case $|E_i \setminus C_{r_{i+1}}| = 0$, we take a representative G_i of E_i contained in $C_{r_{i+1}}$, and we have

$$I_{r_{i+1}}(v) \leqslant P(G_i) = P(E_i). \tag{4.7}$$

In case $|E_i \setminus C_{r_{i+1}}| > 0$, since $|\nabla f| \leqslant L$, the coarea formula implies

$$\frac{1}{L}\int_{r_i}^{r_{i+1}} \mathcal{H}^{m-1}(E_i \cap \partial C_t)\, dt < |E_i| = v.$$

Hence, the set of $r \in [r_i, r_{i+1}]$ such that $\mathcal{H}^{m-1}(E_i \cap \partial C_r) \leqslant Lv/(r_{i+1} - r_i)$ has positive measure, where $\mathcal{H}^{m-1}$ is the $(m-1)$-dimensional Hausdorff measure in M. By Maggi [275, Chap. 18, Ex. 18.3, p. 216], we can choose $\rho(i)$ in this set so that

$$P(E_i \cap C_{\rho(i)}) = P(E_i, \operatorname{int} C_{\rho(i)}) + \mathcal{H}^{m-1}(E_i \cap \partial C_{\rho(i)}).$$

By the choice of $\rho(i)$ and the properties of $\{r_i\}_{i\in\mathbb{N}}$, we also have

$$\mathcal{H}^{m-1}(E_i \cap \partial C_{\rho(i)}) \leqslant \frac{Lv}{i}.$$

Take now $t > 0$ such that $|C_t| > v = |E_i| \geqslant |E_i \cap C_{\rho(i)}|$ for all i, and let $\delta(t)$ be the maximum of the sectional curvatures of M in C_t. Let $v_i = |E_i| - |E_i \cap C_{\rho(i)}|$. The sequence $\{v_i\}_{i\in\mathbb{N}}$ converges to 0 since $v_i = |E_i \setminus C_{\rho(i)}| \leqslant |E_i \setminus C_{r_i}| < 1/i$. We take s_i defined by the equality

$$v_i = \frac{|C_t| - |E_i \cap C_{\rho(i)}|}{|C_{2t}|} V_{\delta(t),m}(s_i),$$

for i large enough. Observe that $\lim_{i\to\infty} s_i = 0$ since $\lim_{i\to\infty} |E_i \cap C_{\rho(i)}| = v < |C_t|$ and $\lim_{i\to\infty} v_i = 0$. From Lemma 4.8 we can find, for every $i \in \mathbb{N}$, a point $x_i \in C_{2t}$ such that $|B(x_i, s_i) \setminus (E_i \cap C_{\rho(i)})| \geqslant v_i$. By the continuity of the functions $s \mapsto |B(x_i, s) \setminus E_i|$, we can find a sequence of radii $s_i^* \in (0, s_i]$ so that $B_i^* = B(x_i, s_i^*)$ satisfies $|B_i^* \setminus (E_i \cap C_{\rho(i)})| = v_i$ for all i. For large i, we have the inclusions $B_i^* \subset C_{\rho(i)}$, the set $F_i = (E_i \cap C_{\rho(i)}) \cup B_i^*$ has volume v, and we get

$$\begin{aligned} I_{r_{i+1}}(v) &\leqslant P(F_i) \leqslant P(E_i \cap C_{\rho(i)}) + P(B_i^*) \\ &\leqslant P(E_i, \operatorname{int} C_{\rho(i)}) + \mathcal{H}^{m-1}(E_i \cap \partial C_{\rho(i)}) + P(B_i^*) \\ &\leqslant P(E_i) + \frac{Lv}{i} + P(B_i^*). \end{aligned} \tag{4.8}$$

Since the balls B_i^* are centered at points of the bounded subset C_{2t} with radii s_i^* converging to 0, Bishop's comparison result for the area of geodesic spheres when the Ricci curvature is bounded below [102, Thm. III.4.3] implies that $\lim_{i\to\infty} P(B_i^*) = 0$. Taking limits in (4.7) and (4.8) when $i \to \infty$, we obtain $\inf_{r>\inf f} I_r(v) \leqslant I_M(v)$. □

Remark 4.10 From the proof of Lemma 4.9, it is clear that the center of the balls B_i^* must be taken in a bounded set of M to have $\lim_{i\to\infty} P(B_i^*) = 0$. Indeed, it is easy to produce a family of geodesic balls, each one in a hyperbolic space, with radii going to 0 and perimeters converging to $+\infty$.

Lemma 4.9 can be applied to the particular case of the distance function to a fixed point in a complete Riemannian manifold and provides the following corollary.

Corollary 4.11 *Let M be a complete Riemannian manifold, and $0 < v < |M|$. Then, for every $\varepsilon > 0$, there exists a bounded measurable set E_ε of finite perimeter such that $I_M(v) \leqslant P(E_\varepsilon) \leqslant I_M(v) + \varepsilon$.*

Proof Fix a point $p \in M$ and let $f : M \to \mathbb{R}$ be the distance function to p, which is 1-Lipschitz'. For every $\varepsilon > 0$, Lemma 4.9 implies the existence of $r(\varepsilon) > 0$ such that $I_M(v) \leqslant I_{r(\varepsilon)}(v) \leqslant I_M(v) + \varepsilon$. As $C_{r(\varepsilon)} = \{f \geqslant r(\varepsilon)\}$ is bounded we take an isoperimetric set E_ε in $C_{r(\varepsilon)}$ of volume v (i.e., a minimizer of the total perimeter $P(\cdot)$ inside $C_{r(\varepsilon)}$ of volume v). Since $I_{r(\varepsilon)}(v) = P(E_\varepsilon)$ we obtain the desired inequalities. □

4.2.4 The Isoperimetric Profile of the Sublevel Sets

The existence of a an exhaustion function on M implies that the hypersurfaces $\partial C_r = \{x \in M : f(x) = r\}$ foliate $M \setminus \{x_0\}$, where x_0 is the only minimum of f. The vector field $\nabla f/|\nabla f|$, defined on $M \setminus \{x_0\}$, is the outer unit normal to the hypersurfaces ∂C_r. If f is strictly convex then, for any $x \in \partial C_r$ and e tangent to ∂C_r at x, we have

$$g\Big(\nabla_e\Big(\frac{\nabla f}{|\nabla f|}\Big), e\Big) = \frac{1}{|\nabla f|}\,\nabla^2 f(e,e) > 0.$$

Hence, the hypersurfaces ∂C_r are strictly convex in the sense that they have strictly positive sectional curvatures. The positive function $\operatorname{div}(\nabla f/|\nabla f|)$ is defined on $M \setminus \{x_0\}$. Its value at $x \in \partial C_r$ is the mean curvature of the hypersurface ∂C_r at x.

Lemma 4.12 *Let M be an m-dimensional complete manifold M possessing a strictly convex Lipschitz continuous exhaustion function $f \in C^\infty(M)$. Then the isoperimetric profile I_r of the sublevel set C_r is a continuous and strictly increasing function for $r >$* $\inf f$.

Proof Continuity follows from the compactness of C_r and the lower semicontinuity of perimeter since a limit of isoperimetric regions of volumes converging to $v \in (0, |C_r|)$ is an isoperimetric region of volume v.

To check that I_r is strictly increasing, consider a restricted isoperimetric set $E \subset C_r$ of volume $v \in (0, |C_r|)$. Let $0 < w < v$ and take $s \in (0, r)$ such that the set $E_s = E \cap C_s$ has volume w. Choose a sequence of radii s_i converging to s such that $P(E \cap C_{s_i}) = P(E, \operatorname{int} C_{s_i}) + \mathcal{H}^{m-1}(E \cap \partial C_{s_i})$ and

$$\int_{E \setminus C_{s_i}} \operatorname{div} X \, dM = -\int_{E \cap \partial C_{s_i}} \langle X, |\nabla f|^{-1} \nabla f\rangle \, d\mathcal{H}^{m-1} + \int_{\partial^* E \setminus C_{s_i}} \langle X, \nu_E\rangle \, d|\partial E|,$$

for any vector field X of class C^1 with compact support in an open neighborhood of $C_r \setminus \operatorname{int} C_{s_i}$. In the above formula, $\partial^* E$ is the reduced boundary of E, and $d|\partial E|$ is the perimeter measure. We apply this formula to $X = \nabla f/|\nabla f|$. Since $\operatorname{div} X > 0$ on $M \setminus \{x_0\}$, and $\langle X, \nu_E\rangle \leqslant 1$, we have

$$\int_{E\setminus C_{s_i}} \operatorname{div} X\, dM + \mathcal{H}^{m-1}(E \cap \partial C_{s_i}) \leqslant P(E, M \setminus C_{s_i}).$$

Adding $P(E, \operatorname{int} C_{s_i})$ to both sides of the above inequality and estimating $P(E, \operatorname{int} C_{s_i}) + P(E, M \setminus C_{s_i}) \leqslant P(E)$, we get

$$\int_{E\setminus C_{s_i}} \operatorname{div} X\, dM + P(E \cap C_{s_i}) \leqslant P(E).$$

Taking inferior limits when $i \to \infty$, using the lower semicontinuity of perimeter, and that $\operatorname{div} X > 0$ on the set $E \setminus C_s$ of positive measure, we obtain

$$P(E_s) < \int_{E\setminus C_s} \operatorname{div} X\, dM + P(E_s) \leqslant P(E),$$

and so

$$I_r(w) \leqslant P(E_s) < P(E) = I_r(v).$$

Thus I_r is a strictly increasing function. □

Remark 4.13 We point out that only the condition $\operatorname{div} X > 0$ on the set $E \setminus C_{s_i}$ has been used in the proof of Lemma 4.12, instead of the stronger property that $\nabla^2 f$ is definite positive. Hence, this proof works if we merely assume that the level sets of the exhaustion function f have positive mean curvature and that the set of critical points of f has measure zero.

4.2.5 *Proof of the Continuity Result*

The following elementary lemma will be needed to prove our main result.

Lemma 4.14 *Let $\{f_i\}_{i\in\mathbb{N}}$ be a non-increasing ($f_i \geqslant f_{i+1}$) sequence of continuous non-decreasing functions defined on an open interval $J \subset \mathbb{R}$. Assume the limit $f(x) = \lim_{i\to\infty} f_i(x)$ exists for every $x \in J$. Then f is a right-continuous function.*

Remark 4.15 The hypotheses in Lemma 4.14 do not imply the left-continuity of f, as shown by the following example. Taking

$$f_i(x) = \begin{cases} 1, & 0 \leqslant x, \\ 1 + i\,x, & -1/i \leqslant x \leqslant 0, \\ 0, & x \leqslant -1/i, \end{cases}$$

we immediately see that the limit of the sequence $\{f_i\}_{i\in\mathbb{N}}$ is the characteristic function of the interval $[0, \infty)$, which is not left-continuous.

Proof of Lemma 4.14 Fix $x \in J$. Let $\{x_i\}_{i\in\mathbb{N}}$ be any sequence in J such that $x_i \geqslant x$. Since f is a non-decreasing function, we get $f(x) \leqslant f(x_i)$ for all i. Hence

$$f(x) \leqslant \liminf_{i\to\infty} f(x_i). \tag{4.9}$$

Assume now, in addition to $x_i \geqslant x$, that $x = \lim_{i\to\infty} x_i$. Let us build an auxiliary strictly decreasing sequence $\{z_i\}_{i\in\mathbb{N}}$ in J converging to x and satisfying

$$\limsup_{i\to\infty} f(z_i) \leqslant f(x). \tag{4.10}$$

To this aim, starting from an arbitrary $x < z_1 \in J$, we inductively choose a point z_i satisfying $x < z_i < \min\{z_{i-1}, x + i^{-1}\}$ and

$$0 \leqslant f_i(z_i) - f_i(x) \leqslant \frac{1}{i}.$$

This last condition follows from the continuity of f_i. By construction, $\{z_i\}_{i\in\mathbb{N}}$ is decreasing and converges to x. Since $f_i \geqslant f$ we get

$$f(z_i) \leqslant f_i(z_i) \leqslant f_i(x) + \frac{1}{i},$$

and taking lim sup, we obtain (4.10). Now choose a subsequence $\{y_i\}_{i\in\mathbb{N}}$ of $\{x_i\}_{i\in\mathbb{N}}$ such that $\lim_{i\to\infty} f(y_i) = \limsup_{i\to\infty} f(x_i)$. Since the sequence $\{y_i\}_{i\in\mathbb{N}}$ converges to x, for every $i \in \mathbb{N}$, we can choose $y_{j(i)}$, with $j(i)$ increasing in i, such that $x \leqslant y_{j(i)} < z_i$. As f is non-decreasing,

$$\limsup_{i\to\infty} f(x_i) = \lim_{i\to\infty} f(y_i) = \lim_{i\to\infty} f(y_{j(i)}) \leqslant \limsup_{i\to\infty} f(z_i) \leqslant f(x) \tag{4.11}$$

by (4.10). Inequalities (4.9) and (4.11) then yield the continuity of f to the right. □

Using the previous results, we give the proof of our main result and their consequences.

Theorem 4.16 *Let M be an m-dimensional complete manifold M possessing a strictly convex Lipschitz continuous exhaustion function $f \in C^\infty(M)$. Then the isoperimetric profile I_M of M is non-decreasing and continuous.*

Proof Lemmas 4.9 and 4.12 imply that the profile I_M is the limit of the non-increasing sequence $\{I_r\}_{r>\inf f}$ of continuous non-decreasing isoperimetric profiles. So I_M is trivially non-decreasing, and Lemma 4.14 implies that I_M is right-continuous.

To prove the left-continuity of I_M at $v > 0$, we take a sequence $\{v_i\}_{i\in\mathbb{N}}$ such that $v_i \uparrow v$. Since I_M is non-decreasing, $I_M(v_i) \leqslant I_M(v)$. Taking limits when $i \to \infty$, we get $\limsup_{i\to\infty} I_M(v_i) \leqslant I_M(v)$. To complete the proof, we shall show

$$I_M(v) \leqslant \liminf_{i\to\infty} I_M(v_i). \tag{4.12}$$

Consider a sequence $\{E_i\}_{i\in\mathbb{N}}$ of sets satisfying $|E_i| = v_i$ and $P(E_i) \leqslant I_M(v_i)+1/i$. By Lemma 4.8, we can find a bounded sequence $\{x_i\}_{i\in\mathbb{N}}$ and a sequence of radii $\{s_i\}_{i\in\mathbb{N}}$ converging to 0 so that

$$|B(x_i, s_i) \setminus E_i| \geqslant v - v_i > 0.$$

We argue now as in the final part of the proof of Lemma 4.9: since the function $s \in [0, s_i] \mapsto |B(x_i, s) \setminus E_i|$ is continuous, there exists, for large i, some $s_i^* \in (0, s_i]$ such that $|B(x_i, s_i^*) \setminus E_i| = v - v_i$. Taking $F_i = E_i \cup B(x_i, s_i^*)$, we have $|F_i| = |E_i| + |B(x_i, s_i^*) \setminus E_i| = v$, and

$$I_M(v) \leqslant P(F_i) \leqslant P(E_i) + P(B(x_i, s_i^*)) \leqslant I_M(v_i) + (1/i) + P(B(x_i, s_i^*)).$$

Taking limits we get (4.12). □

Theorem 4.17 *The isoperimetric profile I_M of a Cartan–Hadamard manifold M is a continuous and non-decreasing function.*

Proof We only need to construct a strictly convex Lipschitz continuous exhaustion function. Fix $x_0 \in M$ and let $h = d^2/2$, where d is the distance function to x_0. Standard comparison results for the Laplacian of the squared distance function imply $\nabla^2 h \geqslant 1$ (see Theorem 1.29 and Remark 1.32 or Chapter 3 in [344]). However, h is not (globally) Lipschitz on M. We consider instead the C^∞ function $\mu : (-1, +\infty) \to \mathbb{R}^+$ defined by $\mu(x) = (1+x)^{1/2}$ and the composition $f = \mu \circ h$. Take some tangent vector e of modulus 1 at some point of M. Then

$$\nabla(\mu \circ h) = (\mu' \circ h)\, \nabla h,$$
$$\nabla^2(\mu \circ h)(e, e) = (\mu'' \circ h)\, g(\nabla h, e)^2 + (\mu' \circ h)\, \nabla^2 h(e, e).$$

From the first formula, we obtain

$$\nabla f = \frac{d}{(1 + \frac{1}{2} d^2)^{1/2}} \nabla d.$$

Hence, $|\nabla f|$ is uniformly bounded from above, and so the function f is Lipschitz continuous on M. On the other hand, from the formula for the Hessian of $(\mu \circ h)$, we get

$$\nabla^2 f(e,e) = -\frac{1}{4}\frac{1}{(1+\frac{1}{2}d^2)^{3/2}}\,g(\nabla h, e)^2 + \frac{1}{2}\frac{1}{(1+\frac{1}{2}d^2)^{1/2}}\,\nabla^2 h(e,e).$$

By Schwarz's inequality $g(\nabla h, e) \leqslant d$, and we have

$$\nabla^2 f(e,e) \geqslant \frac{1}{2}\frac{1}{(1+\frac{1}{2}d^2)^{3/2}} > 0.$$

Hence, f is strictly convex. Since the sublevel sets of f are geodesic balls, f is an exhaustion function on M. Theorem 4.16 then implies that the isoperimetric profile of M is a continuous and non-decreasing function. □

Theorem 4.18 *The isoperimetric profile I_M of a complete non-compact manifold M with strictly positive sectional curvatures is a continuous and non-decreasing function.*

Proof The existence of a strictly convex Lipschitz continuous exhaustion function follows from Theorem 1(a) in the paper by Greene and Wu [191]. The properties of the isoperimetric profile from Theorem 4.16. □

Remark 4.19 A continuous monotone function can be decomposed as the sum of an absolutely continuous function and a continuous singular function (such as the Cantor function or Minkowski's question mark function). It would be desirable to find conditions ensuring the absolute continuity of I_M.

4.3 Minimizing Sequences

In the previous sections, we have considered the continuity of the isoperimetric profile I_M of a non-compact Riemannian manifold M. We already know not only that I_M is not necessarily continuous but also that we may impose some geometric conditions implying continuity. In this section, we start to deal with the problem of existence of isoperimetric sets.

4.3.1 Structure of Minimizing Sequences

A standard way of proving existence of solutions in some variational problems is the *direct method*: given a functional, a "minimizing" sequence converging to the infimum of the functional is taken. Under suitable compactness assumptions, convergence of the minimizing sequence or of a subsequence can be ensured to produce a minimizer of the functional. However, for the perimeter functional, this convergence is not guaranteed when M is a non-compact manifold as we have already seen in Theorem 2.27, where an example of a 2-dimensional manifold without isoperimetric sets for any given area is given. In this particular example, any minimizing sequence leaves any compact set and does not converge in any topology.

Definition 4.20 Let M be a complete Riemannian manifold. We say that $\{E_i\}_{i\in\mathbb{N}}$ is a *minimizing sequence* for volume $v \in (0, |M|)$ if

- $|E_i| = v$ for all i and,
- $\lim_{i\to\infty} P(E_i) = I_M(v)$.

In an unbounded manifold, the compactness of a minimizing sequence in some topology (the natural one would be the $L^1(M)$ topology) cannot be guaranteed as in the compact case. However, we may prove the following result concerning the structure of minimizing sequences. The easier two-dimensional case was considered in Lemma 2.21.

Theorem 4.21 (Structure of Minimizing Sequences [367, Theorem 2.1], [175, 369]) *Let M be a complete unbounded Riemannian manifold. Consider a minimizing sequence $\{E_i\}_{i\in\mathbb{N}}$ for volume $v \in (0, |M|)$.*

Replacing $\{E_i\}_{i\in\mathbb{N}}$ by a subsequence, we can find a finite perimeter set E, possibly empty, and sequences $\{E_i^c\}_{i\in\mathbb{N}}$, $\{E_i^d\}_{i\in\mathbb{N}}$ of finite perimeter sets such that

1. $\lim_{i\to\infty} E_i = \lim_{i\to\infty} E_i^c = E$ *in* $L^1_{\text{loc}}(M)$*, the sequence* $\{E_i^d\}_{i\in\mathbb{N}}$ *diverges, and we have* $|E_i^c| + |E_i^d| = v$ *for all* i.
2. $\lim_{i\to\infty} \big(P(E_i^c) + P(E_i^d)\big) = I_M(v)$.
3. $\lim_{i\to\infty} P(E_i^c) = P(E)$.
4. *If* $|E| \neq 0$*, then* E *is an isoperimetric region of volume* $|E|$.

Proof We fix a point $p_0 \in M$. For any $r > 0$, let $B(r)$ be the ball of radius r centered at p_0 and $S(r) = \partial B(r)$ its boundary.

Let $\mathbf{1}_{E_i}$ be the characteristic function of E_i. For any $j \in \mathbb{N}$, the perimeters $P(E_i, B(j))$, $i \in \mathbb{N}$, are uniformly bounded. We apply the Compactness Theorem 1.42 on $B(1)$ to get a subsequence of $\mathbf{1}_{E_i\cap B(1)}$ converging in $L^1(B(1))$ to the characteristic function of a set in $B(1)$. Now we apply the compactness Theorem 1.42 recursively in the balls $B(j)$ to the subsequence obtained in the previous step. By a diagonal argument, we deduce the existence of the characteristic function $\mathbf{1}_E \in L^1_{\text{loc}}(M)$ of a set E and of a non-relabelled subsequence of $\{E_i\}_{i\in\mathbb{N}}$ such that $\lim_{i\to\infty} \mathbf{1}_{E_i} = \mathbf{1}_E$ in $L^1_{\text{loc}}(M)$. From Fatou's Lemma and the lower semicontinuity of perimeter in Proposition 1.41, we get $|E| \leqslant v$ and $P(E) \leqslant \liminf_{i\to\infty} P(E_i) = I_M(v)$.

We claim that we can choose a sequence of divergent increasing radii $r_i > 0$ so that, after passing to a subsequence, we had

$$\int_{B(r_i)} |\mathbf{1}_E - \mathbf{1}_{E_i}|\, dM \leqslant \frac{1}{i}, \tag{4.13}$$

$$P(E_i \setminus B(r_i), \partial B(r_i)) \leqslant \frac{v}{i}, \tag{4.14}$$

for all $i \in \mathbb{N}$. To prove the claim, we consider a sequence of positive numbers $\{s_i\}_{i\in\mathbb{N}}$ so that $s_{i+1} - s_i \geqslant i$ for all i. Taking a subsequence of $\{E_i\}_{i\in\mathbb{N}}$, we may assume that inequality

$$\int_{B(s_{i+1})} |\mathbf{1}_E - \mathbf{1}_{E_i}|\, dM \leqslant \frac{1}{i}$$

holds for all i. Hence, (4.13) holds for any $r \in (0, s_{i+1})$. To prove (4.14) we observe that the function $\mu_i(r) = |E_i \cap B(r)|$ is an increasing function, and so

$$\int_{s_i}^{s_{i+1}} \mu_i'(r)dr \leqslant \mu_i(s_{i+1}) - \mu_i(s_i) \leqslant |E_i| = v.$$

Hence, for almost everywhere $r > 0$,

$$P(E_i \setminus B(r), \partial B(r)) \leqslant \mu_i'(r)$$

which implies that there is a set J_i of positive measure in $[s_i, s_{i+1}]$ so that $\mu_i'(r) \leqslant v/i$. Hence, the inequality in (4.14) holds for some r_i chosen in J_i.

We define

$$E_i^c = E \cap B(r_i), \quad E_i^d = E \setminus B(r_i).$$

We prove now 1. Since E has finite volume and (4.13) holds, we conclude that $\{E_i^c\}_{i\in\mathbb{N}}$ converges in $L^1_{\text{loc}}(M)$ to E. From the definitions of E_i^c and E_i^d, there follows that $\{E_i^d\}_{i\in\mathbb{N}}$ is a divergent sequence and that equality $|E_i^c| + |E_i^d| = v$ holds.

In order to prove 2, we take into account

$$P(E_i^c) \leqslant P(E_i, B(r_i)) + P(E_i \setminus B(r_i), \partial B(r_i)),$$

$$P(E_i^d) \leqslant P(E_i, M \setminus \overline{B}(r_i)) + P(E_i \setminus B(r_i), \partial B(r_i)).$$

Since $E_i = E_i^c \cup E_i^d$, inequality (4.14) implies

$$P(E_i) \leqslant P(E_i^c) + P(E_i^d) \leqslant P(E_i) + \frac{2v}{i}.$$

Taking limits when $i \to \infty$ we get 2.

To prove 3, we show first that

$$P(E) = \liminf_{i\to\infty} P(E_i^c) \tag{4.15}$$

holds. Since E_i^c converges to E in $L^1_{\text{loc}}(M)$ we have $P(E) \leqslant \liminf_{i\to\infty} P(E_i^c)$. Let us reason by contradiction and assume that the strict inequality $P(E) < \liminf_{i\to\infty} P(E_i^c)$ holds. Reasoning as above, we can find a non-decreasing sequence of diverging radii $\{\rho_i\}_{i\in\mathbb{N}}$ with $\rho_i < r_i$ and

$$P(E \setminus B(\rho_i), \partial B(\rho_i)) \leqslant \frac{v}{i}$$

for all i. Define the set $E_i' = E \cap B(\rho_i)$. The perimeter of E_i' satisfies

$$P(E_i') \leqslant P(E, B(\rho_i)) + P(E \setminus B(\rho_i), \partial B(\rho_i)) \leqslant P(E) + \frac{v}{i},$$

and the volume

$$\lim_{i\to\infty} |E_i'| = |E| = v - \lim_{i\to\infty} |E_i^d|.$$

We can now use Theorem 1.50 to produce a small deformation of E_i', localized in a fixed ball, to obtain a sequence of sets E_i'' so that E_i'' and E_i^d are disjoint and $|E_i''| + |E_i^d| = v$. The perimeter of E_i'' satisfies

$$P(E_i'') \leqslant P(E_i') + C\,\big||E_i''| - |E_i'|\big| \leqslant P(E_i) + \frac{v}{i} + C\,\big||E_i''| - |E_i'|\big|$$

for some constant $C > 0$ that does not depend on i. Hence we have

$$\liminf_{i\to\infty} P(E_i'') \leqslant P(E).$$

Then $F_i = E_i'' \cup E_i^d$ is a sequence of sets of volume v satisfying

$$\liminf_{i\to\infty} P(F_i) \leqslant P(E) + \liminf_{i\to\infty} P(E_i^d) < \liminf_{i\to\infty} \big(P(E_i^c) + P(E_i^d)\big) = I_M(v),$$

which clearly gives a contradiction and proves (4.15). To complete the proof of 3, we remark that we can replace the inferior limit in (4.15) by a true limit since every subsequence of a minimizing sequence is also minimizing.

Finally, to prove 4, we consider a set $F \subset M$ of finite perimeter such that $|F| = |E|$ and $P(F) < P(E)$, and we reason as in the first part of the proof of 3 with F instead of E. In particular, if $|E| = v$, then E is an isoperimetric region of volume v. □

Observe that in the proof of Theorem 4.21(1), we have only used that the perimeters are uniformly bounded to get the $L^1_{\mathrm{loc}}(M)$ convergence of the minimizing sequence. Hence, with the same proof, we have

Proposition 4.22 *Let M be a complete unbounded Riemannian manifold. Let $\{E_i\}_{i\in\mathbb{N}}$ be a sequence of sets with positive volume and perimeter, both uniformly bounded from above. Then, after passing to a subsequence, there exists a set of finite perimeter $E \subset M$, possibly empty, and sequences $\{E_i^c\}_{i\in\mathbb{N}}$, $\{E_i^d\}_{i\in\mathbb{N}}$ so that*

1. $\lim_{i\to\infty} E_i = \lim_{i\to\infty} E_i^c = E$ *in* $L^1_{\mathrm{loc}}(M)$,
2. *the sequence* $\{E_i^d\}_{i\in\mathbb{N}}$ *is divergent,*
3. $|E_i^c| + |E_i^d| = |E_i|$ *for all* $i \in \mathbb{N}$.

4.3.2 Existence of Isoperimetric Regions on Manifolds with Finite Volume

An easy consequence of Theorem 4.21 is the existence of isoperimetric regions in manifolds of finite volume.

Theorem 4.23 *Let M be a complete unbounded manifold of finite volume. Then isoperimetric regions exist for any* $0 < v < |M|$.

Proof Fix some $0 < v < |M|$, and take a minimizing sequence $\{E_i\}_{i\in\mathbb{N}}$ for volume v. Using Theorem 4.21 we obtain two sequences $\{E_i^c\}_{i\in\mathbb{N}}$ and $\{E_i^d\}_{i\in\mathbb{N}}$ so that $\{E_i^c\}_{i\in\mathbb{N}}$ converges in $L^1_{\mathrm{loc}}(M)$ to some set E of finite perimeter, $\{E_i^d\}_{i\in\mathbb{N}}$ is diverging, and $|E_i^c|+|E_i^d| = v$. As M has finite volume, we have $\lim_{i\to\infty} |E_i^d| = 0$, and so $\lim_{i\to\infty} |E_i^c| = |E| = v$. By 2 and 3 in Theorem 4.21, we get

$$P(E) = \lim_{i\to\infty} P(E_i^c) \leqslant \lim_{i\to\infty} \big(P(E_i^c) + P(E_i^d)\big) = I_M(v).$$

Hence, E is an isoperimetric region of volume v. □

4.4 Existence of Isoperimetric Sets Under a Cocompact Action of the Isometry Group

Let $\mathrm{Isom}(M)$ be the group of isometries of the Riemannian manifold M. We denote by $M/\mathrm{Isom}(M)$ the quotient topological space. When this quotient is compact, we say that the action of $\mathrm{Isom}(M)$ on M is *cocompact*. When the action of $\mathrm{Isom}(M)$ on M is cocompact, there exists a compact subset $K \subset M$ such that every point in M can be mapped to a point in K by means of an isometry of M. We prove in this section that isoperimetric sets always exist under the assumption of a cocompact action of the isometry group. This result can be found on page 133 in Morgan [307].

Definition 4.24 A Riemannian manifold M is *cocompact* if there exists a compact set $K \subset M$ such that

$$M = \bigcup_{f \in \mathrm{Isom}(M)} f(K).$$

A similar notion for metric spaces is defined by Bridson and Haefliger, see p. 202 in [79].

Theorem 4.25 *Let M be a complete cocompact Riemannian manifold. Then isoperimetric regions exist on M for any $0 < v < |M|$.*

Our proof of Theorem 4.25 follows the approach by Leonardi and Rigot [266] as in Galli and Ritoré [175].

For the proof of Theorem 4.25, some preliminaries are needed. First of all, we realize that, given a compact set $K \subset M$, there are positive constants ℓ, L, r_0, such that M is *Ahlfors-regular*. This means

$$\ell r^m \leqslant |B(x,r)| \leqslant L r^m, \tag{4.16}$$

for all $x \in K$, $0 < r < r_0$. This follows from standard comparison results for the volume of geodesic balls in M such as in Sect. 1.3.5, or even from the more elementary approach in Sect. 1.2.5. In particular, inequalities (4.16) imply the *doubling property* for the volume of balls in M: given a compact set $K \subset M$, there are positive constants C, r_0 such that

$$|B(x,2r)| \leqslant C|B(x,r)|, \tag{4.17}$$

for all $x \in K$, $0 < r < r_0$.

We shall also need the relative isoperimetric inequality for balls in Lemma 1.46: given a compact subset $K \subset M$, there exists constants $C_I > 0$, $r_0 > 0$, only depending on K, so that, for any set $E \subset M$ with locally finite perimeter, we have

$$P(E, B(x,r)) \geqslant C_I \min\left\{|E \cap B(x,r)|, |B(x,r) \setminus E|\right\}^{(m-1)/m},$$

for any $x \in K$.

We observe that, since $M/\operatorname{Isom}(M)$ is compact, there is no dependence on K so that we have an Ahlfors property, a doubling constant, and a local relative isoperimetric inequality in all of M although only for balls of radii no larger than $r_0 > 0$.

4.4.1 An Isoperimetric Inequality for Small Volumes

One of the first consequences of the existence of a doubling constant and a relative isoperimetric inequality is the following isoperimetric inequality for small volumes

Lemma 4.26 (Isoperimetric Inequality for Small Volumes) *Let M be a complete unbounded cocompact Riemannian manifold. Then there exists $v_0 > 0$ and $C_I > 0$ such that*

$$P(E) \geqslant C_I |E|^{(m-1)/m}, \tag{4.18}$$

for any finite perimeter set $E \subset M$ with $|E| < v_0$.

Proof This is a classical argument (see Lemma 4.1 in Leonardi and Rigot [266]). We fix $r_0 > 0$ small enough so that a uniform relative isoperimetric inequality holds for balls of radius smaller than or equal to r_0. Since M is cocompact, the Ahlfors property implies the existence of $v_0 > 0$ so that $|B(x, r_0)| \geqslant 2v_0$ for all $x \in M$. Let $E \subset M$ be a set of finite perimeter with $|E| < v_0$. We fix a maximal family of points $\{x_i\}_{i \in \mathbb{N}}$ with the properties

$$d(x_i, x_j) \geqslant \frac{r_0}{2} \quad \text{for } i \neq j, \qquad E \subset \bigcup_{i \in \mathbb{N}} B(x_i, r_0). \tag{4.19}$$

Letting $q = (m-1)/m$ we have

$$\begin{aligned} |E|^q &\leqslant \left(\sum_{i \in \mathbb{N}} |B(x_i, r_0) \cap E|\right)^q \leqslant \sum_{i \in \mathbb{N}} |B(x_i, r_0) \cap E|^q \\ &\leqslant C_1 \sum_{i \in \mathbb{N}} P(E, B(x_i, r_0)) \end{aligned} \tag{4.20}$$

from (4.19), the concavity of the function $x \mapsto x^q$, and the relative isoperimetric inequality in Lemma 1.46. Note that $|B(x_i, r_0) \cap E| \leqslant |B(x_i, r_0) \setminus E|$ since $|E| < v_0$ and $|B(x_i, r_0)| \geqslant 2v_0$.

For $z \in M$, we define $A(z) = \{x_i : z \in B(x_i, r_0)\}$, so that $B(x, r_0/4) \subset B(z, 2r_0)$ and $B(z, r_0/4) \subset B(x, 2r_0)$ for $x \in A(z)$. Since the balls $B(x_i, r_0/4)$ are disjoint by (4.19), we get

$$\#A(z) \min_{x \in A(z)} |B(x, r_0/4)| \leqslant \left| \bigcup_{x \in A(z)} B(x, r_0/4) \right| \leqslant |B(z, 2r_0)|. \tag{4.21}$$

On the other hand, since $B(z, r_0/4) \subset B(x, 2r_0)$, we have

$$|B(x, r_0/4)| \geqslant C_D^{-3} |B(z, r_0/4)|, \tag{4.22}$$

where $C_D > 0$ is the doubling constant. We conclude from (4.21) and (4.22) that

$$\#A(z) \leqslant C_D^6,$$

and so

$$\sum_{i \in \mathbb{N}} P(E, B(x_i, r_0)) \leqslant C\, P(E).$$

This inequality together with (4.20) implies (4.18). □

4.4.2 Boundedness of Isoperimetric Sets

One of the main consequences of the existence of an isoperimetric inequality for small volumes is the boundedness of isoperimetric regions. We follow here an argument by F. Morgan [307]. A key role in the proof is also played by the deformation result in Theorem 1.50.

Lemma 4.27 (Boundedness of Isoperimetric Regions) *Let M be a complete unbounded Riemannian manifold so that an isoperimetric inequality*

$$P(E) \geqslant C_I |E|^{(m-1)/m},$$

holds for any finite perimeter set $E \subset M$ of volume $|E|$ less that some given constant v_0. Let $E \subset M$ be a set minimizing perimeter under a volume constraint. Then E is bounded.

Proof We fix $p \in M$ and, for all $r > 0$, denote the ball $B(p, r)$ by $B(r)$. We let $V(r) = |E \setminus B(r)|$, so that $V(r) \to 0$ when $r \to \infty$ since E has finite volume.

Let us assume that $V(r) > 0$ for all $r > 0$. Applying the isoperimetric inequality for small volumes when r is large enough to the set $E \cap (M \setminus B(r))$, we get, taking $q = (m-1)/m$,

$$\begin{aligned} C_I V(r)^q &\leqslant P(E \setminus B(r)) \\ &\leqslant P(E, M \setminus \overline{B}_r) + P(E \cap B(r), \partial B(r)) \\ &\leqslant P(E, M \setminus \overline{B}_r) + |V'(r)| \\ &\leqslant P(E) - P(E, B(r)) + |V'(r)|. \end{aligned} \tag{4.23}$$

We now fix some $r_0 > 0$. For $r > r_0$, the deformation result in Theorem 1.50 implies the existence of a set E_r so that E_r is a small deformation of $E \cap B(r)$, $E_r \setminus (E \cap B(r))$ is properly contained in $B(r_0)$, $|E_r| = |E|$ (which implies $|E \setminus E_r| = V(r)$), and $P(E_r, B(r)) \leqslant P(E, B(r)) + C\,V(r)$. So we have

$$\begin{aligned} P(E_r) &\leqslant P(E_r, B(r)) + P(E_r \cap B(r), \partial B(r)) \\ &= P(E_r, B(r)) + P(E \cap B(r), \partial B(r)) \\ &\leqslant P(E_r, B(r)) + |V'(r)|. \end{aligned} \tag{4.24}$$

By the isoperimetric property of E, we also have

$$P(E) \leqslant P(E_r), \tag{4.25}$$

for all $r \geqslant r_0$.

From (4.23), (4.24), and (4.25), we finally get

$$C_I V(r)^q \leqslant C\,V(r) + 2\,|V'(r)|. \tag{4.26}$$

Since $V(r) = V(r)^{1-q} V(r)^q \leqslant (C_I/2C)\,V(r)^q$ for r large enough, we get

$$-\frac{C_I}{2}\,V(r)^q \geqslant 2\,V'(r),$$

as $V'(r) \leqslant 0$ or, equivalently,

$$(V^{1/m})' = \frac{V'}{m V^{(m-1)/m}} \leqslant -\frac{C_I}{4m} < 0,$$

which forces $V(r)$ to be negative for r large enough. This contradiction proves the result. □

4.4.3 Concentration of Mass

Finally, in the proof of Theorem 4.25, we shall need the following result. Essentially means that a set with finite perimeter and measure not concentrated in small balls of fixed radius must satisfy a linear isoperimetric inequality.

Lemma 4.28 (Concentration Lemma) *Let M be a complete unbounded cocompact Riemannian manifold. Let $E \subset M$ be a set with finite perimeter and positive measure. Let $\mu \in (0, \inf_{x\in M} |B(x, r_0)|/2)$, where $r_0 > 0$ is the radius for which the relative isoperimetric inequality holds. Assume that $|E \cap B(x, r_0)| < \mu$ for all $x \in M$. Then*

$$C|E|^m \leqslant \mu P(E)^m, \tag{4.27}$$

for some constant $C > 0$ that only depends on the local isoperimetric C_I and the doubling constant C_D.

Proof We closely follow the proof of Lemma 4.1 in [266]. We consider a maximal family of points $\mathcal{A}$ in M so that

- $d(x, x') \geqslant r_0/2$ for all different points $x, x' \in \mathcal{A}$,
- $|E \cap B(x, r_0/2)| > 0$ for all $x \in \mathcal{A}$.

Then $\bigcup_{x\in\mathcal{A}} B(x, r_0)$ cover almost all of E, and we have, taking $q = (m-1)/m$,

$$\begin{aligned} |E| &\leqslant \sum_{x\in\mathcal{A}} |E \cap B(x, r_0)| \leqslant \mu^{1/m} \sum_{x\in\mathcal{A}} |E \cap B(x, r_0)|^q \\ &\leqslant \mu^{1/m} C_I \sum_{x\in\mathcal{A}} P(E, B(x, r_0)). \end{aligned}$$

The second inequality follows since $(1/m) + q = 1$ and $|E \cap B(x, r_0)| < \mu$. The last one from the relative isoperimetric inequality since $|E \cap B(x, r_0)| < \mu \leqslant |B(x, r_0)|/2$. The overlapping of the balls $B(x, r_0)$, with $x \in \mathcal{A}$, is controlled in the same way as in the proof of Lemma 4.26 to conclude the proof. □

4.4.4 Proof of Existence

We give in this section the proof of Theorem 4.25.

Proof of Theorem 4.25 We fix a volume $0 < v < |M|$, and we consider a minimizing sequence $\{E_i\}_{i\in\mathbb{N}}$ of sets of volume v whose perimeters approach $I_M(v)$. By comparing with balls, it is not difficult to prove that, given $v > 0$, there exists a constant $C(v) > 0$ so that

$$I_M(w) \leqslant C(v)\, w^{(m-1)/m} \tag{4.28}$$

for all $w \in (0, v]$.

By Lemma 4.28, for any $\mu > 0$ such that $\mu v < \inf_{x\in M} |B(x, r_0)|/2$, there exists a constant $C > 0$, depending on the local isoperimetric and doubling constants, such that

$$C|E|^m \leqslant (\mu|E|)P(E)^m$$

for any finite perimeter set $E \subset M$ of volume at most v satisfying $|E \cap B(x, r_0)| < \mu|E|$ for all $x \in M$. Hence,

$$P(E) \geqslant \left(\frac{C}{\mu}\right)^{1/m} |E|^{(m-1)/m}. \tag{4.29}$$

Taking $\mu_0 > 0$ small enough so that

$$\left(\frac{C}{\mu_0}\right)^{1/m} > 2C(v)$$

we conclude from (4.29) and (4.28) that for any E with $0 < |E| \leqslant v$,

$$P(E) \geqslant 2I_M(|E|). \tag{4.30}$$

Inequality (4.30) implies that, for i large enough, the sets in the minimizing sequence $\{E_i\}_{i\in\mathbb{N}}$ cannot satisfy the property $|E \cap B(x, r_0)| < \mu_0|E|$ for all $x \in M$. So we can choose points $x_i \in M$ such that

$$|E_i \cap B(x_i, r_0)| \geqslant \mu_0|E_i| = \mu_0 v,$$

for i large enough.

We may assume that the sequence $\{x_i\}_{i\in\mathbb{N}}$ is contained in a compact subset of M and that it converges to some point $x_0 \in M$. This can be done simply by replacing the original minimizing sequence $\{E_i\}_{i\in\mathbb{N}}$ by another one $\{f_i(E_i)\}_{i\in\mathbb{N}}$, with $f_i \in \operatorname{Isom}(M)$, so that the points $f_i(x_i)$ are bounded, and by passing to a subsequence.

By Theorem 4.21, we may assume that $\{E_i\}_{i\in\mathbb{N}}$ converges in $L^1_{\mathrm{loc}}(M)$ to some finite perimeter set E. Moreover

$$\mu_0 v \leqslant \liminf_{i\to\infty} |E_i \cap B(x_0, r_0)| = |E \cap B(x_0, r_0)|,$$

and

$$|E| \leqslant \liminf_{i\to\infty} |E_i| = v.$$

So we have proven the following fact: from every minimizing sequence of sets of volume $v > 0$, one can produce, suitably applying isometries of M to each member of the sequence, a new minimizing sequence which converges in $L^1_{\mathrm{loc}}(M)$ to some finite perimeter set E with $\mu_0 v \leqslant |E| \leqslant v$, where $\mu_0 > 0$ is a universal constant that only depends on v, C_I, and C_D. Hence, a fraction of the total volume is captured by the minimizing sequence.

By Theorem 4.21, the set E is isoperimetric for volume $|E|$ and hence bounded by Lemma 4.27. We consider the sequences $\{E_i^c\}_{i\in\mathbb{N}}$, $\{E_i^d\}_{i\in\mathbb{N}}$ obtained in Theorem 4.21 so that $E_i^c \to E$ and E_i^d diverges.

If $|E| = v$ we are done since $P(E) \leqslant \liminf_{i\to\infty} P(E_i) = I_M(|E|)$, and hence E is an isoperimetric region. So assume that $|E| < v$. In this case, we claim that

$$\lim_{i\to\infty} P(E_i^d) = I_M(v - |E|). \tag{4.31}$$

Otherwise $\lim_{i\to\infty} P(E_i^d) > I_M(v - |E|)$. This is obtained by making small adjustments so that the volume of the sets E_i^d is equal to v and passing to the limit. Consider a minimizing sequence $\{E_i'\}_{i\in\mathbb{N}}$ for volume $v - |E|$. By applying isometries of M to the sets E_i', we assume that $E \cap E_i' = \emptyset$ for all i. Then the sequence $\{E \cup E_i'\}_{i\in\mathbb{N}}$ is composed of sets of volume v and

$$\begin{aligned}\lim_{i\to\infty} P(E \cup E_i') &= P(E) + \lim_{i\to\infty} P(E_i') \\ &= I_M(v) + I_M(v - |E|) < P(E) + \lim_{i\to\infty} P(E_i^d) = \lim_{i\to\infty} P(E_i).\end{aligned}$$

This is a contradiction to the fact that $\{E_i\}_{i\in\mathbb{N}}$ is a minimizing sequence and proves (4.31). Hence,

$$I_M(|E|) + I_M(v - |E|) = I_M(v).$$

We have completed the first step of an induction process. We let $F_0 = E$ and modify slightly E_i^d so that the sets E_i^d have volume $v - |E|$. Now we apply the previous argument to the minimizing sequence $\{E_i^d\}_{i\in\mathbb{N}}$ for volume $v - |E|$. Suitably translating the sets E_i^d we obtain another isoperimetric region F_1 of volume

$$v - |F_0| \geqslant |F_1| \geqslant \mu_0(v - |F_0|),$$

and a new diverging minimizing sequence for volume $v - |F_0| - |F_1|$.

By induction we get a sequence of isoperimetric regions $\{F_i\}_{i\in\mathbb{N}}$ so that the volume of F_i satisfies

$$|F_i| \geqslant \mu_0\Big(v - \sum_{j=0}^{i-1} |F_j|\Big).$$

Hence we have

$$v \geqslant \sum_{i=0}^{k} |F_i| \geqslant (k+1)\mu_0 v - k\mu_0 \sum_{i=0}^{k-1} |F_i| \geqslant (k+1)\mu_0 v - k\mu_0 \sum_{i=0}^{k} |F_i|,$$

and so

$$v \geqslant \sum_{i=0}^{k} |F_i| \geqslant \frac{(k+1)\mu_0 v}{1+k\mu_0}.$$

Taking limits when $k \to \infty$, we get

$$\lim_{k\to\infty} \sum_{i=0}^{k} |F_i| = v.$$

Moreover,

$$\sum_{i=0}^{\infty} P(F_i) = I_M(v).$$

Each region F_i is bounded, so that we can place them in M using isometries so that they are at positive distance (each one contained in an annulus centered at some given point). Hence, $F = \bigcup_{i=0}^{\infty} F_i$ is an isoperimetric region of volume v. In fact, F must be bounded by Lemma 4.27, so we only need a finite number of steps to recover all the volume. □

4.4.5 Some Properties of the Isoperimetric Profile

The existence Theorem 4.25 implies that many of the properties of the isoperimetric profile in compact manifolds proven in Lemma 3.19 and Theorem 3.20 are also valid when the action of the isometry group is cocompact. We first observe that the isoperimetric profile of complete unbounded cocompact manifold is continuous.

Theorem 4.29 *Let M be a complete unbounded cocompact manifold. Then I_M is locally Hölder continuous with exponent $(m-1)/m$.*

Proof We take $r_0 > 0$ so that a uniform local isoperimetric inequality, uniform Ahlfors estimates, and the uniform isoperimetric inequality $P(B(z,r)) \leqslant C|B(z,r)|^{(m-1)/m}$ hold for balls of radius $0 < r \leqslant r_0$ with $z \in M$ arbitrary.

For any $v > 0$ we take an isoperimetric set E_v of volume v, and since E_v is bounded, a ball $B(x, r_0)$ disjoint from E. Let

$$\varepsilon_1 < \inf_{y \in M} |B(y, r_0)|.$$

For $0 < \varepsilon \leqslant \varepsilon_1$, there exists $0 < s \leqslant r_0$ such that $|B(x, s)| = \varepsilon$. Hence

$$\begin{aligned} I_M(v + \varepsilon) &\leqslant P(E_v \cup B(x, s)) \\ &= P(E) + P(B(x, s)) \\ &= I_M(v) + P(B(x, s)) \leqslant I_M(v) + C\varepsilon^{(m-1)/m}. \end{aligned} \tag{4.32}$$

This inequality is valid for any $v > 0$ and $0 < \varepsilon \leqslant \varepsilon_1$.

By Eq. (4.28), given $v_1 > 0$, there exists a constant $C(v_1)$ depending on v_1 such that

$$I_M(v) \leqslant C(v_1) v^{(m-1)/m}$$

for all $0 < v \leqslant v_1$.

By Lemma 4.28, for any $\mu > 0$ such that $\mu v < \inf_{x \in M} |B(x, r_0)|/2$, there exists a constant $C' > 0$, depending on the local isoperimetric and doubling constants, such that

$$C'|E|^m \leqslant (\mu|E|) P(E)^m$$

for any finite perimeter set $E \subset M$ of volume at most v_1 satisfying $|E \cap B(x, r_0)| < \mu|E|$ for all $x \in M$. Hence

$$P(E) \geqslant \left(\frac{C'}{\mu}\right)^{1/m} |E|^{(m-1)/m}.$$

If $E \subset M$ is an isoperimetric set of volume $0 < v \leqslant v_1$, then we have

$$C'(v_1) \, |E|^{(m-1)/m} \geqslant \left(\frac{C'}{\mu}\right)^{1/m} |E|^{(m-1)/m}.$$

Taking $\mu_0 > 0$ small enough so that $\mu_0 v < \inf_{x \in M} |B(x, r_0)|/2$ and

$$\left(\frac{C'}{\mu_0}\right)^{1/m} > C'(v_1)$$

we conclude that, for any isoperimetric set E with $0 < |E| \leqslant v_1$, there exists a point $x \in M$, depending on E, such that

$$|E \cap B(x, r_0)| \geqslant \mu_0 |E|.$$

Fix some $v_0 > 0$ and take $\varepsilon_0 > 0$ such that $\varepsilon_0 \in (0, \mu_0 v_0]$. For any isoperimetric set E_v of volume $v_0 < v < v_1$, there exists a point $x_v \in M$ so that $|E_v \cap B(x_v, r_0)| \geqslant \mu_0 v_0 \geqslant \varepsilon_0$. Then, for every $0 < \varepsilon \leqslant \varepsilon_0$, there exists $0 < s_v \leqslant r_0$ such that $|E \setminus B(x_v, s_v)| = \varepsilon$. Hence

$$I_M(v - \varepsilon) \leqslant P(E \setminus B(x_v, s_v)) = P(E) + P(B(x_v, s_v)) \leqslant I_M(v) + C\varepsilon^{(m-1)/m} \tag{4.33}$$

for all $v_0 < v < v_1$, $0 < \varepsilon \leqslant \varepsilon_0$.

Inequalities (4.32) and (4.33) imply the local Hölder continuity of the profile. □

Theorem 4.30 *Let (M, g) be a complete unbounded cocompact Riemannian manifold. Let $I = I_M$ be the isoperimetric profile of M. Let $\delta \in \mathbb{R}$ such that* $\operatorname{Ric} \geqslant (m-1)\delta$. *Then we have*

1. *If $1 \leqslant \alpha \leqslant m/(m-1)$, then I^α satisfies the differential inequality*

$$(I^\alpha)'' \leqslant -\alpha(m-1)\delta(I^\alpha)^{(\alpha-2)/\alpha}, \tag{4.34}$$

   in weak sense in $(0, |M|)$.
2. *The isoperimetric profile is locally the sum of a concave function and a smooth function. Hence, it is differentiable almost everywhere, left and right derivatives I'_-, I'_+ exist everywhere, and $I'_- \geqslant I'_+$. Moreover*

$$I(w) - I(v) = \int_v^w I'_+(\xi)\, d\xi = \int_v^w I'_-(\xi)\, d\xi. \tag{4.35}$$

3. *If E is an isoperimetric region of volume v_0 with boundary mean curvature H then*

$$I'_+(v_0) \leqslant H \leqslant I'_-(v_0).$$

   In particular, if I is regular at v_0, then all isoperimetric regions of volume v_0 have boundary mean curvature $H = I'(v_0)$.
4. *If I is not regular at v_0, then there exist two isoperimetric regions in M of volume v_0 with boundary mean curvatures $I'_+(v_0)$ and $I'_-(v_0)$.*
5. *If v is a regular value of I, let $H(v)$ be the boundary mean curvature of any isoperimetric set of volume v. Then*

$$I(w) - I(v) = \int_v^w H(\xi)\, d\xi \tag{4.36}$$

   for all $0 < v < w < |M|$.

Proof The differential inequality (4.34) follows since isoperimetric sets exist by Theorem 4.25, the Ricci curvature is bounded below on M, and the isoperimetric profile is locally bounded above by some positive constant (we simply estimate the profile by the perimeter of metric balls). Then we can follow the steps of the proof of Theorem 3.13 word for word.

To prove 2 we use exactly the same arguments as in the proof of Lemma 3.19.

Items 3 and 5 are proven like items 1 and 3 in Theorem 3.20.

Finally, we prove item 4. Arguing as in the beginning of the proof of Theorem 4.25 in Sect. 4.4.4 we obtain the following fact: given $v > 0$, there exists $\mu_0 > 0$ depending on v such that for any isoperimetric region E of volume $|E| \leqslant v$, there exists a point $x \in M$ satisfying

$$|E \cap B(x, r_0)| \geqslant \mu_0 \, |E|.$$

As a consequence of this fact, if $\{E_i\}_{i\in\mathbb{N}}$ is a sequence of isoperimetric sets of volumes v_i so that $v_0 = \lim_{i\to\infty} v_i$, suitably translating the sets E_i, we may assume that the $L^1_{\mathrm{loc}}(M)$ limit E of E_i has positive volume. If we can prove that E is part of an isoperimetric region of volume v_0, then 4 follows since some regular point at ∂E would be a limit of regular points of ∂E_i, and we argue as in the proof of Theorem 3.20(2). Of course it is enough to consider the cases $v_i < v_0$ for all $i \in \mathbb{N}$ and $v_i > v_0$ for all $i \in \mathbb{N}$.

To prove that E is part of an isoperimetric region of volume v_0 in case $v_i < v_0$ for all $i \in \mathbb{N}$, we use that the isoperimetric sets E_i are bounded to obtain a sequence of balls $B_i \subset M$ so that

- B_i is disjoint from E_i,
- $|E_i \cup B_i| = v_0$
- $\{B_i\}_{i\in\mathbb{N}}$ is a divergent sequence.

Observe that $|B_i| = v_0 - v_i \to 0$ so that $P(B_i) \to 0$. Let $F_i = E_i \cup B_i$. Then

$$P(F_i) = P(E_i) + P(B_i) = I(v_i) + P(B_i)$$

and as $P(B_i) \to 0$ and I is continuous by Theorem 4.29, we obtain that $P(F_i) \to I(v_0)$. Hence, $\{F_i\}_{i\in\mathbb{N}}$ is a minimizing sequence for volume v_0. The proof of Theorem 4.25 then implies that E is part of an isoperimetric region of volume v_0.

We finally assume $v_i > v_0$ for all $i \in \mathbb{N}$. If $|E| = v_0$, then we know from Theorem 4.25 that E is an isoperimetric set of volume v_0, and we are done. So we assume $|E| < v_0$. As E is the $L^1_{\mathrm{loc}}(M)$ limit of $\{E_i\}_{i\in\mathbb{N}}$, we use Proposition 4.22 to decompose each E_i as $E_i^c \cup E_i^d$, where E is the $L^1_{\mathrm{loc}}(M)$ limit of $\{E_i^c\}_{i\in\mathbb{N}}$ and $\{E_i^d\}_{i\in\mathbb{N}}$ is a divergent sequence. We have

$$|E_i^d| = v_i - |E_i^c| > v_i - v_0 > 0$$

for i large since $|E_i^c| \to |E| < v_0$. For i large enough, we choose a sequence $\{B_i\}_{i\in\mathbb{N}}$ of concentric balls with $|B_i| \to 0$ so that $(E \setminus B_i) \cup E_i^d$ has volume v_0. Let $F_i = (E \setminus B_i) \cup E_i^d$. As

$$\begin{aligned}\lim_{i\to\infty} P(F_i) &= \lim_{i\to\infty} P(E \setminus B_i) + P(E_i^d)\\ &= \lim_{i\to\infty} P(E) + P(E_i^d) = \lim_{i\to\infty} P(E_i) = \lim_{i\to\infty} I(v_i) = I(v_0),\end{aligned}$$

the sequence $\{F_i\}_{i\in\mathbb{N}}$ is minimizing for volume v_0. As E is the L^1_{loc} limit of $\{F_i\}_{i\in\mathbb{N}}$, we get from Theorem 4.25 that E is part of an isoperimetric set of volume v_0.

Finally (4.36) follows from (4.35) and 3. □

4.5 The Isoperimetric Profile of the Euclidean and Hyperbolic Spaces

In this section, we prove that the geodesic balls are the only isoperimetric sets in Euclidean and hyperbolic spaces. The proof follows from the existence Theorem 4.25 and a slight adaptation of Montiel and Ros proof of Alexandrov's Theorem (i.e., that geodesic spheres are the only compact embedded hypersurfaces with constant mean curvature in Euclidean space) (see [300] and also [298]). The corresponding result for the sphere was proven in Sect. 3.4.2.

We first recall the well-known relation between the Laplacian on a Riemannian manifold M and on a hypersurface S of M.

Lemma 4.31 *Let $S \subset M$ be a hypersurface in a Riemannian manifold M and $f \in C^\infty(M)$. Then*

$$\Delta_S f|_S = \Delta f - \nabla^2 f(N, N) - N(f)H, \tag{4.37}$$

where N is a unit normal to S and H is the mean curvature of S with respect to N.

Proof On S we have $\nabla_S f|_S = \nabla f - N(f)N$. Taking $p \in S$ and an orthornormal basis $\{e_i\}_{i=1,\dots,m-1}$ of T_pS, we obtain

$$\begin{aligned}(\Delta_S f|_S)(p) &= \sum_{i=1}^{m-1} \langle \nabla_{e_i} \nabla_S f|_S, e_i\rangle = \sum_{i=1}^{m-1} \langle \nabla_{e_i} \big(\nabla f - N(f)N\big), e_i\rangle\\ &= \big(\Delta f - \nabla^2 f(N, N) - N(f)H\big)(p).\end{aligned}$$

□

4.5.1 The Isoperimetric Profile of $\mathbb{R}^m$

If d is the distance function to some fixed point $p = (p_1, \ldots, p_m) \in \mathbb{R}^m$ then

$$\nabla \tfrac{1}{2} d^2 = \sum_{i=1}^{m} (x_i - p_i) \frac{\partial}{\partial x_i}$$

is the radial vector field with center p. Moreover $\nabla \frac{1}{2} d^2 = d \nabla d$ satisfies the property

$$\nabla_e \big(\nabla \tfrac{1}{2} d^2 \big) = e \tag{4.38}$$

for any $e \in T_q \mathbb{R}^m$ and $q \in \mathbb{R}^m$. This is trivial for $e = \nabla d$ since $\nabla d(d) = 1$ and $\nabla_{\nabla d} \nabla d = 0$. This implies (4.38) at p. For a point different from p and e orthogonal to ∇d, we have $e(d) = 0$ and $\nabla_e \nabla d = (1/d)\, e$, as $1/d$ is the principal curvature of a sphere of radius d. This implies (4.38) at points different from p. Hence, $\nabla(\frac{1}{2} d^2)$ is a *conformal vector field* in Montiel's terminology [298]. In particular, this implies that the Hessian of $\frac{1}{2} d^2$ is proportional to the Riemannian metric, that is,

$$\nabla^2 \big(\tfrac{1}{2} d^2 \big)(e, v) = \langle e, v \rangle, \tag{4.39}$$

and we also have

$$\tfrac{1}{2} \Delta d^2 = m.$$

We prove now a Minkowski type formula for isoperimetric boundaries.

Lemma 4.32 (Generalized First Minkowski Formula in $\mathbb{R}^m$) *Let $\Omega \subset \mathbb{R}^m$ be a bounded set with boundary $\partial \Omega$ composed of a smooth hypersurface S of finite area and mean curvature H with respect to the outer unit normal N and a singular set S_0 with $\mathcal{H}^{m-3}(S_0) = 0$. Then we have*

$$\int_S \big\{ (m-1) - H \,\langle \nabla \tfrac{1}{2} d^2, N \rangle \big\}\, dS = 0, \tag{4.40}$$

where d is the distance function to an arbitrary fixed point in $\mathbb{R}^m$.

Proof of Lemma 4.32 Take $\varphi \in C_0^\infty(S)$, a smooth function with compact support in S and let $f = \frac{1}{2} d^2$. By the divergence theorem,

$$\begin{aligned} 0 = \int_S \Delta_S(\varphi f)\, dS &= \int_S \big(f \Delta_S \varphi + \varphi \Delta_S f + 2 \langle \nabla_S f, \nabla_S \varphi \rangle \big)\, dS \\ &= \int_S \big(\varphi \Delta_S f + \langle \nabla_S f, \nabla_S \varphi \rangle \big)\, dS. \end{aligned} \tag{4.41}$$

Moreover

$$\begin{aligned}\left|\int_S \langle\nabla_S f, \nabla_S\varphi\rangle\, dS\right| &\leqslant \left(\int_S |\nabla_S f|^2 dS\right)^{1/2}\left(\int_S |\nabla_S\varphi|^2 dS\right)^{1/2}\\ &\leqslant ||\nabla_S f||_\infty A(S)^{1/2}\left(\int_S |\nabla_S\varphi|^2 dS\right)^{1/2}.\end{aligned} \tag{4.42}$$

Now we take a sequence of functions $\{\varphi_i\}_{i\in\mathbb{N}}$ as in Lemma 1.61: the functions $\varphi_i : S \to [0, 1]$ form an increasing sequence converging pointwise to the constant function 1 on S, and $\int_S |\nabla_S\varphi_i|^2 dS$ converges to 0 when $i \to \infty$. Inserting the function φ_i in place of φ in (4.41), and taking limits when $i \to \infty$ using the properties of the sequence $\{\varphi_i\}_{i\in\mathbb{N}}$, the dominated convergence theorem, and (4.42), we get

$$\int_S \Delta_S f\, dS = 0.$$

Now we use formula (4.37) relating the Laplacian on $\mathbb{R}^m$ and on S and formula (4.39) to conclude

$$\int_S \left\{(m-1) - H\,\langle\nabla\tfrac{1}{2}d^2, N\rangle\right\} dS = 0,$$

thus providing (4.40). □

Minkowski formula (4.40) implies the following relation between the volume of E and its boundary area when S has constant mean curvature.

Corollary 4.33 *Let $\Omega \subset \mathbb{R}^m$ be a bounded set. Assume that its boundary $\partial\Omega$ is composed of a smooth hypersurface S of finite area and constant mean curvature H with respect to the outer unit normal N, and a singular set S_0 with $\mathcal{H}^{m-3}(S_0) = 0$. Then we have*

$$A(S) - \tfrac{H}{m-1}\, m|\Omega| = 0. \tag{4.43}$$

Proof It follows from (4.40) the fact that H is constant and the divergence formula

$$m|\Omega| = \int_\Omega \Delta\Big(\frac{1}{2}d^2\Big)\, d\mathbb{R}^m = \int_S \langle\nabla\tfrac{1}{2}d^2, N\rangle\, dS.$$ □

Now we prove the main result in this subsection following [300]. Observe that isoperimetric sets in $\mathbb{R}^m$ are bounded by Theorem 4.25. The mean curvature H of the regular part of boundary of an isoperimetric set must be positive with respect to the outer unit normal since we may find a ball B containing E with $p \in \partial B \cap \partial E$. We conclude that p must be a regular point of ∂E since the tangent cone of E at p is contained in a half-space. The maximum principle for the mean curvature equation then implies that H must be no smaller than the mean curvature of ∂B, which is strictly positive.

Theorem 4.34 *Let $E \subset \mathbb{R}^m$ an isoperimetric set, $S \subset \partial E$ the regular part of its boundary, and H the mean curvature of S with respect to the outer unit normal N. Then we have*

$$m|E| \leqslant \tfrac{m-1}{H} A(S). \tag{4.44}$$

Equality holds in (4.44) if and only if E is a geodesic ball.

Proof For every point $q \in E$, we can find a length-minimizing geodesic of unit speed $\gamma : [0, r] \to \mathbb{R}^m$ realizing the distance to ∂E. The ball $B(q, r)$ is contained in E, and its boundary is regular at $p = \gamma(0)$. Then the tangent cone of E at p contains a half-space, and so $p \in S$, and the geodesic γ is normal to S. Moreover r is no larger than the cut distance $c(p)$ along the geodesic, which is no larger than the focal distance, equal to

$$f(p) = \frac{1}{\kappa_m(p)},$$

where $\kappa_m(p)$ is the maximum of the principal curvatures of S at p with respect to the outer normal. Observe that

$$f(p) \leqslant \frac{m-1}{H}.$$

Since the Jacobian of the normal map in Euclidean space is given by $\prod_{i=1}^{m-1}(1-t\kappa_i)$, where κ_i are the principal curvatures of S, we have

$$|E| = \int_S \left\{ \int_0^{c(p)} \prod_{i=1}^{m-1} \left(1 - t\kappa_i(p)\right) dt \right\} dS(p).$$

The arithmetic-geometric inequality implies

$$\prod_{i=1}^{m-1} (1 - t\kappa_i) \leqslant \left(1 - t\tfrac{H}{m-1}\right)^{m-1}$$

and since $c(p) \leqslant (m-1)/H$ we have

$$|E| \leqslant \int_S \left\{ \int_0^{(m-1)/H} \left(1 - t\frac{H}{m-1}\right)^{m-1} dt \right\} dS(p) = \frac{m-1}{mH} A(S).$$

This implies (4.44).

If equality holds in (4.44), then S is totally umbilical almost everywhere (and so everywhere by the continuity of the principal curvatures), and we have equality almost everywhere of $c(p) = (m-1)/H$. Hence, each connected component of S is part of a sphere of radius $(m-1)/H$. Hence, S_0 is the empty set, and each connected component of E is a ball of the same radius $(m-1)/H$. But a configuration of more than one sphere cannot be an isoperimetric set since the perimeter of a ball in terms of its volume is a strictly concave function. □

Theorem 4.35 *The only isoperimetric sets in $\mathbb{R}^m$ are the geodesic balls.*

Proof Given an isoperimetric set $E \subset \mathbb{R}$, formula (4.43) implies that equality holds in inequality (4.44). By Theorem 4.34, the set E is a geodesic ball. □

4.5.2 The Isoperimetric Profile of $\mathbb{H}^m$

In this section, we show that the geodesic balls are the only isoperimetric sets in hyperbolic space. As in the Euclidean case, the proof follows from the Existence Theorem 4.25 and an adaptation of Montiel and Ros' arguments in [300]. The main difference with the Euclidean case is the use of the function $\cosh d$ instead of $\frac{1}{2}d^2$.

To simplify the notation and the computations, we shall assume that $\mathbb{H}^m$ is the hyperbolic space of constant sectional curvatures equal to -1. We fix a point $p \in \mathbb{H}^m$ and consider the distance d to this point. Then

$$\nabla_e\big(\nabla \cosh d\big) = (\cosh d)\, e \tag{4.45}$$

for any $e \in T_q\mathbb{H}^m$ and $q \in \mathbb{H}^m$. Note that this equality holds for $e = \nabla d$ since $\nabla \cosh d = (\sinh d)\nabla d$ and $\nabla d(\sinh d) = \cosh d$ and $\nabla_{\nabla d}\nabla d = 0$. If e is orthogonal to ∇d at some point different from p, then e is tangent to the totally umbilical geodesic sphere of principal curvatures $\coth d$. Hence, $e(\sinh d) = 0$ and $\nabla_e\nabla d = (\coth d)\, e$. This implies (4.45). In particular, the Hessian of $\cosh d$ is proportional to the Riemannian metric of $\mathbb{H}^m$, that is

$$\nabla^2\big(\cosh d\big)(e, v) = (\cosh d)\langle e, v\rangle. \tag{4.46}$$

As in the Euclidean case, we have a Minkowski type formula.

Lemma 4.36 (Generalized First Minkowski Formula in $\mathbb{H}^m$) *Let $\Omega \subset \mathbb{H}^m$ be a bounded set with boundary $\partial\Omega$ composed of a smooth hypersurface S of finite area and mean curvature H with respect to the outer unit normal N and a singular set S_0 with $\mathcal{H}^{m-3}(S_0) = 0$. Then we have*

$$\int_S \big\{(m-1)\cosh d - H\,\langle \nabla \cosh d, N\rangle\big\}\, dS = 0, \tag{4.47}$$

where d is the distance function to an arbitrary point in $\mathbb{H}^m$.

Proof The proof follows the steps of the Euclidean one in Lemma 4.32 taking $f = \cosh d$. In the final step, we use (4.46) instead of (4.39). □

From Lemma 4.36 we obtain

Corollary 4.37 *Let $\Omega \subset \mathbb{H}^m$ be a bounded set with boundary $\partial\Omega$ composed of a smooth hypersurface S of finite area and constant mean curvature H with respect to the outer unit normal N and a singular set S_0 with $\mathcal{H}^{m-3}(S_0) = 0$. Then we have*

$$\int_S (\cosh d)\, dS - \tfrac{H}{m-1}\, m \int_\Omega (\cosh d)\, d\mathbb{H}^m = 0, \tag{4.48}$$

where d is the distance function to an arbitrary point in $\mathbb{H}^m$.

Proof It follows from (4.47) the fact that H is constant and formula

$$\int_S \langle \nabla \cosh d, N\rangle\, dS = \int_\Omega (\Delta \cosh d)\, d\mathbb{H}^m = m \int_\Omega (\cosh d)\, d\mathbb{H}^m. \qquad \square$$

Now we prove the main result in this subsection. Observe that isoperimetric sets in $\mathbb{H}^m$ are bounded by Lemma 4.26. Since ball of radius $r > 0$ in $\mathbb{H}^m$ have positive mean curvature $\coth r > 0$ with respect to the outer unit normal, we can reason as in the Euclidean case to prove that the mean curvature of the regular part of an isoperimetric region is strictly positive.

Theorem 4.38 *Let $E \subset \mathbb{H}^m$ an isoperimetric set, $S \subset \partial E$ the regular part of its boundary, and H the mean curvature of S with respect to the outer unit normal N. Then we have*

$$m \int_E (\cosh d)\, d\mathbb{H}^m \leqslant \tfrac{m-1}{H} \int_S (\cosh d)\, dS, \tag{4.49}$$

where d is the distance function to an arbitrary point in $\mathbb{H}^m$.

Equality holds in (4.49) if and only if E is a geodesic ball in $\mathbb{H}^m$.

Proof For every point $q \in E$, we can find a length-minimizing geodesic of unit speed $\gamma : [0, r] \to \mathbb{R}^m$ realizing the distance to ∂E. The ball $B(q, r)$ is contained in E, and its boundary is regular at $p = \gamma(0)$. Then the tangent cone of E at p contains a half-space, and so $p \in S$, and the geodesic γ is normal to S. Moreover, r is no larger than the cut distance $c(p)$, which is no larger than the focal distance of the geodesic, equal to

$$f(p) = \coth^{-1}(\kappa_{\max}(p)),$$

where $\kappa_{\max}(p)$ is the maximum of the principal curvatures of S at p with respect to the outer normal. Observe that

$$f(p) \leqslant \coth^{-1}\left(\tfrac{H}{m-1}\right).$$

Since the Jacobian of the normal map in the hyperbolic space is given by

$$\prod_{i=1}^{m-1}(\cosh t-\kappa_i\sinh t),$$

where κ_i are the principal curvatures of S, we have

$$\int_E \big(\cosh d\big)\,d\mathbb{H}^m=\int_S\left\{\int_0^{c(p)}(\cosh d)\prod_{i=1}^{m-1}(\cosh t-\kappa_i\sinh t)\big)dt\right\}dS(p).$$

The arithmetic-geometric inequality implies

$$\prod_{i=1}^{m-1}\big(\cosh t-\kappa_i\sinh t)\leqslant\left(\cosh t-\tfrac{H}{m-1}\sinh t\right)^{m-1}$$

and since $c(p)\leqslant\coth^{-1}(\frac{H}{m-1})$, we have

$$\begin{aligned}&\int_E\big(\cosh d\big)\,d\mathbb{H}^m\\&\leqslant\int_S\left\{\int_0^{\coth^{-1}(\frac{H}{m-1})}\big(\cosh d\big)\left(\cosh t-\tfrac{H}{m-1}\sinh t\right)^{m-1}dt\right\}dS(p).\end{aligned}$$

Let us express the function $\cosh d$ in terms of the parameter t. Let $\gamma:[0,c(p)]\to\mathbb{H}^m$ be the normal geodesic (parameterized by arc-length) starting from p. Let $f(t)=\cosh d(\gamma(t))$. We compute the derivatives

$$f'=\langle\nabla\cosh d,\gamma'\rangle,\qquad f''=\nabla^2\big(\cosh d\big)(\gamma',\gamma')=\cosh d(\gamma)=f.$$

Hence

$$\cosh d(\gamma(t))=f(t)=A\,\cosh t+B\,\sinh t,$$

where

$$A=\big(\cosh d\big)(p),\qquad B=\langle\nabla\cosh d,\gamma'(0)\rangle=-\langle\nabla\cosh d,N\rangle(p).$$

Writing

$$A\cosh t+B\sinh t=\big(A+\tfrac{H}{m-1}B\big)\cosh t+B\left(\sinh t-\tfrac{H}{m-1}\cosh t\right)$$

we express

$$\int_0^{\coth^{-1}(\frac{H}{m-1})} (\cosh d)\Big(\cosh t - \tfrac{H}{m-1}\sinh t\Big)^{m-1} dt = C(H)\big(A + \tfrac{H}{m-1}B\big) - \frac{B}{m},$$

where

$$C(H) = \int_0^{\coth^{-1}(\frac{H}{m-1})} \cosh t \left(\cosh t - \tfrac{H}{m-1}\sinh t\right)^{m-1} dt$$

is a positive constant only depending on H. The second term is obtained from the equality

$$B\int_0^{\coth^{-1}(\frac{H}{m-1})} \big(\sinh t - \tfrac{H}{m-1}\cosh t\big)\Big(\cosh t - \tfrac{H}{m-1}\sinh t\Big)^{m-1} dt = -\frac{B}{m}.$$

Since

$$\int_S -\frac{B}{m}\, dS = \int_E (\cosh d)\, d\mathbb{H}^m,$$

we finally obtain

$$0 \leqslant \int_S \big(A + \frac{H}{m-1}B\big)\, ds = \int_S \big\{(\cosh d) - \tfrac{H}{m-1}\langle \nabla \cosh d, N\rangle\big\}\, dS,$$

which is equivalent to (4.49).

If equality holds in (4.49), then S is totally umbilical almost everywhere (and so everywhere by the continuity of the principal curvatures), and we have equality $c(p) = \coth^{-1} H/(m-1)$ almost everywhere. Hence, each connected component of S is part of a sphere of radius $\coth^{-1} H/(m-1)$. Hence, the singular part S_0 of the isoperimetric boundary is empty, and each connected component of E is a ball of the same radius $\coth^{-1} H/(m-1)$.

But a configuration of more than one sphere cannot be an isoperimetric set since the function that assigns to a given volume v the perimeter $P(B(v))$ of the ball $B(v)$ of volume v is strictly concave. To check this, we let $f(v) = P(B(v))$. Note that $|B(v)|$ is also a smooth function of the radius t of the ball and $dv/dt = P(B(v))$. Then

$$\frac{df}{dv} = H(v), \qquad \frac{d^2 f}{dv^2} = \frac{dH}{dv} = \frac{dH/dt}{dv/dt} = \frac{dH/dt}{f(v)}.$$

By formula (1.41) for the variation of the mean curvature (or simply taking into account that $H(t) = (m-1)\coth(t)$), we have

$$\frac{dH}{dt} = (m-1) - (m-1)\frac{\cosh^2 t}{\sinh^2 t} = -\frac{(m-1)}{\sinh^2 t} < 0.$$

Hence $f(v) = P(B(v))$ is strictly concave with $f(0) = 0$, which implies $f(v_1 + v_2) < f(v_1) + f(v_2)$. Hence, a ball of volume $v_1 + v_2$ has strictly less perimeter than the disjoint union of two balls of volumes v_1 and v_2. This implies that E must be a single ball. □

We finally obtain the following.

Theorem 4.39 *The only isoperimetric sets in $\mathbb{H}^m$ are the geodesic balls.*

Proof Given an isoperimetric set $E \subset \mathbb{H}$, formula (4.48) implies that equality holds in inequality (4.49). By Theorem 4.38, the set E is a geodesic ball. □

4.6 Generalized Existence of Isoperimetric Sets

In this section, we prove a generalized existence result of isoperimetric sets in a certain class of manifolds. Here generalized means that the isoperimetric set may exist on a manifold related to the asymptotic geometry of M, but not on M.

4.6.1 *Manifolds with Bounded Lipschitz Geometry*

A *pointed* Riemannian manifold (M, g, p) is a Riemannian manifold (M, g) together with a fixed point $p \in M$.

Definition 4.40 A sequence of complete pointed Riemannian manifolds (M_i, g_i, p_i) is said to *converge in the pointed Lipschitz topology* to a pointed Riemannian manifold (M, g, p) if, for every $R > 0$, there exists an open set $\Omega \subset M$ containing $\overline{B}(p, R)$ and bilipschitz maps $f_i : \Omega \to f_i(\Omega) \subset M_i$ such that

1. $\lim_{i\to\infty} f_i^{-1}(p_i) = p$,
2. $\mathrm{dil}(f_i)$, $\mathrm{dil}(f_i^{-1})$ converge to 1.

A notion of convergence with higher regularity is presented by Petersen [345, §3.2]: a sequence of complete pointed Riemannian manifolds (M_i, p_i, g) is said to *converge in the pointed $C^{k,\alpha}$-topology*, $k \in \mathbb{N}$, $\alpha \in (0, 1)$, to a pointed Riemannian manifold (M, p, g) if for every $R > 0$, we can find a domain Ω in M containing $\overline{B}(p, R)$ and embeddings $f_i : \Omega \to M_i$ such that $f_i^{-1}(p_i) \to p$ and $f_i^* g_i$ converges to g in the $C^{k,\alpha}$ topology.

Definition 4.41 A complete non-compact Riemannian manifold (M, g) is of *bounded Lipschitz geometry* if the following conditions are satisfied:

1. There exists $r_0 > 0$ such that
   - a relative isoperimetric inequality in balls of radius smaller than r_0 holds,
   - a uniform doubling constant exists for balls of radius smaller than r_0.
   - $b(r_0) = \inf_{x \in M} |B(x, r_0)| > 0$.
2. For every divergent sequence of points $\{p_i\}_{i \in \mathbb{N}}$ in M, there exists a pointed Riemannian manifold $(M_\infty, g_\infty, p_\infty)$ such that a subsequence of (M, g, p) converges in pointed Lipschitz topology to $(M_\infty, g_\infty, p_\infty)$.

Any of the limits $(M_\infty, g_\infty, p_\infty)$ obtained this way is called an *asymptotic manifold* to M. The set of asymptotic manifolds to M is denoted by $\mathcal{A}(M)$.

Example 4.42 If (M, g) is a complete unbounded cocompact Riemannian manifold and $\{p_i\}_{i \in \mathbb{N}}$ is a diverging sequence in M, then there is a point $p_0 \in M$ such that (M, g, p_i) converges in pointed Lipschitz topology to (M, g, p). To prove this, we simply take a sequence of isometries $f_i : M \to M$ so that $f_i^{-1}(p_i)$ is a bounded sequence, and we extract a convergent subsequence to some point $p_0 \in M$.

Example 4.43 If (M, g) is a complete Riemannian manifold with cylindrical ends, then the asymptotic manifolds to M are Riemannian cylinders. Roughly speaking, an end of M is a connected component of the complement of a large compact set (Fig. 4.1).

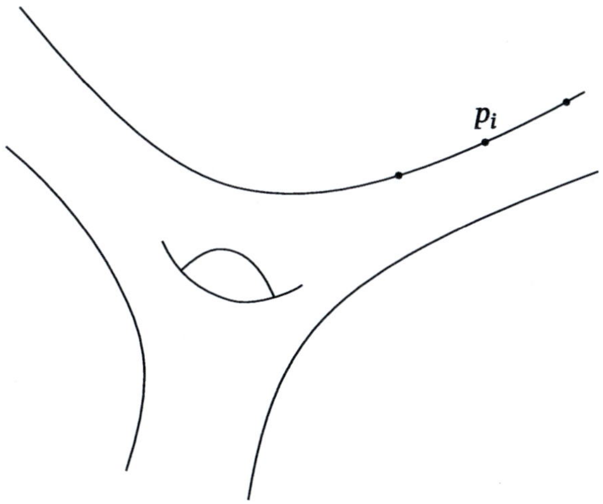

Fig. 4.1 A complete non-compact manifold with cylindrical ends. Its asymptotic manifolds are cylinders

Remark 4.44 If M is of bounded Lipschitz geometry and M_∞ is an asymptotic manifold to M, then a uniform relative isoperimetric inequality and a doubling constant hold in M_∞, possibly with a radius slightly smaller than r_0. Also the positive lower bound on the volume of balls of small radius holds. In particular, isoperimetric regions are bounded not only in M, but on any asymptotic manifold to M.

Definition 4.45 Let (M, g) be a complete non-compact Riemannian manifold of bounded Lipschitz geometry. We say that a finite family $E^0, E^1, \dots, E^k$ of sets of finite perimeter is a *generalized isoperimetric region in M if* $E^0 \subset M = M^0$, $E^i \subset M^i \in \mathcal{A}(M)$ for $i \geqslant 1$ and, for any family of sets $F^0, F^1, \dots, F^r$ such that $F^0 \subset M = N^0$, $F^i \subset N^i \in \mathcal{A}(M)$, for $i \geqslant 1$, and

$$\sum_{i=0}^{k} |E^i| = \sum_{i=0}^{r} |F^i|,$$

we have

$$\sum_{i=0}^{k} P_{M^i}(E^i) \leqslant \sum_{i=0}^{r} P_{N^i}(F^i).$$

From this definition, it follows easily that each set E^i is an isoperimetric region of volume $|E^i|$ in M^i.

For the proof of the main result in this section, Theorem 4.48, we need the following technical result.

Lemma 4.46 *Let (M_i, g_i, p_i) be a sequence of complete non-compact Riemannian manifolds converging in pointed Lipschitz topology to a pointed Riemannian manifold (M_0, g_0, p_0). Let $E_i \subset M_i$ be a sequence of sets with uniformly bounded perimeter and volumes v_i satisfying $v = \lim_{i\to\infty} v_i$.*

Then there exists a set $E_0 \subset M_0$ obtained as limit in $L^1_{\text{loc}}(M_0)$ of images of E_i under Lipschitz maps with constant close to 1 *and a sequence of increasing radii $\{r_i\}_{i\in\mathbb{N}}$ such that $E_i^d = E_i \setminus B(p_i, r_i)$ satisfy*

1. $|E_0| + \lim_{i\to\infty} |E_i^d| = v$,
2. $P_0(E_0) + \liminf_{i\to\infty} P_i(E_i^d) \leqslant \liminf_{i\to\infty} P_i(E_i)$.

Moreover, given any diverging sequence $\{r_i'\}_{i\in\mathbb{N}}$, the sequence $\{r_i\}_{i\in\mathbb{N}}$ can be taken so that each $r_i \geqslant r_i'$ for all i.

Remark 4.47 By abuse of terminology, we say that the sequence $E_i \subset M_i$ converges to $E \subset M_0$.

Proof of Lemma 4.46 For any $R > 0$, there is a sequence of bilipschitz maps $f_i^R : B_0(p_0, R) \to f_i^R(B_0(p_0, R)) \subset M_i$, where B_0 denotes the ball with respect to the metric g_0. Let $0 < \lambda_i^R \leqslant \Lambda_i^R$ be such that, for all $x, y \in B_0(p_0, R)$,

$$\lambda_i^R d_0(x, y) \leqslant d_i(f_i^R(x), f_i^R(y)) \leqslant \Lambda_i^R d_0(x, y).$$

By hypothesis, λ_i^R, Λ_i^R converge to 1 when i converges to ∞.

Now we consider a sequence of increasing radii $\{s_i\}_{i\in\mathbb{N}}$ converging to ∞ so that the balls $B_0(p_0, s_i)$ have Lipschitz boundary. This sequence can be taken larger than any other diverging sequence. Passing to a non-relabeled subsequence, we may assume that

$$d_i(p_i, f_i^{s_i}(p_0)) \leqslant 1/i, \quad \Lambda_i^{s_i} \leqslant 1 + 1/i, \quad \lambda_i^{s_i} \geqslant 1 - 1/i, \tag{4.50}$$

for all i. From now on, we let $f_i = f_i^{s_i}$ to simplify the notation. A direct application of the triangle inequality taking (4.50) into account implies

$$\begin{aligned} f_i(B_0(p_0, s_i)) &\subset B_i(p_i, s_i + \tfrac{1}{i}(1 + s_i)) = B_i(p_i, s_i''). \\ f_i(B_0(p_0, s_i)) &\supset B_i(p_i, (1 - \tfrac{1}{i}) s_i - \tfrac{i-1}{i^2}) = B_i(p_i, s_i'). \end{aligned} \tag{4.51}$$

Observe that $s_i' \leqslant s_i \leqslant s_i''$.

As $\{E_i\}_{i\in\mathbb{N}}$ has uniformly bounded perimeter, also the sequence of relative perimeters $P_0(f_j^{-1}(E_j), B_0(p_0, s_i))_j$ is uniformly bounded in j for all i. By a standard diagonal argument, we can then find a convergent non-relabeled subsequence of $\{E_i\}_{i\in\mathbb{N}}$ and a set $E_0 \subset M_0$ of locally finite perimeter so that

$$\int_{B_0(p_0, s_i)} |\mathbf{1}_{E_0} - \mathbf{1}_{f_i^{-1}(E_i)}| \, dM_0 \leqslant 1/i. \tag{4.52}$$

By the lower semicontinuity of the perimeter,

$$P_0(E_0) \leqslant \liminf_{i\to\infty} P_0(f_i^{-1}(E_i), B_0(p_0, s_i)).$$

Now we assume that the original sequence is taken so that

$$s_{i+1} - s_i \geqslant i$$

for all $i \in \mathbb{N}$. By the coarea formula

$$\int_{s_i'}^{s_{i+1}''} \mathcal{H}^{m-1}(E_i \cap \partial B_i(p_i, t)) \, dt \leqslant v_i,$$

where s_i', s_i'' are defined in (4.51). As $s_{i+1}'' - s_i' \geqslant s_{i+1} - s_i \geqslant i$, there exists $r_i \in [s_i', s_{i+1}'']$ such that

$$\mathcal{H}^{m-1}(E_i \cap \partial B_i(p_i, r_i)) \leqslant \frac{v_i}{i}.$$

We let $E_i^d = E_i \setminus B_i(p_i, r_i)$. Then we have

$$|E_i| = |E_i \cap B_i(p_i, r_i)| + |E_i^d|.$$

Taking into account (4.52) and the fact that the Lipschitz constants $\lambda_i^{s_i}$, $\Lambda_i^{s_i}$ converge to 1 we get $\lim_{i\to\infty} |E_i \cap B_i(p_i, r_i)| = |E_0|$. This way we prove 1.

To prove the inequality for perimeters, we observe that

$$P_i(E_i^d) \leqslant P_i(E_i, M \setminus B_i(p_i, r_i)) + \mathcal{H}^{m-1}(E_i \cap \partial B_i(p_i, r_i))$$

and so

$$\begin{aligned} P_i(E_i) &= P_i(E_i, B_i(p_i, r_i)) + P_i(E_i, M_i \setminus B_i(p_i, r_i)) \\ &\geqslant P_i(E_i, B_i(p_i, r_i)) + P_i(E_i^d) - \mathcal{H}^{m-1}(E_i \cap \partial B_i(p_i, r_i)). \end{aligned}$$

Taking inferior limits, we get

$$\liminf_{i\to\infty} P_i(E_i, B_i(p_i, r_i)) + \liminf_{i\to\infty} P_i(E_i^d) \leqslant \liminf_{i\to\infty} P_i(E_i).$$

Finally we observe that

$$P_0(E_0) \leqslant \liminf_{i\to\infty} P_i(E_i, B_i(p_i, r_i))$$

implies 2. □

4.6.2 *Existence of Isoperimetric Sets in Manifolds with Bounded Lipschitz Geometry*

We now prove our main result guaranteeing the existence of isoperimetric regions in manifolds of bounded Lipschitz geometry in a generalized sense: any miminizing sequence converges to a union of minimizing sets, one of which may be contained in M and the others in asymptotic manifolds of M. Item 3 in the statement is interpreted in the sense of Remark 4.47.

Theorem 4.48 (Existence of Generalized Isoperimetric Regions) *Let (M, g) be a complete non-compact Riemannian manifold of bounded Lipschitz geometry. Let $v_0 > 0$.*

Then there exists $\beta > 0$ and $\ell \in \mathbb{N}$, only depending on M and v_0, with the following property: for any minimizing sequence $\{F_i\}_{i\in\mathbb{N}}$ for volume v_0, one can find a (non-relabeled) subsequence such that there exist

- *a divergent sequence $\{p_i^j\}_{i\in\mathbb{N}}$ in M, for $j \in \{1, \dots, \ell\}$,*
- *a divergent sequence of sets $\{F_i^j\}_{i\in\mathbb{N}}$, for $j \in \{1, \dots, \ell\}$,*
- *an asymptotic manifold $M_\infty^j \in \mathcal{A}(M)$, for $j \in \{1, \dots, \ell\}$. We let $M_\infty^0 = M$.*
- *an isoperimetric region $E^j \subset M_\infty^j$ for all $j \in \{0, \dots, \ell\}$, with volume larger than β, for $j \geqslant 2$,*

such that

1. *$F_i^{j+1} \subset F_i^j \subset F_i$ for all $i \in \mathbb{N}$ and $j \in \{1, \dots, \ell - 1\}$;*
2. *$\{(M, g, p_i^j)\}_{i\in\mathbb{N}}$ converges to $(M_\infty^j, g_\infty^j, p_\infty^j)$ in pointed Lipschitz topology for all $j \in \{1, \dots, \ell\}$;*
3. *The sequence $\{F_i^j\}_{i\in\mathbb{N}}$, where F_i^j is in the pointed manifold (M, g, p_i^j), converges to $E^j \subset M_\infty^j$ in L^1_{loc} for all $j \in \{1, \dots, \ell\}$;*
4. *$E^0, E^1, \dots, E^\ell$ is a generalized isoperimetric region in M of volume $\sum_{j=0}^{\ell} |E^j| = v_0$;*
5. *$I_C(v_0) = \sum_{j=0}^{\ell} P_j(E_j)$, where $P = P_0$ is the perimeter in M and P_j is the perimeter in M_∞^j, $j \in \{1, \dots, \ell\}$.*

Proof We split the proof into several steps.

Step One. Let $E^0 \subset M$ be the (possibly empty) limit in $L^1_{\mathrm{loc}}(M)$ of a non-relabeled subsequence of $\{F_i\}_{i\in\mathbb{N}}$. Given a fixed point $p \in M$, Lemma 4.46, applied with $(M_i, g_i) = (M, g)$ and $p_i = p$ for all i, implies the existence of a sequence of diverging radii $r_i^0 > 0$ so that the set $F_i^1 = F_i \setminus B(p, r_i^0)$ satisfies

$$|E^0| + \lim_{i\to\infty} |F_i^1| = v_0, \tag{4.53}$$

and

$$P_M(E^0) + \liminf_{i\to\infty} P_M(F_i^1) \leqslant \liminf_{i\to\infty} P_M(F_i).$$

In case $|E^0| > 0$ the set E^0 is isoperimetric for its volume: otherwise there would exist a bounded measurable set $G^0 \subset M$ satisfying $|G^0| = |E^0|$ and $P(G^0) < P(E^0)$. This set can be approximated by a sequence $\{G_i\}_{i\in\mathbb{N}}$ of uniformly bounded sets of finite perimeter satisfying $|G_i| + |F_i^1| = v_0$ and $\lim_{i\to\infty} P(G_i) = P(G^0)$.

For large i, the sets G_i and F_i^1 are disjoint, $|G_i \cup F_i^1| = v_0$, and

$$\begin{aligned} I_M(v_0) &\leqslant \liminf_{i\to\infty} P(G_i \cup F_i^1) \\ &= \liminf_{i\to\infty} (P(G_i) + P(F_i^1)) = P(G^0) + \liminf_{i\to\infty} P(F_i^1) \\ &< P(E^0) + \liminf_{i\to\infty} P(F_i^1) \leqslant \liminf_{i\to\infty} P(F_i) = I_M(v_0), \end{aligned}$$

yielding a contradiction. A similar argument proves the equality

$$P(E^0) + \liminf_{i\to\infty} P(F_i^1) = I_M(v_0). \tag{4.54}$$

In particular, 4 is trivially satisfied for $q = 0$. Note that E^0 is a bounded set by Lemma 4.27.

To complete the proof of step 1, let us show the existence of a constant $C(v_0) > 0$ so that

$$\liminf_{i\to\infty} P(F_i^d) \leqslant C(v_0)(v_0 - |E^0|)^{(m-1)/m}. \tag{4.55}$$

The constant $C(v_0)$ will be obtained from the following construction, which we fix henceforth. We consider a finite family of mutually disjoint balls $\{B(q_i, r_0) : i = 1, \ldots, N\}$ outside a tubular neighborhood of E^0 and inside a bounded region of M so that $\sum_{i=1}^N |B(q_i, r_0)| \geqslant v_0$. Then there is a constant $C'(v_0) > 0$ such that

$$\frac{P(B(q_i, r))}{|B(q_i, r)|^{(m-1)/m}} \leqslant C'(v_0)$$

for all $i \in \{1, \ldots, N\}$ and $r \in (0, r_0]$. Hence, for any $0 < \lambda \leqslant 1$, we have

$$P\Big(\bigcup_{i=1}^N B(q_i, \lambda r_0)\Big) \leqslant C(v_0) \Big|\bigcup_{i=1}^N B(q_i, \lambda r_0)\Big|^{(m-1)/m} \tag{4.56}$$

with $C(v_0) = NC'(v_0)$.

In case $\liminf_{i\to\infty} P(F_i^d) > C(v_0)(v_0 - |E^0|)^{(m-1)/m}$, we take λ in the previous construction so that $B = \bigcup_{i=1}^N B(q_i, \lambda r_0)$ has volume $|B| = v_0 - |E^0|$. Hence

$$\begin{aligned} P(E^0 \cup B) &= P(E^0) + P(B) \\ &\leqslant P(E^0) + C(v_0)(v_0 - |E^0|)^{(m-1)/m} \\ &< P(E^0) + \liminf P(F_i^d) = I_M(v_0). \end{aligned}$$

This is a contradiction to the definition of $I_M(v_0)$ and proves (4.55).

Step Two. The set E_0 obtained in the previous step satisfies $|E_0| \leqslant v_0$. If $|E^0| = v_0$, then E^0 is an isoperimetric region of volume v_0, and the theorem holds with $\ell = 0$.

Let us treat the case $|E^0| < v_0$, which corresponds to a "volume loss at infinity." We claim that in this case, we have a divergent sequence of points $\{p_i^1\}_{i\in\mathbb{N}}$ so that

$$|F_i^1 \cap B(p_i^1, r_0)| \geqslant \mu |F_i^1| \tag{4.57}$$

for all $i \in \mathbb{N}$, where

$$\mu = \min\left\{\frac{C}{2^m C(v_0)^m}, \inf_{x\in M} \frac{|B(x, r_0)|}{4v_0}\right\}.$$

Here C is the constant that appears in the concentration Lemma 4.28 and $C(v_0)$ is the constant in (4.55). To prove the claim, we reason by contradiction assuming

$$|F_i^1 \cap B(x, r_0)| < \mu|F_i^1| < \frac{|B(x, r_0)|}{2}.$$

for all $i \in \mathbb{N}$ and $x \in M$. Then Lemma 4.28 (with $\mu|F_i^1|$ in the role of μ) implies

$$P(F_i^1) \geqslant \left(\frac{C}{\mu}\right)^{1/m} |F_i^1|^{(m-1)/m} > 2C(v_0)(v_0 - |E^0|)^{(m-1)/m}$$

for i large enough. Taking inferior limits, we have

$$\liminf_{i\to\infty} P(F_i^1) > C(v_0)(v_0 - |E^0|)^{(m-1)/m}.$$

This clearly contradicts (4.55) and proves the existence of a divergent sequence $\{p_i^1\}_{i\in\mathbb{N}}$ satisfying (4.57).

By hypotheses, a non-relabeled subsequence of the pointed Riemannian manifolds (M, g, p_i^1) converges in pointed Lipschitz topology to a pointed Riemannian manifold $(M_\infty^1, g_\infty^1, p_\infty^1)$. By Lemma 4.46 a non-relabeled subsequence of the sets $\{F_i^1\}_{i\in\mathbb{N}}$ converge to a set $E^1 \subset M_\infty^1$ of volume $|E^1| \geqslant \mu(v_0 - |E^0|)$, and there exists a sequence of radii $r_i^1 > r_i$ such that $F_i^2 = F_i^1 \setminus B(p_i^1, r_i^1)$ satisfies

$$\begin{aligned} |E^1| + \lim_{i\to\infty} |F_i^2| &= v_0 - |E_0|, \\ P_1(E^1) + \liminf_{i\to\infty} P(F_i^2) &\leqslant \liminf_{i\to\infty} P(F_i^1). \end{aligned} \tag{4.58}$$

We claim that the following properties are satisfied:

1. $|E^0| + |E^1| + \lim_{i\to\infty} |F_i^2| = v_0$,
2. E^0, E^1 is a generalized isoperimetric region in M of volume $|E^0| + |E^1|$,

3. $P(E^0) + P_1(E^1) + \liminf_{i\to\infty} P(F_i^2) = I_M(v_0)$,
4. $\liminf_{i\to\infty} P(F_i^2) \leqslant C(v_0)(v_0 - |E^0| - |E^1|)^{(m-1)/m}$,

where $C(v_0)$ is the constant appearing in (4.56).

The first property follows from the first equation in (4.58). If E^0, E^1 is not a generalized isoperimetric region in M, then we can find $G^0, G^1, \ldots, G^k$, where $G^0 \subset M$ and $G^j \subset M^j$, $M_j \in \mathcal{A}(M)$ for $j \geqslant 2$, such that $\sum_{i=0}^k |G^i| = |E^0|+|E^1|$ and

$$P(E^0) + \sum_{i=1}^k P_{M_i}(G^i) < P(E^0) + P_1(E^1).$$

As the sets G^j are bounded, we can approximate them by disjoint unions of bounded sets K_i^j such that $\sum_{j=0}^k |K_i^j| + |F_i^2| = v_0$, and

$$P(K_i^0) + \sum_{j=1}^k P(K_i^j) < P(E^0) + P_1(E^1) - \varepsilon,$$

for some $\varepsilon > 0$. This implies

$$\liminf_{i\to\infty} \big(P(K_i^0) + \sum_{j=1}^k P(K_i^j)\big) < P(E^0) + P_1(E^1),$$

and so

$$\begin{aligned}\liminf_{i\to\infty} P\Big(\Big(\bigcup_{j=0}^k K_i^j\Big) \cup F_i^2\Big) &\leqslant \liminf_{i\to\infty} \Big(\sum_{j=0}^k P(K_i^j) + P(F_i^2)\Big) \\ &< P(E^0) + P(E^1) + \liminf_{i\to\infty} P(F_i^2) \leqslant I_M(v_0).\end{aligned}$$

This is a contradiction to the definition of $I_M(v_0)$ and proves that E^0, E^1 is a generalized isoperimetric region of volume $|E^0| + |E^1|$. A similar argument shows that 3 holds. To prove the last property, we consider the same configuration of balls $\{B(q_i, r_0) : i = 1, \ldots, N\}$ as in step 1, and we adjust the parameter λ to get a union of balls B so that $|B| = v_0 - |E^0| - |E^1|$. We approximate E^1 in volume and perimeter by bounded sets with finite perimeter $G_i^1 \subset M$ disjoint from E^0 and B so that $E^0 \cup B \cup G_i^1$ has volume v_0. In case $\liminf_{i\to\infty} P(F_i^2) > C(v_0)(v_0 - |E^0| - |E^1|)^{(m-1)/m}$, we have

$$\begin{aligned}\lim_{i\to\infty} P(E^0 \cup G_i^1 \cup B) &= P(E_0) + \lim_{i\to\infty} P(G_i) + P(B)\\ &\leqslant P(E^0) + P_1(E^1) + C(v_0)(v_0 - |E^0| - |E^1|)^{(m-1)/m}\\ &< P(E^0) + P_1(E^1) + \liminf_{1\to\infty} P(F_i^2) = I_M(v_0).\end{aligned}$$

This contradiction proves 4.

Step Three (Induction). Assume we have obtained, for $j = 1, \dots, k$, asymptotic manifolds M_∞^j, sets $E^j \subset M_\infty^j$ of positive volume, non-relabeled subsequences of increasing radii $\{r_i^j\}_{i\in\mathbb{N}}$ with $r_i^{j+1} > r_i^j$ for all i, j, and associated sets $F_i^{j+1} = F_i \setminus B(p_i^j, r_i^j)$ so that

1. $\sum_{i=0}^k |E^i| + \lim_{i\to\infty} |F_i^{k+1}| = v_0$,
2. $E^0, E^1, \dots E^k$ is a generalized isoperimetric region in M of volume $\sum_{i=0}^k |E^i|$,
3. $\sum_{i=0}^k P_i(E^i) + \liminf_{i\to\infty} P(F_i^{k+1}) = I_M(v_0)$,
4. $\liminf_{i\to\infty} P(F_i^{k+1}) \leqslant C(v_0)\big(v_0 - \sum_{i=0}^k |E^i|\big)^{(m-1)/m}$.

In case $\lim_{i\to\infty} |F_i^{k+1}| = 0$ we are done. If $\lim_{i\to\infty} |F_i^{k+1}| > 0$, we use property 4 to reason as in step 1 to get a divergent sequence of points $\{p_i^{k+1}\}_{i\in\mathbb{N}}$ so that

$$|F_i^k \cap B(p_i^{k+1}, r_0)| \geqslant \mu |F_i^{k+1}|.$$

By hypotheses, a non-relabeled subsequence of the pointed Riemannian manifolds (M, g, p_i^{k+1}) converge in pointed Lipschitz topology to a pointed Riemannian manifold $(M_\infty^{k+1}, g_\infty^{k+1}, p_\infty^{k+1})$. By Lemma 4.46 a non-relabeled subsequence of the sets $\{F_i^{k+1}\}_{i\in\mathbb{N}}$ converge to a set $E^{k+1} \subset M_\infty^{k+1}$ of volume

$$|E^{k+1}| \geqslant \mu\Big(v_0 - \sum_{i=0}^k |E^i|\Big),$$

and there exists a sequence of radii $r_i^{k+1} > r_i^k$ such that $F_i^{k+2} = F_i^{k+1} \setminus B(p_i^{k+1}, r_i^{k+1})$ satisfies

$$|E^{k+1}| + \lim_{i\to\infty} |F_i^{k+2}| = v_0 - \sum_{i=0}^k |E_i|,$$

$$P_{k+1}(E^{k+1}) + \liminf_{i\to\infty} P(F_i^{k+2}) \leqslant \liminf_{i\to\infty} P(F_i^{k+1}).$$

We reason now exactly as in step 2 to obtain the following properties:

1. $\sum_{i=0}^{k+1} |E^i| + \lim_{i\to\infty} |F_i^{k+2}| = v_0$,
2. $E_0, E^1, \ldots, E^{k+1}$ is a generalized isoperimetric region in M of volume $\sum_{i=0}^{k+1} |E^i|$
3. $\sum_{i=0}^{k+1} P_i(E^i) + \liminf_{i\to\infty} P(F_i^{k+2}) = I_M(v_0)$,
4. $\liminf_{i\to\infty} P(F_i^{k+2}) \leqslant C(v_0)\big(v_0 - \sum_{i=0}^{k+1} |E^i|\big)^{(m-1)/m}$.

This concludes the induction process.

Step Four (Finiteness). We prove now that only a finite number of steps is required. We know from Lemma 4.26 that an isoperimetric inequality for small volumes holds in M, namely, $P(F) \geqslant C_I |F|^{(m-1)/m}$ whenever $|F| < v_{\min}$.

In case $|E^0| = v_0$, we are done. Otherwise $|E^1| > 0$, and we take a deformation of E^1 by sets E_1^t, where $t \in [-\delta, \delta]$ for some $\delta > 0$, so that

$$|E_t^1| = |E^1| + t, \qquad P(E_t^1) \leqslant P(E^1) + C|t|,$$

for any $t \in [-\delta, \delta]$. If

$$|F| < \min\left\{\delta, v_{\min}, \left(\frac{C_I}{2C}\right)^m\right\} = \beta,$$

where C is the constant in the inequality $P(E_t^1) \leqslant P(E^1) + C|t|$, C_I is an isoperimetric constant on M for volumes $0 < v < v_{\min}$; then

$$P(F) \geqslant C_I |F|^{(m-1)/m}.$$

Let us assume that $v_0 - \sum_{i=0}^{k} |E^i| < \beta$. This implies $|F_i^{k+1}| < \beta \leqslant v_{\min}$ for i large, and hence

$$\liminf_{i\to\infty} P(F_i^{k+1}) \geqslant C_I \lim_{i\to\infty} |F_i^{k+1}|^{(m-1)/m}.$$

We take $t_i \in [-\delta, \delta]$ so that $|E_{t_i}^1| = |E^1| + |F_i^{k+1}|$. This means $t_i = |F_i^{k+1}|$, and we have

$$P_1(E_{t_i}^1) \leqslant P_1(E^1) + C|t_i| = P_1(E^1) + C|F_i^{k+1}| \leqslant P_1(E^1) + \frac{C_I}{2}|F_i^{k+1}|^{(m-1)/m}$$

since $|F_i^{k+1}| \leqslant (C_I/2C)^m$. Taking inferior limits, we get

$$\liminf_{i\to\infty} P_1(E_{t_i}^1) < P_1(E^1) + \frac{1}{2} \liminf_{i\to\infty} P(F_i^{k+1}).$$

Approximating the sets $E^1_{t_i}, E^2, \dots, E^k$, we can build a sequence $\{G_i\}_{i\in\mathbb{N}}$ in M of sets of volume v_0 so that $\liminf_{i\to\infty} P(G_i) < I_M(v_0)$, a contradiction.

Note that the same argument implies

$$|E^j| \geqslant \beta \tag{4.59}$$

when $j = 2, \dots, k$. As an immediate consequence of (4.59), one obtains

$$\ell \leqslant 2 + \lfloor v/\beta \rfloor, \tag{4.60}$$

where $\lfloor x \rfloor$ denotes the largest integer $\leqslant x$. The property expressed by (4.60) is actually stronger than a generic finiteness of ℓ, as the right-hand side of (4.60) does not depend upon the specific choices made during each application of step three.

Step Five (Uniformity of β). The fact that β only depends on M and v_0, but not on the minimizing sequence $\{F_i\}_i$, can be proved as follows. Assume by contradiction that there exists a countable family $E_i^0, \dots, E_i^{\ell_i}$ of generalized isoperimetric regions for the prescribed volume v_0, such that the volume of the smallest component of each region, $E_i^{s_i}$, is infinitesimal as $i \to \infty$.

We have two possibilities, either $|E_i^0| \geqslant v_0/2$ for a subsequence or $|E_i^1| \geqslant \mu v_0/2$ also for a subsequence. We only consider the first case. The second one is similar with obvious modifications. Passing to a subsequence, we assume that the volumes $|E_i^0|$ converge to some volume in the interval $[v_0/2, v_0]$. By Lemma 4.46 applied to the sequence E_i^0 we can find an asymptotic manifold N and sets $\tilde{E}^0 \subset M$, $\tilde{E}_1 \subset N$ obtained as L^1_{loc} limits of E_i^0. Now, for i large so that $|E_i^{s_i}|$ is very small, we proceed to build a competitor to the generalized isoperimetric region $E_i^0, \dots, E_i^{\ell_i}$ replacing E_i^0 and $E_i^{s(i)}$ by $\tilde{E}^0$ and $\tilde{E}_t^1$ (here $\tilde{E}_t^1$ is obtained from the deformation Theorem 1.50 as in step 4), and we reach a contradiction using the same arguments as in step 4. □

4.6.3 Some Consequences

One of the consequences of Theorem 4.48 is the following regularity result for the isoperimetric profile.

Theorem 4.49 *Let (M, g) be a complete non-compact manifold of bounded Lipschitz geometry. Then I_M is locally the sum of a concave function and a smooth function. In particular, it possesses the same regularity properties as a concave function.*

Proof The proof is done by considering a generalized isoperimetric region and applying a deformation as in the proof of Theorem 3.13. This way we obtain locally the concavity of $I_M(v) + Cv^2$. The reader is referred to Lemma 5.5 in Leonardi et al. [267] for complete details. □

4.7 Notes

Notes for Sect. 4.1 An example of a manifold with density with discontinuous isoperimetric profile was described by Adams et al. [6, Prop. 2]. As indicated in the remark after Proposition 1 in [6], the authors tried to produce an example of a Riemannian manifold with discontinuous isoperimetric profile by using pieces of increasing negative curvature. By Theorem 4.17 in Sect. 4.2, such a construction is not possible if the resulting manifold M is simply connected with non-positive sectional curvatures.

An example of a two-dimensional connected Riemannian manifold with discontinuous isoperimetric profile was given by Papasoglu and Swenson [333]. Its construction is based on the notion of *expander* graph. Assume that $\Gamma(V, E)$ is a graph given by sets of vertexes V and edges E. For any subset S of vertexes, define the edge boundary of S as the set of edges connecting a vertex in S to a vertex in $V \setminus S$. The volume $|S|$ of S is the cardinal of S, and the boundary area $|\partial S|$ of S is the cardinal of its edge boundary. A k-regular graph (i.e., exactly k edges meet at every vertex) is a c-expander graph if for all $S \subset V$ with $|S| \leqslant |V|/2$, we have $|\partial S| \geqslant c|S|$. It is known that there is a constant $c > 0$ such that for any i large enough, there is a 3-regular c-expander graph with i-vertices. The manifolds M_i considered by the authors are connected sums of spheres of radius $1/i$ placed at the vertexes of a 3-regular c-expander graph of $i^2 + i$ vertexes. These spheres are joined by small cylinders corresponding to the edges of the graph glued on three small equally separated discs on an equator of each sphere.

Notes for Sect. 4.2 We already know that the continuity of the isoperimetric profile of a compact manifold follows from standard compactness results for sets of finite perimeter and the lower semicontinuity of perimeter [275]. Alternative proofs are obtained from concavity arguments [45, §7(i)], [47, 48, 311], or from the metric arguments by Gallot [176, Lemme 6.2].

When the ambient manifold is a non-compact homogeneous space, Hsiang showed that its isoperimetric profile is a non-decreasing and absolutely continuous function [237, Lemma 3, Thm. 6]. Hass [221] obtained examples of disconnected isoperimetric regions in Cartan–Hadamard manifolds, thus showing that the corresponding isoperimetric profiles are not strictly concave.

In Carnot groups or in cones, the existence of a one-parameter group of dilations implies that the isoperimetric profile is a concave function of the form $I(v) = C\, v^{q/(q+1)}$, where $C > 0$ and $q \in \mathbb{N}$, and so it is a continuous function [266, 313, 367].

Nardulli [318, Cor. 1] showed the absolute continuity of the isoperimetric profile under the assumption of bounded $C^{2,\alpha}$ geometry. A manifold N is of $C^{2,\alpha}$ bounded geometry if there is a lower bound on the Ricci curvature, a lower bound on the volume of geodesic balls of radius 1, and for every diverging sequence $\{p_i\}_{i\in\mathbb{N}}$, the pointed Riemannian manifolds (M, p_i) subconverge in the $C^{2,\alpha}$ topology to a pointed manifold. Muñoz Flores and Nardulli [315] prove continuity of the isoperimetric profile of a complete non-compact manifold M with Ricci curvature bounded below and volume of balls of radius one uniformly bounded below. The Hölder continuity property has been extended to a class of non-compact Riemannian manifold by Muñoz Flores and Nardulli [316]. More precisely, they show that the isoperimetric profile is locally $(m-1)/m$ Hölder continuous when there is a uniform bound from below of the Ricci curvature and the volume of unit balls.

Partial results for cylindrically bounded convex sets have been obtained by Ritoré and Vernadakis [370, Prop. 4.4]. For general unbounded convex bodies, in $\mathbb{R}^m$, the concavity of the power $I^{(m-1)}$ of the isoperimetric profile has been proven recently by Leonardi et al. [267].

Notes for Sect. 4.3 The use of minimizing sequences to treat existence issues in non-compact manifolds goes back at least up to Fiala [158]. The structure result for minimizing sequences in general Riemannian manifolds (Theorem 4.21) was introduced by Ritoré and Rosales (see Theorem 2.1 in [367]). A version for surfaces was given by Ritoré (see Lemma 2.2 in [361]).

Notes for Sect. 4.4 The validity of the existence result of isoperimetric sets in manifolds with cocompact action of the isometry group was provided by Morgan (see page 133 in [307]), where he indicates that the proof of existence of minimal clusters in $\mathbb{R}^m$, §13.4 in [307] or [301], holds in any smooth Riemannian manifold with compact quotient by the isometry group. Our proof of Theorem 4.25 is based on Galli and Ritoré [175], which is in turn based on ideas of Leonardi and Rigot [266]. In [266] it is proven the existence of isoperimetric sets in sub-Riemannian Carnot groups, where the quotient under the action of left translations of the group, which preserve the sub-Riemannian perimeter and the volume (the Haar measure), is a single point. The proof in [266] is simplified by the fact that the isoperimetric profile of a Carnot group is of the form $I(v) = Cv^q$, with $C > 0$ and $0 < q < 1$, and so it is a strictly concave function.

The properties of the isoperimetric profile stated in Theorem 4.30 are valid in the particular cases of homogeneous Riemannian manifolds (where the isometry group acts transitively) and symmetric spaces. Many of the properties in Theorem 4.25 were obtained by Hsiang for symmetric spaces (see §5 in [237]). In addition, since isoperimetric sets have positive mean curvature in symmetric spaces, the isoperimetric profile is proven to be an increasing function (see Lemma 3 in [237]), with derivative bounded from below by a geometric constant (see Corollary on page 172 in [237]).

Existence on manifolds under Ricci or scalar curvature conditions has been proved by Nardulli and Mondino [294]. Antonelli et al. [24] show existence results for isoperimetric sets of large volume in Riemannian manifolds with non-negative Ricci curvature and Euclidean volume growth. As a by-product, they show that isoperimetric sets of big volume always exist on manifolds with non-negative sectional curvature and Euclidean volume growth. Antonelli et al. [25] prove existence of isoperimetric regions of any volume in a Riemannian manifold with Ricci bounded below under the assumption that it is Gromov-Hausdorff asymptotic to simply connected models of constant sectional curvature. The Gromov-Hausdorff asymptotic analysis allows them to show, in low dimensions, nonexistence of isoperimetric regions in Cartan–Hadamard manifolds that are Gromov-Hausdorff asymptotic to the Euclidean space. They combine techniques of smooth and non-smooth spaces.

A detailed study of the questions considered in this section is being carried out recently on RCD spaces. Essentially a RCD space is a metric measure space satisfying a weak version of the Bochner inequality for Riemannian manifolds when the Ricci curvature is bounded from below. We refer the reader to Sturm [413, 414], Lott and Villani [273], Ambrosio [18], and Gigli [182] for background on this notion. Concerning RCD spaces, with some applications to Riemannian isoperimetric problems, Antonelli et al. [27] prove regularity and topological properties of volume constrained minimizers of quasi-perimeters in RCD spaces with the Hausdorff measure; Antonelli et al. [28] study sharp and rigid isoperimetric comparison theorems and sharp dimensional concavity properties of the isoperimetric profile for non-smooth spaces RCD spaces with the Hausdorff measure; Antonelli et al. [26] establish a structure theorem for minimizing sequences for the isoperimetric problem on non-compact RCD spaces with the Hausdorff measure under the only assumption that the measure of unit balls is uniformly bounded away from zero.

The behavior of the isoperimetric profile of a non-compact manifold for large volumes in terms of the volume growth of balls has been studied by Saloff-Coste and Coulhon [125]. See Leonardi et al. [267] for a similar study in Euclidean convex sets. On homogeneous Riemannian manifolds, the asymptotic behavior of the isoperimetric profile corresponds to three possible situations, $I_M(v) \sim v, v/\log/v)$ or $v^{(d-1)/d}$, depending on the structure of the identity components of the isometry group of the manifold (see Theorem 2.1 in Pittet [346]).

Notes for Sect. 4.5 While existence of isoperimetric sets in $\mathbb{R}^m$ and $\mathbb{H}^m$ follows from Theorem 4.25, the characterization of such sets as geodesic balls follows from the proof of Alexandrov's Theorem by Montiel and Ros [300] based on Heintze-Karcher inequalities [225]. Of course, it is necessary to take into account the singularities at the boundary of the isoperimetric set. It is worth noting that a similar use of Heintze-Karcher inequality by Gromov (see §VI.5 in Chavel [102]) provides the uniqueness of geodesic balls as isoperimetric sets.

Notes for Sect. 4.6 Generalized existence of isoperimetric sets in a Riemannian manifold M was first proven by Nardulli [318] under the assumption of bounded geometry: this means that the Ricci curvature is bounded below by a constant, the volume of unit balls in uniformly bounded from below, and the pointed manifolds (M, g, p_i) converge in the $C^{k,\alpha}$ topology to a pointed Riemannian manifold $(M_\infty, g_\infty, p_\infty)$ for any divergent sequence $\{p_i\}_{i\in\mathbb{N}}$. The limit manifold depends on the sequence. In Carnot groups, existence of isoperimetric sets for the sub-Riemannian perimeter was shown by Leonardi and Rigot [266]. Their proof uses a concentration lemma to capture a fraction of the volume of the minimizing sequence and the concavity of the isoperimetric profile of a Carnot group to avoid the loss of volume when part of the minimizing sequence diverges.

Our proof of Theorem 4.48 is based on the proof of a similar compactness result for unbounded convex bodies in Euclidean space by Leonardi et al. [267]. In this case, under the hypothesis that an unbounded convex body $C \subset \mathbb{R}^m$ satisfy the non-degeneracy condition $|B(x, 1) \cap C| \geqslant C_0 > 0$ for all $x \in C$, it is proven the existence of generalized isoperimetric sets in the convex body. The role of asymptotic manifolds is replaced by the one of asymptotic cylinder: the Hausdorff limit of the translations $-p_i + C$ when $\{p_i\}_{i \in \mathbb{N}} \subset C$ is a divergent sequence in C. The non-degeneracy condition is equivalent to that the isoperimetric profile of the convex set is strictly positive.

Chapter 5
Symmetrization and Classical Results

In the theory of isoperimetric inequalities, a *symmetrization of a set E* is a procedure to obtain another set E^* with the same volume as E, no larger perimeter, and additional symmetries. Symmetrization is a very powerful method to reduce the number of isoperimetric candidates in a given space and generally requires that the ambient manifold itself possesses symmetries.

The modern use of symmetrization methods goes back to Steiner [407–409] and Schwarz [402]. Steiner introduced many ingenious symmetrization procedures for the classical planar isoperimetric problem, including the one that is widely known as the *Steiner symmetrization* with respect to a line L: given a set E, we obtain its *symmetral* E^* by replacing the intersection of E with every line L' orthogonal to L by a segment of the same length centered at $L \cap L'$; see Blåsjö [64] for a good exposition of some of the Steiner methods and Talenti [419] for a modern presentation of the Steiner symmetrization with respect to a hyperplane. The *Schwarz symmetrization* in $\mathbb{R}^3$ with respect to a line L consists of taking all planes P orthogonal to L and replacing $E \cap P$ with the disk in P of the same area centered at $L \cap P$. We refer the reader to Kawohl [243], Burago and Zalgaller [83], Chavel [101], and Baernstein [32] for an account of symmetrization methods related to isoperimetric inequalities.

In this chapter, we first consider symmetrization in warped products. This is a general method that allows us to derive at once many of the classical symmetrizations line Steiner's, Schwarz's, and spherical symmetrization; see Burago and Maz'ya [82], Bokowski and Sperner [67], and Almgren [13]. Symmetrization of Riemannian submersions can also be considered by similar methods. For the presentation, we follow Morgan et al. [309], who derived the result for manifolds with density.

Afterward, we consider a symmetrization method by Hsiang [234, 236]. This one allows us to obtain symmetries in products of Riemannian manifolds when one of the factors is a simply connected space form or, more generally, when there exists

M. Ritoré, *Isoperimetric Inequalities in Riemannian Manifolds*,
Progress in Mathematics 348, https://doi.org/10.1007/978-3-031-37901-7_5

an isometric action of the isometry group of a simply connected space form with a given isotropy group.

In the final part of this chapter, we consider some classical ways to obtain the isoperimetric inequality in Euclidean space: using the Brunn-Minkowski inequality, the linear Brunn-Minkowski inequality, mass transport techniques, the Alexandrov theorem for embedded hypersurfaces with constant mean curvature in $\mathbb{R}^m$, and Reilly-Ros' and Cabré's proofs of the isoperimetric inequality.

5.1 Symmetrization in Warped Products

In this section, we present a symmetrization result for warped product proven by Morgan et al. [309] for manifolds with special densities. Most of the classical symmetrization techniques can be recovered from this result. We recall first some basic properties of the geometry of warped products.

5.1.1 *Geometry of Warped Products*

Given two Riemannian manifolds (B, g_B) and (Q, g_Q) with $\dim B, \dim Q \geqslant 1$, and a positive smooth function $f : B \to \mathbb{R}$, the *warped product* $B \times_f Q$ is the product manifold $B \times Q$ with the Riemannian metric

$$g = \pi_B^* g_B + (f \circ \pi_B)^2 \pi_Q^* g_Q,$$

where π_B and π_Q are the projections over the factors. B is usually called the *base manifold* and Q the *fiber*. Standard references for warped products are O'Neill's monograph [323, p. 204–205] and also Bishop and O'Neill's [61] and Alexander and Bishop's [9]. Given $b \in B$, we denote by Q_b the set $\pi_B^{-1}(\{b\}) = \{b\} \times Q$ and, given $q \in Q$, by B_q the set $\pi_Q^{-1}(\{q\}) = B \times \{q\}$. The sets B_q are totally geodesic and the sets Q_b totally umbilical.

Any curve γ in $B \times Q$ can be expressed as (γ_B, γ_Q), where $\gamma_B = \pi_B \circ \gamma$ and $\gamma_Q = \pi_Q \circ \gamma$. From elementary properties of the Levi-Civita connection on $B \times_f Q$ (see Proposition 35 on page 206 of [323]), an arc-length parameterized curve $\gamma = (\gamma_B, \gamma_Q)$ is a geodesic if and only if

$$\begin{aligned} 2\gamma_B'' &= |\gamma_Q'|^2 (f \circ \gamma_B) \nabla f, && \text{in } B, \\ \gamma_Q'' &= -2 \frac{(f \circ \gamma_B)'}{f \circ \gamma_B} \gamma_Q', && \text{in } Q, \end{aligned} \tag{5.1}$$

where primes denote the covariant derivative along γ and $|\gamma_Q'|$ is the modulus in (Q, g_Q); see Proposition 38 in p. 208 of [323]. The second property in (5.1) implies that γ_Q is a reparameterization of a geodesic in Q and that the function

$$(f \circ \gamma_B)^2 \, |\gamma_Q'|$$

is constant along γ. Hence, there exists a constant C such that

$$|\gamma_Q'| = \frac{C}{(f \circ \gamma_B)^2}. \tag{5.2}$$

Obviously, $C = f(\gamma_Q(0))^2 \, |\gamma_Q'(0)|$. From the first equation in (5.1), we get

$$\gamma_B'' = \frac{C^2}{(f \circ \gamma_B)^3} \nabla f = -C^2 \, \nabla\left(\frac{1}{f^2}\right).$$

The following metric property in $B \times_f Q$ is essential for what follows. It establishes that the intersection of a metric ball in $B \times_f Q$ intersected with a slice Q_b is a metric ball in Q (Fig. 5.1).

Lemma 5.1 *If $B((b_0, q_0), r) \cap Q_b \neq \emptyset$, then*

$$\overline{B}((b_0, q_0), r) \cap Q_b = \{b\} \times \overline{B}(q_0, r'), \tag{5.3}$$

where $B((b_0, q_0), r)$ is a metric ball in $B \times_f Q$ and $B(q_0, r')$ a metric ball in (Q, g_Q). Moreover, the constant r' only depends on b_0 and is independent of q_0.

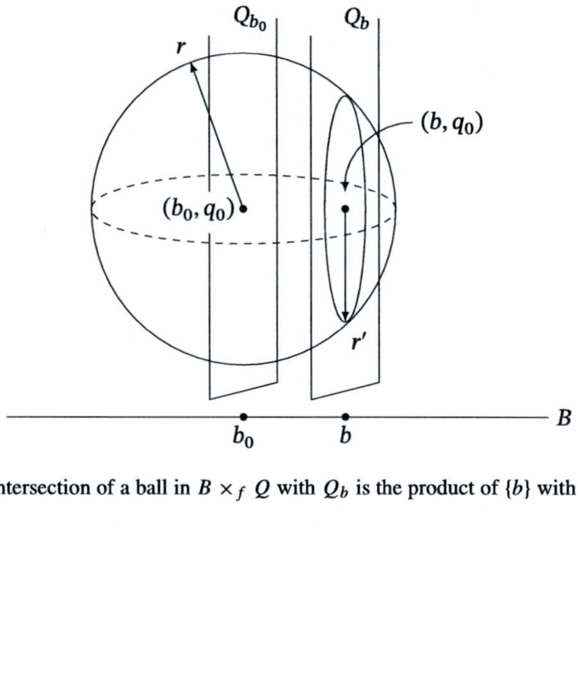

Fig. 5.1 The intersection of a ball in $B \times_f Q$ with Q_b is the product of $\{b\}$ with a metric ball in (Q, g_Q)

Proof Let

$$r' = \sup\left\{d_Q(q, q_0) : (b, q) \in \overline{B}((b_0, q_0), r)\right\},$$

where d_Q is the Riemannian distance in (Q, g_Q). Then we clearly have

$$\overline{B}((b_0, q_0), r) \cap Q_b \subseteq \{b\} \times \overline{B}(q_0, r').$$

On the other hand, take $q \in \overline{B}(q_0, r')$. By the definition of r', there exists $\tilde{q} \in \overline{B}(q_0, r')$ such that $(b, \tilde{q}) \in \overline{B}((b_0, q_0), r)$ and $d_Q(q_0, q) \leqslant d_Q(q_0, \tilde{q})$. Take a unit speed length-minimizing geodesic $\gamma = (\gamma_B, \gamma_Q)$ in $B \times_f Q$ of length $d \leqslant r$ joining (b_0, q_0) and $(b, \tilde{q})$. Take also a minimizing geodesic in Q joining q_0 and $\tilde{q}$, and reparameterize it to get a curve α defined in the interval $[0, d]$ and satisfying $|\alpha'| = C_\alpha/(f \circ \gamma_B)^2$. As length$(\alpha) \leqslant$ length(γ_Q), we get $C_\alpha \leqslant C$, where C is the constant obtained in (5.2) satisfying $|\gamma_Q'| = C/(f \circ \gamma_B)^2$. Hence, the curve (γ_B, α) connects (b_0, q_0) and (b, q) and

$$\begin{aligned}
\text{length}((\gamma_B, \alpha)) &= \int_0^d \Big(|\gamma_B'|^2 + \frac{C_\alpha^2}{(f \circ \gamma_B)^2}\Big)^{1/2} ds \\
&\leqslant \int_0^d \Big(|\gamma_B'|^2 + \frac{C^2}{(f \circ \gamma_B)^2}\Big)^{1/2} ds = \text{length}(\gamma) = d \leqslant r.
\end{aligned}$$

Hence, $(b, q) \in \overline{B}((b_0, q_0), r)$, and equality holds in (5.3).

Finally, let us prove that r' is independent of q_0. Take another point $\tilde{q}_0$ and the corresponding radius $\tilde{r}'$. Let $\bar{q}, \tilde{q} \in Q$ such that $d_Q(q_0, q) = r'$ and $d_Q(\tilde{q}_0, \tilde{q}) = \tilde{r}'$. Then $d((b_0, q_0), (b, q))$ and $d((b_0, \tilde{q}_0), (b, \tilde{q}))$ are both equal to r (otherwise, the definition of $r', \tilde{r}'$ as suprema yields a contradiction). Take a length-minimizing geodesic $\tilde{\gamma} = (\tilde{\gamma}_B, \tilde{\gamma}_Q)$ joining $(b_0, \tilde{q}_0)$ and $(b, \tilde{q})$. Assume that $\tilde{r}' < r'$, and reparameterize $\tilde{\gamma}_Q$ to get a curve α defined in the interval $[0, r]$ and such that $|\alpha'| = \tilde{C}/(f \circ \gamma_B)^2$. Let C be the constant defined in (5.2). As α As γ_Q and $\tilde{\gamma}_Q$ are reparameterizations of geodesics and $r' > \tilde{r}'$, we have $C > \tilde{C}$. Hence,

$$\begin{aligned}
r \leqslant \text{length}((\gamma_B, \alpha)) &= \int_0^r \Big(|\gamma_B'|^2 + \frac{\tilde{C}^2}{(f \circ \gamma_B)^2}\Big)^{1/2} ds \\
&< \int_0^r \Big(|\gamma_B'|^2 + \frac{C^2}{(f \circ \gamma_B)^2}\Big)^{1/2} ds = \text{length}(\gamma) = r,
\end{aligned}$$

a contradiction. The case $r' < \tilde{r}'$ is handled in a similar way. □

5.1.2 Symmetrization and Basic Properties

We now define a symmetrization procedure for the Minkowski content in warped products. A slightly more general version for manifolds with certain densities was proven in Proposition 3 of [309]. The main assumption is that, for some $q_0 \in Q$, the balls about q_0 are isoperimetric for the Minkowski content. Given any set $E \subset B \times_f Q$ and $r > 0$, we denote by E_r the closed tubular neighborhood of radius r. The following result is well-known.

Lemma 5.2 *Let (M, g) be a complete Riemannian manifold. Assume there exists a point $p_0 \in M$ such that the geodesic balls of center p_0 are isoperimetric sets for the Minkowski content. Given $E \subset M$ of finite volume, let D be the ball centered at p_0 of the same volume. Then we have*

$$|D_r| \leqslant |E_r| \tag{5.4}$$

for all $r > 0$.

Proof If $|E_r| = |M|$, then inequality (5.4) holds, so we restrict ourselves to the case $|E_r| < |M|$. Observe first that E_r is measurable since it is a closed set. The functions $r \mapsto |D_r|$ and $r \mapsto |E_r|$ are continuous, strictly increasing while $|D_r|, |E_r| < |M|$, and differentiable except in a set at most countable. For every $r > 0$ such that $|E_r| < |M|$, let $s(r)$ be the unique value such that $|D_{s(r)}| = |E_r|$. The function s is monotone increasing. Assume that r is a differentiability point of s and of $t \mapsto |E_t|$ and that $s(r)$ is a differentiability point of $t \mapsto |D_t|$. Then we get $s'(r)\,\mathrm{Mink}(D_{s(r)}) = \mathrm{Mink}(E_r)$, and, as $D_{s(r)}$ is isoperimetric and has the same volume as E_r, we have

$$s'(r) = \frac{\mathrm{Mink}(E_r)}{\mathrm{Mink}(D_{s(r)})} \geqslant 1$$

for a.e. r. Since $s(0) = 0$, we have $s(r) \geqslant r$ for all $r \geqslant 0$ and so $|D_r| \leqslant |D_{s(r)}| = |E_r|$ for all $r \geqslant 0$. □

Consider a warped product $B \times_f Q$, and assume that Q satisfies the hypothesis of Lemma 5.2: that there exists a point $q_0 \in Q$ such that the geodesic balls about q_0 are isoperimetric sets for the Minkowski content. Given a measurable set $E \subset B \times Q$, we define, for any $b \in B$, the set

$$E^b = \{q \in Q : (b, q) \in E\} \subset Q$$

and the closed disk $D^b \subset Q$ around q_0 of the same g_Q-volume as E^b. We set $D^b = \emptyset$ if $E_b = \emptyset$. Observe that the set E^b is the projection of the slice $E_b =$

$\{b\} \times E^b = E \cap Q_b$ to Q. The *symmetral* sym(E) of E is defined by

$$\operatorname{sym}(E) = \bigcup_{b \in B} \{b\} \times D^b.$$

We have the following important property for sym(E).

Proposition 5.3 *The set* sym(E) *is measurable and* $|E| = |\operatorname{sym}(E)|$.

Proof By Fubini-Tonelli's theorem, the function

$$\varphi(b) = \int_Q \mathbf{1}_E(b, q)\, dQ(q) = \int_Q \mathbf{1}_{E^b}\, dQ = |E^b|$$

is measurable in B, where $|E^b|$ denotes the volume of E^b in (Q, g_Q). By the definition of sym(E), we have $\varphi(b) = |E^b| = |D^b|$, so that φ assigns to $b \in B$ the volume in (Q, g_Q) of the disk D^b.

For any $b \in B$, let $\rho(b)$ be the radius of the disk D^b. In case $D^b = \emptyset$, we set $\rho(b) = 0$. We know that the function $v(\rho) = |D(q_0, \rho)|$, where $D(q_0, \rho)$ is the closed disk in Q of center q_0 and radius $\rho \geqslant 0$ is monotone increasing and continuous. Hence, its inverse v^{-1} is also continuous. As $\rho = v^{-1} \circ \varphi$, we conclude that the function ρ is measurable on Q.

Finally, we observe that

$$\operatorname{sym}(E) \cup \big\{(b, q_0) : E_b = \emptyset\big\} = \big\{(b, q) : d_{B \times Q}\big((b, q), B \times \{q_0\}\big) \leqslant \rho(b)\big\},$$

where $d_{B \times Q}$ is the product distance in $B \times Q$. As ρ is measurable, the set on the right side of the equality is clearly measurable on $B \times Q$. On the other hand, sym(E) and $\big\{(b, q_0) : E_b = \emptyset\big\}$ are disjoint, and

$$\big\{(b, q_0) : E_b = \emptyset\big\} = \big(B \setminus \pi_B(E)\big) \times \{q_0\}$$

is measurable. Hence, sym(E) is measurable.

To prove that $|E| = |\operatorname{sym}(E)|$, we observe that, for any measurable set $F \subset B \times Q$, the area formula and Fubini-Tonelli's theorem in $B \times_f Q$ imply

$$|F| = \int_B f^{\dim(Q)} |F^b|\, dB,$$

where $|F^b|$ is computed in (Q, g_Q). The equality of volumes follows since we have $|E^b| = |\operatorname{sym}(E)^b|$ for all $b \in B$. □

For the Steiner symmetrization, it was proven in Lemma 2(ii) in §2.2 of Evans and Gariepy [149] that sym(E) is measurable. If we assume that E is compact, it is not difficult to prove that sym(E) is compact; see §2.10.30 in Federer [154].

Proposition 5.4 *If $E \subset B \times Q$ is compact, then* sym(E) *is compact.*

Proof Take a sequence $\{(b_i, q_i)\}_{i\in\mathbb{N}}$ contained in sym(E). Since E is compact, the set $\pi_B(E)$ is compact, and there exists $r_0 > 0$ such that

$$E \subset \pi_B(E) \times D(q_0, r_0),$$

where $D(q_0, r_0)$ is the metric closed disk of center q_0 and radius $r_0 > 0$ in (Q, g_Q). As $\pi_B(E)$ is compact, we may assume, passing to a subsequence if necessary, that $\{b_i\}_{i\in\mathbb{N}}$ converges to some point $b \in \pi_B(E)$. On the other hand, as the sequence $\{q_i\}_{i\in\mathbb{N}}$ is contained in $D(q_0, r_0)$, we can also extract a non-relabeled convergent subsequence to some $q \in D(q_0, r_0)$.

It remains to prove that $q \in D^b$. As $(b_i, q_i) \in$ sym(E), we have $d_Q(q_i, q_0) \leqslant \rho(b_i)$ for all $i \in \mathbb{N}$, where $\rho(b_i)$ is the radius of the ball D^{b_i}. From the inclusion $E \subset \pi_B(E) \times D(q_0, r_0)$, we get $\mathbf{1}_{E^{b_i}}, \mathbf{1}_{E^b} \leqslant \mathbf{1}_{D(q_0,r_0)}$. As E is closed, we have $\mathbf{1}_{E^b} \geqslant \limsup_{i\to\infty} \mathbf{1}_{E^{b_i}}$, and Fatou's lemma implies

$$|E^b| = \int_Q \mathbf{1}_{E^b} dQ \geqslant \int_Q \limsup_{i\to\infty} \mathbf{1}_{E_{b_i}} dQ \geqslant \limsup_{i\to\infty} \int_Q \mathbf{1}_{E_{b_i}} dQ = \limsup_{i\to\infty} |E^{b_i}|.$$

Then we have $\rho(b) \geqslant \limsup_{i\to\infty} \rho(b_i)$ and so

$$d_Q(q, q_0) = \lim_{i\to\infty} d_Q(q_i, q_0) \leqslant \limsup_{i\to\infty} \rho(b_i) \leqslant \rho(b).$$

Hence, $q \in D(q_0, \rho(b)) = D^b$. □

We end this section with a result which will play a key role later.

Lemma 5.5 *Let $\{E_i\}_{i\in\mathbb{N}}$ be a sequence of compact sets converging to E in Hausdorff distance. Assume that the sequence* $\{\text{sym}(E_i)\}_{i\in\mathbb{N}}$ *converges in Hausdorff distance to E'. Then $E' \subset$* sym(E).

Proof We take $(b, q) \in E'$. As E' is the Hausdorff limit of sym(E_i), there exists a sequence of points $(b_i, q_i) \in$ sym(E_i) such that $(b, q) = \lim_{i\to\infty}(b_i, q_i)$. Let $\rho_i(b_i)$ be the radius of the disk sym$(E_i)^{b_i}$, of volume $|E_i^{b_i}|$. Hence,

$$d_Q(q, q_0) = \lim_{i\to\infty} d_Q(q, q_i) \leqslant \liminf_{i\to\infty} \rho_i(b_i).$$

On the other hand, as E is the Hausdorff limit of E_i, we have

$$\limsup_{i\to\infty} \mathbf{1}_{E_i^{b_i}} \leqslant \mathbf{1}_{E^b}.$$

Hence, $\limsup_{i\to\infty} |E_i^{b_i}| \leqslant |E^b|$ and so

$$\limsup_{i\to\infty} \rho_i(b_i) \leqslant \rho(b),$$

where $\rho(b)$ is the radius of the disk $\mathrm{sym}(E)^b$ of volume $|E^b|$. We conclude that $d_Q(q, q_0) \leqslant \rho(b)$. Hence, $q \in \mathrm{sym}(E)^b$ and so $(b, q) \in \mathrm{sym}(E)$. □

5.1.3 Proof of the Symmetrization Result

Now we prove the symmetrization result for the lower Minkowski content in warped products.

Theorem 5.6 (Symmetrization for Warped Products) *Let $B \times_f Q$ be a warped product and $q_0 \in Q$ such that balls about q_0 are isoperimetric sets for the Minkowski content. Let $E \subset B \times Q$ be a measurable set of finite volume. Then*

$$|\,\mathrm{sym}(E)_r| \leqslant |E_r| \tag{5.5}$$

for all $r > 0$ and

$$\mathrm{Mink}_-(\mathrm{sym}(E)) \leqslant \mathrm{Mink}_-(E). \tag{5.6}$$

Proof The strategy of the proof is to show that

$$\mathrm{sym}(E)_r \subseteq \mathrm{sym}(E_r) \tag{5.7}$$

for any $r > 0$. This would imply that $|\,\mathrm{sym}(E)_r| \leqslant |\,\mathrm{sym}(E_r)| = |E_r|$, thus proving (5.5). As $|E| = |\,\mathrm{sym}(E)|$, inequality (5.6) would follow from the definition of the lower Minkowski content.

So let us prove (5.7). We fix $r > 0$ and let

$$\mathrm{sym}(E) = \bigcup_{b \in \pi_B(E)} \{b\} \times D^b,$$

where D^b is the closed disk in Q centered at q_0 with the same volume as E^b. Since such disks are isoperimetric, Lemma 5.2 implies

$$|(D^b)_r| \leqslant |(E^b)_r| \tag{5.8}$$

for all $b \in \pi_B(E)$ and any $r > 0$.

As $\overline{B}((b, q), r) = (\{b\} \times \{q\})_r$ and $\overline{B}(q, r') = \{q\}_{r'(b)}$, Lemma 5.1 implies

$$(\{b\} \times \{q\})_r \cap Q_{b'} = \{b'\} \times \{q\}_{r'(b)},$$

implicitly assuming $\{q\}_{r(b')} = \emptyset$ if $B((b,q),r) \cap Q_{b'} = \emptyset$. Writing D^b as the union of its points, we have

$$(\{b\} \times D^b)_r \cap Q_{b'} = \{b'\} \times (D^b)_{r'(b)}.$$

Hence,

$$\mathrm{sym}(E)_r \cap Q_{b'} = \{b'\} \times \bigcup_{b \in \pi_B(E)} (D^b)_{r'(b)},$$

and so

$$(\mathrm{sym}(E)_r)^{b'} = \bigcup_{b \in \pi_B(E)} (D^b)_{r'(b)}. \tag{5.9}$$

On the other hand,

$$\mathrm{sym}(E_r)^{b'} = \tilde{D}^{b'}$$

for all $b' \in \pi_B(E_r)$, where $\tilde{D}^{b'} \subset Q$ is a disk centered at q_0 with volume $|\,\mathrm{sym}(E_r)^{b'}| = |(E_r)^{b'}|$. Replacing $\mathrm{sym}(E)$ by E in (5.9), we get

$$(E_r)^{b'} = \bigcup_{b \in \pi_B(E)} (E^b)_{r'(b)}. \tag{5.10}$$

So we obtain, from (5.8), $|\tilde{D}^{b'}| = |(E_r)^{b'}|$ and, from (5.10),

$$|(D^b)_{r'(b)}| \leqslant |(E^b)_{r'(b)}| \leqslant |\tilde{D}^{b'}|.$$

As $(D^b)_{r'(b)}$ and $\tilde{D}^{b'}$ are concentric disks, we have $(D^b)_{r'(b)} \subset \tilde{D}^{b'}$ for all $b \in B$ and so, by (5.9),

$$(\mathrm{sym}(E)_r)^{b'} = \bigcup_{b \in \pi_B(E)} (D^b)_{r'(b)} \subset \tilde{D}^{b'} = \mathrm{sym}(E_r)^{b'}$$

for all b', which implies $\mathrm{sym}(E)_r \subset \mathrm{sym}(E_r)$. □

Theorem 5.6 implies the following comparison for the perimeter.

Corollary 5.7 *Let $B \times_f Q$ be a warped product and $q_0 \in Q$ such that balls about q_0 are isoperimetric sets for the Minkowski content. Let $E \subset B \times Q$ be a measurable set of finite volume. Then*

$$P(\mathrm{sym}(E)) \leqslant P(E). \tag{5.11}$$

Proof We showed in Theorem 1.52 that the perimeter of a set E of finite volume is equal to the relaxed lower Minkowski content

$$\operatorname{Mink}_{-}^{*}(E) = \inf\Big\{ \liminf_{i\to\infty} \operatorname{Mink}_{-}(E_i) : E_i \to E \text{ in measure}\Big\}.$$

Take any sequence $\{E_i\}_{i\in\mathbb{N}}$ of measurable sets converging to E in measure. By Theorem 5.6, we have

$$\operatorname{Mink}_{-}(\operatorname{sym}(E_i)) \leqslant \operatorname{Mink}_{-}(E_i) \tag{5.12}$$

for all $i \in \mathbb{N}$.

The sequence $\{\operatorname{sym}(E_i)\}_{i\in\mathbb{N}}$ converges to $\operatorname{sym}(E)$ in measure. This is easy to check since, for any pair of measurable sets A, C, we have $(A\Delta C)^b = A^b\Delta C^b$, and

$$|A\Delta C| = \int_B f(b)^{\dim(Q)}|A^b\Delta C^b|\, dB(b),$$

where the volume of $A^b\Delta C^b$ is computed in (Q, g_Q). As $\operatorname{sym}(E_i)^b$, $\operatorname{sym}(E)^b$ are concentric disks, we have

$$\begin{aligned}
|\operatorname{sym}(E_i)^b \,\Delta\, \operatorname{sym}(E)^b| &= \big||\operatorname{sym}(E_i)^b| - |\operatorname{sym}(E)^b|\big| \\
&= \big||E_i^b| - |E^b|\big| \leqslant |E_i^b \,\Delta\, E^b|.
\end{aligned}$$

The last inequality holds since $A \subset (A\Delta C) \cup C$ for any pair of measurable sets. Hence,

$$|\operatorname{sym}(E_i)\Delta \operatorname{sym}(E)| \leqslant |E_i \Delta E|$$

and we conclude that $\operatorname{sym}(E_i) \to \operatorname{sym}(E)$ in measure if $E_i \to E$ in measure.

Taking inferior limits in Eq. (5.12), we get

$$\operatorname{Mink}_{-}^{*}(\operatorname{sym}(E)) \leqslant \liminf_{i\to\infty} \operatorname{Mink}_{-}(\operatorname{sym}(E_i)) \leqslant \liminf_{i\to\infty} \operatorname{Mink}_{-}(E_i).$$

Taking now infimum over all sequences $\{E_i\}_{i\in\mathbb{N}}$ converging to E in measure, we get

$$\operatorname{Mink}_{-}^{*}(\operatorname{sym}(E)) \leqslant \operatorname{Mink}_{-}^{*}(E)$$

and so $P(\operatorname{sym}(E)) \leqslant P(E)$. □

Remark 5.8 Theorem 5.6 and Corollary 5.7 can be applied to different situations.

When $B = \mathbb{R}^{m-1}$, $Q = \mathbb{R}$, and $f \equiv 1$, we get the classical *Steiner symmetrization*. In this case, we are simply writing the Euclidean space as a product $\mathbb{R}^m = \mathbb{R}^{m-1} \times \mathbb{R}$ and replacing the intersection of a set with a line parallel to the

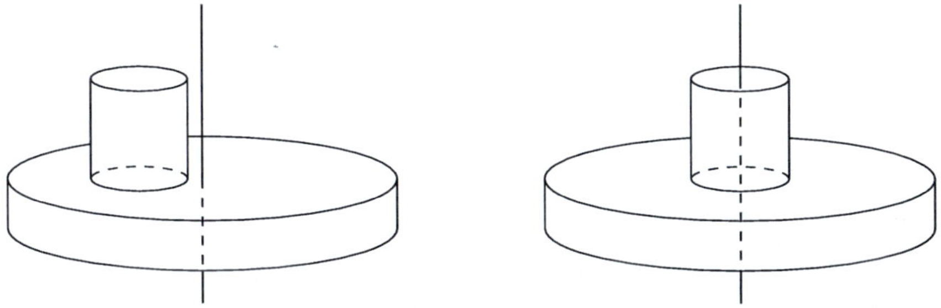

Fig. 5.2 A set E whose Schwarz symmetral sym(E) with respect to a vertical line has the same perimeter as E but it is not congruent to E. Here, B corresponds to a vertical line and Q to a horizontal plane

x_m-axis by the segment of the same length centered at the intersection of the line with the hyperplane $x_m = 0$.

If $B = \mathbb{R}$, $Q = \mathbb{R}^{m-1}$, and $f \equiv 1$, we recover the *Schwarz symmetrization.* In this case, we replace the intersection of a set with the hyperplane $x_1 = c$ by a disk of the same area centered at the intersection of the x_1-axis with the hyperplane.

In case we represent $\mathbb{R}^m \setminus \{0\}$ by the isometric warped product $(0, +\infty) \times_f \mathbb{S}^{m-1}$, where $f(t) = t$, we obtain the *spherical symmetrization.* Here, the intersection of a set with a sphere centered at 0 is replaced by a spherical disk of the same area centered at the intersection of the sphere with a given half-line.

We have seen in Corollary 5.7 that inequality $P(\text{sym}(E)) \leqslant P(E)$ holds for any set E of finite perimeter and finite volume in the warped product $B \times_f Q$. However, the proof does not offer any information in case equality holds.

It is not difficult to produce examples of sets E such that $P(\text{sym}(E)) = P(E)$ but sym(E) and E are not isometric. This might happen when there are pieces of ∂E which contain large portions of fibers $\{b\} \times Q$, which add a positive amount to the perimeter of E. In Fig. 5.2, we present an example for Schwarz symmetrization in $\mathbb{R}^3$.

5.1.4 An Inequality for the Perimeter

In this section, we estimate the perimeter of a set with C^1 boundary in a warped product in terms of the Hausdorff measure of the intersection of the set and its boundary with the fibers and the mean curvature vector of the fibers.

We consider the Riemannian warped product $B \times_f Q$ with Riemannian metric $g = g_B + f^2 g_Q$. Let $b \in B$ be fixed and $p \in Q_b$. To compute the mean curvature vector $\vec{H}_b$ of Q_b at p and its projection $\mathbf{H}_b$ to B, we take an orthonormal frame of vector fields $E_1, \ldots, E_q$ tangent to Q_b in an open neighborhood U of p in Q_b containing p. We complete an orthonormal frame of TM by adding vector fields

$N_1, \ldots, N_\beta$, with $\beta = \dim(B)$, whose projections to B are orthonormal vectors $n_1, \ldots, n_\beta$ in $T_b B$. Since

$$\nabla_{E_i} N_j = \frac{n_j(f)}{f} E_i,$$

by Prop. 35(2) in [323], we have

$$\vec{H}_b = \sum_{i=1}^{\beta} q \frac{n_j(f)}{f} N_j, \tag{5.13}$$

and so

$$\mathbf{H}_b = \sum_{i=1}^{\beta} q \frac{n_j(f)}{f} n_j. \tag{5.14}$$

Observe that the projection $\mathbf{H}_b$ does not depend on $p \in S_b$.

We now prove a lower estimate for the area of a C^1 hypersurface enclosing a bounded domain in a warped product. All Hausdorff measures are taken with respect to the Riemannian distance in the warped product.

Proposition 5.9 *Let (M, g) be the warped product $B \times_f Q$. Let $E \subset M$ a bounded set of finite perimeter, positive volume, and C^1 boundary $S = \partial E$. Assume also that the set of regular points $\mathcal{C}(F)$ of $F = \pi_B|_S$ has $\mathcal{H}^{m-1}$ measure 0. Then*

$$P(E) = \mathcal{H}^{m-1}(S) \geqslant \int_B \sqrt{\mathcal{H}^{q-1}(S_b)^2 + |(\nabla A - A\mathbf{H})_b|^2}\, dB(b), \tag{5.15}$$

where $\mathbf{H}_b$ is the projection of the mean curvature vector of $Q_b = \{b\} \times Q$ to B, $S_b = S \cap Q_b$, $A(b) = \mathcal{H}^q(E \cap Q_b)$, and ∇ is the gradient on (B, g_B).

Equality holds in (5.15) *if and only if the scalar product $g(N, \nu_b)$ of the outer unit normal N to S and the outer unit normal ν_b of $E \cap Q_b$ on Q_b is a constant function on S_b for all regular values $b \in B$ of F.*

Proof The set of regular points of F is composed of those $p \in S$ where S is transverse to the fiber $Q_{\pi_B(p)}$ at p. Around these points, S_b is a smooth submanifold of dimension $q - 1 = \dim(Q) - 1$, a hypersurface in $Q_{\pi_B(p)}$. Let $E_b = E \cap Q_b$ for all $b \in B$. For each $b \in B$, E_b is a measurable set. Fubini-Tonelli's theorem implies that $\mathcal{H}^b(E_b) < +\infty$ for a.e. $b \in B$. Sard's theorem implies that S_b is a smooth submanifold for a.e. $b \in B$.

If we assume that the set of regular points $\mathcal{C}(F)$ of F *has full measure in S*, then the coarea formula, Theorem 1.3, implies

$$\mathcal{H}^{m-1}(S) = \int_B \left\{ \int_{S_b} \frac{1}{\mathrm{Jac}^{\perp}(F)}\, dS_b \right\} dB(b),$$

where $\mathrm{Jac}(h)^{\perp}$ is the normal Jacobian, the determinant of dF restricted to the subspace orthogonal to $\ker(dF)$. Writing

$$\frac{1}{\mathrm{Jac}^{\perp}(F)} = \left(1 + \left(\frac{1}{\mathrm{Jac}^{\perp}(F)^2} - 1\right)\right)^{1/2},$$

and taking into account Minkowski's inequality

$$\left(\int (f+g)^{1/2}\right)^2 \geqslant \left(\int f^{1/2}\right)^2 + \left(\int g^{1/2}\right)^2$$

for $k = 1/2$ (see §6.13 in [217]), we obtain

$$\mathcal{H}^{m-1}(S) \geqslant \int_B \sqrt{\mathcal{H}^{q-1}(S_b)^2 + \left(\int_{S_b} \left(\frac{1}{\mathrm{Jac}^{\perp}(F)^2} - 1\right)^{1/2} dS_b\right)^2}\, dB(b), \tag{5.16}$$

where $\mathcal{H}^{q-1}(S_b)$ is the $(q-1)$-dimensional Hausdorff measure of S_b. Equality holds in (5.16) if and only if f, g are proportional a.e. on S_b for a.e. $b \in B$. This is equivalent to that $\mathrm{Jac}^{\perp}(F)^{-2}$ is constant a.e. in S_b for a.e. $b \in B$.

Let us show that

$$\int_{S_b} \left(\frac{1}{\mathrm{Jac}^{\perp}(F)^2} - 1\right)^{1/2} dS_b = |(\nabla A - A\mathbf{H})_b|, \tag{5.17}$$

where $A(b) = \mathcal{H}^q(E_b)$ and $\mathbf{H}_b$ is the projection of the mean curvature vector $\vec{H}_b$ of Q_b to B.

For any vector field U on $B \times_f Q$ such that $d\pi_B(U) = u$, (5.13) and (5.14) imply

$$g(U, \vec{H}_b) = \sum_{i=1}^{\beta} q\, \frac{n_j(f)}{f}\, g_B(u, n_j) = g_B(u, \mathbf{H}_b)$$

So we get

$$\int_{E_b} g(U, \vec{H}_b)\, dQ_b = A(b)\, g_B(u, \mathbf{H}_b). \tag{5.18}$$

Let us compute ∇A at some regular value $b \in B$. Since S is a compact submanifold, $S_{b'}$ is also a smooth submanifold for b' close enough to b. Consider a smooth curve $\sigma : (-\varepsilon, \varepsilon) \to B$ such that $\sigma(0) = b$, $\sigma'(0) = u$. Take a smooth vector field U on M so that $d\pi_B(U) = u$ and the flow of U takes E_b onto $E_{\sigma(t)}$

for t small. The first variation formulas and (5.18) imply that the derivative of the function $s \mapsto \mathcal{H}^q(E_{\sigma(s)})$ is given by

$$\begin{aligned} g_B((\nabla A)_b, u) &= \left.\frac{d}{ds}\right|_{s=0} \mathcal{H}^q(E_{\sigma(s)}) \\ &= \int_{E_b} g(U, \vec{H}_b)\, dQ_b + \int_{S_b} g(U, \nu_b)\, dS_b \\ &= A(b)\, g_B(u, \mathbf{H}_b) + \int_{S_b} g(U, \nu_b)\, dS_b, \end{aligned}$$

where ν_b is the outer unit normal to S_b inside Q_b. Hence, we have

$$g_B((\nabla A)_b - A(b)\mathbf{H}_b, u) = \int_{S_b} g(U, \nu_b)\, dS_b. \tag{5.19}$$

Let us prove finally that

$$\int_{S_b} g(U, \nu_b)\, dS_b = g_B(u, e) \int_{S_b} \left(\frac{1}{\mathrm{Jac}^{\perp}(F)^2} - 1\right)^{1/2} dS_b \tag{5.20}$$

for some vector unit vector $e \in T_b B$ that could depend on $p \in S_b$. Equations (5.20) and (5.19) then imply (5.17).

To prove (5.20), let us compute $\mathrm{Jac}^{\perp}(F)$ at some point $p \in S_b$. We first observe that $T_p S_b = \ker(dF_p)$ since $T_p S_b \subseteq \ker(dF_p)$ and both subspaces have the same dimension as b is a regular value of F. Recall that $T_p S_b$ is the subspace orthogonal to $(\nu_b)_p$, where ν_b is the outer unit normal to S_b at p. We choose a g_B-orthonormal basis $e_1, \dots, e_\beta$ so that $(e_i, 0) \in T_p M$ is tangent to S for $i = 1, \dots, \beta - 1$, and $(\lambda e_\beta, \mu \nu_b)$ with $\lambda^2 + \mu^2 = 1$ is tangent to $T_p S$. Hence,

$$(e_1, 0), \dots, (e_{\beta-1}, 0), (\lambda e_\beta, \mu \nu_b)$$

is an orthonormal basis of $\ker(dF_p)^{\perp}$ and so

$$\mathrm{Jac}^{\perp}(F)(p) = |\lambda|. \tag{5.21}$$

This implies in particular that $\lambda \neq 0$ as F is regular at p. A unit normal vector N to S is then $N = (\mu e_\beta, -\lambda \nu_b)$. As U is tangent to S, we have

$$0 = g(U, N) = \mu g_B(u, e_\beta) - \lambda g(U, \nu_b).$$

Hence,

$$g(U, \nu_b) = \frac{\mu}{\lambda} g_B(u, e_\beta) = \pm\left(\frac{1}{\lambda^2} - 1\right)^{1/2} g_B(u, e_\beta)$$
$$= \pm\left(\frac{1}{\mathrm{Jac}^{\perp}(F)^2(p)} - 1\right)^{1/2} g_B(u, e_\beta).$$

Integrating on S_b, we get (5.20) with $e = \pm e_\beta$. This completes the proof of (5.15).

Equality holds in (5.15) if and only if equality holds in the Minkowski inequality (5.16). This only happens when $\mathrm{Jac}^{\perp}(F)$ is constant a.e. in S_b for all regular values $b \in B$. This means that the constant λ in Eq. (5.21) is constant on S_b and so the scalar product $g(N, \nu_b)$ is constant on S_b. □

Inequality (5.15) explains why we should expect smaller perimeter after symmetrization in warped products. Note that in the symmetrization procedure, the term $|\nabla A - A\mathbf{H}|$ remains constant, while $\mathcal{H}^{q-1}((\partial E)_b) \geqslant \mathcal{H}^{q-1}(\partial D_b)$, where $D_b = \mathrm{sym}(E) \cap Q_b$ is the isoperimetric disk in Q_b with the same volume as E_b.

Remark 5.10 A closer look to formula (5.15) would permit to characterize the cases when equality holds in the symmetrization procedure. We refer the reader to Chlebík et al. [110] and Morgan et al. [309] for a detailed analysis of the equality case.

Remark 5.11 When B is one-dimensional, Proposition 5.9 was derived in §1.3.2 in [369]. In this particular case, the derivation is much easier, but enough for most of the applications.

5.2 Spaces with Constant Sectional Curvature

In this section, we obtain isoperimetric inequalities in the Euclidean spaces, hyperbolic spaces, and spheres. The proofs will use the symmetrization results for warped products of Sect. 5.1. We start treating the one-dimensional case that does not require any symmetrization.

5.2.1 The One-Dimensional Case

In this section, we consider the one-dimensional problem isoperimetric problem for sets of finite perimeter. Since our proof of the isoperimetric inequality in Euclidean space will be done by induction, this would provide the first step of the induction process. As stated in the following lemma, the complete solution is that the perimeter of a set is larger than or equal to 2, with equality for essentially

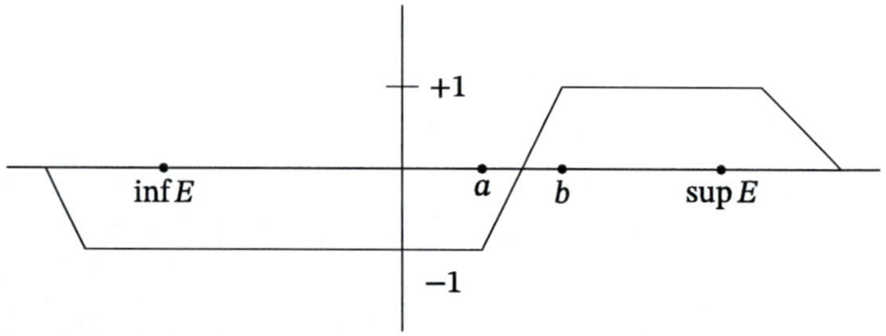

Fig. 5.3 The function $f_{a,b}$

connected sets. The reader should compare this proof with the one by Talenti; see §3.6 in [419].

Lemma 5.12 *Let $E \subset \mathbb{R}$ or $\mathbb{S}^1$ be a bounded measurable set of finite perimeter and positive volume. In case $E \subset \mathbb{S}^1$, assume also that $|\mathbb{S}^1 \setminus E| \neq 0$. Then $P(E) \geqslant 2$. Equality holds if and only if E is L^1 equivalent to a connected set.*

Proof We recall that the perimeter, in the one-dimensional case, is given by

$$P(E) = \sup\left\{ \int_E f' : f \text{ Lipschitz with compact support, } ||f||_\infty \leqslant 1 \right\}.$$

We consider first the case $E \subset \mathbb{R}$. For $a, b \in \mathbb{R}$, $a < b$, we consider the function $f_{a,b}$ defined as in the following Fig. 5.3:

Since E is a measurable set with positive volume, there exists a point $x \in \mathbb{R}$ of density 1. Taking the function $f = f_{x-r,x+r}$, we have

$$\int_E f' = \frac{2}{2r}|E \cap (x - r, x + r)| \to 2.$$

Hence, $P(E) \geqslant 2$.

Assume now that there is a point y of density $0 \leqslant \rho < 1$ between two points $x_1 < y < x_2$ of density 1. For $r > 0$ small enough, we consider the function f_r defined as in the following Fig. 5.4:

Then we have

$$\int_E f_r' = \frac{1}{2r}\big(|E \cap (x_1 - r, x_1 + r)| - |E \cap (y_r, y + r)| + |E \cap (x_2 - r, x_2 + r)|\big)$$
$$= 2(2 - \rho) > 2.$$

We conclude that, in case $P(E) = 2$, there are no points of density $0 \leqslant \rho < 1$ between two points of density 1. Then the set of Lebesgue points is connected and hence an interval. Since E is L^1 equivalent to its set of Lebesgue points, we obtain the desired result.

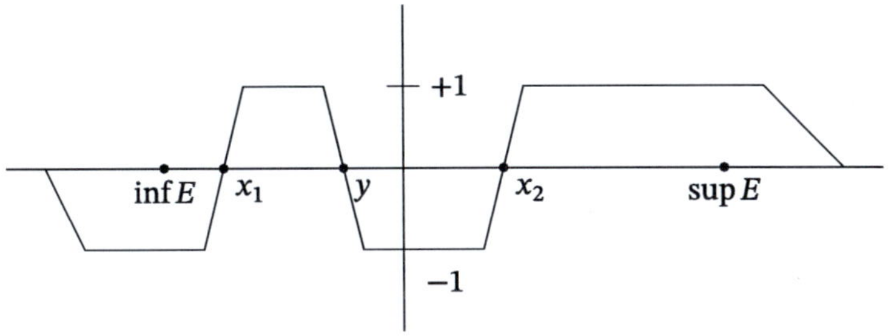

Fig. 5.4 The function f_r. The slope around x_1, x_2, y is $1/(2r)$

In case $E \subset \mathbb{S}^1$, we only need to modify slightly the functions near a point of density 0 of f. □

5.2.2 *Decomposition of Simply Connected Space Forms as Warped Products*

The isoperimetric inequality in the complete m-dimensional simply connected spaces of constant sectional curvature, the Euclidean space $\mathbb{R}^m$, the hyperbolic space $\mathbb{H}^m$, and the sphere $\mathbb{S}^m$, will be proved by induction, expressing each of these spaces as a warped product involving spaces of less dimension. This strategy appears in the classical proof by Schwarz [402] of the isoperimetric inequality in the three-dimensional Euclidean space $\mathbb{R}^3$, where the Euclidean space $\mathbb{R}^3$ is expressed as the Riemannian product of the line $\mathbb{R}$ with the plane $\mathbb{R}^2$. Given a set $E \subset \mathbb{R}^3$ of finite perimeter, if we know that geodesic balls in $\mathbb{R}^2$ are isoperimetric balls, we can use the symmetrization Theorem 5.6 to produce a body of revolution about a line with the same volume and no larger perimeter than E. In a second step, we shall characterize the sets of revolution minimizing the perimeter. This strategy was further developed by Schmidt in a series of papers [394, 395, 397] to complete the proof of the isoperimetric inequality in $\mathbb{R}^m$, $\mathbb{H}^m$, and $\mathbb{S}^m$.

Actually, we have several options for the induction process. In the Euclidean space, it seems natural to consider the decomposition $\mathbb{R}^m = \mathbb{R}^{m-1} \times_f \mathbb{R}$, with $f \equiv 1$. A similar one is possible in the hyperbolic space; see p. 290 in [102]. In the warped product $M = \mathbb{R} \times_f \mathbb{H}^{m-1}$, where $f(t) = \cosh(t)$, the formulas for the sectional curvatures of a warped product (see pp. 210 ff. in O'Neill [323]) imply that M has constant sectional curvatures equal to -1. As M is simply connected, it is isometric to the m-dimensional hyperbolic space $\mathbb{H}^m$.

Another approach is to consider a polar decomposition using a fixed point and the geodesic spheres centered at this point. This way, the Euclidean space $\mathbb{R}^m$ minus the origin is isometric to the warped product $(0, +\infty) \times_f \mathbb{S}^{m-1}$, where $\mathbb{S}^{m-1}$ is the $(m-1)$-dimensional unit sphere, with warping function $f(t) = t$. The hyperbolic space $\mathbb{H}^m(\kappa)$ of constant sectional curvature $\kappa < 0$ minus a point is isometric to the warped product $(0, \infty) \times_f \mathbb{S}^{m-1}$ with warping function

$$f(t) = \frac{1}{\sqrt{-\kappa}} \sinh(\sqrt{-\kappa} t).$$

And the sphere $\mathbb{S}^m(\kappa)$ of constant sectional curvature $\kappa > 0$ minus two antipodal points is isometric to the warped product $(0, \frac{\pi}{\sqrt{\kappa}}) \times \mathbb{S}^{m-1}$ with warping function

$$f(t) = \frac{1}{\sqrt{\kappa}} \sin(\sqrt{\kappa} t).$$

The sets where t is constant are concentric geodesic spheres of radius t.

Of course, these are not the only possible symmetrizations in space forms. We refer the reader to §9.2.1 in Burago and Zalgaller [83] for a different type of symmetrization, in particular, for the one called by the authors *spherical Steiner symmetrization*. This symmetrization is also a particular case of symmetrization in warped products.

5.2.3 The Isoperimetric Inequality in Euclidean Space

We sketch in this section a proof of the isoperimetric inequality in Euclidean space using symmetrization. The main ingredients in the proof are the symmetrization Theorem 5.6 and the calibration argument in Theorem 5.14 (Fig. 5.5).

Theorem 5.13 *The geodesic balls are isoperimetric sets in $\mathbb{R}^m$ for all $m \geqslant 1$.*

Proof The proof is by induction on the dimension m of the Euclidean space. The case $m = 1$ was considered in Lemma 5.12. Assume by induction that geodesic balls are isoperimetric sets in $\mathbb{R}^{m-1}$.

Let $B = \mathbb{R}$ and $Q = \mathbb{R}^{m-1}$, and we fix $q_0 = 0 \in \mathbb{R}^{m-1}$. By induction, we know that geodesic balls centered at $0 \in \mathbb{R}^{m-1}$ are isoperimetric sets. Let $E \subset \mathbb{R}^m$ be a bounded set of finite volume and perimeter. We apply Theorem 5.6 to conclude that $P(\text{sym}(E)) \leqslant P(E)$, where $\text{sym}(E)$ is the symmetrized of E in the Riemannian product $\mathbb{R} \times \mathbb{R}^{m-1}$. The set $\text{sym}(E)$ is of revolution around the line $\mathbb{R} \times \{0\}$ and contains a section disk $\{t\} \times D(0, r)$ of maximal radius. Then Theorem 5.14 implies that $P(\text{sym}(E)) \geqslant P(B)$, where B is a ball in $\mathbb{R}^m$ of the same volume as $\text{sym}(E)$. So we have $|E| = |\text{sym}(E)| = |B|$ and $P(E) \geqslant P(\text{sym}(E)) \geqslant P(B)$. If E is not bounded, we approximate its perimeter by the one of a bounded set with the same volume, and we are done. □

Theorem 5.14 *Let $E \subset \mathbb{R}^m$ be a compact set of finite volume and perimeter. Let C be a closed right circular cylinder with axis L and base D so that*

$$D \subset E \subset C. \tag{5.22}$$

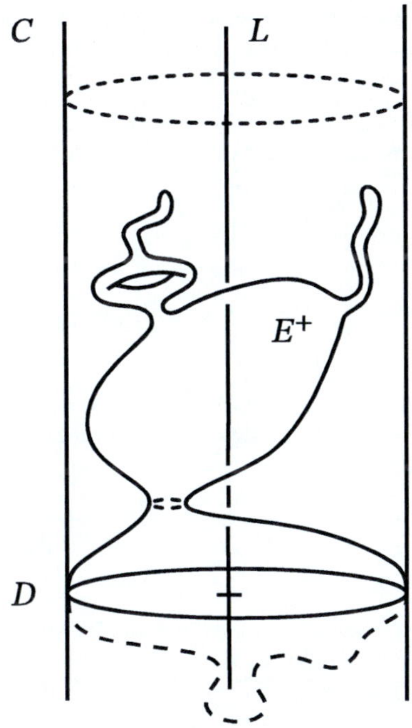

Fig. 5.5 Representation of the geometric configuration in Theorem 5.14

Then we have

$$P(E) \geqslant P(B), \tag{5.23}$$

where B is a ball in $\mathbb{R}^m$ *satisfying with* $|E| = |B|$.

Moreover, equality holds in (5.23) *if and only if* $E = B$ *and* B *and* D *have the same radius.*

Proof For simplicity, we assume that L is the x_m-axis and that D is the disk in the hyperplane $x_m = 0$ centered at 0 of radius $r_0 > 0$. Let S_λ be the sphere of radius r_0 and mean curvature $\lambda = (m-1)/r_0$ centered at $0 \in \mathbb{R}^m$. Its intersection with the hyperplane $x_m = 0$ is the disk D.

Let S_λ^+ be the closed upper half-sphere

$$S_\lambda^+ = \{p \in S_\lambda : p_m \geqslant 0\}.$$

We translate vertically S_λ^+ to produce a foliation of the cylinder C. Let X be the unit normal vector (pointing upward) to the leaves of the foliation. It is immediate to check that

$$\operatorname{div} X = \lambda.$$

We integrate the divergence of X in the set of finite perimeter

$$E^+ = \{p \in E : p_m \geqslant 0\}$$

and use the divergence theorem to obtain

$$\begin{aligned}\lambda|E^+| &= \int_{\partial^* E \cap \{x_m > 0\}} \langle X, N_{\partial E}\rangle \, dP - \int_D \langle X, N_D\rangle \, dD \\ &\leqslant P(E, \{x_m > 0\}) - \int_D \langle X, N_D\rangle \, dD,\end{aligned}$$

where $\partial^* E$ is the reduced boundary of E, $N_{\partial E}$ is the measurable outer unit normal, and N_D is the downward pointing unit normal to the hyperplane $x_m = 0$. Hence, we have

$$P(E, \{x_m > 0\}) - \lambda|E^+| \geqslant \int_D \langle X, N_D\rangle \, dD.$$

If equality holds, then $N_{\partial E}$ coincides P-a.e. on $\partial^* E \cap \{x_m > 0\}$ with the continuous vector field X. Then $\partial^* E \cap \{x_m > 0\}$ is of class C^1 (see Theorem 4.11 in Giusti [184]), and it is equal to a vertical translation of the half-sphere $S_\lambda^+ \cap \{x_m > 0\}$.

Let B_λ be the ball enclosed by S_λ. The same application of the divergence theorem to $B_\lambda^+ = \{p \in B_\lambda : p_m \geqslant 0\}$ provides the equality

$$P(B_\lambda^+, \{x_m > 0\}) - \lambda|B_\lambda^+| = \int_D \langle X, N_D\rangle \, dD.$$

So we get

$$P(E, \{x_m > 0\}) - \lambda|E^+| \geqslant P(B_\lambda^+, \{x_m > 0\} - \lambda|B_\lambda^+| \tag{5.24}$$

with equality if and only if $\partial^* E \cap \{x_m > 0\}$ is a vertical translation of the half-sphere $S_\lambda \cap \{x_m > 0\}$.

With a similar reasoning, we get, for $E^- = \{p \in E : p_m \leqslant 0\}$, $B_\lambda^- = \{p \in B_\lambda : p_m \leqslant 0\}$, the inequality

$$P(E, \{x_m < 0\}) - \lambda|E^-| \geqslant P(B_\lambda, \{x_m < 0\}) - \lambda|B_\lambda^-|, \tag{5.25}$$

with equality if and only if $\partial^* E \cap \{x_m < 0\}$ is a vertical translation of the half-sphere $S_\lambda \cap \{x_m < 0\}$.

Adding inequalities (5.24) and (5.25), we get

$$P(E) - \lambda|E| \geqslant P(B_\lambda) - \lambda|B_\lambda|, \tag{5.26}$$

with equality if and only if ∂E consists of two vertical translations of $\mathbb{S}_\lambda^+$ and $\mathbb{S}_\lambda^-$ and a portion of the boundary of the cylinder in between.

For $\mu > 0$, we define

$$f(\mu) = P(B_\mu) + \mu\big(|E| - |B_\mu|\big),$$

where B_μ is a ball of radius $(m-1)/\mu$ whose boundary is a sphere with constant mean curvature μ. Since spheres of mean curvature μ_0 are critical points of the functional $P(\cdot) - \mu_0|\cdot|$, we have

$$f'(\mu) = |E| - |B_\mu|.$$

Moreover, $f''(\mu) > 0$. Hence, $f(\mu)$ has a unique minimum λ_0 for which $f'(\lambda_0) = 0$ and so $|E| = |B_{\lambda_0}|$. From (5.26), we get

$$P(E) \geqslant f(\lambda) \geqslant f(\lambda_0) = P(B_{\lambda_0}), \tag{5.27}$$

and so E has more perimeter than a ball B_{λ_0} of the same volume.

Equality holds in (5.27) if and only of $B_\lambda = B_{\lambda_0}$ and $P(E) = P(B_\lambda)$. Then $|E| = |B_\lambda|$ and E must be coincide with a vertical translation of the sphere B_λ. □

In Theorem 5.13, we did not obtain uniqueness of isoperimetric sets. However, once we know that balls are isoperimetric sets, we can use an argument by Gromov (see §VI.5 in Chavel [102]) to prove uniqueness. The argument is based on Heintze-Karcher's inequality and resembles the arguments in the proof of Theorem 4.34.

Theorem 5.15 *Let E be an isoperimetric set in $\mathbb{R}^m$. Then E is a geodesic ball.*

Proof If geodesic balls are isoperimetric, then the isoperimetric profile of $\mathbb{R}^m$ is the smooth function

$$I(v) = I_{\mathbb{R}^m}(v) = \mathbf{c}(m)\, v^{(m-1)/m},$$

where

$$\mathbf{c}(m) = m\omega_m^{1/m}$$

is the isoperimetric constant in $\mathbb{R}^m$. As the isoperimetric profile is differentiable at every point, all isoperimetric sets of volume v_0 have the same mean curvature $I'(v_0)$.

Now we reason exactly as in the proof of Theorem 4.34 to get

$$|E| \leqslant \int_S \left\{ \int_0^{(m-1)/H} \left(1 - t\frac{H}{m-1}\right)^{m-1} dt \right\} dS(p),$$

where H is the mean curvature of the regular part of ∂E. If B is a ball with the same volume as E, then B is an isoperimetric set and has the same mean curvature H as E, so we have

$$\begin{aligned} |E| &\leqslant P(E) \int_0^{(m-1)/H} \left(1 - t\frac{H}{m-1}\right)^{m-1} dt \\ &= P(B) \int_0^{(m-1)/H} \left(1 - t\frac{H}{m-1}\right)^{m-1} dt = |B|. \end{aligned}$$

Hence, we have equality in the arithmetic-geometric inequality, and E must be a geodesic ball since the regular part of ∂E is totally umbilical. □

5.2.4 *The Isoperimetric Inequality in Hyperbolic Space*

We briefly sketch the proof of the isoperimetric characterization of balls in $\mathbb{H}^m$ since the arguments are similar to the ones in Euclidean case in Sect. 5.2.3.

Theorem 5.16 *Let E be an isoperimetric set in the hyperbolic space $\mathbb{H}^m$, $m \geqslant 2$. Then E is a geodesic ball.*

Proof The proof is by induction on the dimension m of the hyperbolic space $\mathbb{H}^m$ of constant sectional curvature -1. For $m = 2$, the result holds because of Theorem 2.24 since the only curves with constant geodesic curvature are the geodesic circles.

We assume that the result holds for the $(m-1)$-dimensional hyperbolic space $\mathbb{H}^{m-1}$ of constant sectional curvature -1. We consider the warped product $M = \mathbb{R} \times_f \mathbb{H}^{m-1}$, where $f(t) = \cosh(t)$. The formulas for the sectional curvatures of a warped product (see pp. 210 ff. in O'Neill [323]) imply that M has constant sectional curvatures equal to -1. Since M is simply connected, it is isometric to the m-dimensional hyperbolic space $\mathbb{H}^m$.

Consider a bounded isoperimetric set $E \subset \mathbb{H}^m \equiv M$. We fix a point $q_0 \in \mathbb{H}^{m-1}$, and we symmetrized E using Theorem 5.6 to get a set $\operatorname{sym}(E)$ with $P(\operatorname{sym}(E)) \leqslant P(E)$. Since every reflection with respect to a totally geodesic submanifold containing the geodesic $\mathbb{R} \times \{q_0\}$ preserves the set $\operatorname{sym}(E)$, it follows that $\operatorname{sym}(E)$ is contained in a tubular neighborhood C of positive radius of the geodesic $\Gamma = \mathbb{R} \times \{q_0\}$.

If $r_0 > 0$ is the smallest of such tubular neighborhoods, it follows that $\operatorname{sym}(E)$ contains the intersection D of a ball of radius r_0 with a totally geodesic submanifold orthogonal to Γ. We can place a ball B of radius r_0 centered at a fixed point of Γ. The ball can be translated by moving its center along Γ, and it is still contained in C. We can split ∂B into two half-spheres and reason as in the proof of Theorem 5.14 to conclude that $P(E) \geqslant P(B_0)$, where $B_0 \subset \mathbb{H}^m$ is the metric ball with $|\operatorname{sym}(E)| =$

$|B_0|$ (we only need to show that the volume of a geodesic ball is a concave function of its mean curvature). Hence, we obtain $P(E) \geqslant P(\mathrm{sym}(E) \geqslant |B_0|$.

If E is not bounded, we approximate the perimeter of E by the one of a bounded set with the same volume as E.

For uniqueness, we reason as in the proof of Theorem 5.15 using instead the formulas in the proof of Theorem 4.38. □

5.2.5 The Isoperimetric Inequality in Euclidean Space by Multiple Symmetrization

We give another proof of the isoperimetric property of balls in Euclidean space, in this case using only Steiner's symmetrization.

Theorem 5.17 *Let $E \subset \mathbb{R}^m$ be a measurable set of positive volume. Let B be the closed ball in $\mathbb{R}^m$ of the same volume as E. Then*

$$|E_r| \geqslant |B_r|, \tag{5.28}$$

for all $r > 0$, and hence,

$$\mathrm{Mink}_-(E) \geqslant \mathrm{Mink}(B). \tag{5.29}$$

Proof We assume that the result holds on $\mathbb{R}^{m-1}$ by induction on m. Theorem 5.6 then implies that Steiner's symmetrization in $\mathbb{R}^m$ with respect to a hyperplane does not increase the lower Minkowski content of the symmetral of a set.

We first assume that $E \subset \mathbb{R}^m$ is a compact set and consider the family $\mathcal{S}(E)$ of compact sets obtained by applying to E a finite number of Steiner symmetrizations with respect to hyperplanes passing through 0. Observe that $|F| = |E|$ for any $F \in \mathcal{S}(E)$. For every compact set $F \subset \mathbb{R}^m$, we define

$$\rho(F) = \inf\{r > 0 : F \subseteq \overline{B}(0, r)\}.$$

Let

$$\rho = \inf\{\rho(F) : F \in \mathcal{S}(E)\}.$$

Observe that $\rho > 0$ since, for all $F \in \mathcal{S}(E)$, we have $|\overline{B}(0, r)| \geqslant |F| = |E|$.

The Steiner symmetral $\mathrm{sym}(\cdot)$ of a set with respect to a hyperplane H has the following properties:

- If $G_1 \subset G_2$, then $\mathrm{sym}(G_1) \subset \mathrm{sym}(G_2)$.
- If $0 \in H$, then $\mathrm{sym}(\overline{B}(0, r)) = \overline{B}(0, r)$ for all $r > 0$.

Both properties and the compactness of E imply that $\mathcal{S}(E)$ is bounded in Hausdorff distance.

Take a sequence $\{F_i\}_{i\in\mathbb{N}}$ in $\mathcal{S}(E)$ such that $\rho(F_i)$ converges to ρ. By Blaschke's selection theorem, there is a non-relabeled subsequence converging to a compact set F such that $\rho = \rho(F)$. As $\mathbf{1}_F \geqslant \limsup_{i\to\infty} \mathbf{1}_{F_i}$, Fatou's lemma implies

$$|F| \geqslant \limsup_{i\to\infty} |F_i| = |E|.$$

On the other hand, since F_i is obtained from E by a finite number of Steiner symmetrizations, we get from (5.6) in Theorem 5.6 that $|(F_i)_r| \leqslant |E_r|$ for all $i \in \mathbb{N}$ and $r > 0$. As $\{F_i\}_{i\in\mathbb{N}}$ converges to F in Hausdorff distance, for every pair of positive numbers $0 < r' < r$, we have $F_{r'} \subset (F_i)_r$ for i large enough and so

$$|E_r| \geqslant |(F_i)_r| \geqslant |F_{r'}| \geqslant |F|.$$

Letting $r \to 0$, we get $|E| \geqslant |F|$ and so

$$|E| = |F|. \tag{5.30}$$

Letting $r' \to r$, we get

$$|E_r| \geqslant |F_r| \tag{5.31}$$

for all $r > 0$, and, since $|E| = |F|$, we obtain

$$\text{Mink}_-(E) \geqslant \text{Mink}_-(F). \tag{5.32}$$

Finally, let us see that F coincides with the ball $\overline{B}(0,\rho)$. We need the following property of the Steiner symmetral. Let G be a compact set with $G \subset \overline{B}(0,r)$. Let S be a compact set contained in the boundary of the ball $\partial\overline{B}(0,r)$ satisfying $S\cap G = \emptyset$. Then the Steiner symmetral of G with respect to a hyperplane H passing through 0 satisfies

$$S \cap \text{sym}(G) = \emptyset, \qquad \sigma(S) \cap \text{sym}(G) = \emptyset, \tag{5.33}$$

where σ is the reflection with respect to H. See Fig. 5.6.

Let us assume that $F \neq \partial\overline{B}(0,\rho)$. Then there exists $x_0 \in \partial\overline{B}(0,\rho) \setminus F$ and $r_0 > 0$ such that $B(x_0, r_0) \cap F = \emptyset$. Let us cover $\partial\overline{B}(0,\rho)$ by a finite number of different balls $\overline{B}(x_0,r_0), \overline{B}(x_1,r_0), \ldots, \overline{B}(x_k,r_0)$. For every $i \in \{1,\ldots,k\}$, we consider the Steiner symmetral with respect to the hyperplane H_i passing through 0 with normal vector $x_i - x_0$. If we apply successively Steiner's symmetrization with respect to $H_1,\ldots,H_k$, we get a set F_0 satisfying $\rho(F_0) < \rho$ because of (5.33). This provides a contradiction to the definition of ρ by the following argument. For every $i \in \mathbb{N}$, let F_i' be the set obtained by applying successive symmetrizations to F_i with

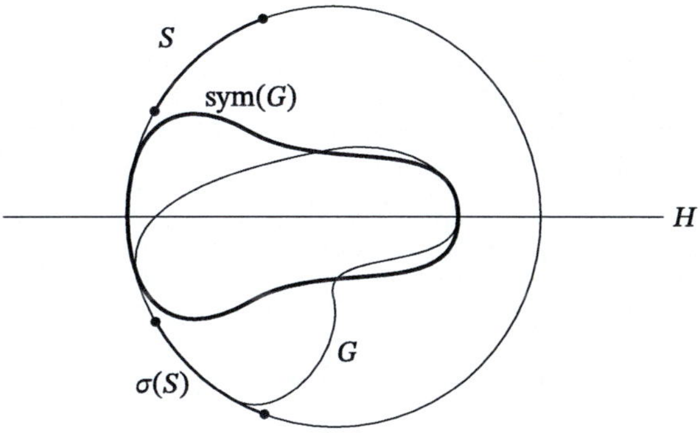

Fig. 5.6 Detachment of the Steiner symmetral sym G from the boundary of a ball when the symmetry hyperplane passes through the center of the ball

respect to $H_1, \ldots, H_k$. Then $F_i \in \mathcal{S}(E)$ for all $i \in \mathbb{N}$. Let F' be the Hausdorff limit of a non-relabeled subsequence of $\{F_i'\}_{i\in\mathbb{N}}$. Then Lemma 5.5 implies $F' \subseteq F_0$ and so $\rho(F') \leqslant \rho(F_0) < \rho(F) = \rho$. This contradiction shows that F is a ball and (5.31) and (5.32) imply (5.28) and (5.29).

This proves (5.28) and (5.29) for compact sets. If $E \subset \mathbb{R}^m$ is measurable, we can find a sequence of compact sets $E_i \subset E$ such that $E_i \to |E|$. Hence,

$$|E_r| \geqslant |(E_i)_r| \geqslant |(B_i)_r|$$

for all $r > 0$ and $i \in \mathbb{N}$, where B_i is a ball with $|B_i| = |E_i|$. Taking limits, we get $|E_r| \geqslant |B_r|$ for all $r > 0$, where B is a ball with $|E| = |B|$, and so $\text{Mink}_-(E) \geqslant \text{Mink}(B)$. □

We could have used Schwarz or spherical symmetrization to prove Theorem 5.17 instead of Steiner's. The strategy of the proof is similar. Since any measurable set can be approximated in volume by compact sets, it is enough to prove inequality (5.29) when E is compact. We consider the class $\mathcal{S}(E)$ of sets that can be obtained from E by a finite number of Schwarz or spherical symmetrizations. In the first case, we require that the axis of symmetry passes through 0. In the second one, we require that the line containing the half-line of symmetry contains 0. Both symmetrizations preserve the closed balls $\overline{B}(0, r)$ for all $r > 0$. We consider the quantity

$$\rho(F) = \inf\{r > 0 : F \subset \overline{B}(0, r)\}$$

and take a sequence $\{F_i\}_{i\in\mathbb{N}}$ so that $\rho(F_i)$ converges to the infimum ρ of the set $\{\rho(F) : F \in \mathcal{S}(E)\}$. Passing to a subsequence, we may assume that F_i converges in Hausdorff distance to some set F such that $\rho = \rho(F)$. Like in the proof of Theorem 5.17, we show that $|F| = |E|$ and $|F_r| \leqslant |E_r|$ for all $r > 0$ and $\text{Mink}_-(F) \leqslant \text{Mink}_-(E)$.

To prove that F coincides with the ball $\overline{B}(0, \rho)$, we see that a detachment property similar to (5.33) holds for Schwarz and spherical symmetrizations. In the latter case, when the origin of the half-line lies outside the open ball $B(0, \rho)$.

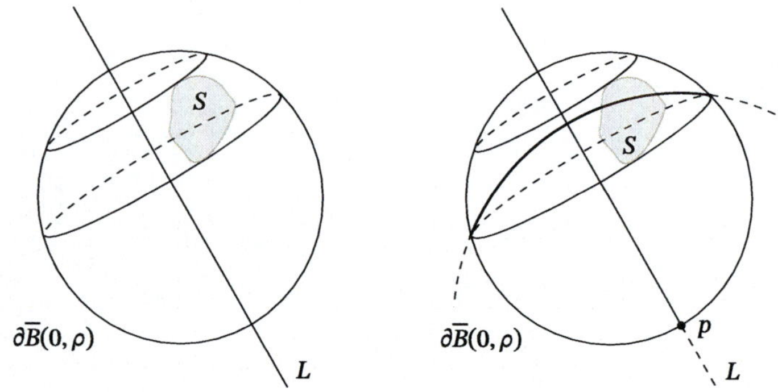

Fig. 5.7 The detachment property for Schwarz and spherical symmetrization. Let E be any compact set at positive distance from $S \subset \partial \overline{B}(0, \rho)$, and let H be a hyperplane orthogonal to L or a sphere centered at p with the property $H \cap S \neq \emptyset$. Then we have $\mathrm{sym}(E) \cap H \cap \partial \overline{B}(0, \rho) = \emptyset$ since $\mathrm{sym}(E) \cap H$ is a disk in H, concentric to $\overline{B}(0, r) \cap H$, and with strictly smaller volume

In this case, the obtained property is the following: let G be a compact set with $G \subset \overline{B}(0, r)$. Let S be a compact set contained in the boundary of the ball $\partial \overline{B}(0, r)$ satisfying $S \cap G = \emptyset$. Then the Schwarz or spherical symmetral of G with respect to a line L passing through 0 satisfies

$$\pi_L^{-1}(\pi_L(S)) \cap \mathrm{sym}(G) = \emptyset, \tag{5.34}$$

where π_L is the orthogonal projection to L. The set $\pi_L^{-1}(\pi_L(S))$ is the smallest slab bounded by two parallel hyperplanes orthogonal to L containing S; see Fig. 5.7.

The proof is completed applying Lemma 5.5.

5.2.6 *The Isoperimetric Inequality in the Round Sphere*

We sketch the proof of the isoperimetric property of geodesic balls in the round sphere $\mathbb{S}^m$. We apply the method of multiple symmetrizations with respect to geodesic balls.

Theorem 5.18 *Let E be a measurable set in the round sphere $\mathbb{S}^m$, $m \geqslant 2$. Then*

$$\mathrm{Mink}_-(E) \geqslant \mathrm{Mink}(B),$$

where B is a geodesic ball of the same volume as E.

Proof The proof is by induction on the dimension m of the sphere $\mathbb{S}^m$ of constant sectional curvature 1. For $m = 2$, the result holds because of Theorem 2.24 since the only curves with constant geodesic curvature are the geodesic circles.

Assume by induction that geodesic balls are isoperimetric sets on the sphere $\mathbb{S}^{m-1}$. We express $\mathbb{S}^m$ minus two antipodal points as a warped product with fiber $\mathbb{S}^{m-1}$. Fix an arbitrary point $p_0 \in \mathbb{S}^m$. By Gauss lemma, the Riemannian metric of $\mathbb{S}^m$ can be written on $\mathbb{S}^m \setminus \{p_0, -p_0\}$ as $dt^2 + \sin(t)^2\, d\sigma_{m-1}^2$, where t is the distance to p_0 and $d\sigma_{m-1}^2$ is the Riemannian metric on $\mathbb{S}^{m-1}$. Thus, $\mathbb{S}^m \setminus \{p_0, -p_0\}$ is isometric to the warped product $M = (0, \pi) \times_f \mathbb{S}^{m-1}$, where $f(t) = \sin(t)$.

Given any compact set $E \subset \mathbb{S}^m$, we denote by $\operatorname{sym}(E)$ the symmetral of E with respect to this warped product structure through the symmetrization introduced in Sect. 5.1. In this particular case, this spherical symmetrization consists of fixing some $p_0 \in \mathbb{S}^m$ and a geodesic segment Γ connecting p_0 to its antipodal point and replacing the intersection $E \cap S(p_0, t)$ of E with the geodesic sphere $S(p_0, t) \subset \mathbb{S}^m$ by the geodesic disk in $S(p_0, t)$ centered at $\Gamma \cap S(p_0, t)$ with the same $(m-1)$-volume. The point (t, q_0) moves, when $t \in (0, \pi)$, along a fixed geodesic connecting p_0 and $-p_0$. Thus, this symmetrization depends on a fixed point $p_0 \in \mathbb{S}^m$ and a geodesic connecting p_0 and its antipodal point $-p_0$. We can preserve the compactness of the symmetral in $\mathbb{S}^m$ by adding the points $\pm p_0$ in case they lie in the closure of the symmetral.

We consider a compact set $E \subset \mathbb{S}^m$ with volume $0 < |E| < |\mathbb{S}^m|$. The compactness of E and condition $|E| < |\mathbb{S}^m|$ imply the existence of $x_0 \in \mathbb{S}^m$ and $0 < r_0 < \pi$ such that $E \cap \overline{B}(-x_0, \pi - r_0) = \emptyset$. This means $E \subset \overline{B}(x_0, r_0)$. An *admissible symmetrization* of E is one obtained taking as base point any $p_0 \in S(x_0, \pi/2)$ and as geodesic segment the one connecting p_0 and $-p_0$ passing through x_0. See Fig. 5.8.

Now we reason as in Sect. 5.2.5. We consider the family $\mathcal{S}(E)$ of compact sets obtained by applying to E a finite number of *admissible* symmetrizations. Observe that $|F| = |E|$ for any $F \in \mathcal{S}(E)$. For any admissible symmetrization, we also have $\operatorname{sym}(\overline{B}(x_0, r)) = \overline{B}(x_0, r)$ for all $0 < r < \pi$. For every compact set $F \subset \mathbb{S}^m$, we define

$$\rho(F) = \inf\{r > 0 : F \subset \overline{B}(x_0, r)\}$$

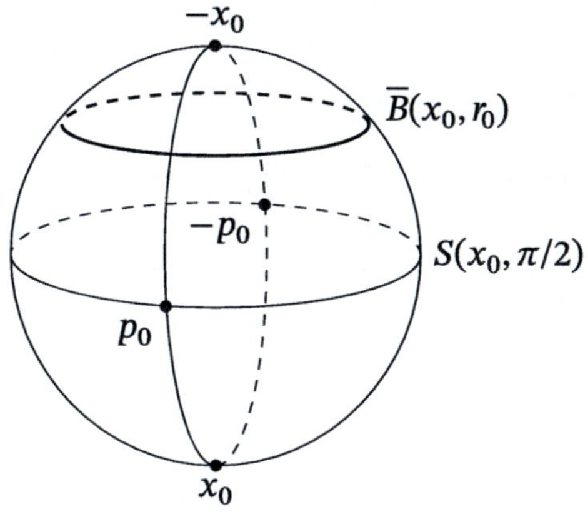

Fig. 5.8 Admissible symmetrizations

and

$$\rho = \inf\{\rho(F) : F \in \mathcal{S}(E)\}.$$

Observe that $\rho > 0$ since, for all $F \in \mathcal{S}(E)$, we have $|\overline{B}(x_0, \rho(F))| \geqslant |F| = |E|$. Note also that, given a compact $F \subset \mathbb{S}^m$, there exists $q(F) \in \mathbb{S}^m$ such that $F \subset \overline{B}(q(F), \rho(F))$.

We take a sequence $\{F_i\}_{i \in \mathbb{N}}$ such that $\rho(F_i)$ converges to ρ. By Blaschke's selection theorem, there is a non-relabeled subsequence converging to a compact set F such that $\rho = \rho(F)$. Reasoning as in the proof of Theorem 5.17, we get

$$|F| = |E|, \qquad \text{Mink}_-(E) \geqslant \text{Mink}_-(F).$$

Let us prove that $F = \overline{B}(x_0, \rho)$. We reason by contradiction assuming that $F \neq \overline{B}(x_0, \rho)$. Then there exists $z_0 \in S(x_0, \rho) \setminus F$ and $s_0 > 0$ such that $\overline{B}(z_0, s_0) \cap F = \emptyset$. Let us cover $S(x_0, \rho)$ by a finite number of different balls $\overline{B}(z_0, s_0), \overline{B}(z_1, s_0), \ldots, \overline{B}(z_k, s_0)$. For every $i \in \{1, \ldots, k\}$, we consider the admissible symmetrizations with respect to any point p_i in the intersection of the hyperplane H_i passing through x_0 with normal vector $z_i - z_0$ and the equator $S(x_0, \pi/2)$. If we apply successively admissible symmetrizations with respect to $p_1, \ldots, p_k$, we get a set F_0 satisfying $\rho(F_0) < \rho$ because of the detachment property. This provides a contradiction to the definition of ρ as in the proof of Theorem 5.17. For every $i \in \mathbb{N}$, let F_i' be the set obtained by applying successive symmetrizations to F_i with respect to $H_1, \ldots, H_k$. Then $F_i \in \mathcal{S}(E)$ for all $i \in \mathbb{N}$. Let F' be the Hausdorff limit of a non-relabeled subsequence of $\{F_i'\}_{i \in \mathbb{N}}$. Then Lemma 5.5 implies $F' \subset F_0$ and so $\rho(F') \leqslant \rho(F_0) < \rho(F) = \rho$.

Finally, if E is measurable, we approximate its volume by compact sets contained in E to get $|E_r| \geqslant |B_r|$ for all $r > 0$ and $\text{Mink}_-(E) \geqslant \text{Mink}(B)$, where B is the geodesic ball with the same volume as E. □

5.3 Hsiang Symmetrization

We present in this section a symmetrization result from W.-T. Hsiang and W.-Y. Hsiang [234], who proved it for products of Euclidean and hyperbolic spaces. A more general version was given by W.-Y. Hsiang in [235, 237].

5.3.1 *The Symmetrization Result*

We consider the product manifold $M = N \times Q$, where N is any Riemannian manifold and Q is the Euclidean space $\mathbb{R}^q$, a round sphere $\mathbb{S}^q$, or a hyperbolic space $\mathbb{H}^q$, $q \geqslant 1$. Note that Q is simply connected except when $Q = \mathbb{S}^1$. If we endow

M with the product metric, then the Lie group $G = \text{Isom}(Q)$ acts isometrically on $M = N \times Q$ through the second factor. Observe that, given $q \in Q$, the isotropy group $G_q = \{f \in G : f(q) = q\}$ is isomorphic to $= O(m-1)$ when $m \geqslant 1$ and isomorphic to $\mathbb{Z}_2$ when $m = 1$.

Let us prove first the following result:

Lemma 5.19 *Let $E \subset M = N \times Q$ be a bounded isoperimetric set and ℓ a geodesic in Q, where N is a Riemannian manifold and Q a Euclidean, spherical, or hyperbolic space. Then every connected component of E is symmetric with respect to the reflection across a totally geodesic hypersurface perpendicular to ℓ.*

Moreover, we may move the connected components of E by a one-parameter group of isometries in the direction of ℓ to assume that E is symmetric with respect to the reflection across a totally geodesic hypersurface P perpendicular to ℓ.

In particular, if E is connected, E is symmetric with respect to the geodesic reflection across P.

Proof The geodesic ℓ is either a line when Q is Euclidean or hyperbolic or a closed geodesic when Q is a sphere. For any $x \in \ell$, consider the reflection r in Q across a totally geodesic hypersurface $P' \subset Q$ containing x and orthogonal to ℓ. Let P_x be the set of fixed points of the action of r in M. Then P_x is a totally geodesic hypersurface with one or two components (the latter in case $Q = \mathbb{S}^1$). The hypersurface P_x contains $N \times P'$ and separates M into two connected components. We parameterize ℓ and denote the connected components of $M \setminus P_x$ by H_x^+ and H_x^- accordingly.

We claim that there exists a point $x_0 \in \ell$ such that $|E \cap H_{x_0}^+| = |E \cap H_{x_0}^-|$. We consider two cases according to ℓ being closed or a line.

Case 1: ℓ is Closed. This happens when Q is a sphere. We take $x \in \ell$. If $|E \cap H_x^+| = |E \cap H_x^-|$, then there is nothing to prove. Otherwise, we may assume

$$|E \cap H_x^+| > |E \cap H_x^-|.$$

Observe that $P_x \cap \ell = \{x, z\}$, where z is the antipodal of x and we have

$$H_x^+ = H_z^-, \qquad H_x^- = H_z^+.$$

Hence,

$$|E \cap H_z^+| < |E \cap H_z^-|.$$

By the continuity of the function $x \mapsto |E \cap H_x^+| - |E \cap H_x^-|$, we can find a point $x_0 \in \ell$ such that

$$|E \cap H_{x_0}^+| = |E \cap H_{x_0}^-|.$$

Case 2: ℓ is a Line. This happens when Q is Euclidean or hyperbolic. Since E has finite volume, we can find $x_1, x_2 \in \ell$ such that

$$|E \cap H_{x_1}^+|, |E \cap H_{x_2}^-| < \frac{|E|}{2}.$$

Therefore, $|E \cap H_{x_1}^+| - |E \cap H_{x_1}^-| < 0$ and $|E \cap H_{x_2}^+| - |E \cap H_{x_2}^-| < 0$. By continuity, there is a point $x_0 \in \ell$ such that

$$|E \cap H_{x_0}^+| = |E \cap H_{x_0}^-|.$$

Since E is an isoperimetric set, $P(E, H_{x_0}^+)$ and $P(E, H_{x_0}^-)$ must be equal since otherwise the set

$$E' = (E \cap H_{x_0}^+) \cup (E \cap P_{x_0}) \cup (E \cap H_{x_0}^+)^*, \tag{5.35}$$

where $(E \cap H_{x_0}^+)^*$ is the reflection of $E \cap H_{x_0}^+$ across P_{x_0} and has smaller perimeter as E and the same volume. This also proves that E' is an isoperimetric region.

We claim that P_{x_0} intersects every connected component of E. Otherwise, move a connected component $\tilde{E}$ of E such that $\tilde{E} \cap P_{x_0} = \emptyset$ by the action of a one-parameter group of isometries in the direction of ℓ until P_{x_0} separates $\tilde{E}$ into two regions of equal volumes. Note that $\tilde{E}$ cannot touch any other component of E by the regularity of isoperimetric sets. We repeat the process for every connected component of E using the same one-parameter group of isometries in the direction of ℓ.

We know from Theorem 1.60 that $\partial E = S \cup S_0$, where S is a smooth hypersurface with constant mean curvature and S_0 is the singular set satisfying $\mathcal{H}^{m-7}(S_0) = 0$. Hence, $P_{x_0} \cap S$ contains points of every connected component of E.

Observe that E' defined in (5.35) is an isoperimetric set. Let S' be the regular part of E'. The sets S and S' coincide on $H_{x_0}^+$ and have common points on the intersection of P_{x_0} with any connected component of E. We apply now the unique continuation principle for hypersurfaces with constant mean curvature, which follows by writing locally S and S' as graphs in Fermi coordinates, noticing that the difference of the graphs satisfies a linear elliptic equation (e.g., §10.4 in [183]), and applying the weak unique continuation principle for linear elliptic operators (see Hardt et al. [216]) to conclude that $S = S'$ and so $E = E'$. This same argument implies that each connected component of E is symmetric with respect to some totally geodesic hyperplane orthogonal to ℓ. □

The proof of the following result is based on Theorem 1 in W.-T. Hsiang and W.-Y. Hsiang [234].

Theorem 5.20 *Let $E \subset M = N \times Q$ be a bounded isoperimetric set, where N is a Riemannian manifold and Q a Euclidean, spherical, or hyperbolic space of dimension $q = \dim(Q)$. Then every connected component of E is invariant under* *the action of the isotropy group* $G_{q_0} \subset \text{Isom}(Q)$ *of some* $q_0 \in Q$*. Hence,* F *is* $O(q-1)$ *symmetric if* $q \geqslant 2$ *or* $\mathbb{Z}_2$ *symmetric if* $q = 1$.

Proof Let F be a connected component of E. Take a geodesic ℓ_1 in Q. By Lemma 5.19, there exists a totally geodesic hypersurface P_1 perpendicular to ℓ_1 such that F is symmetric with respect to P_1. If $q = 1$, then we are done.

If $q \geqslant 2$, we take a second geodesic ℓ_2 contained in P_1 and apply again Lemma 5.19 to get another totally geodesic hypersurface P_2 perpendicular to ℓ_2 such that F is symmetric with respect to P_2. The intersection $P_1 \cap P_2$ is a totally geodesic submanifold, and P_1, P_2 meet orthogonally along $P_1 \cap P_2$. We take now a geodesic $\ell_3 \subset P_1 \cap P_2$ and use again Lemma 5.19 to get a totally geodesic submanifold P_3 such that F is symmetric with respect to P_3. We continue inductively the process until we get q totally geodesic orthogonal hypersurfaces $P_1, \ldots, P_q$ so that F is symmetric with respect to the reflections across $P_1, \ldots, P_q$. Hence, there is a point $\{p_0\}$ in Q such that $P_1 \cap \cdots \cap P_q = N \times \{p_0\}$.

In case Q is the Euclidean or hyperbolic space, then the boundedness of E implies that any other reflection must contain $N \times \{p_0\}$ in its set of fixed points. Hence, F is invariant by the group generated by these reflections, that is, the isotropy group of p_0, and hence isomorphic to $O(q-1)$.

In case Q is the sphere, we refer the reader to Lemma 1 in §2.2 in [235] for the proof. □

W.-Y. Hsiang stated a more general version of the symmetry principle of Theorem 5.20 in §2 of [236] and [235]. An immediate corollary of Theorem 5.20 is the extension to a finite number of products with Euclidean, spherical, or hyperbolic spaces.

Corollary 5.21 *Let* $E \subset M = N \times Q_1 \times \cdots \times Q_r$, $r \geqslant 1$, *be a bounded isoperimetric set, where* N *is a Riemannian manifold and* Q_i *a Euclidean, spherical, or hyperbolic space of dimension* $q_i = \dim(Q_i)$*. Then every connected component of* E *is invariant under the action of the isotropy groups* $G_{q_i} \subset \text{Isom}(Q_i)$ *of some points* $q_i \in Q_i$*, for* $i = 1, \ldots, r$.

Let us finally remark the essential role played by the symmetries across totally geodesic hypersurfaces in the proof of Theorem 5.20. In particular, this result is not applicable in spaces with no orientation reversing isometries, like the Riemannian Heisenberg groups.

5.3.2 Some Consequences

Theorem 5.20 and its Corollary 5.21 are powerful results that allow us to reduce drastically the isoperimetric candidates in the presence of symmetry. Let us begin with an easy application.

Proposition 5.22 *Let M be a compact Riemannian manifold. Then every isoperimetric region $E \subset M \times \mathbb{R}^q$ is invariant under the action of the isotropy group of some point $q_0 \in \mathbb{R}^q$.*

In particular, isoperimetric sets in $M \times \mathbb{R}$ are invariant under the reflection across a totally geodesic hypersurface $M \times \{q_0\}$.

Proof We only need to check that isoperimetric sets are bounded, which follows from Theorem 4.25, and apply Theorem 5.20. □

Proposition 5.23 *Let $\mathbb{S}^1(r_1) \times \cdots \times \mathbb{S}^1(r_m)$ be a rectangular m-dimensional torus. Then any isoperimetric set is invariant by m reflections with respect to totally geodesic $(m-1)$-dimensional tori.*

Proof We simply apply Corollary 5.21 to conclude that an isoperimetric set E is invariant by the reflection across the totally geodesic hypersurfaces $\mathbb{S}^1(r_1) \times \cdots \times \{q_i\} \times \cdots \times \mathbb{S}^1(r_m)$, for $q_i \in \mathbb{S}^1(r_i)$, $i = 1, \ldots, q$. Observe that in this case, the set of fixed points of every reflection is disconnected. □

5.3.3 *Products of Euclidean and Hyperbolic Spaces*

In case $M = \mathbb{R}^n \times \mathbb{H}^q(-\kappa^2)$ or $M = \mathbb{H}^n(-1) \times \mathbb{H}^q(-\kappa^2)$, we have the following result. For the computations of the ordinary differential equations obtained after the symmetrization process, we refer to §3 in W.-T. Hsiang and W.-Y. Hsiang [234].

Proposition 5.24 *Any hypersurface in $\mathbb{R}^n \times \mathbb{H}^q(-\kappa^2)$ or $\mathbb{H}^n(-1) \times \mathbb{H}^q(-\kappa^2)$ with constant mean curvature H can be obtained from a generating curve $\gamma(s) = (x(s), y(s))$ in the planar quadrant $\{(x, y) \in \mathbb{R}^2 : x, y \geqslant 0\}$ which satisfies the system of ordinary differential equations*

$$x' = \cos\sigma, \qquad y' = \sin\sigma,$$

and

$$\sigma' = -H - (n-1)\frac{\sin\sigma}{x} + (q-1)\,\kappa\cos\sigma\coth(\kappa y),$$

in the Euclidean case, or

$$\sigma' = -H - (n-1)\sin\sigma\coth(x) + (q-1)\,\kappa\cos\sigma\coth(\kappa y),$$

in the hyperbolic case. The normal vector, with respect to which the mean curvature is computed, is in both cases $\mathbf{n} = (-\sin\sigma, \cos\sigma)$.

In addition, the functions

$$I = x^{n-1}\sin\sigma + \frac{H}{n}x^n, \qquad J = \sinh(\kappa y)^{q-1}\cos\sigma - H\int_0^y \sinh(\kappa t)^{q-1}\,dt$$

are increasingly monotone along γ in the sense that $dI/ds, dJ/ds \geqslant 0$.

Thus, Proposition 5.24 reduces the study of isoperimetric sets in these spaces to the classification of closed curves satisfying the relevant differential equations. A qualitative study of such curves was given in §5 in [234].

In the particular case of $\mathbb{R} \times \mathbb{H}^q(-\kappa^2)$, the generating curves must satisfy the differential equations

$$\begin{aligned} x' &= \cos\sigma, \\ y' &= \sin\sigma, \\ \sigma' &= -H + (q-1)\kappa\coth(\kappa y)\cos\sigma. \end{aligned} \tag{5.36}$$

Moreover, in this case, the quantity

$$J = \sinh(\kappa y)^{q-1}\cos\sigma - H\int_0^y \sinh(\kappa t)^{q-1}\,dt$$

is a first integral of the system (5.37). An analysis of the system implies that two solutions with the same mean curvature generating closed embedded hypersurfaces must be congruent; see Theorem 3 in [234]. Moreover, the correspondence between the enclosed volume and the mean curvature is bijective. This would imply that they are the only isoperimetric solutions. The discussion of the system (5.37) is similar to that of the system (2.26) we encountered when considering constant geodesic curvature curves in surfaces of revolution.

5.3.4 *Spherical Cylinders*

Another interesting case which can be treated with the symmetrization Theorem 5.20 is the product. This was considered by Pedrosa in his Ph.D. thesis [336]; see also [337]. In this case, the relevant system of ordinary differential equations is given by

$$\begin{aligned} x' &= \cos\sigma, \\ y' &= \sin\sigma, \\ \sigma' &= -H + (q-1)\cot y\cos\sigma, \end{aligned} \tag{5.37}$$

with first integral

$$J = \sin(y)^{q-1}\cos\sigma - H\int_0^y \sin(t)^{q-1}\,dt.$$

As we shall see later, isoperimetric sets for large volume are necessarily slices of the form $[a, b] \times \mathbb{S}^q$. See also Theorem 4 in Pedrosa [337]. In addition, there is only a family of embedded topological spheres with constant mean curvature. However, there is not a bijective correspondence between volumes and mean curvatures of the spheres in this family. This could imply non-uniqueness of isoperimetric sets for some volumes, although Pedrosa showed in Theorem 5 in [337] that there is indeed uniqueness for small volumes.

5.3.5 *Product of a Circle with a Simply Connected Two-Dimensional Space Form*

As a more elaborated application of Theorem 5.20 we consider the product of the circle $\mathbb{S}^1(r)$ with a two-dimensional space form Q^2 of constant curvature 0, +1, or −1, and we apply Theorem 5.20 only on the second factor, although we could obtain also an additional symmetry with respect to a totally geodesic {point} × Q from the $\mathbb{S}^1(r)$ factor.

After symmetrization, the candidate isoperimetric sets are obtained from curves $\gamma(s) = (x(s), y(s))$ in the space $\{(x, y) : x \in \mathbb{R}, y \geqslant 0\}$ in the Euclidean and hyperbolic cases and $\{(x, y) : x \in \mathbb{R}, y \in (0, \pi)\}$ in the spherical case. In all cases, we identify points whose x-coordinates are integer multiples of $2\pi r$. Such curves satisfy the equations

$$\begin{aligned} x' &= \cos\sigma, \\ y' &= \sin\sigma, \\ \sigma' &= -H + \cot_\kappa(y)\cos\sigma. \end{aligned} \tag{5.38}$$

Solutions of these equations satisfy the first integral

$$J_\kappa = \cos\sigma\,\sin_\kappa(y) - H\int_0^y \sin_\kappa(y)\,dt. \tag{5.39}$$

Here, $\sin_\kappa$ is the function defined in (1.48) and $\cot_\kappa = \cos_\kappa / \sin_\kappa$, where $\cos_\kappa = (\sin_\kappa)'$.

The system (5.38) has the same qualitative behavior as (2.26), the system of ordinary differential equations we found when computing curves with constant geodesic curvature in surfaces of revolution. In particular, the existence of the first integral allows us to classify the solutions in four types: curves touching the x-axis,

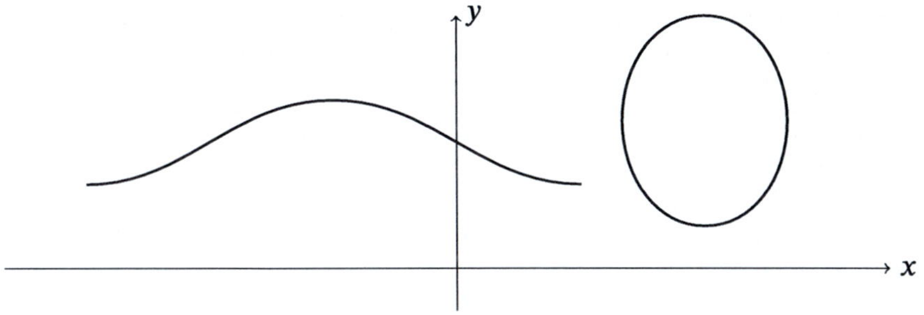

Fig. 5.9 Unduloid- and closed nodoid-type solutions. In the first case, the function σ' changes its sign. In the second one, the function $\cos\sigma$ changes its sign

which produce spheres of revolution; unduloids, which are non-constant periodic x-graphs; solutions with y constant, which produce embedded tubes around a closed geodesic $\mathbb{S}^(r) \times \{\text{point}\}$; and nodoids, which are not graphs over the x-axis but can produce closed curves in the spherical case. In addition, in the spherical case, we also have the curves with x constant, which produce totally geodesic submanifolds $\{\text{point}\} \times \mathbb{S}^2$. See Proposition 2.16 or §1 in Pedrosa and Ritoré [338] (Fig. 5.9).

Let us see that we can discard unduloids and closed nodoids as isoperimetric candidates by proving that they are *unstable*. This means that there exist functions just depending on s with mean zero so that the second variation is negative. This notion and the one of *stable* hypersurface will be considered in depth in Sect. 6.1. The second variation for a function u with compact support on S is given by

$$-\int_S u\,\mathcal{L}u\,dS$$

(see Theorem 1.27 and §1 in [338]), where $\mathcal{L} = \Delta_S + (|\sigma|^2 + \mathrm{Ric}(N, N))$. If we take a function $u = u(s)$ depending only on s, we get

$$\mathcal{L}u = u'' + [\cot_\kappa(y)\sin\sigma]\,u' + [(\sigma')^2 + \sin_\kappa^{-2}(y)\cos^2\sigma]\,u$$

We have the following results:

Proposition 5.25 ([338, Lemma 2.2 and Proposition 2.3]) *Closed nodoid-type solutions to* (5.38) *generate unstable surfaces in* $\mathbb{S}^1(r) \times \mathbb{S}^2$.

Proof In a closed nodoid-type curve, the function $u = \cos\sigma$ changes its sign. A direct computation using (5.38) implies $\mathcal{L}u = \sin_\kappa^{-2}(y)u$. Hence, $-u\mathcal{L}u < 0$. □

Proposition 5.26 ([338, Proposition 4.1]) *Unduloid-type solutions to* (5.38) *generate unstable surfaces in* $\mathbb{S}^1 \times Q$.

Proof We consider the test function $u = \sin_\kappa(y)\,\sigma'$ and compute $\mathcal{L}u$ using (5.38), to obtain $\mathcal{L}u = fu$, where $f = ((\sigma')^2 + \sin_\kappa^{-2}\sin^2\sigma) > 0$. Hence, $-u\,\mathcal{L}u = -f\,u^2 < 0$. Since on an unduloid the function σ' changes its sign, we obtain instability by considering the positive and negative parts of u. □

As a consequence of Propositions 5.25 and 5.26, we conclude that the only isoperimetric candidates are balls (touching the axis of revolution), tubes around the closed geodesics $\mathbb{S}^1(r) \times \{\text{point}\}$, and, in the spherical case, also slabs of the form $I \times \mathbb{S}^2$, where $I \subset \mathbb{S}^2$ is a connected set.

We consider first the Euclidean and hyperbolic cases; see Theorem 4.2 in [338] for the hyperbolic case.

Theorem 5.27 *The isoperimetric sets in the Riemannian product* $\mathbb{S}^1 \times \mathbb{R}^2$ *and* $\mathbb{S}^1(r) \times \mathbb{H}^2$ *are:*

1. *Balls*
2. *Tubes around closed geodesics* $\mathbb{S}^1(r) \times \{\textit{point}\}$

Moreover, the first ones are solutions for small values of volume. The second ones solve the problem for large values of volume.

Proof By the previous discussion, balls and tubes are the only isoperimetric candidates. The curve $y = y_0$ generating a tube around a geodesic $\mathbb{S}^1(r) \times \{\text{point}\}$ has an operator $\mathcal{L}$ given by

$$\mathcal{L}u = u'' + \sin_\kappa^{-2}(y_0)u$$

and is applied to periodic functions with period $2\pi r$. So the tube generated by $y = y_0$ is unstable for small y_0. Hence, balls are isoperimetric sets for small volume. On the other hand, balls are not embedded in $\mathbb{S}^1(r) \times Q$ for large volume, and so tubes are the isoperimetric solutions. □

For the spherical case, we can make a similar analysis, and we have; see Theorem 4.3 in [338].

Theorem 5.28 *The isoperimetric sets in the Riemannian product* $\mathbb{S}^1(r) \times \mathbb{S}^2$ *are either*

1. *balls or complements of balls, or*
2. *tubular neighborhoods of the closed geodesics* $\mathbb{S}^1(r) \times \{\textit{point}\}$*, which are diffeomorphic to* $\mathbb{S}^1 \times \mathbb{S}^1$*, or*
3. *sections bounded by two totally geodesic* $\{\textit{point}\} \times \mathbb{S}^2$*, which are diffeomorphic to* $[a, b] \times \mathbb{S}^2$.

The balls are solutions for small values of volume. If r *is such that*

$$r > 1, \tag{5.40}$$

then the tubes are not solutions, and, if r is such that

$$r < \frac{\mathbf{c}_2}{\pi \mathbf{c}_1}, \tag{5.41}$$

where $\mathbf{c}_i = |\mathbb{S}^i|$, *then the sections are not solutions.*

5.4 Classical Proofs of the Isoperimetric Inequality in Euclidean Space

In this section, we review some of the classical proofs of the isoperimetric inequality in Euclidean space.

5.4.1 Using Brunn-Minkowski Inequality

In the Euclidean space $\mathbb{R}^m$, the *Minkowski sum* of two sets E, F is defined by

$$E + F = \{x + y : x \in E, y \in F\}.$$

When E is a closed set, $r > 0$, and B the closed unit ball, the set $E + rB$ coincides with the closed tubular neighborhood E_r of points $q \in \mathbb{R}^m$ satisfying $d(q, E) \leqslant r$. Hence, the lower Minkowski content of a closed set is equal to

$$\mathrm{Mink}_-(E) = \liminf_{r \to 0} \frac{|E + rB| - |E|}{r}.$$

For measurable sets $E, F \subset \mathbb{R}^m$ such that $\lambda E + (1 - \lambda)F$ is measurable for all $\lambda \in [0, 1]$, we have the *Brunn-Minkowski inequality*

$$|(1 - \lambda)E + \lambda F|^{1/m} \geqslant (1 - \lambda)|E|^{1/m} + \lambda |F|^{1/m}. \tag{5.42}$$

Theorem 5.29 *For any measurable set* $E \subset \mathbb{R}^m$ *of positive volume, we have*

$$\mathrm{Mink}_-(E) \geqslant (m\omega^{1/m})|E|^{(m-1)/m}, \tag{5.43}$$

where B *is the closed unit ball in* $\mathbb{R}^m$. *In particular, if* $r_0 > 0$ *satisfies* $|E| = |r_0 B|$, *then*

$$\mathrm{Mink}_-(E) \geqslant \mathrm{Mink}(r_0 B).$$

Proof We apply (5.42) when $F = B$. Subtracting $(1-\lambda)|E|^{1/m}$ from both sides of (5.42) and dividing by λ, we have

$$\frac{|E + \frac{\lambda}{1-\lambda}F|^{1/m} - |E|^{1/m}}{\frac{\lambda}{1-\lambda}} \geqslant |B|^{1/m}.$$

Taking liminf when λ goes to 0, we get

$$\frac{1}{m}|E|^{(1-m)/m}\,\mathrm{Mink}_{-}(E) \geqslant |B|^{1/m}.$$

This implies (5.43). As $\mathrm{Mink}(B) = m|B|$, we obtain

$$\mathrm{Mink}_{-}(E) \geqslant \frac{\mathrm{Mink}(B)}{|B|^{(m-1)/m}}\,|E|^{(m-1)/m}.$$

Since the quantity $\mathrm{Mink}(B)|B|^{-(m-1)/m}$ is invariant by scaling of the ball choosing $r_0 > 0$ so that $|E| = |r_0 B|$, we obtain $\mathrm{Mink}_{-}(E) \geqslant \mathrm{Mink}(r_0 B)$. □

5.4.2 Using the Linear Brunn-Minkowski Inequality

We show in this section how to obtain the isoperimetric property of balls in the Euclidean space $\mathbb{R}^m$ from the *linear Brunn-Minkowski inequality*

$$|(1-\lambda)E + \lambda F| \geqslant (1-\lambda)|E| + \lambda|F|, \qquad \lambda \in [0,1], \tag{5.44}$$

valid for measurable sets $E, F \subset \mathbb{R}^m$ such that $\lambda E + (1-\lambda)F$ is measurable. Inequality (5.44) holds if the projections of E and F to a given hyperplane have the same $\mathcal{H}^{m-1}$ area or they have a maximal sections of equal $\mathcal{H}^{m-1}$ areas with respect to parallel hyperplanes.

The following result can be found in Remark 4.5 in Yepes-Nicolás and Ritoré [374].

Theorem 5.30 *If the linear Brunn-Minkowski inequality* (5.44) *holds on* $\mathbb{R}^m$*, then, for any bounded measurable set* $E \subset \mathbb{R}^m$ *of positive volume, we have*

$$\mathrm{Mink}_{-}(E) \geqslant \mathrm{Mink}(r_0 B),$$

where B *is the closed unit ball in* $\mathbb{R}^m$ *and* $r_0 > 0$ *satisfies* $|E| = |r_0 B|$.

Proof As E is bounded, we have

$$\sup_{t\in\mathbb{R}} \mathcal{H}_{m-1}(E(t)) < +\infty,$$

where $E(t) = \{x \in \mathbb{R}^{m-1} : (t, x) \in E\}$. Take $r > 0$ so that

$$\sup_{t\in\mathbb{R}} \mathcal{H}^{m-1}(E(t)) = \sup_{t\in\mathbb{R}} \mathcal{H}^{m-1}(rB(t)),$$

where $B \subset \mathbb{R}^n$ is the closed unit ball. Inequality (5.44) then implies

$$|(1-\lambda)E + \lambda(rB)| \geqslant (1-\lambda)|E| + \lambda|rB|.$$

Subtracting $(1-\lambda)^m|E|$ from both sides and dividing by λ, we obtain

$$(1-\lambda)^{m-1}\frac{|E + \frac{\lambda}{1-\lambda}(rB)| - |E|}{\frac{\lambda}{1-\lambda}} \geqslant (1-\lambda)\frac{(1-(1-\lambda)^{m-1})|E|}{\lambda} + |rB|.$$

Taking lim inf when λ goes to 0, we get

$$\text{Mink}_-(E) - \frac{(m-1)}{r}|E| \geqslant r^{n-1}|B|.$$

As $\text{Mink}(B) = m|B|$, we obtain

$$\text{Mink}_-(E) \geqslant \text{Mink}_-(rB) + \frac{(m-1)}{r}\big(|E| - |rB|\big). \tag{5.45}$$

Now define the function

$$f(s) = \text{Mink}(sB) + \frac{(m-1)}{s}\big(|E| - |sB|\big).$$

Its derivative is given by

$$f'(s) = \frac{(m-1)}{s^2}\big(|sB| - |E|\big).$$

Hence, for the unique $r_0 > 0$ such that $|r_0B| = |E|$, we have $f'(r_0) = 0$. Moreover, $f'(s) > 0$ when $s > r_0$ and $f'(s) < 0$ for $s < r_0$. This implies that r_0 is a global minimum for f. From (5.45), we get

$$\text{Mink}_-(E) \geqslant f(r) \geqslant f(r_0) = \text{Mink}(r_0B),$$

as desired. □

5.4.3 *Using Mass Transport: Knothe and Brenier-McCann Maps*

Another proof of the isoperimetric property of the ball in Euclidean space was given by Gromov in Appendix I in Milman and Schechtman [288]. Given the closed m-dimensional unit ball B in $\mathbb{R}^m$ centered at 0 and a set $E \subset \mathbb{R}^m$ with the same volume as B, a volume-preserving transformation from E to B is considered and used as a vector field to estimate the perimeter of E. Formally, the proof is as follows: consider a map $\Phi : E \to B$ with Jacobian equal to 1. Since $|\Phi| \leqslant 1$, the vector field $X = \sum_{i=1}^m \Phi_i \partial/\partial x_i$ satisfies $||X||_\infty \leqslant 1$. Hence, we have

$$P(E) \geqslant \int_E \operatorname{div} X \, d\mathbb{R}^m.$$

If the matrix $(\partial\Phi_i/\partial x_j)_{i,j=1,\ldots,m}$ is symmetric or triangular, then $\operatorname{div} X \geqslant m\,(\operatorname{Jac}\Phi)^{1/m} = m$. So we have

$$P(E) \geqslant m\,|E| = m|B| = P(B).$$

An analysis of the equality case taking into account the properties of the map Φ implies that E is indeed isometric to the ball B.

Actually, Gromov used Knothe's map (see [251]) defined as follows. We fix an orthonormal basis $e_1, \ldots, e_m$ on $\mathbb{R}^m$ and the associated coordinates $x_1, \ldots, x_m$. Given any set $F \subset \mathbb{R}^m$, we consider the sections

$$F_{(x_1,\ldots,x_k)} = \{y \in F : y_1 = x_1, \ldots, y_k = x_k\}.$$

For $x = (x_1, \ldots, x_m)$, the coordinates of the map

$$\Phi(x) = (\Phi_1(x_1), \Phi_2(x_1, x_2), \ldots, \Phi_m(x_1, \ldots, x_m))$$

are constructed recursively. We first define Φ_1 by the equality

$$\mathcal{H}^m(\{y \in E : y_1 < x_1\}) = \mathcal{H}^m(\{y \in B : y_1 < \Phi_1\}),$$

and, for $1 \leqslant k \leqslant m-1$, we define Φ_{k+1} inductively by

$$\frac{\mathcal{H}^{m-k}(\{y \in E_{(x_1,\ldots,x_k)} : y_{k+1} < x_{k+1}\})}{\mathcal{H}^{m-k}(E_{(x_1,\ldots,x_k)})}$$
$$= \frac{\mathcal{H}^{m-k}(\{y \in B_{(\Phi_1,\ldots,\Phi_k)} : y_{k+1} < \Phi_{k+1}\})}{\mathcal{H}^{m-k}(B_{(\Phi_1,\ldots,\Phi_k)})}.$$

The Jacobian matrix of the map Φ is triangular with determinant 1; see Knothe [251] and Brothers and Morgan [80]. An analysis of Gromov's arguments when E has smooth boundary is given in Berger; see §12.11.4 in [55].

Of course, the use of a different volume-preserving map is possible. Brenier [77] introduced a canonical volume-preserving map between two measures in Euclidean space. McCann [280] extended Brenier's results to a larger class of measures.

Theorem 5.31 ([280, Main Theorem]) *Let μ and ν be Borel probability measures in $\mathbb{R}^m$. Assume that μ is absolutely continuous with respect to Lebesgue measure $\mathcal{L}^m$. Then there exists a convex function φ on $\mathbb{R}^m$ whose gradient pushes μ forward to ν. The map $\Phi = \nabla\varphi$ is determined by μ almost everywhere.*

Given two Borel sets E, F of finite volume, we may apply Theorem 5.31 to the Borel probability measures

$$\mu(A) = \frac{\mathcal{L}^m(A \cap E)}{\mathcal{L}^m(E)}, \qquad \nu(A) = \frac{\mathcal{L}^m(A \cap F)}{\mathcal{L}^m(F)}.$$

Note that they are both absolutely continuous with respect to Lebesgue measure $\mathcal{L}^m$.

See §2 in Figalli et al. [160] for a detailed discussion of the use of Brenier's map to obtain the Euclidean isoperimetric inequality. If E has smooth boundary, Caffarelli's results [91] imply that Brenier's map is of class C^∞. See also Villani [432] as a general reference for mass transportation problems.

5.4.4 Reilly-Ros' Proof of Alexandrov Theorem

Once we know that isoperimetric sets exist in Euclidean space and that their boundaries have constant mean curvature except on a small singular set, we can try to characterize compact embedded hypersurfaces with constant mean curvature and try to check whether the proof extends in the presence of singularities.

The classical characterization of embedded hypersurfaces with constant mean curvature was given by Alexandrov in 1958; see [7, 8].

Theorem 5.32 (Alexandrov) *The only compact, connected, embedded hypersurfaces with constant mean curvature in $\mathbb{R}^m$ are the round spheres.*

Alexandrov introduced in his proof the method of moving hyperplanes. One considers a direction in $\mathbb{R}^m$ and the hyperplanes orthogonal to this direction. Reflecting the hypersurface with respect to the hyperplanes and making use of the maximum principle for elliptic operators, one can prove that the original hypersurface is symmetric across one of the hyperplanes. Since this can be done in every direction, the surface must be a round sphere. We refer the reader to Chapter 9 (addendum 3) in volume V of Spivak [406] for a complete detailed proof.

Many proofs of Alexandrov's theorem 5.32 have been given with different techniques. We provide one given by Reilly [356] extended to hypersurfaces with constant mean curvature of higher order by Ros [377]. The techniques developed by Reilly (see [354, 355]) allowed Ros to prove the following result.

Theorem 5.33 ([377, Theorem 1]) *Let E be a domain with smooth boundary $S = \partial E$ in a Riemannian manifold with non-negative Ricci curvature. If the mean curvature H of S is strictly positive, then we have*

$$m|E| \leqslant \int_S \frac{m-1}{H}\, dS. \tag{5.46}$$

Moreover, if equality holds, then E is isometric to a Euclidean ball.

Proof We consider the solution of the Dirichlet problem

$$\begin{aligned} 2\Delta u &= 1, && \text{on } E,\\ u &= 0, && \text{on } S. \end{aligned} \tag{5.47}$$

Existence follows from Fredholm's alternative and regularity from classical regularity theory for elliptic operators. The maximum principle for elliptic operators implies that the maximum of u is achieved at $S = \partial E$. Hopf's boundary maximum principle implies that $\nabla u \neq 0$ at S. Hence, the outer normal to E is given by

$$N = \frac{\nabla u}{|\nabla u|}.$$

The relation between the Laplacians Δu and $\Delta_S u|_S$ On S is given by Eq. (4.37). Since (5.47) holds and $u|_S = 0$, we have

$$1 = \nabla^2 u\big(N, N\big) + N(u)H \quad \text{on } S.$$

Multiplying by $N(u)$, we have

$$N(u) = N(u)\nabla^2 u(N, N) + N(u)^2 H. \tag{5.48}$$

Since $N(u) = |\nabla u|$ and $\nabla^2 u(\nabla u, \nabla u) = \frac{1}{2}\nabla u(|\nabla u|^2)$,

$$N(u)\nabla^2 u(N, N) = N\big(\tfrac{1}{2}|\nabla u|^2\big).$$

Integrating over S, using the divergence theorem, and taking into account Bochner's formula

$$\Delta\big(\tfrac{1}{2}|\nabla u|^2\big) = \nabla u(\Delta u) + |\nabla^2 u|^2 + \mathrm{Ric}(\nabla u, \nabla u)$$

we obtain

$$\int_S N(u)\nabla^2 u(N,N)\,dS = \int_S N\big(\tfrac{1}{2}\,|\nabla u|^2\big)$$
$$= \int_E \Delta(\tfrac{1}{2}|\nabla u|^2)\,dM = \int_E \big\{|\nabla^2 u|^2 + \mathrm{Ric}(\nabla u,\nabla u)\big\}\,dM.$$

As $\mathrm{Ric} \geqslant 0$ and $|\nabla^2 u|^2 \geqslant \frac{1}{m}(\Delta u)^2 = \frac{1}{m}$, we finally get

$$\int_S N(u)\nabla^2 u(N,N)\,dS \geqslant \frac{1}{m}\,|E|. \tag{5.49}$$

As

$$\int_S N(u)\,dS = \int_E \Delta u\,dM = |E|$$

we obtain from (5.48) and (5.49) the inequality

$$\frac{m-1}{m}\,|E| \geqslant \int_S N(u)^2 H\,dS. \tag{5.50}$$

So we get, from Cauchy-Schwarz's inequality,

$$\begin{aligned} |E|^2 &= \left(\int_E \Delta u\,dM\right)^2 = \left(\int_S N(u)\,dS\right)^2 \\ &\leqslant \left(\int_S N(u)^2 H\,dS\right)\left(\int_S H^{-1} dS\right). \end{aligned} \tag{5.51}$$

Hence, from (5.50) and (5.51), we get (5.46).

If equality holds in (5.46), then we have $\mathrm{Ric} \equiv 0$ in E and $|\nabla^2 u|^2 = m^{-1}(\Delta u)^2$. The last equality happens when $\nabla^2 u$ is proportional to the Riemannian metric g (i.e., there exists a smooth function ϕ on M such that $\nabla^2 u = \phi g$). As $\Delta u = 1$, we have

$$\nabla^2 u = m^{-1} g. \tag{5.52}$$

Let x_0 be a minimum of u in the interior of E. For any unit speed geodesic $\gamma : [0,a] \to M$, we have $(u \circ \gamma)''(t) = \nabla^u(\gamma'(t),\gamma'(t)) = m^{-1}$ by (5.52). Hence, $(u \circ \gamma)(t)$ is a quadratic function of t with leading term $(2m)^{-1}$. If $\gamma(0) = x_0$, as $(\nabla u)_{x_0} = 0$, we have

$$u(\gamma(t)) - u(x_0) = \frac{t^2}{2m} = \frac{1}{2m} d^2(\gamma(t))$$

where d is the distance function to x_0. This implies that $u - u(x_0) = (2m)^{-1}d^2$. In particular, $\Delta d^2 = 2m$ and $\Delta d = m - 1$. As Ric $\geqslant 0$, comparison results for the Laplacian of the distance function imply that E is flat and a geodesic ball. □

5.4.5 Cabré's Proof: The Alexandrov-Bakelman-Pucci Technique

In this section, another proof the Euclidean isoperimetric inequality is given following Cabré [87]. In contrast to Reilly-Ros' proof, this one relies on the solvability of a Neumann problem.

Theorem 5.34 *Let $E \subset \mathbb{R}^m$ be a domain with C^2 boundary. Then*

$$A(\partial E) \geqslant m\omega_m^{1/m} |E|^{(m-1)/m}. \tag{5.53}$$

Equality holds in (5.53) if and only if E is a round ball.

Proof We consider a solution u to the Neumann problem

$$\begin{aligned} 2\Delta u &= 1, && E, \\ \frac{\partial u}{\partial \nu} &= c, && S, \end{aligned} \tag{5.54}$$

where ν is the outer unit normal to $S = \partial E$. A necessary condition for the existence of a solution of (5.54) is given by the divergence theorem

$$|E| = \int_E \Delta u \, dM = \int_S \frac{\partial u}{\partial \nu} \, dS = c\, A(S).$$

Hence,

$$c = \frac{|E|}{A(S)}.$$

This condition is indeed sufficient for existence by standard elliptic theory.

Let Γ be the graph of u in $E \times \mathbb{R} \subset \mathbb{R}^{m+1}$, and let

$$\Gamma^+ = \big\{x \in E : u(z) \geqslant u(x) + \nabla u(x) \cdot (z - x) \ \ \forall z \in E\big\}.$$

The set Γ^+ is composed of the points $x \in E$ such that the tangent hyperplane to the graph of u at $(x, u(x))$ is a supporting hyperplane of the epigraph of u. If $x \in \Gamma^+$, then the function $F : E \to \mathbb{R}$ defined by

$$F(z) = u(x) - u(z) + \nabla u(x) \cdot (z - x)$$

has a global maximum at $z = x$, and so the Hessian of u is positive semidefinite at x. This implies, by the arithmetic-geometric mean inequality and the area formula,

$$\begin{aligned} |E| \geqslant |\Gamma^+| &= \int_{\Gamma^+} (\Delta u)^m d\mathcal{H}^m \\ &\geqslant \int_{\gamma^+} m^m \det(\nabla^2 u)\, d\mathcal{H}^m \geqslant m^m |(\nabla u)(\Gamma^+)|, \end{aligned} \tag{5.55}$$

where $(\nabla u)(\Gamma^+)$ is the image of Γ^+ by the map $\nabla u : D \to \mathbb{R}^m$. Let us see that

$$B(0, c) \subset (\nabla u)(\Gamma^+) \tag{5.56}$$

holds. This inclusion, together with (5.55), implies (5.53).

To prove (5.56), let $e \in \mathbb{R}^m$. Take a hyperplane in $\mathbb{R}^{m+1}$ orthogonal to the vector $(e, -1)$ disjoint from the graph of u in the negative direction of the x_{m+1} axis. We translate it vertically until it becomes a supporting hyperplane to the epigraph of u at some point $(x, u(x))$. If $x \in \partial E$, then we take a curve in the graph with tangent vector $-(\nu, \partial u/\partial \nu) = -(\nu, c)$. Taking scalar product with $(e, -1)$, we get $-|e| + c \leqslant 0$ as ν, e are proportional. Hence, in case $|e| < c$, there is a supporting hyperplane with normal vector $(e, -1)$ to the graph of u. This implies $e = (\nabla u)(x)$ for some $x \in \Gamma^+$ and so $e \in (\nabla u)(\Gamma^+)$. This proves (5.56).

If equality holds in (5.53), then we have $\Gamma^+ = E$, and equality holds in the arithmetic-geometric mean inequality. This implies $\nabla^2 u = (1/m)\, I_m$, where I_m is the identity matrix of order m. We conclude that E is a ball using the arguments in the final part of the proof of the Reilly-Ros Theorem 5.33. □

5.5 Notes

Notes for Sect. 5.1 The main symmetrization result for warped products in this section, Theorem 5.6, was proven By Morgan et al. [309] as a generalization to higher codimension of a result by Ritoré; see §1.3.2 in [369]. Theorem 5.6 provides a uniform proof for most of the classical symmetrizations, like Steiner's [410], Schwarz's [402], and Almgren's spherical symmetrization [13].

Formula (5.15), estimating the area of a hypersurface in terms of the area of the sections and the mean curvature of the fibers, seems to have appeared first in the classical paper by Schwarz on the isoperimetric property of balls in $\mathbb{R}^3$; see Equation (II.b) on page 9 in [402]. There the base manifold is $\mathbb{R}$ and the fiber $\mathbb{R}^2$. Equation (II.b) in [402] reads

$$S \geqslant \int_{x_0}^{x_1} \sqrt{4\pi\, Q(x) + Q'(x)^2}\, dx,$$

where S is the area of the boundary of a set E and $Q(\xi)$ the area of the intersection of E with the plane $x = \xi$. The term $4\pi Q(x)$ appears after applying the planar isoperimetric inequality to the boundary of the section with the plane.

Notes for Sect. 5.2 The use of symmetrization results to prove isoperimetric inequalities goes back to Steiner [410] in the plane and to Schwarz [424] in the three-dimensional Euclidean space. Steiner's proofs got some criticism since he assumed the existence of an isoperimetric set, as remarked by Weierstraß; see Blåsjö [64]. Edler [144] proved, without assuming existence, that the circle enclosed more area than any other figure with the same perimeter. Carathéodory and Study [94] provided proofs that iterated symmetrizations preserving the perimeter of a planar set converge with increasing area to the disk. Carathéodory provided a proof for curves and Study for polygons. The proofs did not use existence of isoperimetric sets, thus showing the value of Steiner's methods. Tonelli [424] used a sequence of Schwarz symmetrals of a given set E and proved convergence to a sphere. We have used only symmetrization to prove the isoperimetric property of the balls in the Euclidean space and the sphere in Sects. 5.2.5 and 5.2.6. Our proof of Theorem 5.17 is based on the sphericity theorem of Groß; see Groß [200] and Theorem 9.1 in Gruber [202].

The combination of symmetrization with analytic estimates to obtain isoperimetric inequalities was initiated by Schwarz [424]. After symmetrization, any set in $\mathbb{R}^3$ has larger perimeter than a set of revolution about an axis. Schwarz then gave a variational argument to prove that the rotationally symmetric candidate has no smaller perimeter than a sphere. We have replaced the final argument in Schwarz's proof by Theorem 5.14, proven by calibration. The arguments in Theorem 5.14 are closely related to a local minimization property of surfaces with constant mean curvature we shall see in Theorem 6.9. Theorem 5.14 can be applied to different situations, spheres, hyperbolic spaces, Nil manifolds (Riemannian Heisenberg groups), and Berger spheres, among others, and also in spaces where regularity results for solutions of the isoperimetric problem are not yet available, like the sub-Riemannian Heisenberg groups [363]. Calibration arguments have been widely used in the calculus of variations. The reader is referred to Hélein [226, 227], to Lawlor for arguments on calibrations and slicing [259, 260], and to Lawlor and Morgan [261, 262].

Schwarz result in $\mathbb{R}^3$ was extended by Schmidt in a series of papers to all simply connected space forms with similar methods: to Euclidean space in [393], to the sphere in [394, 397], and to hyperbolic space in [395, 397]. The comparison is proven for sets with smooth boundary.

We refer the reader to Bonnesen [71], Bonnesen and Fenchel [72], Blaschke [63], Hadwiger [211], Osserman [325], Burago and Zalgaller [83], Blåsjö [64], and Ros [380] for historical references and comments.

Notes for Sect. 5.3 The idea of reflecting the least perimeter portion of an isoperimetric set E with respect to a line separating E into two equal areas goes back to Steiner, see Weierstraß's comments in Blåsjö [64].

Hsiang symmetrization combines this simple geometric procedure with the regularity theory for isoperimetric sets to provide symmetries of isoperimetric regions. Their main symmetrization result is proven in W.-T. Hsiang and W.-Y. Hsiang in [234] and, in a more general setting, in Hsiang [235, 236]. We have presented a proof in Theorem 5.20 in the product of a Riemannian manifold with a Euclidean, spherical, or hyperbolic space.

In the proof of Theorem 5.20, we have used the weak unique continuation principle instead of the Cauchy-Kowalevski theorem used by W.-Y. Hsiang. The weak unique continuation principle asserts that a solution of an elliptic equation vanishing in an open set must be identically zero. There are many applications of the weak continuation principle in the calculus of variations; see p. 201 in Vogel [434], where a uniqueness argument based on the unique continuation property is provided. As for the weak maximum principle for elliptic operators, the literature is enormous; see Miranda [291], pp. 59–61, and the references therein, Garofalo and Lin [177, 178], Hardt et al. [216], and Kazdan [244], among many other references.

Once we have the symmetrization Theorem 5.20, we can reduce the analysis of isoperimetric shapes to the one of solutions of ordinary differential equations. One could be also able to produce an examples of an isoperimetric set with singularities at its boundary.

Notes for Sect. 5.4 In this section, we reviewed some classical proofs of the isoperimetric inequality in Euclidean space. The Brunn-Minkowski inequality allows us not only to prove the classical isoperimetric inequality in Euclidean space but to classify the isoperimetric sets for crystalline norms in $\mathbb{R}^m$, called *Wulff shapes*. The interested reader is referred to pp. 158 ff. in Grüber [202] for an historical account, Taylor [422], Fonseca [164], Fonseca and Müller [165], Brothers and Morgan [80], and Figalli et al. [160]. See also Palmer [329] and Winklmann [447, 448] for a more geometric point of view. A generalization of Wulff shapes to the sub-Riemannian first Heisenberg group has been obtained recently by Pozuelo and Ritoré [348].

The use of the linear Brunn-Minkowski inequality in Sect. 5.4.2 replaces the calibration argument in Theorem 5.14. Both methods allow us to obtain the inequality

$$P(E) - \lambda|E| \geqslant P(B) - \lambda|B|$$

for any finite perimeter set E contained in a right circular cylinder of radius $(m - 1)/\lambda$ and containing a base of the cylinder. Here, B is the ball of radius $(m - 1)/\lambda$.

The proof in Sect. 5.4.5 is due to Cabré [87]; see also §3 in the survey [88] and §VI.2 in Chavel [102]. The proof is based on a method by Alexandrov, Bakelman, and Pucci, usually named the ABP method, initially used to obtain L^∞ bounds for solutions of second-order elliptic operator in divergence form. See Cabré, Ros-Otón, and Serra [89] for applications to isoperimetric inequalities in weighted convex cones and Brendle [76] for a proof of an optimal isoperimetric inequality, up to codimension 2, on minimal submanifolds of Euclidean space.

Chapter 6
Space Forms

In this chapter, isoperimetric inequalities in *space forms*, complete Riemannian manifolds with constant sectional curvatures, are considered. We have already seen in Sect. 4.5 that isoperimetric sets in simply connected space forms are geodesic balls. When considering manifolds with non-trivial topology, we should expect existence of isoperimetric sets of different topological types. This often implies the existence of several explicit isoperimetric inequalities, one for each topological type of isoperimetric solution. A way of dealing with this phenomenon is through a classification of stable hypersurfaces, second-order minima of the perimeter under volume-preserving variations.

In the first part of this chapter, we briefly recall well-known results about stability of surfaces with constant mean curvature, including a classification of stable hypersurfaces in simply connected space forms. We also show how to adapt the results for stable smooth hypersurfaces to hypersurfaces with a singular set of small Hausdorff dimension, such as isoperimetric boundaries.

Then we focus on three-dimensional space forms, where we have several advantages: isoperimetric boundaries are regular everywhere and we can use the full theory of surfaces when considering *orientable* isoperimetric boundaries. In particular, we obtain an estimate on the genus of isoperimetric boundaries when the sectional curvatures are non-negative and characterize them when the genus is 0 or 1. This is enough to classify the isoperimetric sets in some three-dimensional manifolds like the real projective space $\mathbb{RP}^3$ or $\mathbb{S}^1 \times \mathbb{R}^2$. We also show that the spaces of stable surfaces of genus larger than one, and their ambient manifolds, are compact in suitable topologies, a result that implies that in most orientable quotients of $\mathbb{R}^3$, the only isoperimetric sets are geodesic spheres, tubes around closed geodesics, and slabs bounded by two totally geodesic parallel tori. In the case of rectangular tori, we see that the isoperimetric candidates of genus 2 or 3 come in completely described families. The genus 3 Schwarz $\mathcal{P}$ lies in one of these families and is a stable surface, but it is not an isoperimetric surface.

M. Ritoré, *Isoperimetric Inequalities in Riemannian Manifolds*,
Progress in Mathematics 348, https://doi.org/10.1007/978-3-031-37901-7_6

We have also included a proof of Hadwiger's isoperimetric inequality for quotients of $\mathbb{R}^m$ by cubic lattices. This is an optimal isoperimetric inequality when separating the ambient torus $(\mathbb{S}^1)^m$ into two equal volumes. The solution in this case is a slab bounded by two totally geodesic compact tori. Also the product manifold $\mathbb{S}^1(r) \times \mathbb{R}^m$ is discussed in great detail. This manifold exhibits, for $m \geqslant 9$, examples of isoperimetric boundaries which are neither geodesic spheres nor tubes of given radius around a closed geodesic. They are in fact the quotients of Delaunay's unduloids, embedded constant mean curvature hypersurfaces of revolution about an axis in $\mathbb{R}^{m+1}$.

The techniques used in this chapter are adequate when the Ricci curvature of the ambient manifold is non-negative, but offer few or no information when it is negative.

6.1 Stable Constant Mean Curvature Hypersurfaces

Recall (see Theorem 1.27) that a region $S' \subset S$ of a two-sided hypersurface S with unit normal N and constant mean curvature in a Riemannian manifold M is *stable* if

$$\int_S \left\{|\nabla_S u|^2 - (\mathrm{Ric}(N,N) + |\sigma|^2)\, u^2\right\} dS \geqslant 0 \tag{6.1}$$

for any smooth function $u : S \to \mathbb{R}$ with compact support on S' satisfying $\int_S u\, dS = 0$. Here, $|\sigma|^2$ is the squared norm of the second fundamental form. This condition is equivalent to that S' is a second-order minimum of the area for any variation of the interior of S' preserving the volume up to second order. Since the regular part of an isoperimetric boundary is perimeter minimizing under a volume constraint, it is a stable hypersurface.

Stability is a strong condition that implies geometrical and topological restrictions. As an example, we show the following:

Proposition 6.1 *Let S be a compact stable two-sided hypersurface in a Riemannian manifold M with* Ric > 0. *Then S is connected.*

Proof If S is not connected, we take a locally constant nowhere vanishing function u on S with *mean zero*. Then the second variation for u is given by

$$-\int_S \left(\mathrm{Ric}(N,N) + |\sigma|^2\right) u^2\, dS < 0,$$

a contradiction. □

6.1.1 The Jacobi Operator and the Index Form

The divergence theorem applied to $u \in C_0^\infty(S)$ implies $\int_S |\nabla_S u|^2 dS = -\int_S u\Delta_S u\, dS$. Hence, condition (6.1) can be rewritten as

$$\mathcal{I}(u) = -\int_S u\mathcal{L}u\, dS \geqslant 0, \tag{6.2}$$

where $\mathcal{L}$ is the *Jacobi operator*

$$\mathcal{L} = \Delta_S + (\mathrm{Ric}(N, N) + |\sigma|^2). \tag{6.3}$$

The quadratic form $\mathcal{I}$ defined by (6.2) is the *index form* of S.

Remark 6.2 If inequality (6.1) holds for any $u \in C_0^\infty(S)$ with mean zero, it is also valid for any function of mean zero in the Sobolev space

$$H_0^1(S) = \{\text{closure of } C_0^\infty(S) \text{ in } W^{1,2}(S) = H^1(S)\}.$$

In particular, (6.1) holds for Lipschitz functions with mean zero and compact support on S. The Sobolev space $W^{1,2}(S)$ is defined as the set of $L^2(S)$ functions whose first derivatives in weak send lie also in $L^2(S)$. It is a Hilbert space with the inner product

$$\langle f, h\rangle = \int_S \big\{fh + \langle\nabla f, \nabla h\rangle\big\}\, dS. \tag{6.4}$$

See Aubin [31] or Brezis [78].

In space forms, the term $\mathrm{Ric}(N, N)$ is a constant function. When $\mathrm{Ric}(N, N) \geqslant 0$, the obstruction to the existence of stable hypersurfaces is particularly strong. If we remove the mean zero condition, we obtain the following definition.

Definition 6.3 We say that a two-sided hypersurface with constant mean curvature is *strongly stable* if condition (6.1) is satisfied for any function $u \in C_0^\infty(S)$ without assuming any condition on the values of $\int_S u\, dS$.

In case S is a minimal hypersurface, $\mathcal{I}(u)$ is the second variation of the perimeter of S under a variation with normal component u. Hence, the strong stability condition is equivalent to being a second-order minimum of the perimeter for variations with compact support. In particular, if S is an area-minimizing hypersurface, then it is a strongly stable hypersurface.

Of course, a strongly stable hypersurface is stable. The following is a condition ensuring that a hypersurface is strongly stable.

Lemma 6.4 ([163, Thm. 1]) *Let M be a Riemannian manifold eventually with boundary. We consider on M the differential operator $\mathcal{L} = \Delta + q$, where $q \in$*

$C^\infty(M)$*. Assume that there exists a function* $u > 0$ *such that* $\mathcal{L}u \leqslant 0$*. Then*

$$\int_M \{|\nabla v|^2 - qv^2\}\, dM \geqslant 0 \tag{6.5}$$

for any smooth function v *with compact support on* M*.*

Proof If $w = \log u$, then

$$\Delta w \leqslant -q - |\nabla w|^2.$$

Let v be a smooth function with compact support in the interior of M. Multiplying this inequality by $-v^2$, integrating on M, and using that $v = 0$ on ∂M and the divergence theorem, we obtain

$$2\int_M v\langle \nabla v, \nabla w\rangle\, dM \geqslant \int_M qv^2\, dM + \int_M |\nabla w|^2 v^2\, dM. \tag{6.6}$$

Applying the Schwarz inequality and the arithmetic-geometric mean inequality, we have

$$2|v|\langle \nabla v, \nabla w\rangle \leqslant 2|v||\nabla v||\nabla w| \leqslant v^2|\nabla w|^2 + |\nabla v|^2.$$

Integrating over M and inserting in (6.6), we obtain (6.5). □

Definition 6.5 Given a two-sided hypersurface S with constant mean curvature, a *Jacobi function* $u \in C^\infty(S)$ is a solution of the equation $\mathcal{L}u = 0$, where $\mathcal{L}$ is the Jacobi operator defined in (6.3).

A way of obtaining Jacobi functions is by means of *Killing fields*, smooth vector fields on M whose associated one-parameter group of diffeomorphisms is composed of isometries. A vector field X is a Killing vector field if the algebraic condition

$$\langle \nabla_Y X, Z\rangle + \langle Y, \nabla_Z X\rangle = 0$$

is satisfied for all vector fields Y, Z in M.

Lemma 6.6 *Let* S *be a two-sided hypersurface with constant mean curvature and unit normal* N *in a Riemannian manifold* M *and* X *a Killing vector field in* M*. Then the function* $u = \langle X, N\rangle$ *is a Jacobi function on* S*. If* S *bounds a relatively compact set* $\Omega \subset M$*, then* $\int_S u\, dS = 0$*.*

Proof Since isometries preserve the mean curvature of a hypersurface, u is a Jacobi function by Lemma 6.6. If S bounds a relatively compact domain Ω and $\{\varphi_t\}_{t\in\mathbb{R}}$ is the flow associated with X, then $|\varphi_t(\Omega)|$ is constant, and so $0 = \int_\Omega \operatorname{div} X\, dM = \int_S u\, dS$. Hence, u has mean zero on S. □

A consequence of Lemmas 6.4 and 6.6 is that the existence of a Killing vector transversal (not tangent) to a hypersurface S on a region S' automatically implies that the region S' is strongly stable for the Jacobi operator. This holds in particular for graphs with constant mean curvature in Euclidean space.

6.1.2 Eigenvalues of the Jacobi Operator

The Jacobi operator $\mathcal{L} = \Delta_S + \mathrm{Ric}(N, N) + |\sigma|^2$ is an elliptic operator consisting of the Laplacian plus a potential function. This kind of operator appears as Hamiltonian in non-relativistic quantum mechanics and is often called a *Schrödinger operator*; see §XIII.2 in [353].

Consider a Schrödinger operator $\mathcal{L} = \Delta + q$ on a Riemannian manifold (M, g), where Δ is the Laplacian on M and $q \in C^\infty(M)$. Given a bounded open set $\Omega \subset M$, we say that λ is an *eigenvalue* of $\mathcal{L}$ in Ω if there exists a function $u \in H_0^1(\Omega)$ such that

$$\int_\Omega \left\{\langle \nabla u, \nabla f\rangle - quf\right\} dM = \lambda \int_M uf\, dM$$

for all $f \in C_0^\infty(\Omega)$. The function u is called an *eigenfunction* of eigenvalue λ in Ω. If the boundary of Ω is Lipschitz, then u is a smooth function, and so u is an eigenfunction with eigenvalue λ on Ω if and only if

$$\begin{aligned} \mathcal{L}u + \lambda u &= 0, & \Omega,\\ u &= 0, & \partial\Omega. \end{aligned}$$

We summarize well-known results about Schrödinger operators on manifolds; see [163].

Theorem 6.7 *Let (M, g) be a Riemannian manifold, Δ its Laplacian, $q \in C^\infty(M)$, and $\mathcal{L} = \Delta + q$. Let $\Omega \subseteq M$ be a bounded set (that could coincide with M). Then*

1. *$\mathcal{L}$ is a self-adjoint operator with respect to the product (6.4) in $H_0^1(\Omega)$.*
2. *There exists a divergent increasing sequence of eigenvalues $\{\lambda_i(\Omega)\}_{i\in\mathbb{N}}$ with associated eigenfunctions $\{u_i\}_{i\in\mathbb{N}}$ so that*

$$\lambda_1(\Omega) < \lambda_2(\Omega) \leqslant \lambda_3(\Omega) \leqslant \cdots .$$

In particular, the space of eigenfunctions associated with $\lambda_1(\Omega)$ is one-dimensional, and a first eigenfunction $u \neq 0$ does not vanish on Ω.

3. *The eigenvalue $\lambda_i(\Omega)$ satisfies the min-max principle*

$$\lambda_i(\Omega) = \min_{U \subset H_0^1(\Omega)} \left\{ \max_{u \in U\setminus\{0\}} \left\{ \frac{\int_M \left\{|\nabla u|^2 - q\, u^2\right\} dM}{\int_M u^2\, dM} : \dim U = i \right\}\right\},$$

which is equivalent to

$$\lambda_i(\Omega) = \min\Big\{ \textstyle\int_M \{|\nabla u|^2 - qu^2\}\, dM : \int_S u^2 dM = 1,$$
$$\int_M u u_j\, dM = 0, 1 \leqslant j \leqslant i-1\Big\},$$

where u_j *are the first eigenfunctions of* $\mathcal{L}$.

4. *If* $\Omega \subseteq \Omega'$, *then* $\lambda_i(\Omega) \geqslant \lambda_i(\Omega')$ *for all* $i \in \mathbb{N}$. *If, in addition,* $\Omega' \setminus \overline{\Omega} \neq \emptyset$, *then* $\lambda_i(\Omega) > \lambda_i(\Omega')$ *for all* $i \in \mathbb{N}$.
5. *An eigenfunction associated with the k-th eigenvalue has at most k nodal domains.*

The set of eigenvalues is called the *spectrum* of the operator. A *nodal domain* of a function u is a connected component of the set $\{p \in M : u(p) \neq 0\}$. The *index* of $\mathcal{L}$ in Ω is the finite number of negative eigenvalues of $\mathcal{L}$ in Ω. It is denoted by $\mathrm{index}(\mathcal{L}, \Omega)$. Theorem 6.7(4) implies $\mathrm{index}(\mathcal{L}, \Omega') \leqslant \mathrm{index}(\mathcal{L}, \Omega)$ if $\Omega \subseteq \Omega'$. Hence, the index of a non-compact manifold M can be defined as the infimum of $\mathrm{index}(\mathcal{L}, \Omega)$ when Ω is a relatively compact domain in M.

Eigenvalues and eigenfunctions of Schrödinger operators have been intensively considered in Riemannian manifolds, specially the Laplacian [56, 100], and the Jacobi operator associated with the second variation of the area on a minimal hypersurface [162, 163].

The spectrum of the Jacobi operator on a hypersurface with constant mean curvature is closely related to the stability of the hypersurface, as shown by the following result obtained by Koiso; see Theorem 1.3 in [254] and also §2 in [238].

Theorem 6.8 ([254]) *Let* $S \subset M$ *be a two-sided stable hypersurface in a Riemannian manifold. Let* $\mathcal{L} = \Delta + q$ *be the Jacobi operator, where* $q = \mathrm{Ric}(N, N) + |\sigma|^2$. *Consider a relatively compact set* $\Omega \subset M$ *with smooth boundary, and let* $\{\lambda_i\}_{i \in \mathbb{N}}$ *be the spectrum of* $\mathcal{L}$ *on* Ω. *Then we have:*

1. *If* $0 \leqslant \lambda_1$, *then* S *is strongly stable, henceforth stable.*
2. *If* $\lambda_1 < 0 < \lambda_2$, *then there is a unique solution* $u \in C^\infty(\Omega) \cap H_0^1(\Omega)$ *satisfying* $\mathcal{L}u = 1$. *Moreover,*

   (a) *If* $\int_S u\, dS \geqslant 0$, *then* S *is stable.*
   (b) *If* $\int_S u\, dS < 0$, *then* S *is unstable.*

3. *If* $\lambda_1 < \lambda_2 = 0$, *then*

   (a) *If there is an eigenfunction* v *with eigenvalue* λ_2 *such that* $\int_S v\, dS \neq 0$, *then* S *is unstable.*

(b) *If all eigenfunctions v with eigenvalue λ_2 satisfy $\int_S v\,dS = 0$, then there exists a smooth function u vanishing at $\partial\Omega$ which satisfies $\mathcal{L}u = 1$. Moreover,*

i. *If $\int_S u\,dS \geqslant 0$, then S is stable.*
ii. *If $\int_S u\,dS < 0$, then S is unstable.*

4. *If $\lambda_2 < 0$, then S is unstable.*

In the proof of Theorem 6.8, the role of a function satisfying $\mathcal{L}u = 1$ is essential. Geometrically, these functions can be obtained from variations of S by hypersurfaces with constant mean curvature.

Proof

1. If $0 \leqslant \lambda_1$, then S is indeed strongly stable since by the min-max characterization of the first eigenvalue, we have

$$0 \leqslant \lambda_1 \leqslant \frac{\int_S \{|\nabla u|^2 - qu^2\}\,dS}{\int_S u^2 dS} = \frac{\mathcal{I}(u)}{\int_S u^2 dS}$$

for all functions $u \in C_0^\infty(\Omega)$.

2. Condition $\lambda_1 < 0 < \lambda_2$ implies that there are no Jacobi functions on Ω. By Fredholm (Riesz-Schauder) alternative, there exists a unique solution of the equation $\mathcal{L}u = 1$ on Ω satisfying $u = 0$ on $\partial\Omega$. For a first non-trivial eigenfunction u_1 of $\mathcal{L}$, we have $\int_S u_1\,dS \neq 0$. Hence, taking $\alpha = -(\int_S u\,dS)(\int_S u_1\,dS)^{-1}$, we get, for $v = u + \alpha u_1$,

$$\int_S v\,dS = \int_S (u + \alpha u_1)\,dS = 0.$$

So we have $-\alpha \int_S u_1\,dS = \int_S u\,dS$ and

$$\begin{aligned}\mathcal{I}(v) &= \mathcal{I}(u) + \alpha^2 \mathcal{I}(u_1) - 2\alpha \int_S u_1\,\mathcal{L}u\,dS \\ &= \int_S u\,dS + \alpha^2\lambda_1 \int_S u_1^2\,dS.\end{aligned}$$

As $\lambda_1 < 0$, we obtain $\mathcal{I}(v) < 0$ in case $\int_S u\,dS < 0$.

Assume now that $\int_S u\,dS \geqslant 0$. Let E be the orthogonal in $L_2(\Omega)$ of the subspace generated by u_1. Since $\lambda_2 > 0$, we get $\mathcal{I}(w) > 0$ for all $w \in E\setminus\{0\}$. As $\mathcal{I}(u) = -\int_S u\mathcal{L}u\,dS = -\int_S u\,dS \leqslant 0$, we conclude that $u \notin E$. Let $v \in C_0^\infty(\Omega)$ be a smooth function with mean zero. Then v can be expressed as

$$v = w + \beta u,$$

with $w \in E$ and $\beta \in \mathbb{R}$. Hence, $\int_S w\, dS = -\beta \int_S u\, dS$, and we have

$$\begin{aligned}\mathcal{J}(v) &= \mathcal{J}(w) + \beta^2 \mathcal{J}(u) - 2\beta \int_S w\mathcal{L}u\, dS \\ &= \mathcal{J}(w) + \beta^2 \int_S u\, dS \geqslant 0\end{aligned}$$

since $\mathcal{J}(w) \geqslant 0$ and $\int_S u dS \geqslant 0$. This proves that S is stable.

3. To prove case (a), we consider an eigenfunction with eigenvalue $\lambda_2 = 0$ such that $\int_S v\, dS \neq 0$. We take a linear combination $\alpha u_1 + v$ so that $\int_S (\alpha u_1 + v)\, dS = 0$, with $\alpha \neq 0$. As u_1, v are $L_2(\Omega)$ orthogonal eigenfunctions and $\mathcal{L}v = 0$, we get

$$\mathcal{J}(\alpha u_1 + v) = \alpha^2 \mathcal{J}(u_1) < 0.$$

Then S is unstable.

To prove case (b), we use again Fredholm alternative to show the existence of a smooth function u vanishing at $\partial\Omega$ such that $\mathcal{L}u = 1$. Assume that $\int_S u\, dS = 0$. Observe that in this case,

$$\mathcal{J}(u) = -\int_S u\mathcal{L}u\, dS = -\int_S u\, dS = 0.$$

If u is orthogonal to u_1, then u is an eigenfunction with eigenvalue 0, but this contradicts $\mathcal{L}u = 1$. If u is not orthogonal to u_1, then the proof follows as in case $\lambda_2 > 0$. Finally, if $\int_S u\, dS < 0$ or $\int_S u\, dS > 0$, then the proof also follows as in case $\lambda_2 > 0$.

□

6.1.3 Local Minimizing Properties for Stable Hypersurfaces

Let us see that the fact that $\lambda_1(\Omega) > 0$ for some region Ω with smooth boundary on a hypersurface S with constant mean curvature implies that Ω is locally area-minimizing under a volume constraint. We follow the arguments by Große-Brauckmann [201].

Theorem 6.9 ([201, Theorems 1 and 3]) *Let S be a two-sided hypersurface with constant mean curvature H and $\Omega \subset S$ a subset with smooth boundary so that $\lambda_1(\Omega) > 0$. Then there is a foliation of a normal cylinder over $\overline{\Omega} = S_0$ by smooth hypersurfaces $\{S_t\}_{t\in(-\varepsilon,\varepsilon)}$ with boundary and constant mean curvature H.*

Moreover, let S' be any C^1 hypersurface contained in the region foliated by $\{S_t\}_{t\in(-\varepsilon,\varepsilon)}$ with $\partial S' = \partial\Omega$. Let E^+ be the region delimited by S and S' on the *side of S where N points and E^- the region delimited by S and S' on the other side of S. Then*

$$A(S') - H|E^+| \geqslant A(\Omega) - H\,|E^-|. \tag{6.7}$$

Equality holds if and only if $S' = \Omega$.

Proof The proof follows the lines of a classical one for minimal surfaces; see §414 in [321]. For $u \in C^{2,\alpha}(\overline{\Omega})$, we consider the normal graph of u defined as the set

$$\{\exp_p(u(p)\,N_p) : p \in \overline{\Omega}\}.$$

For u small, the graph of u is a regular hypersurface diffeomorphic to $\overline{\Omega}$ of class $C^{2,\alpha}$ with a well-defined mean curvature $H(u)$. We consider the function

$$F : C^{2,\alpha}(\overline{\Omega}) \to C^{0,\alpha}(\overline{\Omega}) \times C^{2,\alpha}(\partial\Omega)$$

defined by

$$F(u) = (H(u), u|_{\partial\Omega}).$$

The derivative of this map between Banach spaces is given by

$$(dF)_0(v) = (\mathcal{L}v, v|_{\partial\Omega}).$$

This map is continuous, injective since there are no eigenfunctions with eigenvalue 0 as $\lambda_1(\Omega) > 0$, and surjective by Fredholm's alternative. The inverse map is continuous by the classical Schauder's estimates. The implicit function theorem for differentiable maps between Banach spaces provides a neighborhood of $(H, 0) \in C^{0,\alpha}(\overline{\Omega}) \times C^{2,\alpha}(\partial\Omega)$ such that F is invertible. In particular, there exists $\varepsilon > 0$ such that (H, t) has a preimage for $|t| < \varepsilon$. We let S_t be the normal graph of the function $u_t = F^{-1}(H, t)$. Reducing ε if necessary, the hypersurface S_t is regular. Let $v = du_t/dt$. Then v satisfies the equation $\mathcal{L}v = 0$ with boundary conditions $v|_{\partial\Omega} = 1$. We claim that $v > 0$ on $\overline{\Omega}$. This would imply that S_t is a foliation for t small enough. If $v \geqslant 0$ and equal to 0 at some point in Ω, Harnack's inequality for elliptic operators gives a contradiction; see Lemma 3 in [271]. If $v < 0$ at some point of Ω, we let $\Omega' = \{p \in \Omega : v(p) < 0\}$. Then $\lambda_1(\Omega') = 0$, but, since $\Omega' \subset \Omega$, we would get a contradiction since $\lambda_1(\Omega) < \lambda_1(\Omega') = 0$.

For the minimization property, we take into account that the sets E^+ and E^- have finite perimeter. We consider the vector field X in the region foliated by S_t, defined at a point $p \in S_t$ as the normal to S_t at p. We make a continuous choice so that X coincides with N on $\overline{\Omega}$. Then

$$\operatorname{div} X = H.$$

Applying the divergence theorem to E^+ and E^- and letting ν be the a choice of unit normal to S', we have $\langle X, \nu\rangle \leqslant 1$, and so

$$H\left(|E^-| - |E^+|\right) = A(\Omega) - \int_{S'} \langle X, \nu\rangle \, dS' \geqslant A(S) - A(S').$$

This implies (6.7). If equality holds in (6.7), then $X = \nu$ and S' coincides with $\overline{\Omega}$.

□

Let us now prove a stronger minimization property for stable hypersurfaces. This was also proven by Große-Brauckmann [201].

Theorem 6.10 ([201]) *Let $S \subset M$ be a strictly stable constant mean curvature hypersurface with constant mean curvature H. Let $S(u)$ be the normal graph of a function $u : S \to \mathbb{R}$. Then there is a constant $C > 0$ such that, if $||u||_{C^{0,1}} < C$, $u = 0$ on ∂S, and S and $S(u)$ enclose the same volume, then $A(S(u)) > A(S)$ unless $u = 0$.*

The result holds if $\partial S = \emptyset$ and also if we replace the stability condition by an equivariant stability condition. This means that S is invariant under the action of some subgroup of isometries of M and the strict stability condition holds for equivariant test functions. The consequence is that S minimizes area in the class of equivariant normal graphs.

Proof of Theorem 6.10 The strict stability condition means the existence of $\lambda > 0$ such that

$$\int_S \left\{ |\nabla_S u|^2 - q\, u^2 \right\} dS \geqslant \lambda \int_S u^2 dS, \tag{6.8}$$

for $q = \mathrm{Ric}(N, N) + |\sigma|^2$ and all $u \in C_0^\infty(S)$ with mean zero.

Let us prove first that there exists a constant $C > 0$ large enough so that

$$\int_S \left\{ |\nabla_S u|^2 - q u^2 \right\} dS + C \left(\int_S u \, dS \right)^2 \geqslant \lambda \int_S u^2 \, dS \tag{6.9}$$

for all $u \in C_0^\infty(S)$, thus removing the mean zero condition. Otherwise, there is a sequence of functions $\{u_i\}_{i\in\mathbb{N}}$ in $C_0^\infty(S)$ with $||u_i||_{L_2(S)} = \int_S u_i^2 dS = 1$ such that

$$\int_S \left\{ |\nabla_S u_i|^2 - q u_i^2 \right\} dS + i \left(\int_S u_i dS \right)^2 < \lambda \int_S u_i^2 \, dS. \tag{6.10}$$

Hence,

$$\int_S |\nabla_S u_i|^2 dS < \int_S (q + \lambda)\, u_i^2 d < C' ||u_i||_{L^2(S)} = C'$$

and so the sequence $\{u_i\}_{i\in\mathbb{N}}$ is bounded in $W_0^{1,2}(S)$. Hence, a non-relabeled subsequence converges weakly in $W_0^{1,2}(S)$ and strongly in $L^2(S)$ to a function u. By (6.10), the function u has mean 0, but this contradicts (6.8) and proves the existence of a constant $C > 0$ so that (6.9) holds.

Let $\delta > 0$ so that in the tubular neighborhood of radius δ, there is a unique metric projection to S. For a function u with $||u||_{L^\infty(S)} < \delta$, the normal graph

$$S(u) = \left\{ \exp_p(u(p)\, N_p) : p \in S \right\}$$

of u is contained in this tubular neighborhood. We define

$$E(u) = \left\{ \exp_p(tu(p)N_p) : p \in S, -\delta \leqslant tu(p) \leqslant 1 \right\}.$$

Then it is easily checked that the left side of (6.9) is the second variation of the functional

$$F(u) = A(S(u)) - H\,|E(u)| + \frac{C}{2}\left(|E(u)| - |E(0)|\right)^2.$$

By the arguments in p. 532 of [201], we have

$$\begin{aligned} F(u) - F(0) &= \int_S \{|\nabla_S u|^2 - qu^2\}\, dS + C\left(\int_S u\, dS\right)^2 + o(||u||_{W_0^{1,2}(S)}) \\ &\geqslant \mu\, ||u||_{W_0^{1,2}(S)} \end{aligned}$$

for all u with sufficiently small $||\cdot||_{C^{0,1}(S)}$ norm and some $\mu > 0$ by (6.9). Hence, $A(S(u)) > A(S)$ if $S(u)$ and S enclose the same volume ($|E(u)| = |E(0)|$) unless $u = 0$. □

6.2 Stable Hypersurfaces in Simply Connected Space Forms

In this section, we consider stable hypersurfaces with constant mean curvature in the simply connected space forms $\mathbb{M}_\kappa^m$ of constant sectional curvature $\kappa \in \mathbb{R}$ and dimension $m \geqslant 2$. We shall use the warped product model for these space forms.

6.2.1 Stability of Geodesic Spheres in Simply Connected Space Forms

Let us see that geodesic spheres in the simply connected space form $\mathbb{M}_\kappa^m$ are stable hypersurfaces with constant mean curvature; see Lemma 3.4 and Proposition 2.13

in [39]. Recall that a geodesic sphere $S(p, r)$ centered at some point $p \in \mathbb{M}^m_\kappa$ of radius $r > 0$ is totally umbilical with principal curvatures

$$\frac{\cos_\kappa(r)}{\sin_\kappa(r)}.$$

The geodesic sphere $S(p, r)$ is isometric to the round sphere $\mathbb{S}^{m-1}(\sin_\kappa(r))$ of radius $\sin_\kappa(r)$; see (1.48). Hence, the Jacobi operator on $S(p, r)$ is given by

$$\mathcal{L} = \Delta_{\mathbb{S}^{m-1}(\sin_\kappa(r))} + (m-1)\left(\kappa + \frac{\cos^2_\kappa(r)}{\sin^2_\kappa(r)}\right) = \Delta_{\mathbb{S}^{m-1}(\sin_\kappa(r))} + \frac{m-1}{\sin^2_\kappa(r)}.$$

As $(m-1)/\sin^2_\kappa(r)$ is constant on $S(p, r)$, for any function $u : S(p, r) \to \mathbb{R}$ with mean zero satisfying $\int_{S(p,r)} u^2 dS(p, r) = 1$, we have

$$-\int_{S(p,r)} u\mathcal{L}u \, dS(p, r) \geqslant \lambda_1(\mathbb{S}^m(\sin_\kappa(r))) - \frac{m-1}{\sin^2_\kappa(r)} = 0$$

as the first eigenvalue of $\Delta_{\mathbb{S}^{m-1}(\sin_\kappa(r))}$ on $\mathbb{S}^{m-1}(\sin_\kappa(r))$ is $(m-1)/\sin^2_\kappa(r)$; see §4 in Chapter II of [100] or page 159 ff. in [56].

6.2.2 *Stable Constant Mean Curvature Surfaces in* $\mathbb{R}^m$

In this section, we prove a classical classification result by Barbosa and do Carmo [38] for stable orientable immersed hypersurfaces in Euclidean space.

Theorem 6.11 *Let $S \subset \mathbb{R}^m$ be a compact orientable stable hypersurface with constant mean curvature. Then S is a geodesic sphere.*

In the proof of Theorem 6.11, we shall use the function

$$u = (m-1) - H\,\langle \nabla(\tfrac{1}{2}d^2), N\rangle, \tag{6.11}$$

where H is the mean curvature with respect to the unit normal N to S and d is the distance function to a fixed arbitrary point in $\mathbb{R}^m$. When S is the boundary of a domain, we shall always choose the outer unit normal. The function u has mean zero since it is the divergence on S of the tangent vector field $\nabla_S(\frac{1}{2}d^2)$. See the first Minkowski formula in Lemma 4.32.

The second Minkowski formula is obtained by taking the derivative of the first one along a deformation by parallel hypersurfaces.

Lemma 6.12 (Second Minkowski Formula in $\mathbb{R}^m$) *Let $S \subset \mathbb{R}^m$ be a compact orientable hypersurface with constant mean curvature H. Then*

$$\int_S \left\{ \langle \nabla(\tfrac{1}{2}d^2), N \rangle \, |\sigma|^2 - H \right\} dS = 0. \tag{6.12}$$

Proof We consider the parallel hypersurfaces S_t in the direction of the normal N (i.e., $S_t = \{\exp_p(tN_p) : p \in S\}$). The first Minkowski formula holds for any S_t. Taking derivatives

$$\left.\frac{d}{dt}\right|_{t=0} \int_{S_t} \big((m-1) - H_t \langle \nabla(\tfrac{1}{2}d^2), N_t \rangle\big)\, dS_t = 0.$$

Since $(dS_t)'_{t=0} = H\,dS$, $(-H_t)'_{t=0} = |\sigma|^2$, $(D/dt)_{t=0}\nabla(d^2/2) = N$, $(D/dt)_{t=0}N_t = 0$, H is constant, and the function u defined in (6.11) has mean zero, we get (6.12). □

We shall need to compute $\mathcal{L}v$ when v is the function

$$v = \langle \nabla(\tfrac{1}{2}d^2), N \rangle.$$

This can be easily done observing that the flow $\{\varphi_t\}_{t\in\mathbb{R}}$ associated with the vector field $\nabla(\frac{1}{2}d^2)$ is a one-parameter group of dilations of ratio e^t. Hence, the mean curvature of $\varphi_t(S)$ is $e^{-t}H$. The variation of the mean curvature along a variation was computed in Lemma 1.26. In this case, we get

$$H = -\left.\frac{d}{dt}\right|_{t=0} e^{-t}H = \mathcal{L}v. \tag{6.13}$$

Hence, for the mean zero function $u = (m-1) - Hv$, since $\mathcal{L} = \Delta_S + |\sigma|^2$ on $\mathbb{R}^m$, we have

$$\begin{aligned}
-\int_S u\mathcal{L}u\, dS &= -\int_S \big((m-1) - Hv\big)\big((m-1)|\sigma|^2 - H^2\big)\, dS \\
&= -(m-1)\int_S |\sigma|^2\big((m-1) - Hv\big)\, dS \\
&\quad - (m-1)\int_S \big((m-1)|\sigma|^2 - H^2\big)\, dS
\end{aligned}$$

The last equality because of the second Minkowski formula (6.12). Hence, by stability, we get

$$\int_S \big((m-1)|\sigma|^2 - H^2\big)\, dS \leqslant 0,$$

but since $(m-1)|\sigma|^2 - H^2 \geqslant 0$, we conclude $(m-1)|\sigma|^2 - H^2 \equiv 0$. This means that S is a totally umbilical geodesic sphere in $\mathbb{R}^m$. This completes the proof of Theorem 6.11.

Wente's Approach

The function $u = (m-1) - H\langle \nabla(\frac{1}{2}d^2), N\rangle$ has an interesting geometric interpretation as shown by Wente [442]. Assume that S is a constant mean curvature hypersurface that encloses a domain E. Steiner's formula in $\mathbb{R}^m$ yields the area $a(t)$ of the parallel hypersurface S_t as

$$a(t) = \int_S \left(\prod_{i=1}^{m-1} (1 + t\kappa_i) \right) dS = \sum_{i=0}^{m-1} a_i t^i.$$

The volume of the enclosed sets E_t is given by

$$v(t) = v_0 + \int_0^t a(\xi)\, d\xi = \sum_{i=0}^{m} v_i t^i,$$

where $v_0 = |E|$ and the area of the parallels is given by

$$a(t) = A(S_t) = \frac{dv(t)}{dt} = \sum_{i=0}^{m-1} a_i t^i.$$

We have the obvious equalities $v_1 = a_0, 2v_2 = a_1 = Ha_0$. For each t, we consider a dilation $h_{s(t)}$ of ratio $s(t)$ so that

$$|h_{s(t)}(E_t)| = |E|.$$

Since $h_{s(t)}(E)| = s(t)^m |E|$, we may compute the asymptotic expansion of the volume

$$\begin{aligned} |h_{s(t)}(E)| &= (s_0 + o(t))^m (v_0 + o(t)). \\ &= (s_0^m + o(t))(v_0 + o(t)) = s_0^m v_0 + o(t). \end{aligned}$$

Hence, $s_0^m v_0 = v_0$, and so

$$s_0 = 1.$$

Since S is a hypersurface with constant mean curvature and $h_{s(t)}(S_t)$ is a volume-preserving variation, we have $\frac{d}{dt}\big|_{t=0} A(h_{s(t)}(S)) = 0$. But a simple computation yields

$$\begin{aligned}A(h_{s(t)}(S)) &= (s_0 + s_1 t + o(t^2))^{m-1}(a_0 + a_1 t + o(t^2))\\ &= (1 + (m-1)s_1 t + o(t^2))(a_0 + a_1 t + o(t^2))\\ &= a_0 + ((m-1)s_1 a_0 + a_1)t + o(t^2),\end{aligned}$$

and so

$$0 = (m-1)s_1 a_0 + a_1 = ((m-1)s_1 + H)a_0.$$

This implies

$$s'(0) = s_1 = -\frac{H}{m-1}.$$

Let f_t be the one-parameter family of deformations by parallels, with associated vector field N, and h_t the one-parameter family of dilations of ratio t, with associated vector field $X_1 = \nabla(\frac{1}{2}d^2)$. Then the vector field associated with the one-parameter family $h_{s(t)} \circ f_t$ is given by

$$s'(0)\nabla(\tfrac{1}{2}d^2) + N = -\frac{H}{m-1}\nabla(\tfrac{1}{2}d^2) + N,$$

whose normal component is, up to a constant, the test function u defined in (6.11) and used in Theorem 6.11 to prove the stability of geodesic spheres in $\mathbb{R}^m$. Hence, the deformation used by Barbosa and do Carmo consists of taking parallel hypersurfaces followed by appropriate dilations so that the enclosed volume is constant.

6.2.3 *Stable Constant Mean Curvature Hypersurfaces in* $\mathbb{S}^m$ *and* $\mathbb{H}^m$

We prove in this section that any compact orientable stable constant mean curvature hypersurface in a round sphere or a hyperbolic space is a geodesic sphere. This result was first proven by Barbosa, do Carmo, and Eschenburg [39]. We follow the computations in [429] assuming for simplicity that the sectional curvature of the ambient manifold is $\kappa = 1$ in the spherical case or $\kappa = -1$ in the hyperbolic case. In these cases, the function $\cos_\kappa$ defined in (1.47) is either cos or cosh.

Observe that in case $\kappa = \pm 1$, we have

$$\nabla_e \nabla \cos_\kappa(d) = (-\kappa \cos_\kappa(d))\, e$$

for all $e \in T_pS$ and $p \in S$. Hence,

$$\nabla^2 \cos_\kappa(d) = -\kappa \cos_\kappa(d) \langle \cdot, \cdot \rangle,$$

when $\kappa = \pm 1$.

In the following result, we compute the Laplacian on a hypersurface S with normal N of the key functions $\cos_\kappa(d)$ and $\langle \nabla \cos_\kappa(d), N \rangle$.

Lemma 6.13 *Let $S \subset \mathbb{M}^m_\kappa$ be a hypersurface with constant mean curvature H with respect to a unit normal N in a space form with $\kappa = \pm 1$. Let d be the distance function in $\mathbb{M}^m_\kappa$ to a fixed point. Then we have:*

1. $\Delta_S \cos_\kappa(d) = -(m-1)\, \kappa \cos_\kappa(d) - H \langle \nabla \cos_\kappa(d), N \rangle$.
2. $\Delta_S \langle \nabla \cos_\kappa(d), N \rangle = -\kappa H \cos_\kappa(d) - |\sigma|^2 \langle \nabla \cos_\kappa(d), N \rangle$.

Proof 1. Since $\nabla^2 \cos_\kappa(d) = -\kappa \cos_\kappa(d) \langle \cdot, \cdot \rangle$, formula (4.31) implies

$$\big(\Delta_S \cos_\kappa(d)\big)(p) = \sum_{i=1}^{m-1} \nabla^2 \cos_\kappa(d)(e_i, e_i) - H \langle \nabla \cos_\kappa(d), N \rangle(p),$$

for $p \in S$ and an orthonormal basis $e_1, \ldots, e_{m-1}$ of T_pS. This implies 1.

2. We fix $p \in S$ and a local orthonormal basis $E_1, \ldots, E_{m-1}$ of TS such that $(\nabla_{e_i} E_i)^\top = 0$ for all $i = 1, \ldots, m-1$ and $e_i = (E_i)_p$. Then we have

$$\begin{aligned}
\big(\Delta_S \langle \nabla \cos_\kappa(d), N \rangle\big)(p) &= \sum_{i=1}^{m-1} e_i (E_i \langle \nabla \cos_\kappa(d), N \rangle) \\
&= \sum_{i=1}^{m-1} e_i \langle \nabla \cos_\kappa(d), \nabla_{E_i} N \rangle \\
&= \sum_{i=1}^{m-1} \langle -\kappa \cos_\kappa(d)\, e_i, \nabla_{e_i} N \rangle + \sum_{i=1}^{m-1} \langle \nabla \cos_\kappa(d), \nabla_{e_i} \nabla_{E_i} N \rangle \\
&= -\kappa H \cos_\kappa(d) + \sum_{i=1}^{m-1} \langle \nabla \cos_\kappa(d), \nabla_{e_i} \nabla_{E_i} N \rangle.
\end{aligned}$$

We compute now $\sum_{i=1}^{m-1} \nabla_{e_i} \nabla_{E_i} N$. Observe that

$$\langle \sum_{i=1}^{m-1} \nabla_{e_i} \nabla_{E_i} N, N_p \rangle = -\sum_{i=1}^{m-1} |\nabla_{e_i} N|^2 = -|\sigma|^2.$$

On the other hand, as H is constant, equality

$$0 = e_j(H) = \sum_{i=1}^{m-1} \langle \nabla_{e_j} \nabla_{E_i} N, e_i \rangle$$

holds for all $j = 1, \ldots, m-1$. This follows since $\langle \nabla_{e_i} N, \nabla_{e_j} E_i \rangle = 0$ as $\nabla_{e_i} N$ is tangent to S and $\nabla_{e_j} E_i$ is orthogonal to S. But $R(e_i, e_j, N_p, e_i) = 0$ as the ambient manifold has constant sectional curvature. As $[E_i, E_j]_p = 0$, we conclude

$$\langle \nabla_{e_j} \nabla_{E_i} N, e_i \rangle = \langle \nabla_{e_i} \nabla_{E_j} N, e_i \rangle,$$

and so

$$\langle \nabla_{e_i} \nabla_{E_j} N, e_i \rangle = e_i \langle \nabla_{E_j} N, E_i \rangle = e_i \langle \nabla_{E_i} N, E_j \rangle = \langle \nabla_{e_i} \nabla_{E_i} N, e_j \rangle$$

because of the symmetry of the second fundamental form. So we have

$$0 = \langle \sum_{i=1}^{m-1} \nabla_{e_i} \nabla_{E_i} N, e_j \rangle$$

for all $j = 1, \ldots, m-1$, and we finally get

$$\sum_{i=1}^{m-1} \nabla_{e_i} \nabla_{E_i} N = -|\sigma|^2 N_p.$$

This completes the computation of 2. □

The Jacobi operator on a hypersurface $S \subset \mathbb{M}^m_\kappa$ is given by

$$\mathcal{L}_\kappa = \Delta_S + (m-1)\kappa + |\sigma|^2.$$

Lemma 6.13 allows to compute $\mathcal{L}_\kappa$ applied to $\cos_\kappa(d)$ and $\langle \nabla \cos_\kappa(d), N \rangle$:

$$\begin{aligned} \mathcal{L}_\kappa \cos_\kappa(d) &= |\sigma|^2 \cos_\kappa(d) - H \langle \nabla \cos_\kappa(d), N \rangle, \\ \mathcal{L}_\kappa \langle \nabla \cos_\kappa(d), N \rangle &= -\kappa H \cos_\kappa(d) + (m-1)\,\kappa \langle \nabla \cos_\kappa(d), N \rangle \end{aligned} \tag{6.14}$$

Let us prove now the characterization of stable hypersurfaces in spheres and hyperbolic spaces.

Theorem 6.14 *Let $S \subset \mathbb{S}^m$ or $S \subset \mathbb{H}^m$ be a compact orientable stable hypersurface with constant mean curvature. Then S is a geodesic sphere.*

Proof Consider the smooth function $u : S \to \mathbb{R}$ defined by

$$u = (m-1)\,\kappa \cos_\kappa(d) + H\langle \nabla \cos_\kappa(d), N\rangle,$$

where d is the distance in the ambient manifold to a fixed point p. Since $u = -\Delta_S \cos_\kappa(d)$ by Lemma 6.13(1), the divergence theorem implies that u has mean zero on S. Formulas (6.14) imply

$$\mathcal{L}_\kappa u = \kappa\,((m-1)|\sigma|^2 - H^2)\, \cos_\kappa(d).$$

So we have

$$u\mathcal{L}_\kappa u = \big[(m-1)|\sigma|^2 - H^2\big]\big[(m-1)\kappa^2 \cos_\kappa(d)^2 + H\kappa \cos_\kappa(d)\, \langle \nabla \cos_\kappa(d), N\rangle\big]. \tag{6.15}$$

To complete the proof, we consider separately the spherical and hyperbolic cases.

In the spherical case $\kappa = 1$, we observe that $(\cos_\kappa(d))(x) = \langle x, p\rangle$ for all $x \in \mathbb{S}^m$, where $\langle \cdot, \cdot\rangle$ is the scalar product in $\mathbb{R}^{m+1}$. Taking an orthonormal basis $p_1, \ldots, p_{m+1}$ in $\mathbb{R}^{m+1}$, and calling d_i to the distance in $\mathbb{S}^m$ to the point $p_i \in \mathbb{S}^m$, we consider the functions

$$u_i = (m-1)\cos_\kappa(d_i) + H\langle \nabla \cos_\kappa(d_i), N\rangle, \qquad i = 1, \ldots, m+1.$$

Since

$$\sum_{i=1}^{m+1} \cos_\kappa(d_i)^2 = 1, \qquad \sum_{i=1}^{m+1} \cos_\kappa(d_i)\nabla \cos_\kappa(d_i) = 0,$$

we get from (6.15)

$$\sum_{i=1}^{m+1} \int_S u_i\, \mathcal{L}_\kappa u_i\, dS = (m-1) \int_S \big((m-1)|\sigma|^2 - H^2\big)\, dS \leqslant 0.$$

As $(m-1)|\sigma|^2 - H^2 \geqslant 0$ on S, we obtain $(m-1)|\sigma|^2 = H^2$, and we conclude that S is totally umbilical and hence a geodesic sphere. This completes the proof when the ambient manifold is $\mathbb{S}^m$.

In the hyperbolic case $\kappa = -1$, we consider the function

$$h = H\, \cos_\kappa(d) - (m-1)\, \langle \nabla \cos_\kappa(d), N\rangle.$$

From Lemma 6.13, we get

$$\Delta_S h = \big((m-1)|\sigma|^2 - H^2\big)\, \langle \nabla \cos_k(d), N\rangle.$$

Since $\Delta_S(\frac{h^2}{2}) = h\Delta_S h + |\nabla_S h|^2$, formula (6.15) implies

$$u\mathcal{L}_k u + \Delta_S(\tfrac{h^2}{2}) = |\nabla_S h|^2 + (m-1)\big[(m-1)|\sigma|^2 - H^2\big]\big[\cos_\kappa(d)^2 - \langle \nabla \cos_\kappa(d), N\rangle^2\big].$$

As we have

$$\cos_\kappa(d)^2 - \langle \nabla \cos_\kappa(d), N\rangle^2 \geqslant \cosh(d)^2 - \sinh(d)^2 = 1,$$

the stability condition yields

$$0 \geqslant \int_S u\mathcal{L}_\kappa u\, dS = \int_S \big(u\mathcal{L}_\kappa u + \Delta_S(\tfrac{h^2}{2})\big)\, dS \geqslant (m-1)\int_S \big((m-1)|\sigma|^2 - H^2\big)\, dS.$$

This implies that S is totally umbilical in $\mathbb{H}^m$ and hence a geodesic sphere. □

6.2.4 The Stability Condition in the Presence of a Singular Set

The results in Sects. 6.2.2 and 6.2.3 apply to smooth hypersurfaces in simply connected space forms. We show in this section that those results are also valid for boundaries which are smooth except on a singular set of small Hausdorff dimension. In particular, they are valid for isoperimetric boundaries, thus providing another proof of the characterization of geodesic balls as the only isoperimetric sets in simply connected space forms.

Lemma 6.15 *Let $E \subset M$ be a bounded isoperimetric set in a Riemannian manifold M. Assume that $\partial E = S \cup S_0$ where S, S_0 are the regular and singular parts of ∂E. Then*

$$\int_S |\sigma|^2 dS < +\infty. \tag{6.16}$$

Moreover, if $u \in C^\infty(S)$ is a mean zero function such that u is bounded and $|\nabla u| \in L^2(S)$, then the stability inequality (6.1) holds for the function u. If, in addition, $\Delta_S u \in L^1(S)$, then also (6.2) holds.

Proof To check that (6.16) holds, we take a sequence $\{\varphi_i\}_{i\in\mathbb{N}}$ as in Lemma 1.61 ($\varphi_i \geqslant 0$, $\varphi_i \uparrow 1$ pointwise, and $\int_S |\nabla_S \varphi_i|^2 dS \to 1$) and a fixed function $v \in C_0^\infty(S)$ so that $\int_S v\, dS > 0$. Note that supp(v) is a compact set in the interior of S. We take constants $a_i > 0$ so that $\varphi_i - a_i v$ has mean zero for all $i \in \mathbb{N}$. As $\int_S v dS > 0$ and

$\int_S \varphi_i dS$ converges to the area $A(S)$ of S, the sequence a_i converges to a constant $a > 0$.

Since S is a stable hypersurface, we have

$$\int_S |\sigma|^2 (\varphi_i - a_i v)^2 dS \leqslant \int_S \left(|\nabla_S(\varphi_i - a_i v)|^2 - (\mathrm{Ric}(N, N)(\varphi_i - a_i v)^2\right) dS. \tag{6.17}$$

As $\lim_{i\to\infty} \int_S |\nabla \varphi_i|^2 dS = 0$, we get

$$\lim_{i\to 0} \int_S |\nabla_S(\varphi_i - a_i v)|^2 dS = a^2 \int_S |\nabla_S v|^2 dS.$$

Since $\mathrm{Ric}(N, N)(\varphi_i - a_i v)^2$ is uniformly bounded on S and S has finite measure, we have

$$\lim_{i\to\infty} \int_S \mathrm{Ric}(N, N)(\varphi_i - a_i v)^2 dS = \int_S \mathrm{Ric}(N, N)(1 - av)^2 dS$$

Finally, we estimate

$$\int_S |\sigma|^2 (\varphi_i - a_i v)^2) dS \geqslant \int_{S\setminus \mathrm{supp}(v)} |\sigma|^2 \varphi_i^2 dS.$$

So we get from Fatou's lemma and (6.17)

$$\int_{S\setminus \mathrm{supp}(v)} |\sigma|^2 dS \leqslant a^2 \int_S |\nabla_S v|^2 dS - \int_S \mathrm{Ric}(N, N)(1 - av)^2 dS < +\infty.$$

This proves (6.17).

We consider now a bounded function $u \in C^\infty(S)$ satisfying $|\nabla_S u| \in L^2(S)$, $\Delta_S \in L^1(S)$. We may assume that $u \not\equiv 0$. Again, we consider a sequence of functions $\{\varphi_i\}_{i\in\mathbb{N}}$ as in Lemma 1.61. For every $i \in \mathbb{N}$, we take

$$u_i = \varphi_i (u^+ - a_i u^-),$$

where $u^+ = \max\{u, 0\}$ and $u^- = \max\{-u, 0\}$. The constant a_i is chosen so that u_i has mean zero on S. As $u \not\equiv 0$ and $\varphi_i \to 1$, the sequence $\{a_i\}_{i\in\mathbb{N}}$ converges to 1. Since u_i has compact support on S, the stability of S implies

$$\int_S \{|\nabla_S u_i|^2 - (\mathrm{Ric}(N, N) + |\sigma|^2) u_i^2\} dS \geqslant 0. \tag{6.18}$$

Since $\lim_{i\to\infty}\int_S |\nabla_S u_i|^2 dS = 0$, φ_i, u are uniformly bounded, and $a_i \to 1$, we get

$$\lim_{i\to\infty}\int_S |\nabla_S u_i|^2 dS = \int_S |\nabla u|^2 dS.$$

As $\mathrm{Ric}(N,N)+|\sigma|^2 \in L^1(S)$, φ_i, u are uniformly bounded, and $a_i \to 1$, we obtain

$$\lim_{i\to\infty}\int_S (\mathrm{Ric}(N,N)+|\sigma|^2)u_i^2\, dS = \int_S (\mathrm{Ric}(N,N)+|\sigma|^2)u^2\, dS$$

Taking limits in (6.18), we get that the stability inequality (6.1) holds for u.

To check that the stability condition (6.2) involving the operator $\mathcal{L}$ is satisfied, we only need to check that equality

$$\int_S |\nabla_S u|^2 dS = -\int_S u\Delta_S u dS \tag{6.19}$$

holds. To prove (6.19), we notice that

$$\int_S \varphi_i u\Delta_S u dS = -\int_S \langle \nabla_S \varphi_i u, \nabla_S u\rangle\, dS = -\int_S \big(\varphi_i |\nabla_S u|^2 + u\langle \nabla\varphi_i, \nabla_S u\rangle\big)\, dS.$$

Taking into account that φ_i, u are uniformly bounded, that $\varphi_i \to 1$ and $\int_S |\nabla_S u_i|^2 dS \to 1$, and that $|\nabla_S u|^2$, $\Delta_S u \in L^1(S)$, we may take limits in this equality to get (6.19). Equations (6.18) and the stability condition (6.1) for u imply that (6.2) holds for u. □

A consequence of Lemma 6.15 is that the results in Sects. 6.2.2 and 6.2.3 also hold for isoperimetric boundaries. We simply notice that the test functions used,

$$u = (m-1) + H\langle \nabla(\tfrac{1}{2}d^2), N\rangle$$

in the Euclidean case and

$$u = (m-1)\,\kappa\cos_\kappa(d) + H\langle \nabla\cos_\kappa(d), N\rangle$$

in the cases $\kappa = \pm 1$, are bounded on an isoperimetric boundary S by a constant depending on H and the supremum of d on S. Their gradients on S are

$$H\sum_{i=1}^{m-1}\langle \nabla(\tfrac{1}{2}d^2), \nabla_{e_i}N\rangle\, e_i, \qquad (m-1)\kappa\nabla_S\cos_\kappa(d) + H\sum_{i=1}^{m-1}\langle \nabla\cos_\kappa(d), \nabla_{e_i}N\rangle\, e_i$$

so that $\int_S |\nabla_S u|^2 dS$ can be estimated in terms of H, $\sup_S d$, and the $L^1(S)$-norm of $|\sigma|^2$.

To estimate $\Delta_S u$, we take into account that

$$\Delta_S u = H^2 - |\sigma|^2 H u$$

in the Euclidean case (we have used (6.13)). In the hyperbolic and spherical cases, we have

$$\Delta_S u = \big(-(m-1)^2 - \kappa H^2 \big) \cos_\kappa(d) - \big((m-1)\kappa + H|\sigma|^2 \big) \langle \nabla \cos_\kappa(d), N \rangle$$

from Lemma 6.13. In this case, we also obtain that $\int_S |\Delta_S u| dS$ is bounded in terms of H, $\sup_S d$, and the $L^1(S)$-norm of $|\sigma|^2$.

6.3 Three-Dimensional Space Forms

In this section, we consider isoperimetric sets in three-dimensional space forms. Here, there are two main differences with the general theory: The first one is that isoperimetric boundaries are always regular by Theorem 1.60. The second one is the existence on every orientable surface S of an atlas whose transition maps are holomorphic, thus defining on S a structure of *Riemann surface* or *one*-dimensional complex manifold.

The existence of this holomorphic atlas is based on the existence of local *isothermal coordinates* near every point of the surface. A local chart (U, φ) with coordinates (x, y) in a Riemannian surface (S, g) is *isothermal* if the metric is conformal to the flat metric $dx^2 + dy^2$. This is equivalent to the existence of a smooth function $u : U \to \mathbb{R}$ such that

$$g = e^{2u}(dx^2 + dy^2).$$

Letting $z = x + iy$, we denote $dx^2 + dy^2$ by $|dz|^2$. A quick proof of existence of isothermal coordinates around any point in S is obtained from elementary elliptic theory. Given two Riemannian metrics g, h on S so that $h = e^{2u} g$, we have the well-known relation between the Gauss curvatures K_g, K_h given by

$$K_h = e^{-2u}(K_g - \Delta_g u).$$

Hence, given a Riemannian metric g on S, there exists a flat metric h conformal to g on $U \subset S$ if and only if there exists $u \in C^\infty(U)$ such that

$$\Delta_g u = K_g.$$

This elliptic equation has a solution if U is a small disk where the first eigenvalue of the Laplacian Δ_g is positive. Hence, (U, h) is flat and so locally isometric to the

Euclidean plane by Cartan's theorem; see §8.2 in [141]. This proves the existence of isothermal coordinates around any given point.

In order to have a structure of Riemann surface on S, we need to assume that S is orientable. Then the system of isothermal charts (U, φ) so that φ is orientation preserving provides an atlas so that the transition maps are orientation preserving and conformal and so they are holomorphic.

We refer the reader to Farkas and Kra's monograph [152] for background on Riemann surfaces.

6.3.1 Bounds on the Genus of an Isoperimetric Boundary

In this section, we provide upper bounds on the genus of a connected stable surface with constant mean curvature. Let S be a two-sided surface in $\mathbb{M}^3_c$. Let N be a unit normal and K, H the Gauss and constant mean curvature of S. If κ_1, κ_2 are the principal curvatures of S, the Gauss equation implies $K = c + \kappa_1\kappa_2$, and we have

$$\operatorname{Ric}(N, N) + |\sigma|^2 = 4c + H^2 - 2K.$$

When $c + H^2$ is positive (e.g., when $c \geqslant 0$), we let $b^2 = 4c + H^2$. Hence, the Jacobi operator is given by

$$\mathcal{L} = \Delta + b^2 - 2K.$$

Given an orientable two-sided compact Riemannian surface M, we shall consider conformal maps $\phi : M \to \mathbb{S}^2$ to the unit *two*-dimensional sphere $\mathbb{S}^2$. These are holomorphic maps from S with its associated Riemann surface structure to M to the sphere $\mathbb{S}^2$ with its usual conformal structure. Writing $\phi = (\phi_1, \phi_2, \phi_3)$, we can use ϕ_i as test functions in the index form $\mathcal{I}$ assuming they have mean zero.

The key result to obtain a bound on the genus of an isoperimetric surface with orientable boundary is the following.

Lemma 6.16 *Let $S \subset M^3$ be a connected stable orientable two-sided compact surface in a three-dimensional space form. Let $\phi : S \to \mathbb{S}^2$ be a conformal map such that $\int_S \phi \, dS = 0$. Assume $4c + H^2 \geqslant 0$. Then*

$$\textit{genus}\, S \leqslant 1 + \deg \phi. \tag{6.20}$$

If equality holds in (6.20), then $4c + H^2 = 0$. If equality holds in (6.20) and $c \geqslant 0$, then $c = H = 0$, $|\nabla\phi|^2 = -2K$, and the stability operator is given by $\mathcal{L} = \Delta + |\nabla\phi|^2$.

Proof Let $b^2 = 4c + H^2$. We use the coordinates of ϕ as test functions in the index form $\mathcal{J}$. Since ϕ is conformal, given an orthonormal basis e_1, e_2 of T_pS, we get

$$\mathrm{Jac}(\phi)(p) = \frac{1}{2}\sum_{j=1}^{2} |d\phi_p(e_j)|^2 = \frac{1}{2}\sum_{i=1}^{3} |\nabla\phi_i|^2(p) = \frac{1}{2}|\nabla\phi|^2(p)$$

and, by the area formula,

$$\frac{1}{2}\int_S \sum_{i=1}^{3} |\nabla\phi_i|^2 dS = 4\pi \deg\phi.$$

Inserting now the test functions ϕ_i in the index form $\mathcal{J}$ and using the stability of S and Gauss-Bonnet theorem, we have

$$0 \leqslant \sum_{i=1}^{3} \mathcal{J}(\phi_i) = \sum_{i=1}^{3} \int_S \{|\nabla\phi_i|^2 - (b^2 - 2K)\,\phi_i^2\}\, dS \tag{6.21}$$

and so we obtain

$$0 \leqslant \int_S b^2\, dS \leqslant 8\pi \deg\phi + 8\pi(1 - \text{genus } S) = 8\pi\,(1 + \deg\phi - \text{genus } S). \tag{6.22}$$

This implies the estimate (6.20).

If equality holds in (6.21), then $b^2 = 4c + H^2 = 0$. If, in addition, $c \geqslant 0$, then $c = H = 0$. So S is a minimal surface in a flat 3-manifold and $K \leqslant 0$. As equality also holds in (6.21), we get $|\nabla\phi|^2 + 2K = 0$. Hence, the Jacobi operator is given by $\mathcal{L} = \Delta - 2K = \Delta + |\nabla\phi|^2$. □

The operators $\Delta + |\nabla\phi|^2$ are often called *Schrödinger operators associated with a holomorphic map* $\phi : S \to \mathbb{S}^3$. They play an essential role in the study of the index of minimal surfaces in Euclidean space.

It is a well-known fact that any holomorphic function $\phi : S \to \mathbb{S}^2$ can be composed of a conformal diffeomorphism ψ of $\mathbb{S}^2$ so that $\int_S (\psi \circ \phi)\, dS = 0$; see [270]. Using this result, we only need to construct holomorphic functions from a compact Riemann surface to the sphere with controlled degree. For instance, if S is a hyperelliptic Riemann surface which can be expressed as a degree 2 branched holomorphic covering of the sphere, we may find $\phi : S \to \mathbb{S}^2$ of degree 2 with mean 0 so that (6.20) would imply genus $S \leqslant 3$. Summing up, we have the following result, where heavy machinery on Riemann surfaces theory is used.

Theorem 6.17 *Let $S \subset M$ be a compact connected orientable two-sided stable surface in a three-dimensional space form of constant curvature c. Assume that $4c + H^2 \geqslant 0$. Then*

$$\textit{genus } S \leqslant 3.$$

Proof Let $g = \text{genus}\, M$. By [193, p. 261], there exists a holomorphic map $\phi : S \to \mathbb{S}^2$ such that

$$\deg \phi \leqslant 1 + \left[\frac{g+1}{2}\right],$$

where $[x]$ denotes the integer part of x. By (6.20), we obtain

$$g \leqslant 1 + \deg(\phi) \leqslant 2 + \left[\frac{g+1}{2}\right]. \tag{6.23}$$

This inequality only holds for $g \leqslant 5$. Moreover, if $g = 4, 5$, then we have equality in (6.23) and so in (6.20). Lemma 6.16 then implies $c = H = 0$ and $\deg \phi = 1 + [(g+1)/2]$. In particular, in case $g = 5$, we have $\deg \phi = 4$ and, in case $g = 4$, $\deg(\phi) = 3$. We also have that $\mathcal{L} = \Delta + |\nabla \phi|^2$ is the Schrödinger operator associated with the holomorphic function $\phi : S \to \mathbb{S}^2$.

In particular, if $c > 0$, then we already have the bound genus $S \leqslant 3$.

Let us discard first the case $g = 5$ following the arguments in Proposition 1.2(iii) in [359]. Let us see that the number of negative eigenvalues of $\Delta + |\nabla \phi|^2$ is larger than 1. This would imply that S is not a stable surface. We first observe that there are no holomorphic maps $\psi : S \to \mathbb{S}^2$ of degree less than or equal to 3 (S is neither hyperelliptic nor trigonal) since otherwise, after composing with a conformal transformation of $\mathbb{S}^2$ so that $\int_S \psi \, dS = 0$, inequality (6.20) would be violated. Fix two different points $p, q \in S$. Riemann-Roch's theorem implies that the space V of abelian differentials vanishing at p and q has dimension 3 since S is not hyperelliptic. Another application of Riemann-Roch's theorem using that S is not trigonal shows that, for any $x \neq p, q$, there exists an element of V not vanishing at x. Thus, choosing a basis of V, one can define a map from $S \setminus \{p, q\}$ to $\mathbb{CP}^2$ that can be easily extended to a map $f : S \to \mathbb{CP}^2$. A third application of the Riemann-Roch theorem shows that there are one-forms in V vanishing either at p or q with order 1. So the degree d of f is equal to $2g - -2 - - \deg(p+q) = 6$. The curve $f(S)$ must be singular since otherwise the genus formula would imply that g would be equal to $(d - -1)(d - -2)/2 = 10$. Projecting from a singular point with multiplicity m to a line $\mathbb{CP}^1 \simeq \mathbb{S}^2$, we get a holomorphic map to $\mathbb{S}^2$ with degree $d - -m = 6 - -m$. As $6 - -m \geqslant 4$, again by inequality (6.20), we obtain $m = 2$. The genus formula then implies that there are $five$ singular points. An easy argument shows that no more than two singular points can be applied onto the same point by f: we may assume that such a point is $P = [(0, 0, 1)]$. If there are three singular points mapped onto P and $f = [(\omega_0, \omega_1, \omega_2)]$, then ω_0 and ω_1 are linearly independent, and they have the same divisors of zeroes, giving us a contradiction. So projecting from different singular points to a line meeting $f(M)$ at six different points, we obtain at least three holomorphic maps to $\mathbb{S}^2$ which are not related by a Möbius transformation. Theorem 6 in [299] then implies that the number of negative eigenvalues of the operator $\Delta + |\nabla \phi|^2$ is larger than 1.

Let us now discard the case $g = 4$. We follow the arguments in Theorem 14 in [381]. The reader is referred to [152, pp. 108-110] for details on Riemann surfaces of genus 4.

Let us see first that ϕ is the Gauss map of an unbranched multivalued conformal minimal immersion $x : S \to \mathbb{R}^3$ so that dx is globally well-defined on S. Let $\omega_1, \ldots, \omega_4$ be a basis of the space of holomorphic differentials. As the space of holomorphic quadratic differentials on S has dimension 9, the *ten* quadratic differentials $\omega_i\omega_j$, with $1 \leqslant i \leqslant j \leqslant 4$, are linearly dependent. Hence, there exists $a_{ij} \in \mathbb{C}$ such that $\sum_{1\leqslant i\leqslant j\leqslant 4} a_{ij}\omega_i\omega_j = 0$. The rank of this quadratic form is 3 or 4. By a linear transformation, this expression is equivalent to

$$\omega_1^2 + \omega_2^2 + \omega_3^2 = 0, \quad \text{or} \quad \omega_1^2 + \omega_2^2 + \omega_3^2 + \omega_4^2 = 0. \tag{6.24}$$

If the rank is 4, then S admits two degree 3 holomorphic functions to $\mathbb{S}^2$ not related by a Möbius transformation. So this case is discarded. We assume henceforth that the rank of the quadratic expression is 3. In this case, the unique holomorphic map to $\mathbb{S}^2 \simeq \overline{\mathbb{C}}$ of degree 3 on S up to Möbius transformations is the meromorphic map $g = \omega_3/(\omega_1 - i\omega_2) : S \to \overline{C}$ since otherwise the number of negative eigenvalues of $\Delta + |\nabla\phi|^2$ would be larger than one by the results of Montiel and Ros [299]. If the assertion in the claim holds for a meromorphic map φ, then it also holds for the composition of φ followed by a Möbius transformation. So it is enough to prove the claim for the meromorphic map g. If we consider $\omega = \omega_1 - i\omega_2$, we can write

$$\omega_1 = \frac{1}{2}\,(1 - g^2)\,\omega, \qquad \omega_1 = \frac{i}{2}\,(1 + g^2)\,\omega, \qquad \omega_3 = g\omega.$$

If g has a pole of order m at some $p \in S$, then ω vanishes at least with order $2m$ at p. As g has three poles and any holomorphic differential on S has exactly *six* zeroes, it follows that the zeroes of ω coincide with the poles of g and that the order of the zero is twice the order of the pole at those points. Therefore, g and ω are the Weierstrass data of a multivalued minimal immersion $x : S \to R^3$. The differential of x is the real part of $(\omega_1, \omega_2, \omega_3)$. This proves the claim.

Working with the metric induced by x, we have $|\nabla\phi|^2 = |\sigma|^2$, where σ is the second fundamental form of x. The Jacobi operator for the second variation of the area of $x(S)$ is given by $\mathcal{L}_x = \Delta + |\nabla\phi|^2$, where the Laplacian and the modulus of the gradient are the ones associated with the metric of the immersion. The differential dx of the immersion x is globally well-defined on S, and so the map

$$\omega \in H^1(S, \mathbb{R}) \mapsto X_\omega$$

is well-defined. Here, $H^1(S, \mathbb{R})$ is the space of harmonic one-forms in S, and $X_\omega : S \to \mathbb{R}^3$ is defined by

$$X_\omega = (\langle\omega, dx_1\rangle, \langle\omega, dx_2\rangle, \langle\omega, dx_3\rangle);$$

see [328]. The dimension of the space $V = \{\omega \in H^1(S, \mathbb{R}) : \int_S \varphi_1 X_\omega = 0\}$, where φ_1 is the first eigenfunction of $\Delta + |\sigma|^2$, is no smaller than $2 \times 4 - -3 = 5$. As in the proof of Theorem 7 in [381], we get that each $\omega \in V$ satisfies $\mathcal{J}_x(X_\omega) = 0$ and so $\mathcal{L}_x X_\omega = 0$. Hence, $\langle \nabla\omega, \sigma\rangle = 0$ for all $\omega \in V$ (here, σ is the second fundamental form of x). This contradiction shows that in the case $g = 4$ and $\deg(\phi) = 3$, the operator $\mathcal{L}$ has index larger than 1. □

From Proposition 6.17, we get the following result:

Corollary 6.18 *Let $E \subset M$ be an isoperimetric set in an orientable three-dimensional space form with sectional curvature $c \geqslant 0$ with connected boundary $S = \partial E$. Then genus $S \leqslant 3$.*

6.3.2 The Genus 0 Case

In the previous section, we have obtained an upper bound on the genus of an isoperimetric boundary in an orientable three-dimensional space form with sectional curvatures $c \geqslant 0$.

In this section, we consider the case of genus 0. In this case, we have a well-known result by H. Hopf; see Chapter VI in [233].

Theorem 6.19 ([233, Thm. 2.1, p. 138]) *Let $S \subset M$ be an orientable two-sided surface of genus 0 with constant mean curvature in a three-dimensional space form. Then S is totally umbilical and hence a geodesic sphere.*

In order to prove Theorem 6.19, let us introduce first the *Hopf differential*. This is a quadratic holomorphic differential for the Riemann surface structure of S. We consider the second fundamental form σ of S in M. Let W be the self-adjoint Weingarten endomorphism defined by $\sigma(X, Y) = \langle W(X), Y\rangle$. In an isothermal chart (U, φ), we consider the $(2, 0)$ part $\sigma^{2,0} = 4\sigma(\partial_z, \partial_z)\, dz^2$ of the complexification of σ. If x, y are the coordinates in U and $z = x + iy$, then

$$\sigma^{2,0} = \big((\sigma(\partial_x, \partial_x) - \sigma(\partial_y, \partial_y)) - 2i\sigma(\partial_x, \partial_y)\big)\, dz^2.$$

Let $A = \sigma(\partial_x, \partial_x)$, $B = \sigma(\partial_y, \partial_y)$, $C = \sigma(\partial_x, \partial_y)$. The Codazzi equation in space forms (see Proposition 3.4 in Chapter 6 of [141]) is

$$\nabla_X W(Y) = \nabla_Y W(X),$$

where W is the Weingarten endomorphism of S. Writing the Riemannian metric g of S as $e^{2u}(dx^2 + dy^2)$, we get from the Koszul formula

$$2e^{2u}\nabla_{\partial_x}\partial_x = (e^{2u})_x\partial_x - (e^{2u})_y\partial_y,$$
$$2e^{2u}\nabla_{\partial_x}\partial_y = (e^{2u})_y\partial_x + (e^{2u})_x\partial_y,$$
$$2e^{2u}\nabla_{\partial_y}\partial_x = -(e^{2u})_x\partial_x + (e^{2u})_y\partial_y.$$

From Codazzi's equation, we get, letting H be the mean curvature of S,

$$2\,(A_y - C_x) = (e^{2u})_y H,$$
$$2\,(B_x - C_y) = (e^{2u})_x H.$$

On the other hand, taking derivatives in the equality $A + B = e^{2u}H$, we obtain

$$A_x + B_x = (e^{2u})_x H + e^{2u} H_x,$$
$$A_y + B_y = (e^{2u})_y H + e^{2u} H_y.$$

Hence, we obtain

$$(A - B)_x + (2C)_y = e^{2u} H_y$$
$$(A - B)_y - (2C)_x = e^{2u} H_x.$$

So the function $(A - B) - 2iC$ satisfies the Cauchy-Riemann equations if and only if S has constant mean curvature.

Proof of Theorem 6.19 If genus $S = 0$, then any holomorphic quadratic differential vanishes on S. Hence, $\sigma^{2,0} = 0$, and S is totally umbilical and a geodesic sphere. □

Corollary 6.20 *Let $E \subset M$ be an isoperimetric set in an orientable three-dimensional space form with spherical boundary. Then E is a geodesic ball.*

6.3.3 The Genus 1 Case

We consider in this section isoperimetric sets with boundary of genus 1. We follow for the most part the arguments in §1 of [365].

We consider again the Hopf differential $\sigma^{2,0}$ associated with the second fundamental form σ. As S has constant mean curvature H, we know from Sect. 6.3.2 that $\sigma^{2,0}$ is a holomorphic quadratic differential with $4g - 4$ zeroes counted with multiplicity, where g = genus S. In particular, if $g = 1$, the surface S has no umbilical points. Let P be the eventually empty finite set of umbilical points in S.

Assuming $4c + H^2 > 0$, the metric $ds_0^2 = |\sigma^{2,0}|$ (if locally $\sigma^{2,0} = f\,dz^2$, then $|\sigma^{2,0}| = |f|\,|dz|^2$, where $|dz|^2 = dx^2 + dy^2$) is Riemannian on $M \setminus P$. Letting ds^2 be the Riemannian metric on S, we define a smooth function $w : M \setminus P \to \mathbb{R}$ by

$$ds^2 = \frac{e^{2w}}{b^2}\, ds_0^2,$$

where $b^2 = 4c + H^2$. Then we have

$$\Delta_0 w + \cosh(w)\sinh(w) = 0 \tag{6.25}$$

on $M \setminus P$, where Δ_0 is the Laplacian for the flat metric ds_0^2. To check that (6.25) holds, we first observe that

$$\Delta_0 w = -\frac{e^{2w}}{b^2}\, K, \tag{6.26}$$

where K is the Gauss curvature of S with respect to the metric ds^2, because of the relation between the Gauss curvatures of the conformal metrics ds^2 and ds_0^2. On the other hand,

$$|\sigma^{2,0}| = e^{2h}|\kappa_1 - \kappa_2|\,|dz|^2 = e^{2h}(b^2 - 4K)^{1/2}|dz|^2,$$

where $ds^2 = e^{2h}|dz|^2$. Since, on the other hand,

$$ds^2 = \frac{e^{2w}}{b^2}\, ds_0^2 = \frac{e^{2w}}{b}\, e^{2h}(b^2 - 4K)^{1/2}|dz|^2,$$

we get

$$K = \frac{b^2}{4}\left(1 - e^{-4w}\right). \tag{6.27}$$

Inserting (6.27) in (6.26), we get

$$\Delta_0 w = -\frac{1}{2}\sinh(2w) = -\cosh(w)\sinh(w).$$

This implies (6.25). Observe that K can be expressed in terms of w from (6.27) to get

$$K = \frac{b^2}{4}\left(1 - e^{4w}\right). \tag{6.28}$$

In particular, K and w have the same sign at any point and the same nodal set.

Theorem 6.21 ([365, Thm. 3]) *Let $S \subset M$ be a compact orientable two-sided surface of positive genus in a three-dimensional space form. Let H be the mean curvature of S, and assume $4c + H^2 > 0$.*

If $\Omega \subset S$ is a connected component of the set $\{p \in S : K(p) \neq 0\}$ without umbilic points, then $\lambda_1(\Omega) < 0$.

If S is stable and K is not identically 0, then the open set $\{p \in S : K(p) < 0\}$ must be connected, and each connected component of $\{p \in S : K(p) > 0\}$ must contain an umbilical point at least.

Proof Equation (6.26) implies

$$b^2 - 2K = b^2(\cosh^2(w) + \sinh^2(w))\, e^{-2w}.$$

Hence, the index form can be expressed as

$$\mathcal{I}(f) = \int_S \{|\nabla_0 f|^2 - (\cosh^2(w) + \sinh^2(w))\, f^2)\, dS_0 \tag{6.29}$$

for any Lipschitz function f with compact support on $S \setminus P$, where P is the set of umbilical points in S. The gradient ∇_0, the measure dS_0, and the modulus of the gradient are taken with respect to the flat metric ds_0^2.

Let Ω be a connected component of $\{p \in S : K(p) \neq 0\}$ without umbilical points. As K and w have the same sign, the function

$$f = \begin{cases} \sinh(w), & \text{on } \Omega, \\ 0, & \text{on } S \setminus \Omega, \end{cases}$$

is in the Sobolev space $W^{1,2}(S)$. Moreover, umbilical points cannot lie at $\partial\Omega$ since at such points, $K(p) = b^2/4 > 0$. So f has compact support in $M \setminus P$. Integrating by parts using (6.25), we get

$$\begin{aligned} \int_S |\nabla_0 f|^2 dS_0 &= \int_S \langle \nabla_0 f, \nabla_0 \sinh(w)\rangle\, dS_0 = -\int_S f\,\Delta_0 \sinh(w)\, dS_0 \\ &= -\int_\Omega \sinh(w)\,\Delta_0 \sinh(w)\, dS_0 \\ &= -\int_\Omega \{\sinh(w)\{\sinh(w)|\nabla_0 w|^2 + \cosh(w)\Delta_0 w\}\, dS_0 \\ &= \int_\Omega \sinh^2(w)\{\cosh^2(w) - |\nabla_0 w|^2\}\, dS_0. \end{aligned}$$

From this equality and (6.29), we obtain

$$\begin{aligned}\mathcal{I}(f) &= \int_{\Omega} \{\sinh^2(w)(\cosh^2(w) - |\nabla_0 w|^2) - (\cosh^2(w) + \sinh^2(w)) \sinh^2(w)\}\, dA_0 \\ &= -\int_{\Omega} \sinh^2(w)(\sinh^2(w) + |\nabla_0 w|^2) dS_0 < 0.\end{aligned}$$

This implies $\lambda_1(\Omega) < 0$.

If S were stable and we could find two connected components of $\{p \in S : K(p) \neq 0\}$ without umbilical points, we would obtain two functions f_i, f_2 in $W^{1,2}(S)$ with disjoint supports and such that $\mathcal{I}(f_i) < 0$ for $i = 1, 2$. So a certain linear combination of f_1 and f_2 would provide a function f with mean zero and such that $\mathcal{I}(f) < 0$. The result follows because the set $\{p \in S : K(p) < 0\}$ is not empty by the Gauss-Bonnet theorem and does not contain umbilical points. □

From Theorem 6.21, we obtain the following stronger conclusion, without any restriction on the values of c and H.

Corollary 6.22 *Let $S \subset M$ be an orientable two-sided stable torus in a three-dimensional space form. Then the Gauss curvature of S is identically zero.*

Proof The number of umbilical points on an orientable two-sided surface of genus g, counted with multiplicity, is $4g - 4$. Hence, the torus S has no umbilical points.

If $4c + H^2 > 0$ and $K \not\equiv 0$, the Gauss-Bonnet theorem implies the existence of at least two nodal regions of $\{p \in S : K(p) \neq 0\}$, each one free of umbilical points. Hence, Theorem 6.21 implies that S is unstable. We conclude that $K \equiv 0$.

If $4c + H^2 \leqslant 0$, the Gauss equation implies $K = c + \kappa_1\kappa_2 \leqslant c + H^2/4 \leqslant 0$. From the Gauss-Bonnet theorem, $K \equiv 0$. □

A flat constant mean curvature surface in a 3-manifold with sectional curvature is locally congruent to a tube around a geodesic or to a totally geodesic plane in case $c = 0$.

For isoperimetric sets, we have the following:

Corollary 6.23 *Let $E \subset M$ be an isoperimetric set in a three-dimensional space form. If ∂E is a torus, then ∂E is a flat surface.*

6.3.4 Higher Genus: Examples

In the previous sections, we have proven that isoperimetric solutions with spherical or toroidal boundary in three-dimensional space forms are geodesic balls or tubular neighborhoods of closed geodesics. This result is obtained from a classification of stable two-sided orientable surfaces. In the higher genus case, we have indeed many examples, as shown by Ross [386].

Theorem 6.24 ([386, Theorem 1]) *Schwarz's $\mathcal{P}$ and $\mathcal{D}$ surfaces are stable.*

These are embedded minimal surfaces in quotients of $\mathbb{R}^3$ by a three-dimensional lattice. The proof is obtained by a detailed analysis of the behavior of the Jacobi operator with respect to the symmetries of the minimal surface.

6.4 Three-Dimensional Elliptic Space Forms

In this section, we consider elliptic space forms, those with positive sectional curvature $c > 0$. All of them are finitely covered by the sphere $\mathbb{S}^3(c^{-1/2})$. We shall fix $c = 1$ in this section for convenience. We recall that isoperimetric sets in elliptic space forms have connected boundary by Proposition 6.1.

6.4.1 The Real Projective Space $\mathbb{RP}^3$

The real projective space $\mathbb{RP}^3(c)$ of sectional curvature $c > 0$ is the quotient of the sphere $\mathbb{S}^3(c)$ of sectional curvature c by the antipodal map $I(x) = -x$ of $\mathbb{R}^4$. The projective space of constant sectional curvature 1 will be denoted by $\mathbb{RP}^3$. The key to solving the isoperimetric problem in the three-dimensional real projective space, as well as in some lens spaces, is the following result.

Proposition 6.25 ([365, Thm. 6]) *Let M be a three-dimensional space elliptic space form with sectional curvature 1. Assume that the universal covering map $\pi : \mathbb{S}^3 \to M$ has k sheets. Let $S \subset M$ be a compact orientable two-sided stable surface with genus larger than one. Then we have:*

1. $(4 + H^2)\, A(S) \leqslant 8\pi$.
2. $k \geqslant 3$.

Proof Inequalities (6.22) and (6.23) imply

$$\int_S (4 + H^2)\, dS \leqslant 8\pi \left(2 + \left[\frac{g+1}{2}\right] - g\right).$$

This inequality only holds for $g \leqslant 3$. For $g = 2, 3$, we obtain inequality 1.

To obtain 2, we consider the pullback $\tilde{S}$ of S via the covering map π. The surface $\tilde{S}$ has constant mean curvature H in $\mathbb{S}^3$ although it is not necessarily connected. By 1, we have, in case $k \leqslant 2$,

$$\int_{\tilde{S}} (4 + H^2)\, d\tilde{S} = k \int_S (4 + H^2)\, dS \leqslant 8k\pi \leqslant 16\pi.$$

However, we know that

$$\int_{\tilde{S}} (4 + H^2)\, d\tilde{S} \geqslant 4 \int_{\tilde{S}} K\, d\tilde{S} = 16\pi,$$

with equality if and only if $\tilde{S}$ is a totally umbilical sphere. But genus $\tilde{S} \geqslant$ genus $S = g > 1$, and we get a contradiction. □

Proposition 6.25 implies that the quotients of $\mathbb{S}^3$ with a small number of sheets have isoperimetric sets with spherical or toroidal boundary.

Theorem 6.26 ([365, Corollary 7]) *Let $S \subset \mathbb{RP}^3$ be a compact two-sided stable surface. Then either*

1. *S is a totally umbilical sphere, or*
2. *S is a flat tube of radius $\pi/6 \leqslant r \leqslant \pi/3$ about a geodesic.*

Proof The bound on the genus of Theorem 6.17 implies that $g =$ genus $S \leqslant 3$. Since the covering map $\pi : \mathbb{S}^3 \to \mathbb{RP}^3$ has two sheets, Proposition 6.25 implies that $g = 0, 1$. In case $g = 0$, S is a totally umbilical sphere by Theorem 6.19. If $g = 1$, then S is a flat tube about a geodesic by Corollary 6.22.

A tube $\tilde{T}$ of radius $0 < r < \pi/2$ about a geodesic is congruent to the standard embedding of a Clifford torus $\mathbb{S}^1(\cos(r)) \times \mathbb{S}^2(\sin(r)) \subset \mathbb{S}^3$, and the Jacobi operator is given by

$$\Delta + \frac{1}{\cos^2(r)} + \frac{1}{\sin^2(r)}.$$

The corresponding embedded tube in $\mathbb{RP}^3$ is $T = \mathbb{R}^2/\Gamma$, where $\Gamma \subset \mathbb{R}^2$ is the lattice generated by the vectors $(2\pi\cos(r), 0)$ and $(\pi\cos(r), \pi\sin(r))$. Its dual lattice Γ^* is generated by the vectors $((2\pi\cos(r))^{-1}, -(2\pi\sin(r))^{-1})$, $(0, (\pi\sin(r))^{-1})$. Hence, the eigenvalues of the Laplacian on $\mathbb{R}^2/\Gamma$ (see Proposition B.I.2 in [56]) are given by

$$\left\{\frac{m^2}{\cos^2(r)} + \frac{(2n+m)^2}{\sin^2(r)} : n, m \in \mathbb{Z}\right\}.$$

As the Jacobi operator is the Laplacian plus a constant function, the tube is stable if and only if all positive eigenvalues of the Laplacian are larger than or equal to $\cos^{-2}(r) + \sin^{-2}(r)$. This is equivalent to $\cos^2(r), \sin^2(r) \geqslant 1/4$. Hence, $\pi/6 \leqslant r \leqslant \pi/3$. □

Theorem 6.27 ([365, Theorem 8]) *Let $E \subset \mathbb{RP}^3$ be an isoperimetric set. Then E is either a geodesic ball or a tubular neighborhood of a geodesic. Moreover, the isoperimetric profile of $\mathbb{RP}^3$ is given by*

$$I_{\mathbb{RP}^3}(v) = \begin{cases} I_{\mathbb{S}^3}(v), & 0 < v \leqslant \mu, \\ 2v^{1/2}(\pi^2 - v)^{1/2}, & \mu \leqslant v \leqslant \pi^2 - \mu, \\ I_{\mathbb{S}^3}(\pi^2 + v), & \pi^2 - \mu \leqslant v \leqslant \pi^2, \end{cases} \tag{6.30}$$

where $0 < \mu < \pi^2$ and $I_{\mathbb{S}^3}$ is the isoperimetric profile of the three-dimensional sphere $\mathbb{S}^3$.

Proof Let $I_{\mathbb{S}^3}$ be the isoperimetric profile of $\mathbb{S}^3$. For any $v \in (0, |\mathbb{S}^3|/2) = (0, 2\pi^2)$, the function $I_{\mathbb{S}^3}$ assigns to v the area of the geodesic sphere of volume v. On the other hand, a tube of radius $0 < r < \pi$ around a closed geodesic in $\mathbb{RP}^3$ has area $A = 2\pi^2 \cos(r)\sin(r)$, and they separate $\mathbb{RP}^3$ into two connected sets with volumes $v_1 = \pi^2 \sin^2(r)$ and $v_2 = \pi^2 \cos^2(r)$. Hence, the relation between the area and the volume enclosed by the tubes is

$$I_T(v) = 2v^{1/2}(\pi^2 - v)^{1/2}.$$

The range where these tubes are stable is the interval $r \in [\pi/6, \pi/6]$, which corresponds to the volumes in the interval $[\pi^2/, 3\pi^2/4]$. Let us prove that there exists a unique μ in the interval $[\pi^2/4, \pi^2/2]$ such that $I_{\mathbb{S}^3}(\mu) = I_T(\mu)$. This is equivalent to finding the fixed points of the explicitly calculable map $I_{\mathbb{S}^3}^{-1} \circ I_T$. We first observe that $I_{\mathbb{S}^3}(\pi^2/4) < I_T(\pi^2/4)$ and $I_{\mathbb{S}^3}(\pi^2/2) > I_T(\pi^2/2)$. On the other hand, a calculation shows that $0 < (I_{\mathbb{S}^3}^{-1} \circ I_T)' < 1$. This is enough to ensure the existence of a unique $\mu \in [\pi^2/4, \pi^2/2]$ where the function $I_{\mathbb{S}^3}$ and I_T take the same value. A numerical calculation provides the approximation $\mu \simeq 4.1432835$. Observe finally that $I_{\mathbb{S}^3}(\pi^2 - v) = I_{\mathbb{S}^3}(2\pi^2 - (\pi^2 - v)) = I_{\mathbb{S}^3}(\pi^2 + v)$. □

6.4.2 *An Application to the Willmore Problem*

As an application of the isoperimetric inequality in the real projective space $\mathbb{RP}^3$, we present a proof of the Willmore conjecture for surfaces in $\mathbb{S}^3$ invariant by the antipodal map due to Ros [378]. We give first an estimate of the Willmore functional on a 3-manifold M with $\mathrm{Ric} \geqslant 2$ in terms of the isoperimetric profile of M. For simplicity, we state and prove the result for embedded surfaces. See the Remark in page 491 in [378] and also Theorem 2.2 in [225].

Proposition 6.28 ([380, Prop. 2.9]) *Let M be a compact three-dimensional manifold with* Ric $\geqslant 2$ *and $S \subset M$ an embedded surface with mean curvature H that separates M into two sets. Then*

$$\int_S \Big(1+\Big(\frac{H}{2}\Big)^2\Big)\, dS \geqslant I_M(|M|/2). \tag{6.31}$$

Equality holds if and only if S is totally umbilical or an isoperimetric boundary with mean curvature 0 *separating M into two equal volumes.*

Proof We take $E \subset M$ with boundary S and $|E| \leqslant |M|/2$ and consider the tubular neighborhood $E_t = \{p \in M : d(p, E) \leqslant t\}$ of radius $t \geqslant 0$ so that $|E_t| = |E|$. Theorem 1.35 implies that the Jacobian of the normal exponential map is bounded above by the corresponding one in $\mathbb{S}^3$. Since $P(E_t)$ is finite by the coarea formula, we have

$$P(E_t) \leqslant \int_{S_t^+} \prod_{i=1}^{2} (\cos t - \kappa_i \sin t)\, dS,$$

where κ_i are the principal curvatures of S and $S_t^+ \subset S$ is the set of points where the Jacobian $\prod_{i=1}^{2}(\cos s - \kappa_i \sin s)$ is non-negative for all $0 < s \leqslant t$. By the arithmetic-geometric and Schwarz's inequalities, we have

$$\begin{aligned}\prod_{i=1}^{2} (\cos t - \kappa_i \sin t) &\leqslant \Big(\cos t - \frac{H}{2} \sin t\Big)^2 \\ &\leqslant 1 + \Big(\frac{H}{2}\Big)^2.\end{aligned}$$

By the definition of the isoperimetric profile, we have $P(E_t) \geqslant I_M(|M|/2)$. Whence,

$$I_M(|M|/2) \leqslant P(E_t) \leqslant \int_{S_t^+} \Big(1+\Big(\frac{H}{2}\Big)^2\Big)\, dS \leqslant \int_S \Big(1+\Big(\frac{H}{2}\Big)^2\Big)\, dS.$$

This implies (6.31). If equality holds in (6.31), then S_t^+ and S coincide up to a set of measure 0. Equality also holds in the arithmetic-geometric inequality in a set of full measure in S, and so S is totally umbilical if $t > 0$ or minimal if $t = 0$. □

Proposition 6.28 can be applied to the sphere $\mathbb{S}^3$. This way, we obtain the following.

Corollary 6.29 *Let $S \subset \mathbb{S}^3$ be an embedded surface with mean curvature H. Then*

$$\int_S \left(1 + \left(\frac{H}{2}\right)^2\right) dS \geqslant 4\pi. \tag{6.32}$$

Equality holds if and only if S is a totally umbilical sphere in $\mathbb{S}^3$.

Proof The surface S divides $\mathbb{S}^3$ into two components. Since $I_{\mathbb{S}^3}(|\mathbb{S}^3|/2) = 4\pi$, Proposition 6.28 implies

$$4\pi \leqslant P(E_t) \leqslant \int_{S_t^+} \left(1 + \left(\frac{H}{2}\right)^2\right) dS \leqslant \int_S \left(1 + \left(\frac{H}{2}\right)^2\right) dS.$$

This implies (6.32). If equality holds in (6.32), then by Proposition 6.28, S is a totally umbilical geodesic sphere in $\mathbb{S}^3$ or an isoperimetric boundary with $H = 0$ separating $\mathbb{S}^3$ into two regions of equal volumes. By the characterization of isoperimetric sets in $\mathbb{S}^3$, S is a totally umbilical geodesic sphere with $H = 0$. □

We have a similar result in the real projective space $\mathbb{RP}^3$.

Theorem 6.30 ([378, Thm. 3]) *Let $S \subset \mathbb{RP}^3$ be an embedded surface with mean curvature H. Then*

$$\int_S \left(1 + \left(\frac{H}{2}\right)^2\right) dS \geqslant \pi^2. \tag{6.33}$$

Equality holds if and only if S is the projection of the minimal Clifford torus $\mathbb{S}^1(1/\sqrt{2}) \times \mathbb{S}^1(1/\sqrt{2})$ in $\mathbb{S}^3$.

Proof We first assume that S is connected. Let $\pi : \mathbb{S}^3 \to \mathbb{RP}^3$ be the covering map. We consider the pullback $\tilde{S} = \pi^{-1}(S)$, which is an embedded surface in $\mathbb{S}^3$ with constant mean curvature H. If $\tilde{S}$ has two components, then S can be lifted to one of the components $\tilde{S}_1$ in $\mathbb{S}^3$. Hence, Corollary 6.29 implies

$$\int_S \left(1 + \left(\frac{H}{2}\right)^2\right) dS = \int_{\tilde{S}_1} \left(1 + \left(\frac{H}{2}\right)^2\right) d\tilde{S}_1 \geqslant 4\pi > \pi^2.$$

So we may assume that $\tilde{S}$ is connected. Then it separates $\mathbb{S}^3$ into two components, and we consider the one $\tilde{E}$ with the smallest volume. The set $\tilde{E}$ is invariant by the antipodal map. Reasoning as in the proof of Corollary 6.29, we consider a tubular neighborhood $\tilde{E}_t$ so that $|\tilde{E}_t| = |\mathbb{S}^3|/2$. The set $\tilde{E}_t$ is also invariant by the antipodal map. Hence, it projects to a set F in $\mathbb{RP}^3$ of finite perimeter. By the isoperimetric inequality in $\mathbb{RP}^3$, we have

$$P(F) \geqslant \pi^2.$$

Hence, we have

$$\pi^2 \leqslant P(F) = 2\,P(E_t) \leqslant 2\int_{\tilde{S}} \Big(1 + \big(\frac{H}{2}\big)^2\Big)\, d\tilde{S} = \int_S \Big(1 + \big(\frac{H}{2}\big)^2\Big)\, dS,$$

which proves (6.33) for S connected.

For S disconnected, the estimate is not optimal, and it is obtained by adding the inequalities for all connected components.

If equality holds, then S is connected, and F is the only isoperimetric set in $\mathbb{RP}^3$ of volume $|\mathbb{RP}^3|/2$, and we conclude that S is the projection to $\mathbb{RP}^3$ of the minimal Clifford torus in $\mathbb{S}^3$. □

In particular, for surfaces in $\mathbb{S}^3$ which are invariant by the antipodal map, we have the following corollary.

Lemma 6.31 ([378, Thm. 2]) *Let $S \subset \mathbb{S}^3$ be an embedded surface with mean curvature H enclosing a region E invariant by the antipodal map. Then*

$$\int_S \Big(1 + \big(\frac{H}{2}\big)^2\Big)\, dS \geqslant 2\pi^2. \tag{6.34}$$

Equality holds if and only if S is the minimal Clifford torus $\mathbb{S}^1(1/\sqrt{2}) \times \mathbb{S}^1(1/\sqrt{2})$ in $\mathbb{S}^3$.

6.4.3 Lens Spaces

Given two coprime integers p and q with $1 \leqslant q < p$, the three-dimensional *lens space* $L(p, q)$ is the quotient of $\mathbb{S}^3 \subset \mathbb{R}^4 \equiv \mathbb{C}^2$ under the action of the finite group of isometries of $\mathbb{S}^3$ generated by

$$(z_1, z_2) \mapsto (e^{2\pi i/p} z_1, e^{2\pi i q/p} z_2), \quad (z_1, z_2) \in \mathbb{C}^2,\ |z_1|^2 + |z_2|^2 = 1.$$

The lens space $L(2, 1)$ is the real projective space $\mathbb{RP}^3$. Lens spaces are endowed with the Riemannian metric induced from $\mathbb{S}^3$, and so they are elliptic space forms of sectional curvature 1. Ros [380, Thm. 2.11] proved that isoperimetric regions in the lens space $L = L(3, 1)$ are either geodesic balls or tubes around closed geodesics. The strategy of the proof is to discard isoperimetric boundaries of genus 2 or 3 in L. This is done using the inequality

$$\int_S \Big(1 + \big(\frac{H}{2}\big)^2\Big)\, dS \leqslant 2\pi$$

obtained in Proposition 6.25(1) for isoperimetric boundaries of genus larger than one, together with the estimate

$$\int_S \left(1+\left(\frac{H}{2}\right)^2\right) dS \geqslant I_L\left(\tfrac{|L|}{2}\right)$$

obtained in Proposition 6.28. Since $I_L(|L|/2) = 2\pi^2/3 > 2\pi$ as the projection of the minimal Clifford torus is the isoperimetric solution in L for volume $|L|/2$, the result follows; see the comments in §3.1 in [430].

Viana [430] has obtained non-existence of isoperimetric boundaries of genus greater than one in lens spaces $L(p,q)$ with sufficiently large p by compactness arguments. He has also obtained a classification of the isoperimetric sets in $L(3,1)$ and $L(3,2)$ using the solution to the Willmore conjecture by Neves and Marques [277]. We sketch the argument. Let S be an isoperimetric boundary of genus larger than one and $\tilde{S}$ the pullback of S to $\mathbb{S}^3$. Assume that $\tilde{S}$ has ℓ components $\tilde{S}_i$, $i = 1, \ldots, \ell$, each one of genus larger than one. Since $\int_{\tilde{S}_i}(1+(\frac{H}{2})^2)\, d\tilde{S}_i \geqslant 2\pi^2$ by [277], we get

$$2\pi^2\ell \leqslant \sum_{i=1}^{\ell} \int_{\tilde{S}_i} \left(1+\left(\frac{H}{2}\right)^2\right) d\tilde{S}_i = 3\int_S \left(1+\left(\frac{H}{2}\right)^2\right) dS$$

Hence, we obtain

$$\int_S \left(1+\left(\frac{H}{2}\right)^2\right) dS \geqslant \frac{2\pi^2\ell}{3} \geqslant \frac{2\pi^2}{3} > 2\pi,$$

and we reach a contradiction to Proposition 6.25(1).

6.5 Three-Dimensional Flat Space Forms

Let $E(3)$ be the group of affine isometries in $\mathbb{R}^3$, the semidirect product $T(3) \rtimes O(3)$ of the group of translations $T(3)$ of $\mathbb{R}^3$ and the orthogonal group $O(3)$. We denote by $E^+(3) = T(3) \rtimes SO(3)$ the subgroup of $E(3)$ of orientation-preserving elements of $E(3)$.

Every complete flat 3-manifold M is the quotient of $\mathbb{R}^3$ by a discrete subgroup G of $E(3)$ acting properly and discontinuously. The manifold M is orientable if and only if $G \subset E^+(3)$ (i.e., when all isometries in G are orientation preserving).

We denote by $\mathcal{G}$ the set of discrete orientation-preserving subgroups of $\mathbb{R}^3$ acting properly and discontinuously in $\mathbb{R}^3$. If $G \in \mathcal{G}$, then $\Gamma(G)$ is defined as the subgroup of translations in G and $\operatorname{inj}(\mathbb{R}^3/G)$ the injectivity radius of the manifold $\mathbb{R}^3/G$. The rank of $\Gamma(G)$ is the dimension of the subspace generated by $\Gamma(G)$. A *crystallographic group* on $\mathbb{R}^3$ is a discrete subgroup of $E(3)$ acting properly and discontinuously with *compact* quotient manifold $\mathbb{R}^3/G$. When G is crystallographic, the compact flat 3-space form $\mathbb{R}^3/G$ is finitely covered by a three-dimensional torus; see Chapter 3 in [449]. The affine diffeomorphism classes of flat 3-space forms are completely classified; see Theorems 3.5.1 and 3.5.5 in [449].

Given a sequence $\{G_i\}_{i\in\mathbb{N}}$, we say that $\lim_{i\to\infty} G_i = G$ if G is the set of accumulation points of elements of G_i in $E(3)$. A fundamental property is the following.

Lemma 6.32 *Let $\{G_i\}_{i\in\mathbb{N}} \subset \mathcal{G}$ such that inj$(\mathbb{R}^3/G_i) = 1$. Then we can extract a subsequence converging to a subgroup $G \in \mathcal{G}$.*

A flat 3-space form M can always be normalized so that $\mathrm{inj}(M) = 1$ by scaling the Riemannian metric, which is still complete and flat after the scaling. In case (M, g) is the quotient of $\mathbb{R}^3$ by a discrete subgroup G of isometries acting properly and discontinuously in $\mathbb{R}^3$, the manifold $(M, \lambda^2 g)$ is isometric to the quotient $\mathbb{R}^3/h_\lambda^{-1}Gh_\lambda$, where $h_\lambda^{-1}Gh_\lambda$ is the conjugation of G by a dilation h_λ of ratio $\lambda > 0$.

The possible types of affinely diffeomorphic orientable flat three-dimensional space forms can be described from a classification of the elements of $\mathcal{G}$ up to affine conjugation; see Wolf's monograph or Lemma 1.1 in [366] for details.

If $\mathrm{rank}\,\Gamma(G) = 0, 1$ then either $G = \{\mathrm{Id}_{\mathbb{R}^3}\}$ or $G = S_\theta$, with $0 \leqslant \theta \leqslant \pi$, where S_θ is the subgroup generated by a screw motion given by a rotation of angle θ followed by a non-trivial translation in the direction of the axis of rotation. Note that in this case, the affine conjugation classes are parametrized by θ.

If $\mathrm{rank}\,\Gamma(G) = 2$, then there are two affine conjugation classes: either G is generated by two linearly independent translations and the quotient R^3/G is the Riemannian product $T^2 \times \mathbb{R}$, where T^2 is a flat *two*-dimensional torus, or G is generated by a screw motion with angle π and a translation orthogonal to the axis of the screw motion. We shall denote the class of quotient manifolds R^3/G by $\mathcal{K}$. Note that every manifold in $\mathcal{K}$ admits a twofold covering by some $T^2 \times \mathbb{R}$.

The above manifolds describe all the possible types of affinely diffeomorphic complete *non-compact* orientable flat 3-manifolds.

Finally, if $\mathrm{rank}\,\Gamma(G) = 3$, then either G contains only translations and $\mathbb{R}^3/G$ is a flat three-dimensional torus, or G contains screw motions with angle different from 0 and $\mathbb{R}^3/\Gamma(G) \to \mathbb{R}^3/G$ is a Riemannian covering with at most six sheets. This way, five different affine diffeomorphism classes are obtained.

For isoperimetric boundaries in these manifolds, we have the following result.

Theorem 6.33 *Let M be an orientable flat three-dimensional space form, and let $S \subset M$ be a compact orientable stable surface. Then either*

1. *S is connected and genus $S \leqslant 3$, or*
2. *S is a union of totally geodesic tori.*

Proof If S is connected, we know that genus $S \leqslant 3$. In case S is not connected, we take a locally constant nowhere vanishing function $u : S \to \mathbb{R}$ with mean zero. The stability condition implies

$$-\int_S |\sigma|^2 u^2 dS \geqslant 0,$$

and so $|\sigma|^2 \equiv 0$ on S. Hence, S is totally geodesic. The Gauss-Bonnet theorem implies that each connected component of S is a torus. □

6.5.1 *Compactness of the Space of Isoperimetric Boundaries of High Genus*

When $G \in \mathcal{G}$ is a discrete subgroup of orientation-preserving isometries of $\mathbb{R}^3$ acting properly and discontinuously, the manifold $\mathbb{R}^3/G$ is an orientable three-dimensional space form. As we have seen in the previous section, any isoperimetric boundary S is orientable and genus $S \leqslant 3$. When $g = 0$ it is a geodesic sphere and when $g = 0$ a flat torus, which can be a tubular neighborhood of a closed geodesic or a totally geodesic submanifold with mean curvature $H = 0$. We are interested in estimating the size of the set of $G \in \mathcal{G}$ so that $\mathbb{R}^3/G$ contains a compact orientable stable surface with genus $S \geqslant 2$.

C^k Convergence of Surfaces

We consider now the notion of convergence we use in this section.

Definition 6.34 A sequence of surfaces $\{S_i\}_{i\in\mathbb{N}}$ in $\mathbb{R}^3$ converges to a surface S in the C^k topology , $k \in \mathbb{N}$, when for any point $p \in S$ and for i large enough, each S_i is locally a union of graphs over a fixed disk $D \subset T_pS$ and the graphs of S_i converge in the usual C^k topology to the graph of S.

A well-known result in geometric analysis establishes that C^k convergence can be obtained through a uniform curvature estimate and uniform local area bounds.

Definition 6.35 We say that the sequence of surfaces $\{S_i\}_{i\in\mathbb{N}}$ satisfy a *uniform curvature estimate* if there exists a constant $C_0 > 0$ such that $|\sigma_i| < C_0$ for all $i \in \mathbb{N}$, where σ_i is the second fundamental form of S_i.

Definition 6.36 We say that the sequence of surfaces $\{S_i\}_{i\in\mathbb{N}}$ satisfy *uniform local area bounds* if there exists a radius $r > 0$ so that, for each ball $B(x, r) \subset \mathbb{R}^3$, we have $A(S_i \cap B(x, r)) < C_0$ for some positive constant independent of i.

For sequences of surfaces with constant mean curvature, we have the following convergence result in Euclidean space.

Theorem 6.37 **([255])** *Let $\{S_i\}_{i\in\mathbb{N}}$ be a sequence of properly embedded surfaces with constant mean curvature such that $|\sigma_i| < C$ for all $i \in \mathbb{N}$. Assume that the sequence $\{S_i\}_{i\in\mathbb{N}}$ satisfies uniform local area bounds. Then either the sequence $\{S_i\}_{i\in\mathbb{N}}$ has no limit points, or we can extract a subsequence convergent in the C^k topology (for all $k \in \mathbb{N}$) to a properly weakly embedded constant mean curvature surface S.*

We say that an immersed surface $S \subset \mathbb{R}^3$ is *weakly embedded* if it only have tangential self-intersections. In C^k convergence, one cannot exclude a priori the convergence of several graphs to a single one. We say that $\{S_i\}_{i\in\mathbb{N}}$ converges to S with multiplicity k if for every point in S, there is a neighborhood of p in S which is the limit of k graphs contained in S_i.

Let us see now that in some special cases, the existence of a uniform curvature estimate implies the existence of a uniform local area bound. First we recall Blaschke's rolling theorem for surfaces with non-negative mean curvature. It ensures the existence of a one-sided embedded tubular neighborhood of uniform radius of a surface with non-negative mean curvature.

Lemma 6.38 **([366, Thm. 3.1])** *Let $S \subset \mathbb{R}^3$ be a non-totally geodesic embedded surface invariant by $G \in \mathcal{G}$ such that S/G is orientable, compact, and connected and separates $\mathbb{R}^3/G$. Assume that the mean curvature of S with respect to N is non-negative. Let $\kappa_{\max} > 0$ be the maximum of the principal curvatures of S. Then the normal exponential map $p \in S \mapsto p - tN(p)$ is a diffeomorphism of $S \times (0, \kappa_{\max}^{-1})$ onto its image.*

Proof Let $F : S \times (0, \kappa_{\max}^{-1}) \to \mathbb{R}^3$ be the normal exponential map. The parallel hypersurface $S_t = \{p - tN_p : t \in \mathbb{R}\}$ is regular for $0 < t < \kappa_{\max}^{-1}$ and has mean curvature

$$H_t = \sum_{i=1}^{2} \frac{\kappa_i}{1 - t\kappa_i} = \frac{H - 2tK}{(1 - t\kappa_1)(1 - t\kappa_2)},$$

where κ_i are the principal curvatures of S and K is the Gauss curvature of S. Since $H^2/4 \geqslant K$ (with equality at umbilical points of S), we have

$$H_t \geqslant \frac{H(1 - t\frac{H}{2})}{(1 - t\kappa_1)(1 - t\kappa_2)} \geqslant 0$$

since $t < \kappa_{\max}^{-1} \leqslant 2/H$.

Let $t_0 = \sup\{t > 0 : \text{the map } F : S \times (0, t) \to \mathbb{R}^3 \text{ is injective}\}$. If $t_0 < \kappa_{\max}^{-1}$, then there exist two different points (p, t_0) and (q, t_0) such that $F(p, t_0) = F(q, t_0)$. So the regular surface S_{t_0} has a tangential self-intersection. As $p \neq q$, the definition of the normal exponential map implies $N_p \neq N_q$ and so $N_p = -N_q$. As $H_{t_0} \geqslant 0$, an application of the maximum principle for the mean curvature equation (see Addemdum 3 in Chapter 9 of [406]) implies $H_{t_0} \equiv 0$ in a neighborhood of the contact point and so S is totally geodesic. This contradiction shows that $F : S \times (0, \kappa_{\max}^{-1}) \to \mathbb{R}^3$ is a diffeomorphism onto its image. □

An important consequence of Lemma 6.38 is a lower estimate of the injectivity radius of a manifold $\operatorname{inj}(\mathbb{R}^3/G)$ containing a compact embedded surface with non-negative mean curvature and genus different from 1.

Corollary 6.39 *Let $S \subset \mathbb{R}^3$ be a non-totally geodesic embedded surface invariant by $G \in \mathcal{G}$ such that S/G is orientable, compact, and connected and separates $\mathbb{R}^3/G$. Assume that the mean curvature of S with respect to N is non-negative. Let $\kappa_{\max} > 0$ be the maximum of the principal curvatures of S.*

Assume also that the genus of S/G is different from 1 and that G is not cyclic. Then

$$inj(\mathbb{R}^3/G) \geqslant \frac{1}{6\kappa_{\max}}. \tag{6.35}$$

Proof As G is not cyclic, the discussion at the beginning of the section implies that order $G/\Gamma(G) \leqslant 6$. Hence, $S/\Gamma(G)$ is a compact orientable surface with genus different from 1. From the Gauss-Bonnet theorem, the Gauss map of $S/\Gamma(G)$ is surjective. Take $v \in \Gamma(G) \setminus \{0\}$ and $p \in S$ such that $N_p = v/|v|$. Then $q = p + v \in S$ and so $F(p, |v|) = F(q, 0)$. Lemma 6.38 implies $|v| \geqslant \kappa_{\min}^{-1}$ and so $\operatorname{inj}(\mathbb{R}^3/\Gamma(G)) \geqslant \kappa_{\max}^{-1}$. Since the number of sheets of the covering $\mathbb{R}^3/\Gamma(G) \to \mathbb{R}^3/G$ is less than or equal to 6, we get (6.35). □

From Lemma 6.38, we can get a uniform local area bound from a uniform curvature estimate.

Theorem 6.40 *Let $S \subset \mathbb{R}^3$ be a non-totally geodesic embedded surface with non-negative mean curvature invariant by a subgroup $G \in \mathcal{G}$ and such that S/G is an orientable compact surface that separates $\mathbb{R}^3/G$. If there exists a constant $C > 0$ such that the second fundamental form σ of S satisfies the bound $|\sigma| < C$, then, for any $R > 0$ with $0 < R < C^{-1}$, we have*

$$A(S \cap B(x, R)) \leqslant 36\pi R^2,$$

for any $x \in \mathbb{R}^3$.

Proof Let $S_R = S \cap B(x, R)$. The normal exponential map $F : S \times (0, \kappa_{\max}^{-1}) \to \mathbb{R}^3$ is a diffeomorphism onto its image by Lemma 6.38. As $\kappa_{\max} \leqslant \sup |\sigma| < C$, it follows that $R/2 < \kappa_{\max}^{-1}$, and so F maps $S_R \times (0, R/2)$ injectively over a region $\Omega \subset \mathbb{R}^3$. By the change of variables formula,

$$|\Omega| = \int_0^{R/2} \left\{ \int_{S_R} (1 - t\kappa_1)(1 - t\kappa_2)\, dS \right\} dt,$$

where κ_i are the principal curvatures of S. Since $0 < t < R/2$ and $\kappa_1, \kappa_2 \leqslant C$, the volume of Ω satisfies

$$|\Omega| \geqslant \int_0^{R/2} \left\{ \left(1 - \frac{R}{2} C\right)^2 dS \right\} dt$$

As $RC \leqslant 1$, the last integral is larger than or equal to

$$\int_0^{R/2} \left\{ \int_{S_R} \frac{1}{4} dS \right\} dt = \frac{R}{8} A(S_R)$$

and we conclude

$$|\Omega| \geqslant \frac{R}{8} A(S_R).$$

Now the theorem follows from the fact that Ω is contained in a Euclidean ball with radius $3R/2$ and so $|\Omega| \leqslant 9\pi R^3/2$. To see this, take $p \in \Omega$. Then $d(p, S_R) < R/2$. So the distance between the center x of $B(x, R)$ and p is smaller than $3R/2$ as claimed. □

Convergence of Manifolds

Lemma 6.32 implies that we can always extract a convergent subsequence to some $G \in \mathcal{G}$ from any sequence $\{G_i\}_{i\in\mathbb{N}} \subset \mathcal{G}$ satisfying $\operatorname{inj}(\mathbb{R}^3/G_i) \geqslant 1$, $i \in \mathbb{N}$. Of course, the limit manifold $\mathbb{R}/G$ is not necessarily affinely diffeomorphic to the manifolds $\mathbb{R}^3/G_i$ in the sequence. We improve this convergence under the assumptions that the groups G_i have $\operatorname{rank}\Gamma(G_i) \geqslant 2$ and that there is a sequence of G_i-invariant surfaces S_i converging in C^k topology, for all $k \in \mathbb{N}$, to a G-invariant surface S with *S/G compact.*

Lemma 6.41 ([366, Lemma 2.7]) *Consider a sequence $\{G_i\}_{i\in\mathbb{N}} \subset \mathcal{G}$ with $\operatorname{rank}\Gamma(G_i) \geqslant 2$. Take a sequence of connected embedded G_i-invariant surfaces $\{S_i\}_{i\in\mathbb{N}}$ in $\mathbb{R}^3$ such that S_i/G_i is compact for all $i \in \mathbb{N}$. Assume that G_i converges to $G \in \mathcal{G}$ and that S_i converges in the C^k topology to a properly immersed G-invariant surface $S \subset \mathbb{R}^3$. If S/G is compact, then $\operatorname{rank}\Gamma(G_i) = \operatorname{rank}\Gamma(G)$ for sufficiently large i, and $\mathbb{R}^3/G_i$ is affinely diffeomorphic to $\mathbb{R}^3/G$ for i large.*

Proof Set $\Gamma = \Gamma(G)$. We have $\Gamma(G_i) \to \Gamma(G)$. Since $\operatorname{order}(G_i/\Gamma(G_i)) \leqslant 6$, it follows that $\operatorname{order}(G/\Gamma(G)) \leqslant 6$ and so S/Γ is compact since S/G is compact and $S/\Gamma \to S/G$ is a finite covering.

We reason by contradiction assuming that $\operatorname{rank}(\Gamma) < \operatorname{rank}(\Gamma(G_i))$. If $\Gamma \neq \{0\}$, we take a system $\{w_j\}_{j\in J}$ of generators of Γ, where $J = \operatorname{rank}\Gamma(G)$. For i large, we have

$$\operatorname{rank}\Gamma < \operatorname{rank}(\Gamma(G_i)).$$

We take vectors w_j^i in $\Gamma(G_i)$ converging to the vectors w_j. Note that $\{w_j^i\}_{j\in J}$ generates a discrete subgroup Γ_i' with rank $\Gamma_i' <$ rank $\Gamma(G_i)$. So for i large, there exists a vector $v_i \in \Gamma(G_i) \setminus \Gamma_i'$. The sequence $\{v_i\}_{i\in\mathbb{N}}$ is not bounded. Otherwise, it should converge to some vector $v \in \Gamma$, and condition $\mathrm{inj}(\mathbb{R}^3/G_i) = 1$ would not be satisfied. As S_i is connected and G_i-invariant, we conclude that S/Γ is not compact and so S/G_i would be non-compact as the number of sheets of the covering $S/\Gamma(G_i) \to S/G$ is finite.

If $\Gamma = \{0\}$, we simply take a divergent sequence of vectors $\{v_i\}_{i\in\mathbb{N}}$, with $v_i \in \Gamma(G_i)$, and reason as above. □

The Compactness Result

Our main result is the following.

Theorem 6.42 ([366, Thm. 4.2]) *Let $\{S_i\}_{i\in\mathbb{N}}$ be a sequence of connected embedded constant mean curvature surfaces in $\mathbb{R}^3$ which are G_i-invariant, where $G_i \in \mathcal{G}$ is not a cyclic group and inj$(\mathbb{R}^3/G_i) = 1$. Assume that S_i/G_i are compact, orientable, and stable in $\mathbb{R}^3/G_i$, with genus greater than one, and that they separate $\mathbb{R}^3/G_i$ into two regions. Then:*

1. *We can extract a non-relabeled subsequence of $\{G_i\}_{i\in\mathbb{N}}$ converging to a subgroup $G \in \mathcal{G}$ which is not cyclic. For i large enough, $\mathbb{R}^3/G_i$ and $\mathbb{R}^3/G$ are affinely diffeomorphic.*
2. *We can extract a subsequence of $\{S_i\}_{i\in\mathbb{N}}$ converging in the C^k topology to a connected properly weakly embedded constant mean curvature G-invariant surface in $\mathbb{R}^3$ such that S/G is a compact stable surface. For i large enough, S_i/G_i and S/G are diffeomorphic.*

Proof Let us see that the sequence $\{S_i\}_{i\in\mathbb{N}}$ satisfies a uniform curvature estimate. We reason by contradiction assuming the existence of a non-relabeled subsequence of $\{S_i\}_{i\in\mathbb{N}}$ and sequence of points $p_i \in S_i$ such that

$$\lambda_i = \max_{p\in S_i} |\sigma_i| = |\sigma_i|(p_i)$$

is increasing and unbounded. Translating the surfaces $-p_i + S_i$ and scaling by a factor λ_i, we obtain a new sequence $\tilde{S}_i$ so that $|\tilde{\sigma}_i| \leqslant |\tilde{\sigma}_i|(0) = 1$. Using Theorem 6.40, we have uniform local area bounds. Passing to a subsequence, we can assume from Theorem 6.37 that $\tilde{S}_i$ converges in the C^k topology, $k \in \mathbb{N}$, to a non-totally geodesic properly weakly embedded constant mean curvature surface $\tilde{S}$ in $\mathbb{R}^3$. Such surface is orientable since it separates $\mathbb{R}^3$ by Lemma 2.3 in [366]. By Theorem 4.1 in [366], $\tilde{S}$ is either totally geodesic or compact. Since $|\tilde{\sigma}(0)| = 1$, we conclude that it is compact and, by Theorem 6.11, is a round sphere. This is not possible since Lemma 6.41 would imply that $\mathbb{R}^3$ would be affinely diffeomorphic to $\mathbb{R}^3/G_i$ for large i. This contradiction shows that $|\sigma_i|$ is uniformly bounded.

Since the sequence $\{S_i\}_{i\in\mathbb{N}}$ satisfies uniform curvature estimates, we know from Theorem 6.40 that $\{S_i\}_{i\in\mathbb{N}}$ satisfies uniform local area bounds. From the normalization $\mathrm{inj}(\mathbb{R}^3/G) = 1$, we obtain from Lemma 6.32 that a subsequence of the subgroups G_i converges to a subgroup $G \in \mathcal{G}$. We can apply again Theorem 6.37 to conclude that a subsequence of $\{S_i\}_{i\in\mathbb{N}}$ converges to a properly weakly embedded surface S with constant mean curvature in $\mathbb{R}^3$. The surface S is G-invariant, and it is not totally geodesic since $\kappa_{\max} \geqslant (6\,\mathrm{inj}(\mathbb{R}^3/G))^{-1} \geqslant 6^{-1}$ by (6.35).

Observe that S/G is either orientable or a double cover of a non-orientable surface. In both cases, the limit is stable and hence compact. By Lemma 6.41, the manifold $\mathbb{R}^3/G_i$ is affinely diffeomorphic to $\mathbb{R}^3/G$ for i large enough. By Theorem 2.8 in [366], the limit surface S has multiplicity 1. □

Some Consequences for Rectangular Lattices

Two consequences of Theorem 6.42 are the following:

Theorem 6.43 ([357, Thm. 4.2]) *There exists a constant $\varepsilon > 0$ such that, if $\mathbb{T}^2$ is a flat two-dimensional torus with inj($\mathbb{T}^2$) $= 1$ and area($\mathbb{T}^2$) $> \varepsilon$, then the only isoperimetric sets in $\mathbb{T}^2 \times \mathbb{R}$ are:*

1. *Geodesic balls*
2. *Tubular neighborhoods of closed geodesics*
3. *Slices bounded by two totally geodesic parallel tori*

Proof Let $E \subset \mathbb{T}^2 \times \mathbb{R}$ be an isoperimetric set. Theorem 6.33 implies that $S = \partial E$ is connected with either genus $S \leqslant 3$ or the union of totally geodesic tori. Theorem 6.42 implies that the set of manifolds $\mathbb{T}^2 \times \mathbb{R}$ with $1 = \mathrm{inj}(\mathbb{T}^2 \times \mathbb{R}) = \mathrm{inj}(\mathbb{T}^2)$ is compact. This implies the existence of $\varepsilon > 0$ with the required properties. If S is a union of totally geodesic tori, each one must be of the form $\mathbb{T}^2 \times \{c\}$ for some $c \in \mathbb{R}$. Hence, the total area of S is

$$\#\{\text{components of } S\} \times \mathrm{area}(\mathbb{T}^2) \geqslant 2\,\mathrm{area}(\mathbb{T}^2).$$

This implies that S is not isoperimetric unless S has only two components. □

A three-dimensional torus is the quotient of $\mathbb{R}^3$ by a discrete subgroup of translations of rank 3. For three-dimensional tori, we have the following:

Theorem 6.44 *There exists a constant $\delta > 0$ such that, if $\mathbb{T}^3$ is a three-dimensional torus with* $\mathrm{inj}(\mathbb{T}^3) = 1$*and* $|\mathbb{T}^3| > \delta$, *then the only isoperimetric sets in $\mathbb{T}^3$ are:*

1. *Geodesic balls and their complements*
2. *Tubular neighborhoods of closed geodesics and their complements*
3. *Slices bounded by totally geodesic parallel tori*

Proof It is similar to the proof of Theorem 6.43. When S is totally geodesic, their components are the projection of planes in $\mathbb{R}^3$ that must be parallel. □

Definition 6.45 A *rectangular lattice* is a discrete subgroup of translations generated by orthogonal vectors. If the lattice has rank 2, we can explicitly describe the isoperimetric profile for most of the manifolds $\mathbb{R}^3/\Gamma$.

Theorem 6.46 ([357, Thm. 4.3]) *Let Γ be the lattice in $\mathbb{R}^2$ generated by $(1,0)$ and $(0,a)$, where $a > \max\{1,\varepsilon\}$ and $\varepsilon > 0$ is the constant in Theorem 6.43. Then the isoperimetric profile of $\mathbb{T}^2 \times \mathbb{R}$ is given by*

$$I_{\mathbb{T}^2\times\mathbb{R}}(v) = \begin{cases} (4\pi)^{1/3}(3v)^{2/3}, & 0 < v \leqslant 4\pi/3^4, \\ (4\pi v)^{1/2}, & 4\pi/3^4 \leqslant v \leqslant a^2/\pi, \\ 2a, & a^2/\pi \leqslant v. \end{cases} \tag{6.36}$$

In the first case, the isoperimetric sets are geodesic balls; in the second one, tubular neighborhoods of closed geodesics; and in the last case, slices bounded by two totally geodesic parallel tori.

6.5.2 Isoperimetric Sets in Rectangular Lattices

In Theorem 6.44, we have proven that for most three-dimensional tori, the only isoperimetric sets are geodesic balls, tubular neighborhoods of geodesics, and slabs bounded by totally geodesic tori. However, in some three-dimensional tori, there could exist isoperimetric sets with boundary of genus 2 or 3. We describe in this section the family of candidate isoperimetric boundaries of genus 2 or 3 in quotients of $\mathbb{R}^3$ by rectangular lattices.

Recall that a rectangular lattice is a discrete subgroup of translations Γ generated by orthogonal vectors. If $\text{rank}(\Gamma) = 1$, then the manifold $\mathbb{R}^3/\Gamma$ is isometric to $\mathbb{S}^1(r) \times \mathbb{R}^2$ for some $r > 0$; if $\text{rank}(\Gamma) = 2$, to $\mathbb{S}^1(r_1) \times \mathbb{S}^1(r_2) \times \mathbb{R}$ for $r_1, r_2 > 0$; and, if $\text{rank}(\Gamma) = 3$, to $\mathbb{S}^1(r_1) \times \mathbb{S}^1(r_2) \times \mathbb{S}^1(r_3)$ for $r_1, r_2, r_3 > 0$.

Let us consider first the easy case of the quotient of $\mathbb{R}^3$ by an helicoidal motion, of which $\mathbb{S}^1(r) \times \mathbb{R}^2$ is a particular case. We consider the flat manifold $\mathbb{R}^3/S_{\theta,\lambda}$, where $\theta \in [0, 2\pi)$, $\lambda > 0$, and $S_{\theta,\lambda}$ is the discrete subgroup of $\mathbb{R}^3$ generated by the screw motion

$$(x, y, z) \mapsto (x\cos\theta - y\sin\theta, x\sin\theta + y\cos\theta, z + \lambda).$$

Theorem 6.47 *The isoperimetric profile of $\mathbb{R}^3/S_{\theta,\lambda}$ is given by*

$$I_{\mathbb{R}^3/S_{\theta,\lambda}}(v) = \begin{cases} (4\pi)^{1/3}(3v)^{2/3}, & 0 < v \leqslant \frac{4\pi\lambda^3}{3^4}, \\ (4\pi\lambda v)^{1/2}, & \frac{4\pi\lambda^3}{3^4} \leqslant v. \end{cases} \tag{6.37}$$

In the first case, the isoperimetric sets are geodesic balls. In the second one, it is the domain enclosed by the torus $(\mathbb{S}^1(a) \times \mathbb{R})/S_{\theta,\lambda}$ for suitable values of a.

Proof As there are no totally geodesic compact surfaces in $\mathbb{R}^3/S_{\theta,\lambda}$, the boundary S of an isoperimetric set must be compact, connected, and also orientable since $\mathbb{R}^3/S_{\theta,\lambda}$ is orientable. A connected pullback $\tilde{S} \subset \mathbb{R}^3$ is either compact (and hence a sphere) or periodic and cylindrically bounded. In this case, the Alexandrov reflection principle (see Addendum 3 in Chapter 9 of [406] or [255]) implies that S is a torus and so a tubular neighborhood of a closed geodesic by Corollary 6.23. The explicit expression of the profile follows from straightforward computations. □

We consider now the case of rectangular lattices of rank 2 or 3. We shall need the following concept. Given a smooth map $\phi : (M, g) \to (N, h)$ between Riemannian manifolds, its *energy* $E(\phi)$ is defined by

$$E(\phi) = \frac{1}{2} \int_M |d\phi|^2 \, dS, \tag{6.38}$$

where $|d\phi|^2(p) = \sum_{i=1}^m |d\phi_p(e_i)|^2$ for $p \in M$ and an orthonormal basis $e_1, \ldots, e_m$ of T_pM. A smooth map is *harmonic* if it is a critical point of the energy functional E. We refer the interested reader to [146] and [145] for background on harmonic maps between Riemannian manifolds.

A map $\phi : (M, g) \to \mathbb{S}^k$ to the round sphere is harmonic if and only if

$$\Delta\phi + |\nabla\phi|^2\phi = 0; \tag{6.39}$$

see Example (4.14) in [145]. The following result by Ruh [388] provides a link between Euclidean immersions with constant mean curvature and harmonic maps.

Theorem 6.48 ([388, Thm. I]) *An isometric immersed hypersurface in $\mathbb{R}^m$ has constant mean curvature if and only if its Gauss map to the sphere $\mathbb{S}^{m-1}$ is harmonic.*

Theorem 6.49 *Let $E \subset \mathbb{R}^3/\Gamma$ be an isoperimetric set, where Γ is a rectangular lattice of rank 2 or 3 generated by vectors proportional to the ones in the standard basis of $\mathbb{R}^3$. Let $S = \partial E$. Then:*

1. *E and S are symmetric with respect to symmetries r_i, $i = 1, 2, 3$, about three orthogonal totally geodesic submanifolds parallel to the projection of the coordinate planes.*
2. *The Gauss map N of S is symmetric ($r_i \circ N = N \circ r_i|_S$) with respect to the symmetries r_i and is a minimum of the energy functional in the class of symmetric maps $\phi : S \to \mathbb{S}^2$ satisfying $r_i \circ \phi = \phi \circ r_i|_S$.*
3. *Either $N_i \equiv 0$ or the nodal set of N_i coincides with the set of fixed points of u_i. The first possibility only happens when S is a torus.*
4. *genus $S \leqslant 3$.*
5. *If genus $S \geqslant 2$, then there are eight connected components of the set $\{p \in S : N_i(p) \neq 0 \text{ for all } i\}$. The closures of these components are congruent right-angled polygons with (genus S+3) sides and geodesic boundary. Each connected component is a graph over the three coordinate planes.*

Proof Observe that $\mathbb{R}^3/\Gamma$ is isometric to either $\mathbb{S}^1(r_1) \times \mathbb{S}^1(r_2) \times \mathbb{R}$ or $\mathbb{S}^1(r_1) \times \mathbb{S}^1(r_2) \times \mathbb{S}^1(r_3)$ for some positive r_i. Then 1 follows from Hsiang's symmetrization Theorem 5.20 since there are $O(2)$ actions on each $\mathbb{S}^1$ factor and an $E(1)$ action on the $\mathbb{R}$ factor. Observe that the set of fixed points of r_i consists of two parallel totally geodesic submanifolds in the first case and a totally geodesic torus in the last one.

To prove 2, let u_i be the restriction of r_i to S. The set of fixed points of u_i is not empty: otherwise, S would be disconnected as the set of fixed points of r_i disconnects $\mathbb{R}^3/\Gamma$, against the conclusion of Theorem 6.33. Let $N : S \to \mathbb{S}^2$ be the Gauss map, with components (N_1, N_2, N_3). We note that $|\nabla N|^2 = |\sigma|^2$ and so the Jacobi operator on S is given by $\Delta + |\nabla N|^2$.

Consider a map $\phi : S \to \mathbb{S}^2$ satisfying $\phi \circ u_i = r_i \circ \phi$. Then $\phi_i \circ u_i = -\phi_i$. In particular, $\phi_i = 0$ on the set of fixed points of u_i. As u_i is an isometry of S, we have $\int_S \phi_i \, dS = 0$. By the stability of S, we have

$$2\,E(\phi) = \int_S |\nabla\phi|^2 dS = \sum_{i=1}^{3} \int_S |\nabla\phi_i|^2 dS \geqslant \sum_{i=1}^{3} \int_S |\nabla N|^2 \phi_i^2 dS$$

$$= \int_S |\nabla N|^2 dS = 2\,E(N).$$

This proves 2.

To prove 3, let us assume that $i = 1$. If $N_1 \equiv 0$, then the Gauss map of S is contained in an equator of $\mathbb{S}^2$. Hence, the Gauss curvature of S is identically zero, and S is a torus. Let us assume that $N_1 \not\equiv 0$. We notice that the set of fixed points of u_1 separates S into two open sets Ω_1 and Ω_2 with $u_1(\Omega_1) = \Omega_2$. We consider the function

$$\phi = \begin{cases} (|N_1|, N_2, N_3), & \Omega_1, \\ (-|N_1|, N_2, N_3), & \Omega_2. \end{cases}$$

ϕ is a Lipschitz map satisfying $|\nabla\phi|^2 = |\nabla N|^2$ almost everywhere. Hence, ϕ is a harmonic map, which is smooth by Lemaire's results [265]. It follows then from the maximum principle that $\phi_1 > 0$ on Ω_1 (and $\phi_1 < 0$ on Ω_2).

Inequality genus $S \leqslant 3$ can be obtained from the general result in Theorem 6.33. However, we can give a direct proof in this case. We consider the isometries $h_i = u_{i+1} \circ u_{i+2}$, $i \in \mathbb{Z}_3$, and the subgroups

$$G_i = \{\mathrm{Id}, h_i\}, \qquad G = \{\mathrm{Id}, h_1, h_2, h_3\}.$$

The isometries h_i preserve the orientation of S, and so they are conformal transformations for the conformal structure of S. Hence, S/G_i and S/G are Riemann surfaces of genus γ_i and γ, respectively. As h_i are conformal (holomorphic) transformations, their sets of fixed consist of isolated points. Since $h_i^2 = \mathrm{Id}$, the branching order of each fixed point is 1. The degree of the projection maps

$S \to S/G_i$ and $S \to S/G$ is 2 and 4, respectively. Let us call B_i and B their total branching orders, which satisfy $B = \sum_{i=1}^{3} B_i$. So the Riemann-Hurwitz formula [152, Prop. v.1.10] implies

$$2g - 2 = 2\,(2\gamma_i - 2) + B_i,$$
$$2g - 2 = (2\gamma - 2) + B,$$

where $g = \operatorname{genus} S$. From these equations and the relation among the branching orders, we get

$$g + 2\gamma = \gamma_1 + \gamma_2 + \gamma_3. \tag{6.40}$$

Now we estimate $\gamma_1, \gamma_2, \gamma_3$. Let $\{v_1, v_2, v_3\}$ be the standard basis of $\mathbb{R}^3$. The projections $X_i = (v_i)^\top = v_i - \langle v_i, N\rangle N$ to S are well-defined vector fields on S. They vanish only at the fixed points of h_i and so $N = \pm v_i$ at these points. Moreover, the index of X_i is either $+1$ (at local maxima and minima of the height function $x \mapsto \langle x, v_i\rangle$) or -1 at saddle points of the local height function. The vector field X_i projects to S/G_i, it has isolated singularities at the branch values of $S \to S/G_i$, and the index is non-negative. Hence, $\gamma_i \leqslant 1$. We conclude from (6.40) that $\gamma = 0$ and $g \leqslant 3$.

To prove 5, we notice that eight isometric copies of the open set $\Omega = \{p \in S : N_i(p) > 0 \text{ for all } i\}$ cover all of S except the zero measure set of points fixed by some u_i. Observe that the boundary of Ω consists of geodesic segments meeting at inner angles of $\pi/2$ radians. Observe also that $\overline{\Omega}$ is connected since S is connected and symmetric. Hence, the Gauss-Bonnet theorem implies

$$4\pi(1 - g) = \int_S K\,dS = 8\int_\Omega K\,dS = 8\,\big(2\pi\chi(\Omega) - \ell\,\frac{\pi}{2}\big),$$

where ℓ is the number of vertices of $\partial\Omega$. Hence, we get

$$\ell = 4\chi(\Omega) + g - 1.$$

As $\ell \geqslant 3$ and $g \leqslant 3$, we have

$$\chi(\Omega) \geqslant \frac{1}{4}$$

and so Ω is a disk with $\chi(\Omega) = 1$ and $\ell = g + 1$, as claimed. □

Theorem 6.49 provides a quite complete geometric description of the candidate isoperimetric boundaries of genus 2 or 3 in quotients of $\mathbb{R}^3$ by rectangular lattices. Observe that the existence of the anticonformal maps u_i in S provides a restriction on the conformal structure of S. It was proved indeed in [358, p. 21] that these families are parameterized by the family of right angle hyperbolic pentagons

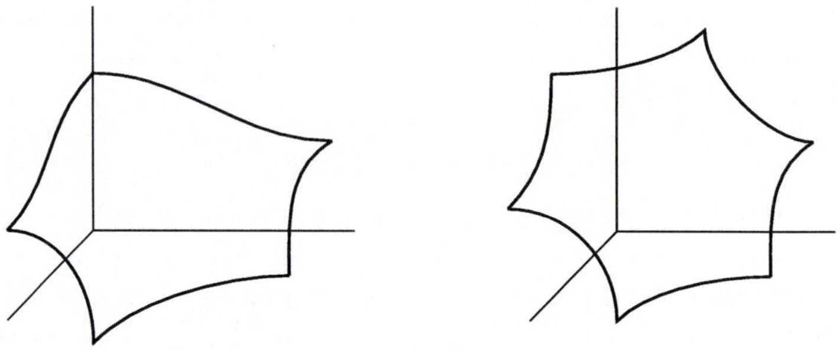

Fig. 6.1 Schematic representation of the fundamental domain $\overline{\Omega}$ of the isoperimetric candidates of genus 2 and 3 in quotients of $\mathbb{R}^3$ by rectangular lattices

and hexagons, which depend on two and three parameters, respectively (Fig. 6.1). Indeed, in the genus 2 case, the surfaces correspond to the Riemann surface associated with the polynomial

$$w^2 = (z^2 + 1)(z^2 + a^2)(z^2 + b^2), \qquad 0 < a < 1 < b.$$

The symmetries u_i correspond in this case to the antiholomorphic involutions $(z, w) \mapsto (\bar{z}, \bar{w})$, $(z, w) \mapsto (-\bar{z}, \bar{w})$, and $(z, w) \mapsto (-\bar{z}, -\bar{w})$.

It was also shown in Theorem 1 in [358] that there is a unique harmonic map, up to composition with r_i, minimizing the energy in the class of symmetric maps. Such harmonic maps can be integrated to obtain a periodic embedding with constant mean curvature in $\mathbb{R}^3$; see Theorem 8 in [358].

6.5.3 Hadwiger's Theorem for Lattices

In this subsection, we prove a classical isoperimetric inequality by Hadwiger [212] on the m-dimensional cubic torus. One of its main consequences is that slabs bounded by two totally geodesic tori separating the cubic three-dimensional torus are isoperimetric sets. In the Euclidean space $\mathbb{R}^m$, we consider the standard orthonormal basis $\{e_1, \ldots, e_m\}$ and the lattice

$$\Gamma = \Big\{\sum_{i=1}^{m} n_i e_i : n_i \in \mathbb{Z}, i = 1, \ldots, m\Big\}.$$

The quotient $T = T^m = \mathbb{R}^m / \Gamma$ is the standard unit cubic m-dimensional torus. We shall consider the fundamental region W in $\mathbb{R}^m$ given by

$$W = \big\{x \in \mathbb{R}^m : -\tfrac{1}{2} \leqslant x_i < \tfrac{1}{2}\big\}.$$

Hadwiger's result is the following:

Theorem 6.50 ([212, Satz 1]) *Let $T = T^m$ be the m-dimensional unit cubic torus. Let $E \subset T$ be a set of finite perimeter and positive volume. Then*

$$P(E) \geqslant 8|E|(1 - |E|). \tag{6.41}$$

Moreover, slabs of width $1/2$ orthogonal to the coordinate axes yield equality in (6.41) for volume $1/2$.

The isoperimetric inequality (6.41) is not optimal for all volumes. However, when $|E| = 1/2$, we obtain $P(E) \geqslant 2$, which is exactly the perimeter of a slab determined by two totally geodesic tori parallel to a coordinate hyperplane. Hence, these slabs are isoperimetric regions in T. It is also worth mentioning that the isoperimetric inequality (6.41) does not depend on the dimension of the torus.

In the following, we need the following definition. Let u be a *unit* vector in $\mathbb{R}^m$. Then the *planar static moment* of W with respect to the plane $\langle x, u\rangle = 0$ is the quantity

$$T(u) = \int_W |\langle u, w\rangle|\, dw. \tag{6.42}$$

It is straightforward to check that $T(e_i) = \frac{1}{4}$ and that

$$T(u) \leqslant \frac{1}{4}. \tag{6.43}$$

In order to prove (6.41), we approximate the set E by polyhedra. Given a polyhedron in T with k faces, we can always lift it to a Γ-periodic polyhedra $A \subset \mathbb{R}^m$ with outer unit normals $u_1, \ldots, u_k$. We define the quantity

$$c = c(A) = \min\{1/T(u_i) : i = 1, \ldots, k\}. \tag{6.44}$$

Theorem 6.50 will then follow from (6.44) and the following isoperimetric inequality for periodic polyhedra.

Theorem 6.51 ([212, Satz 2]) *For a polyhedron in T of perimeter P and volume V, we have*

$$P \geqslant 2c\,V(1 - V), \tag{6.45}$$

where $c \geqslant 4$ is defined in (6.44).

Proof of Theorem 6.51 Let $A \subset \mathbb{R}^m$ be the corresponding Γ-periodic polyhedron and $u_1, \ldots, u_k$ the outer unit normals to the faces. We fix $w \in W$ so that $\langle w, u_i \rangle \neq 0$ for all $i \in \{1, \ldots, k\}$. We define

$$\omega(x, w) = \begin{cases} 1, & x \in A, x + w \in A^c, \\ 0, & \text{otherwise}, \end{cases} \qquad \omega^c(x, w) = \begin{cases} 1, & x \in A^c, x + w \in A, \\ 0, & \text{otherwise}. \end{cases}$$

Here, A^c is the complementary set of A. The functions ω, ω^c are Γ-periodic functions of x. For fixed w, $\omega(\cdot, w)$ is the characteristic function of $A \cap (-w + A^c)$, and $\omega^c(\cdot, w)$ is the one of $A^c \cap (-w + A)$.

We also define

$$\chi(x, w) = \#\big\{\partial A \cap [x, x + w]\big\}.$$

This quantity is always finite since w is never tangent to the faces of A. The function $\chi(\cdot, w)$ is a Γ-periodic locally integrable function of x. For natural i, the set

$$P_i = \{x \in \mathbb{R}^m : \chi(x, w) = i\}$$

is a Γ-periodic polyhedron, and we have

$$\int_W \chi(x, w)\, dx = \sum_i i\, V(P_i),$$

where for a Γ-periodic set $B \subset \mathbb{R}$, $V(B)$ is the volume in T of B/Γ, equal to $\int_W \chi_B(x)\, dx$.

Another way of computing the integral $\int_W \chi(x, w)\, dx$ is as follows: take a face C of the polyhedron with outer unit normal u, and take the oblique cylindrical polyhedron $Q_C = C + [-w, 0]$ with base C and height $|\langle u, w \rangle|$. Here, $+$ denotes the Minkowski sum. Then it is geometrically clear that $\sum_i i\, |P_i/\Gamma|$ is the sum of the volumes of the cylinders Q_C for all faces C. So we have

$$\int_W \chi(x, w)\, dx = \int_{\partial A \cap W} |\langle u, w \rangle|\, d(\partial A). \tag{6.46}$$

Finally, we notice that

$$\chi(x, w) \geqslant \omega(x, w) + \omega^c(x, w)$$

for all $x \in \mathbb{R}^m$ and $w \in W$. Integrating over W

$$\int_W \chi(x, w)\, dx \geqslant \int_W \big(\omega(x, w) + \omega^c(x, w)\big)\, dx.$$

Integrating (6.46) over W and taking (6.44) into account, we get

$$\int_{W\times W} \chi(x, w)\,dx\,dw \leqslant \frac{1}{c}\int_{\partial A\cap W} d(\partial A) = \frac{1}{c}\,P.$$

On the other hand, since $\omega(x, w) = \mathbf{1}_A(x)\mathbf{1}_{A^c}(x + w)$ is the characteristic function of $A \cap (-w + A^c)$, a standard application of Fubini-Tonelli's theorem yields

$$\int_{W\times W} \omega(x, w)\,dx\,dw = V(A)V(A^c).$$

A similar equality holds for the integral of ω^c over $W \times W$. So we finally get

$$\frac{1}{c}\,P \geqslant 2\,V(A)V(A^c) = 2\,V\,(1 - V).$$

This implies (6.41). □

6.5.4 *Product of Circles and Euclidean Spaces*

We consider the flat manifold $N = \mathbb{S}^1(r)\times\mathbb{R}^m$, where $r > 0$. Since the group $E(m)$ of affine isometries of $\mathbb{R}^m$ acts on N through the second factor, the quotient by the isometry group is a compact set. Hence, isoperimetric sets exist on $\mathbb{S}^1(r)\times\mathbb{R}$ for any given volume by Theorem 4.25. Moreover, Hsiang's symmetrization Theorem 5.20 can be applied to this situation and yields the following:

Theorem 6.52 *An isoperimetric set $E \subset \mathbb{S}^1(r) \times \mathbb{R}^m$ is invariant under an $O(m - 1)$ action on the second factor.*

Using the $O(m-1)$ symmetry, the orbit space $N/O(m-1)$ is identified to $\mathbb{S}^1(r) \times [0, \infty)$. We study the equations satisfied by $S = \partial E$ in $\mathbb{R} \times [0, \infty)$, identifying points whose first coordinates differ by integral multiples of $2\pi r$. The axis of revolution corresponds to the subset $\mathbb{R} \times \{0\}$ of the orbit space.

The boundary S is regular out of the axis of revolution since any point of the singular set out of the axis contains a submanifold of dimension $(m-1)$, which is not possible since the Hausdorff dimension of the singular set is no larger than $(m+1) - 8$ by Theorem 1.60.

The projection of S to the orbit space $N/O(m-1)$ is a curve γ that we shall call the *generating curve* of S. We parameterize $\gamma = (x, y)$ by arc-length s and define $\sigma(s)$ by the equality

$$(x'(s), y'(s)) = (\cos\sigma(s), \sin\sigma(s)).$$

Note that σ is the angle between the tangent to $\gamma(s)$ and the positive x direction. By computing the mean curvature with respect to the normal $\mathbf{n} = (-\sin\sigma, \cos\sigma)$, we obtain the following result, the proof of which follows from straightforward analysis.

Lemma 6.53 *The generating curve γ of an $O(m-1)$ invariant hypersurface S in $\mathbb{S}^1(r) \times \mathbb{R}^m$ with constant mean curvature H with respect to the normal $(-\sin\sigma, \cos\sigma)$ satisfies the following system of ordinary differential equations:*

$$\begin{aligned} x' &= \cos\sigma, \\ y' &= \sin\sigma, \\ \sigma' &= -H + (m-1)\frac{\cos\sigma}{y} \end{aligned} \tag{6.47}$$

Moreover, the above ODE system has a first integral: along the solution γ of (6.47), the function

$$J(s) = \cos\sigma\, y^{m-1} - \frac{H}{m}\, y^m \tag{6.48}$$

is constant.

A standard analysis of the solutions to the system (6.47) implies that $2\pi r$-periodic solutions in the x-direction are of three types: of spherical type meeting the x-axis orthogonally; the curves $y(s) = r > 0$ for all s, which correspond to the tubular neighborhoods $\mathbb{S}^1(r) \times B(x,r)$ of the closed geodesic $\mathbb{S}^1(r) \times \{x\}$ (here, $B(x,r) \subset \mathbb{R}^m$ is a geodesic ball); and $2\pi r$-periodic solutions which are graphs over the horizontal x-axis with alternative maxima and minima of the y-coordinate. The first ones are the generating curves of geodesic spheres in $\mathbb{R}^{m+1}$. The second ones are characterized by the condition $\sigma \equiv 0$ or, equivalently, $H + (m-1)\, y^{-1} \equiv 0$. The third ones are called *unduloids*. They have symmetries with respect to vertical lines in the orbit space. They are also characterized as the solutions of (6.47) whose first integral (6.48) is positive and σ is not identically zero; see §1 in [338] for complete details. These curves were first studied by Delaunay [138]. Their behavior is similar to the system (2.26) we reached when we considered curves with constant geodesic curvature in surfaces of revolution (Fig. 6.2).

In particular, the only curves touching the axis of revolution correspond to spheres, which are regular hypersurfaces. Since there are no compact minimal hypersurfaces in $\mathbb{S}^1(r) \times \mathbb{R}^m$, all isoperimetric boundaries are connected. This implies the following:

Theorem 6.54 *All isoperimetric sets in $\mathbb{S}^1(r) \times \mathbb{R}^m$ have regular boundary. The only possible candidates are geodesic spheres, tubular neighborhoods of the closed geodesics $\mathbb{S}^1(r) \times \{x\}$, and $O(m-1)$ invariant domains whose boundary is generated by an unduloid.*

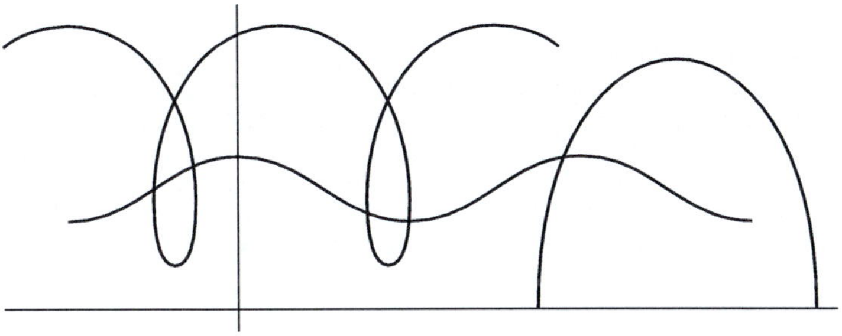

Fig. 6.2 Solutions of the system (6.47). The periodic graphs over the horizontal axis correspond to unduloids

The stability of the boundaries of tubular neighborhoods of the closed geodesics $\mathbb{S}^1(r) \times \{x\}$, $x \in \mathbb{R}^m$, can be completely characterized.

Lemma 6.55 ([338, Lemma 2.5]) *The boundary of the tube $\mathbb{S}^1(r) \times B(\rho)$, where $B(\rho)$ is a geodesic ball of radius $\rho > 0$ in $\mathbb{R}^m$, is a stable hypersurface if and only if*

$$\rho \geqslant r\sqrt{m-1}$$

Proof We just take into account that the Jacobi operator of the tube is the sum of the Laplacian of the tube plus the constant $(m-1)/\rho^2$. The first non-zero eigenvalue of the Laplacian for the product $\mathbb{S}^1(r) \times \partial B(\rho)$, the boundary of the tube $\mathbb{S}^1(r) \times B(\rho)$, is the minimum of

$$\left\{r^{-2}, (m-1)\,\rho^{-2}\right\},$$

so that the criterion for stability given in Proposition 2.13 in [39] is equivalent to

$$r^{-2} \geqslant (m-1)\,\rho^{-2},$$

and the result follows. □

As for the stability of the unduloids, we have the following result. We define the *period* of an unduloid as the horizontal distance between two consecutive maxima (or minima) of the y-coordinate. Fixing the mean curvature H, the period is a smooth function of the maximum point y_0 of the y-coordinate of the unduloid (Fig. 6.3).

Theorem 6.56 ([338, Proposition 2.7]) *Let $S \subset \mathbb{S}^1(r) \times \mathbb{R}^m$ be a hypersurface with constant mean curvature H generated by an unduloid with maximum distance $y_{\max} > 0$ to the axis of revolution. If the derivative of the period (with H fixed) with respect to the maximum of the y-coordinate is strictly negative at y_0, then S is unstable.*

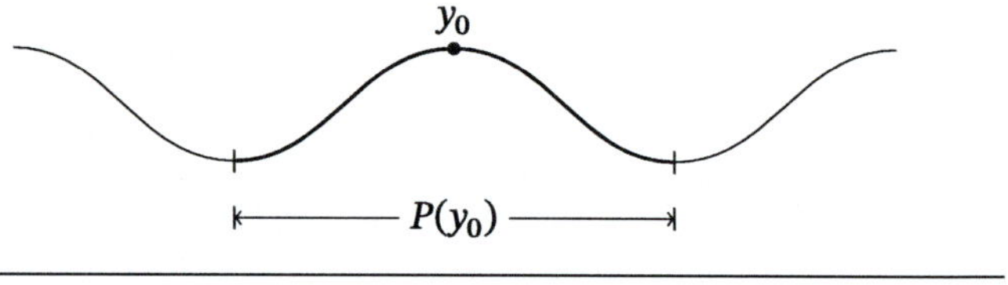

Fig. 6.3 The period of an unduloid as a function of the maximum distance to the axis of revolution

Proof We consider a solution (x, y, σ) of the system (6.47) of unduloid type such that the function y achieves a maximum $y_{\max}$ at $s = s_0$. Let $s_1 > s_0$ be the first instant where $y'(s) = 0$, and let $y_{\min} = y(s_1)$. The function y achieves a minimum at $s = s_1$. For $i = 0, 1$ and $\varepsilon \in \mathbb{R}$, the solution of (6.47) with initial conditions $(x(s_i), y(s_i) + \varepsilon, 0)$ and mean curvature H will be denoted by $(x_i(s, \varepsilon), y_i(s, \varepsilon), \sigma_i(s, \varepsilon))$. For ε close enough to 0, the solutions $(x_i(\cdot, \varepsilon), y_i(\cdot, \varepsilon), \sigma_i(\cdot, \varepsilon))$, $i = 0, 1$, are of unduloid type since the first integral (6.48) is positive and $-H + (m-1)\, y_i(s_i, \varepsilon)^{-1} \neq 0$.

For $\varepsilon > 0$ close to 0, let $s_1^\varepsilon > s_0$ be the first instant where $y_0'(s, \varepsilon) = 0$, and let $y_{\min}(\varepsilon) = y_0(s_1^\varepsilon, \varepsilon)$. The derivative $y_{\min}'(\varepsilon)$ at $\varepsilon = 0$ can be computed from the first integral (6.48) since

$$(y_0 + \varepsilon)^{m-1} - \frac{H}{m}(y_0 + \varepsilon)^m = (y_{\min}(\varepsilon))^{m-1} - \frac{H}{m}(y_{\min}(\varepsilon))^m$$

and so

$$y_{\min}'(0) = \frac{y_0^{m-1}}{y_1^{m-1}} \frac{\sigma'(s_0)}{\sigma'(s_1)} < 0. \tag{6.49}$$

We fix $t \in (s_0, s_1)$. As $y'(t) = \sin\sigma(t) \neq 0$, the implicit function theorem ensures the existence of smooth functions $s_0(\varepsilon), s_1(\varepsilon)$ such that, for ε close to 0,

$$y_0(s_0(\varepsilon), \varepsilon) = y(t), \qquad y_1(s_1(\varepsilon), \varepsilon) = y(t).$$

The derivatives of such functions with respect to ε at $\varepsilon = 0$ are given by

$$\left.\frac{ds_0}{d\varepsilon}\right|_{\varepsilon=0} = \frac{-1}{\sin\sigma(t)} \frac{\partial y_0}{\partial\varepsilon}(t, 0), \qquad \left.\frac{ds_1}{d\varepsilon}\right|_{\varepsilon=0} = \frac{-1}{\sin\sigma(t)} \frac{\partial y_1}{\partial\varepsilon}(t, 0) \tag{6.50}$$

The period $P(\varepsilon)$ of the solution $(x_0(\cdot, \varepsilon), y_0(\cdot, \varepsilon), \sigma_0(\cdot, \varepsilon))$ is given by (see Fig. 6.4)

$$\frac{1}{2} P(\varepsilon) = \frac{1}{2} P(0) + x_0(s_0(\varepsilon), \varepsilon) - x_1(s_1(y_{\min}(\varepsilon)), y_{\min}(\varepsilon))).$$

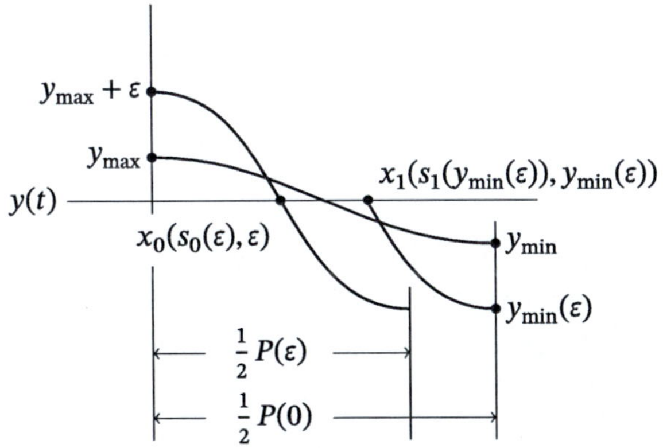

Fig. 6.4 The derivative of the period of an unduloid

So we have, taking (6.50) into account,

$$\frac{1}{2}P'(0) = \frac{-1}{\sin\sigma(t)}\left\{u_0(t) - y'_{\min}(0)\,u_1(t)\right\}, \tag{6.51}$$

where

$$u_i(s) = \left(-\sin\sigma\,\frac{\partial x_i}{\partial\varepsilon} + \cos\sigma\,\frac{\partial y_i}{\partial\varepsilon}\right)(s,0).$$

For i fixed, the solutions $(x_i(\cdot,\varepsilon), y_i(\cdot,\varepsilon), \sigma_i(\cdot,\varepsilon))$ produce a variation in $\mathbb{R}\times\mathbb{R}^m$ of S by rotationally invariant hypersurfaces with the same mean curvature H. Hence, the functions

$$u_i(s) = \left(\frac{\partial x_i}{\partial\varepsilon}, \frac{\partial y_i}{\partial\varepsilon}\right)(s,0)\cdot\mathbf{n}(s) = -\sin\sigma(s)\,\frac{\partial x_i}{\partial\varepsilon}(s,0) + \cos\sigma(s)\,\frac{\partial y_i}{\partial\varepsilon}(s,0)$$

yield rotationally invariant Jacobi fields (i.e., solutions of $\Delta + (\mathrm{Ric}(N,N) + |\sigma|^2 = 0)$ on the lifting $\tilde{S}$ of S to $\mathbb{R}\times\mathbb{R}^m$.

Let us prove now that the nodal region of u_0 around s_0 and the one of u_1 around s_1 cannot overlap, that is, that the first zero t_0 of u_0 in the interval (s_0, s_1) is strictly smaller than the first zero t_1 of u_1 in the same interval. We have $t_0, t_1 \in (s_0, s_1)$ since $u_i(s_i) = 1$ and $u_i(s_j) < 0$ for $i, j \in \{0,1\}, i \neq j$. We reason by contradiction assuming $t_0 \geqslant t_1$. Then there exists $t \in (s_0, s_1)$ such that $u_0(t), u_1(t) \geqslant 0$. Equation (6.51), together with (6.49), implies $P'(0) \geqslant 0$, a contradiction to our hypothesis.

So $t_0 < t_1$. This implies that the nodal region of u_1 around s_0 and the nodal region of u_1 around s_1 do not overlap. Hence, we obtain two regions on S, at positive distance, whose first eigenvalues for the Jacobi operator are equal to 0. Enlarging slightly both regions, we obtain two disjoint regions on S whose first eigenvalues for the Jacobi operator are negative. A linear combination of both eigenfunctions with mean zero ensures the instability of S. □

Hence, Theorem 6.56 provides an instability criterion for unduloids in $\mathbb{S}^1 \times \mathbb{R}^m$. Unfortunately, an asymptotic analysis near cylinders implies the following result.

Proposition 6.57 ([338, Prop. 3.1]) *The derivative of the period of unduloids with mean curvature H and initial conditions close to the cylinder of mean curvature H is strictly negative if $m \leqslant 8$ and strictly positive for $m \geqslant 9$.*

Hence, unduloids near cylinders are unstable in $\mathbb{S}^1(r) \times \mathbb{R}^m$ near cylinders for $2 \leqslant m \leqslant 8$. Theorem 6.56 does not provide any useful information when $m \geqslant 9$ or far from cylinders. However, we can prove the following result.

Proposition 6.58 ([338, Prop. 3.2]) *The derivative of the period of unduloids with respect to the maximum distance to the axis of revolution is strictly negative if $2 \leqslant m \leqslant 7$. Hence, hypersurfaces generated by unduloids are unstable in $\mathbb{S}^1(r) \times \mathbb{R}^m$ for all $r > 0$ and $2 \leqslant m \leqslant 7$.*

Proof By scaling, we assume that $H = (m - 1)$. The first integral (6.48) then becomes

$$J = \cos\sigma \, y^{m-1} - \lambda y^m, \qquad \lambda = \frac{m-1}{m}.$$

For an unduloid with $J > 0$, the maximum and minimum values of y are $y_{\max}(J)$ and $y_{\min}(J)$ and satisfy the equation $J = y^{m-1} - \lambda y^m = y^{m-1}(1 - \lambda y)$. In particular, $y_{\max} < \lambda^{-1} = m/(m+1)$ ($y_{\max}(J)$ corresponds to the sphere with $J = 0$). We also have $y_{\min}(J) < 1 < y_{\max}(J)$ since the second derivative of y is positive or negative.

The period of the unduloid is given by

$$2\int_{y_{\min}(J)}^{y_{\max}(J)} \frac{dx}{dy}\, dy = 2\int_{y_{\min}(J)}^{y_{\max}(J)} \frac{\cos\sigma}{\sin\sigma}\, dy$$
$$= 2\int_{y_{\min}(J)}^{y_{\max}(J)} \frac{J + \lambda y^m}{\sqrt{y^{2(m-1)} - (J + \lambda y^m)^2}}\, dy.$$

We consider the Riemann surface $S_{m,J}$ associated with the algebraic equation

$$w^2 = y^{2(m-1)} - (J + \lambda y^m)^2 = \big(y^{m-1} - (J + \lambda y^m)\big)\big(y^{m-1} + (J + \lambda y^m)\big).$$

Since the polynomials $y^{m-1} \pm (J + \lambda y^m)$ have roots with multiplicity one, the Riemann surface $S_{m,J}$ is hyperelliptic for integer $m \geqslant 2$ and $J \in (0, 1/m)$. Hence, the period can be computed as the integral in $S_{m,J}$ given by

$$P(J) = \int_{\alpha} (J + \lambda y^m) \frac{dy}{w}, \tag{6.52}$$

where α is the lifting to $S_{m,J}$ of a curve in $\mathbb{C}$ enclosing the segment $[y_{\min}(J), y_{\max}(J)]$ but no other roots of the polynomials $y^{m-1} \pm (J + \lambda y^m)$. Computing the derivative of (6.52) with respect to J, we obtain

$$\begin{aligned}\frac{dP}{dJ} &= \int_{\alpha} \frac{y^{2(m-1)}}{w^3}\, dy \\ &= \int_{\alpha} \frac{1}{2}\left(\frac{y^{m-1} dy}{w\,(y^{m-1} + (J + \lambda\, y^m))} + \frac{y^{m-1} dy}{w\,(y^{m-1} + (J - \lambda\, y^m))}\right).\end{aligned}$$

Such integral cannot be reduced to the real axis since the one-form $y^{2(m-1)}dy/w^3$ has poles in $S_{m,J}$ at $y_{\min}(J)$, $y_{\max}(J)$. We consider the meromorphic function

$$\phi = \frac{1}{1-y}\left\{y - \frac{m}{1-mJ}\left(y^{m-1} - (J + \lambda\, y^m)\right)\right\}\frac{1}{w}.$$

A direct computation shows

$$\frac{y^{2(m-1)}\, dy}{w^3} + \frac{d\phi}{m-1} = \frac{\psi(y, J)\, dy}{(m-1)(1-mJ)\, w\,(y^{m-1} + (J + \lambda\, y^m))}, \tag{6.53}$$

where

$$\begin{aligned}\psi(y, J) = \left(y^{m-1} + \left(J + \lambda\, y^m\right)\right)&\left(\frac{1 - m\, y^{m-1} + (m-1)\, y^m}{(1-y)^2}\right) \\ &+ (m-1)\left(\frac{-y^m(1-mJ) - (mJ\, y^{m-2} - y^{2m-2})}{1-y}\right).\end{aligned} \tag{6.54}$$

We have

$$\begin{aligned}1 - m\, y^{m-1} + (m-1)\, y^m &= (1-y)^2 \sum_{k=0}^{m-2} (k+1)\, y^k \\ -y^m(1-mJ) - (mJ\, y^{m-2} - y^{2m-2}) &= (y-1)\, y^{m-2}\left(mJ\,(1+y) + \sum_{2 \leqslant k \leqslant m-1} y^k\right),\end{aligned}$$

where the sum in the last line from $k = 2$ to $m-1$ must be taken equal to zero when $m = 2$.

The integral of $y^{2(m-1)}/w^3 + d\phi/(m-1)$ can be reduced to the real axis since the denominator of the fraction obtained does not vanish in a neighborhood of the segment $[y_{\min}, y_{\max}]$. Moreover, the denominator is positive since $(1 - mJ) > 0$ and $w > 0$.

On the other hand, a direct computation shows that, for any polynomial $p(y)$, we have

$$d\left(p(y)\,\frac{y^{m-1} - (J + \lambda\, y^m)}{w}\right) = p'(y)\,\frac{y^{m-1} - (J + \lambda\, y^m)}{w} + (m-1)\,p(y)\,\frac{J\,y^{m-2} - \frac{1}{m}\,y^{2m-2}}{w\,\bigl(y^{m-1} + (J + \lambda\, y^m)\bigr)}. \tag{6.55}$$

So we conclude that

$$\frac{y^{2(m-1)}\,dy}{w^3} + \frac{d\phi}{m-1} + d\left(\frac{p(y)\,\bigl(y^{m-1} - (J + \lambda\, y^m)\bigr)}{(m-1)(1-mJ)\,w}\right) = \frac{q_m(y, J)\,dy}{(m-1)(1-mJ)\,w\,\bigl(y^{m-1} + (J + \lambda\, y^m)\bigr)},$$

where $q_m(y, J)$ is given by

$$\begin{aligned} q_m(y, J) = {} & \Bigl(y^{m-1} + \bigl(J + \lambda\, y^m\bigr)\Bigr)\Bigl(\sum_{k=0}^{m-2}(k+1)\,y^k + p'(y) \\ & \times \bigl(y^{m-1} - (J + \lambda\, y^m)\bigr)\Bigr) \\ & + (m-1)\,y^{m-2}\Bigl(-mJ\,(1+y) - \sum_{2\leqslant k\leqslant m-1} y^k \\ & + p(y)\bigl(J - \frac{1}{m}\,y^m\bigr)\Bigr), \end{aligned} \tag{6.56}$$

and again, the sum in the last line from $k = 2$ to $m-1$ is ignored when $m = 2$.

We want to see that q_m is positive, for $2 \leqslant m \leqslant 7$, inside the region Ω in the (y, J)-plane defined by the inequalities $0 < J < y^{m-1} - \lambda\, y^m$ and $0 < y < \lambda^{-1}$.

Taking a polynomial $p(y)$ with $p'(y) \geqslant 0$, we obtain that the function $q_m(y, J)$ is concave in J for y fixed. So it is enough to see that it is non-negative at the boundary of Ω to conclude that $q_m \geqslant 0$ in Ω. Restricting q_m to the curve $J = y^{m-1} - \lambda\, y^m$

and to the axis $J = 0$, respectively, we obtain polynomials f_m and g_m given by

$$f_m(y) = y^{m-1}\Big\{2\sum_{k=0}^{m-2}(k+1)\,y^k + (m-1)\Big(-\sum_{k=1}^{m} y^k + (y^{m-2} - y^{m-1})(p(y) - m\,(1+y))\Big)\Big\}, \qquad (6.57)$$

$$g_m(y) = y^{m-1}\Big\{(1+\lambda\,y)\Big(\sum_{k=0}^{m-2}(k+1)\,y^k + p'(y)\,(y^{m-1} - \lambda\,y^m)\Big) + (m-1)\Big(-\sum_{k=1}^{m-2} y^k - \frac{1}{m}\,p(y)\,y^{m-1}\Big)\Big\}, \qquad (6.58)$$

where the sum from $k = 1$ to $m - 1$ does not appear when $m = 2$.

We now make choices of polynomials $p(y)$ with $p'(y) \geqslant 0$ and study the resulting functions f_m and g_m. First we take

$$p(y) = \frac{n\,(n+1)}{6}.$$

The polynomials so obtained, for $n \leqslant 6$, are given by

$$\begin{aligned}
f_2(y) &= (y-1)^2 y,\\
f_3(y) &= (y-1)^2 y^2 (2+4y),\\
f_4(y) &= (y-1)^2 y^3 (2+5y+9y^2),\\
f_5(y) &= (y-1)^2 y^4 (2+4y+8y^2+16y^3),\\
f_6(y) &= (y-1)^2 y^5 (2+3y+5y^2+10y^3+25y^4),
\end{aligned}$$

and

$$\begin{aligned}
g_2(y) &= (1/2)\,2y,\\
g_3(y) &= (1/3)\,y^2(3+2y),\\
g_4(y) &= (1/4)\,y^3(4-y+6y^2-y^3),\\
g_5(y) &= (1/5)\,y^4(5-6y+3y^2+12y^3-4y^4),\\
g_6(y) &= (1/6)\,y^5(6-13y-2y^2+9y^3+20y^4-10y^5).
\end{aligned}$$

Trivially, we have $f_m(y) > 0$ in $(0, m/(m-1)) - \{1\}$ for $2 \leqslant m \leqslant 6$. Also by elementary methods, we can check that $g_m > 0$ for $2 \leqslant m \leqslant 5$ in the intervals $(0, m/(m-1))$. To see that $g_6 > 0$ in the interval $(0, 6/5)$, it is necessary to appeal to Sturm's theorem [241, p. 298].

Theorem 6.59 (Sturm's Theorem) *Let $f(x)$ be a polynomial with real coefficients, and consider the standard sequence $\{f_0(x), f_1(x), \ldots, f_s(x)\}$ for $f(x)$ given by*

$$\begin{aligned} f_0(x) &= f(x), \\ f_1(x) &= f'(x), \\ f_0(x) &= q_1(x)\, f_1(x) - f_2(x), \qquad \text{deg}\, f_2 < \text{deg}\, f_1, \\ &\cdots \\ f_{i-1}(x) &= q_i(x)\, f_i(x) - f_{i+1}(x), \qquad \text{deg}\, f_{i+1} < \text{deg}\, f_i, \\ &\cdots \\ f_{s-1}(x) &= q_s(x)\, f_s(x). \end{aligned}$$

Assume that $[a, b]$ is an interval such that $f(a) \neq 0$ and $f(b) \neq 0$. Then the number of distinct roots of $f(x)$ in (a, b) is $v(a) - v(b)$, where $v(c)$ denotes the number of variations in sign of $\{f_0(c), f_1(c), \ldots, f_s(c)\}$.

A long but straightforward computation shows that, for $6\, g_6/y^5$, the signs of the standard sequence at $y = 0$ are given by

$$\{+, -, -, +, -, -\},$$

and at $y = 6/5$ by

$$\{+, +, -, -, +, -\},$$

so that $v(0) - v(6/5) = 3 - 3 = 0$. Hence, g_5 has no roots, and it is positive in the interval $(0, 6/5)$.

For the case $m = 7$, we consider the polynomial

$$p(y) = \frac{m(m+1)}{6} + 98\,(y-1)^3.$$

We obtain

$$f_7(y) = (y-1)^2 y^6 \big(2 + 2y + 2y^2 + 4y^3 + 10y^4 - 552y^5 + 1176y^6 - 588y^7\big)$$
$$g_7(y) = (1/7)\, y^6 \{7 - 22y - 9y^2 + 4y^3 + 17y^4 + 30y^5 + 2626y^6$$

$$- 5880y^7 + 2310y^8$$
$$+ 2436y^9 - 1512y^{10}\}.$$

An application of Sturm's theorem shows that the signs of the standard sequence for $f_7/(y^6(y-1)^2)$ at $y = 0$ and $y = 7/6$ are given, respectively, by

$$\{+, +, -, +, -, -, +, +\} \quad \text{and} \quad \{+, -, -, -, +, +, -, +\},$$

so that $v(0) - v(7/6) = 4 - 4 = 0$ and f_7 is positive in the open interval $(0, 7/6)$. Another application of Sturm's theorem gives the following variation of signs for the standard sequence of $7g_7/y^6$ at $y = 0$ and $y = 7/6$:

$$\{+, -, -, -, +, -, -, +, -, -, -\} \quad \text{and} \quad \{+, -, -, -, -, , +, +, -, +, +, -\}.$$

Hence, g_7 has no zeroes, and it is positive on the open interval $(0, 7/6)$. □

Proposition 6.58 leaves open the possibility of having isoperimetric solutions with boundaries generated by unduloids in case $m \geqslant 8$. We prove that this is indeed the case for $m \geqslant 9$ (Fig. 6.5).

Proposition 6.60 ([338, Prop. 3.4]) *There exist isoperimetric boundaries in* $\mathbb{S}^1(r) \times \mathbb{R}^m$*,* $m \geqslant 9$*, generated by unduloids.*

Proof By scaling, we assume that $r = 1/\pi$. With this choice, any geodesic sphere of radius 1 in $\mathbb{S}^1(1/\pi) \times \mathbb{R}^m$ has a tangential self-intersection, and so it cannot be an isoperimetric set. We denote by ω_{m+1} the volume of the unit ball in $\mathbb{R}^{m+1}$ and by $\mathbf{c}_m$ its perimeter.

Consider now a tube T of radius $t > 0$ around a closed geodesic $\mathbb{S}^1(1/\pi) \times \{x\}$, for some $x \in \mathbb{R}^m$. The tube is the product $\mathbb{S}^1(1/\pi) \times B(x,t)$, and its volume and perimeter are given by

$$|T| = 2\,\omega_m t^m, \qquad P(T) = 2\,\mathbf{c}_{m-1} t^{m-1}.$$

We compute the radius $t > 0$ so that $|T| = \omega_{m+1}$ to get

$$t = \left(\frac{\omega_{m+1}}{2\omega_m}\right)^{1/m}.$$

Using the equalities $\mathbf{c}_{k-1} = k\,\omega_k$ and $\mathbf{c}_m\mathbf{c}_{m-1}^{-1} = 2\int_0^{\pi/2} \sin^{m-1} s\, ds$, we conclude that $P(T) > \mathbf{c}_m$ if and only if

$$\left(\frac{m}{m+1}\right)^{m-1} > \int_0^{\pi/2} \sin^{m-1} s\, ds. \tag{6.59}$$

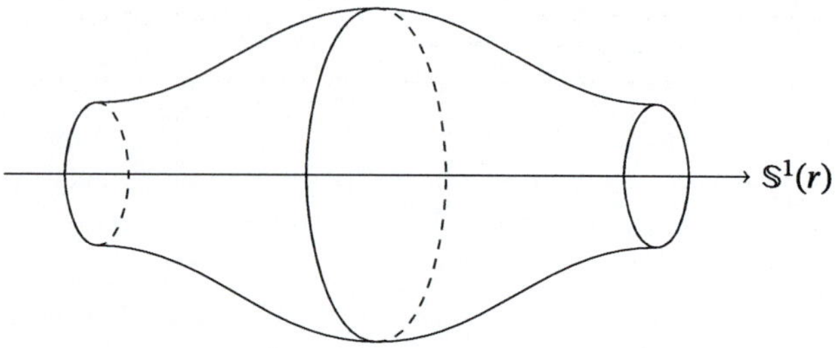

Fig. 6.5 An isoperimetric region bounded by a hypersurface of unduloid type in $\mathbb{S}^1(r) \times \mathbb{R}^m$, $m \geqslant 9$

Calling $I_m = \int_0^{\pi/2} \sin^m s\, ds$, we get $I_{m+2} = (m+1)\, I_m/(m+2)$ and $I_0 = \pi/2$, $I_1 = 1$. So I_m can be explicitly computed. On the other hand, the sequence I_m decreases to 0, and the decreasing sequence $(m/(m+1))^{m-1}$ converges to e^{-1}. A direct computation shows that (6.59) holds for $m \geqslant 9$. So for $m \geqslant 9$, a non-embedded sphere of radius 1 has smaller perimeter than the tube of the same volume. Hence, both sets of candidates are discarded as isoperimetric sets, and the only remaining possibility is a set whose boundary is generated by an unduloid. We recall that isoperimetric sets exist for any given volume by Theorem 4.25. □

The case $m = 8$ is not covered by these arguments. Numerical experiments do not exclude the possibility of existence of stable unduloids, but they indicate that they are not isoperimetric sets; see the remark in page 1384 in [338].

Summing up the results in this section, we have the following:

Theorem 6.61 *The isoperimetric sets in $\mathbb{S}^1(r) \times \mathbb{R}^m$ are geodesic balls or tubular neighborhoods of closed geodesics if $2 \leqslant m \leqslant 7$. In case $m \geqslant 9$, there exist isoperimetric sets whose boundary is generated by an unduloid.*

For $2 \leqslant m \leqslant 7$, the isoperimetric profile of $\mathbb{S}^1(r) \times \mathbb{R}^m$ is given by

$$I_{\mathbb{S}^1(r)\times\mathbb{R}}(v) = \begin{cases} \left((1+m)^{m/(m+1)} \mathbf{c}_m^{1/(m+1)}\right) v^{1-1/(m+1)}, & v \leqslant v_0, \\ \left(m^{1-1/m}(2\pi r \mathbf{c}_{m-1})^{1/m}\right) v^{1-1/m}, & v \geqslant v_0, \end{cases} \tag{6.60}$$

where

$$v_0 = m^{(m-1)(m+1)} (2\pi r \mathbf{c}_{m-1})^{m+1} (1+m)^{-m^2} \mathbf{c}_m^{-m}.$$

The solutions for $v \leqslant v_0$ are geodesic balls and, for $v \geqslant v_0$, tubes around the closed geodesics $\mathbb{S}^1(r) \times \{x\}$, $x \in \mathbb{R}^m$.

6.6 Notes

Notes for Sect. 6.1 From the point of view of partial differential equations, the Jacobi operator is nothing but a Schrödinger operator, the Laplacian plus a potential. It has received considerable attention by differential geometers and geometric analysts because of its relation to the stability problem for minimal surfaces (without a volume-preserving condition) and to the Bernstein problem; see Fischer-Colbrie and Schoen [163], Fischer-Colbrie [162], Barbosa and do Carmo [37], do Carmo and Peng [139], Giusti [184, pp. 203 ff.], and Choe [113], among many other interesting works. The strong stability condition, without a volume constraint, also played a key role in the classification of (strongly) stable cones in $\mathbb{R}^m$, $3 \leqslant m \leqslant 7$; see Simons [405] or Appendix B in Simon's lecture notes [404]

A spectral theory for the Jacobi operator in the space of first-order volume-preserving variations was developed by Barbosa and Berard [40], who obtained existence of eigenvalues and comparison to the eigenvalues of the usual Dirichlet problem, as well as properties of the eigenfunctions. The relation between the stability of a hypersurface with constant mean curvature and the classical eigenvalues of the Jacobi operator was investigated by Koiso [254] and also by Huang and Lin [238]. The local area-minimizing properties in Sect. 6.1.3 were obtained by Große-Brauckmann [201].

Notes for Sect. 6.2 The characterization of geodesic spheres in simply connected space forms as the only stable (orientable) compact hypersurfaces was obtained in the Euclidean case by Barbosa and do Carmo [38] and by Barbosa, do Carmo, and Eschenburg [39] in the spherical and hyperbolic cases. These techniques were extended to some warped products by Veeravalli [429] following ideas by Montiel [297] and to cones over manifolds with non-negative Ricci curvature by Morgan and the author [313].

The results in Wente's paper [442] provide not only a geometric characterization of the deformation used by Barbosa and do Carmo [38] but also a complete independent proof of the stability of geodesic spheres in $\mathbb{R}^m$. It should be noted that the main ingredients in the proof are the use of parallel hypersurfaces and dilations, as in the proof of the isoperimetric inequality in $\mathbb{R}^m$ obtained from the Brunn-Minkowski inequality.

All these results are valid for smooth hypersurfaces. The extension to isoperimetric boundaries in Sect. 6.2.4, which are regular except on a singular set with small Hausdorff dimension, relies on the work by Sternberg and Zumbrum [411]; see also Lemma 1.61 and [313].

Notes for Sect. 6.3 In three-dimensional ambient manifolds, we can apply the full machinery of Riemann surfaces to the study of isoperimetric boundaries. The technique used to obtain the genus estimate in Lemma 6.16 originated from Szegö [417]; see also Hersch [229], Li and Yau [270], and Montiel and Ros [299].

The characterization of constant mean curvature surfaces of genus 0 in $\mathbb{R}^3$ in Sect. 6.3.2 was given by Heinz Hopf [233] using the holomorphic quadratic differential $\sigma^{2,0}$, known nowadays as the Hopf differential, associated with the second fundamental form of the surface. The extension to the three-dimensional sphere and hyperbolic space is immediate. In homogeneous 3-manifolds, a holomorphic quadratic differential was introduced by Abresch and Rosenberg [2, 3]; see also Fernández and Mira [155, 156] and Franceschi et al. [166].

In homogeneous 3-manifolds, a classification of isoperimetric candidates in Sol_3 has been given by Daniel and Mira [130]; see Daniel [129] for the background on isometric immersions into three-dimensional homogeneous manifolds. Meeks et al. [282] have shown the relation

$$\lim_{v\to 0}\frac{I_M(v)}{v}=H_M,$$

in a non-compact simply connected homogeneous 3-manifold. Here, H_M is the infimum of max H_S, where H_S is the maximum of the mean curvature function on S, and S ranges over compact smooth surfaces in M. See Hsiang [237] for similar results in arbitrary dimension.

The test function $\sinh(w)$ used in Sect. 6.3.3 depends on the Gauss and mean curvature of the surface and on the sectional curvature of the ambient manifold. In the Euclidean case, it is given by

$$w=\sinh\left(\log\left(\frac{H^2-4K}{H^2}\right)^{-1/4}\right).$$

The function $\sinh(w)$, as well as $\cosh(w)$, is an eigenfunction of the stability operator $\Delta_0+\cosh^2(w)+\sinh^2(w)$ on the constant mean curvature surfaces of revolution in $\mathbb{R}^3$. It is possible to use some other test functions associated with K. Huang and Lin used the Gauss curvature itself as a test function to obtain the weaker result $\lambda_1(\{K<0\})\leqslant 0$; see Theorem 2 in [238]. They also obtained the interesting consequence that the region with negative Gauss curvature in a constant mean curvature graph in $\mathbb{R}^3$ over an open set cannot be compactly contained in the graph; see Theorem 3 in [238].

Notes for Sect. 6.4 The classification of isoperimetric sets in $\mathbb{RP}^3$ was given by the author and Ros [365]. The application to the Willmore problem by Ros [378]. Recently, Viana [430] has proven that the only isoperimetric sets in the lens spaces $L(p,q)$ for p large are geodesic spheres and tubes around closed geodesics. Compact minimal surfaces with index one and stable constant mean curvature hypersurfaces in real projective spaces $\mathbb{RP}^m$ were studied by do Carmo et al. [140], Perdomo [339], and Alías et al. [10]. Viana [431] has recently announced a proof of the classification of stable constant mean curvature hypersurfaces in $\mathbb{RP}^m$. They are tubular neighborhoods of totally geodesic submanifolds.

Notes for Sect. 6.5 The compactness result for stable surfaces of genus larger than or equal to 2 and their ambient manifolds in Theorem 6.42 were proven by the author and Ros [365]. A different proof in quotients of $\mathbb{R}^3$ by lattices, using convergence results for harmonic maps to the sphere $\mathbb{S}^2$ with bounded energy, was given by the author in [357]. The stability condition allows to deal successfully with the bubbling phenomenon [334] for harmonic maps. The classification results in Sect. 6.5.2 are completed by the examples obtained by the author in [358]. More results, including the limits of these examples, can be found in author's Ph.D. thesis [375].

Some more papers focusing on isoperimetric inequalities in quotients of $\mathbb{R}^3$ are Hauswirth et al. [222]; Ros [382], where it is proven that if Γ is a lattice in $\mathbb{R}^3$ of rank k, any isoperimetric boundary satisfies genus $S \leqslant k$; and Ros [379].

Hadwiger's result in Sect. 6.5.3 provides a non-optimal isoperimetric inequality in the cubic m-dimensional torus; see [212]. It is however optimal for half of the volume of the torus and implies that slabs parallel to the sides of the cube are the only isoperimetric sets. This result can also be obtained by a different technique, comparing the isoperimetric profile of the cube with the one of Gauss space; see Barthe and Maurey [44], Barthe [41], and also Theorem 3.8(c) in Ros [380]. See also Morgan [304].

In Sect. 6.5.4, we have classified the isoperimetric sets in $\mathbb{S}^1(r) \times \mathbb{R}^m$ using the $O(m-1)$ rotational symmetry in the Euclidean factor. There is another isolated symmetry with respect to a totally geodesic $\{x\} \times \mathbb{R}$, with $x \in \mathbb{S}^1(r)$, coming from the $O(2)$ action on $\mathbb{S}^1(r)$. This implies an additional symmetry which shows that the isoperimetric minimization problem is equivalent to the isoperimetric free boundary problem in a slab of $\mathbb{R}^{m+1}$ determined by two parallel hyperplanes; see §5 in [338]. This problem was considered in $\mathbb{R}^3$ by Athanassenas [29] and Vogel [433], who proved that half-spheres and tubes around a segment perpendicular to the boundary planes are the only solutions to the problem.

Chapter 7
The Isoperimetric Profile for Small and Large Volumes

In this chapter, a few results on the isoperimetric profile for small and large volumes are presented. The behavior of the profile for small volumes in compact manifolds was treated in an elementary way in Sect. 3.3 in Chap. 3, where a result by Bérard and Meyer [52] was proven, namely, that the isoperimetric profile for small volumes in compact manifolds is asymptotic to the one of Euclidean space.

In the first part of this chapter, we focus on isoperimetric sets of small volume in compact Riemannian manifolds. As an application of the implicit function theorem, Nardulli proved in [317] the uniqueness of *pseudo bubbles* with given center of mass and small volume in compact manifolds. A pseudo bubble is a radial graph over its center of mass, thus parameterized by $\mathbb{S}^{m-1}$, which mean curvature is a constant plus a eigenfunction of $\Delta_{\mathbb{S}^{m-1}} + 2$. This class includes all isoperimetric boundaries enclosing small volume by a result of Morgan and Johnson [311]. The uniqueness part of the implicit function theorem provides a symmetrization result: all isoperimetric sets of small volume are invariant by the subgroup of isometries of M fixing the center of mass of the set. The fact that isoperimetric sets of small volume are radial graphs allows us to recover an isoperimetric comparison result for small volumes involving the scalar curvature of the manifold proven by Druet [142]. It also yields a Maclaurin-Taylor approximation of the profile at $v = 0$.

In the second part of the chapter, we consider isoperimetric regions of large volume in Riemannian products $M \times \mathbb{R}^k$, where M is a compact Riemannian manifold. The case $k = 1$ was proven by Duzaar and Steffen [143], while the case $k > 1$ was obtained independently by Ritoré and Vernadakis [373] and Gonzalo [187].

M. Ritoré, *Isoperimetric Inequalities in Riemannian Manifolds*,
Progress in Mathematics 348, https://doi.org/10.1007/978-3-031-37901-7_7

7.1 A Symmetrization Result for Small Volumes

In this section, we prove that an isoperimetric set of sufficiently small volume in a compact Riemannian manifold M is invariant by the isotropy subgroup of $\mathrm{Isom}(M)$ fixing its center of mass.

7.1.1 The Center of Mass

The reference for this section is Karcher [242]. Let (X, μ) be a probability space, a measure space with volume 1, and $u : X \to B \subset M$ be a measurable map into an open convex ball B in a Riemannian manifold M. Here, the convexity of B means that two points in B can be joined by a unique length-minimizing geodesic in B. We assume that the radius of B is small enough so that $d_q^2(\cdot) = d^2(\cdot, q)$ is a strictly convex function on $\overline{B}$ for all $q \in \overline{B}$ (see, e.g., Lemma 4.1 in do Carmo [141]). In Cartan-Hadamard manifolds, we have convexity of balls for all radii and global strict convexity of the function d_q^2 for any point q.

We define a function $P_u : \overline{B} \to \mathbb{R}$ by

$$P_u(q) = \frac{1}{2} \int_X d^2(u(x), q)\, d\mu(x). \tag{7.1}$$

The function P_u is strictly convex. To check it, we take two different points $p, q \in \overline{B}$ and connect p to q with a geodesic $\gamma : [0, a] \to \overline{B}$. As $d^2(u(x), \cdot)$ is strictly convex for all $x \in X$, we have

$$d^2(u(x), \gamma(t)) < (1 - t)\, d^2(u(x), p) + t\, d^2(u(x), q)$$

for all $0 < t < 1$. Integrating over X, we get

$$P_u(\gamma(t)) < (1 - t)\, P_u(p) + t\, P_u(q)$$

for $0 < t < 1$, thus proving the strict convexity of P_u, that implies that the minimum of P_u in $\overline{B}$ is unique. This minimum is called the *center of mass* of u and is denoted by $c(u)$.

To check that $c(u)$ lies in the interior of the ball B, we take an arbitrary point $c \in \overline{B}$ and a smooth curve $\alpha : [0, \varepsilon) \to \overline{B}$ such that $\alpha(0) = c$. Let $\gamma : [0, d(u(x), c)] \to \overline{B}$ be the unit speed geodesic connecting $u(x)$ and q. Then

$$\begin{aligned} \frac{d}{dt}\bigg|_{t=0^+} \tfrac{1}{2}\, d^2(u(x), \alpha(t)) &= d(u(x), c)\, \langle \gamma'(d(u(x), c)), \alpha'(0)\rangle \\ &= -\langle \exp_c^{-1}(u(x)), \alpha'(0)\rangle. \end{aligned}$$

This implies that a minimum c of P_u cannot lie at the boundary ∂B since otherwise, choosing a curve α so that $\alpha(0) = c$ and $\alpha'(0)$ is the unit inner normal at c, we would had

$$\langle \exp_c^{-1}(u(x)), \alpha'(0)\rangle > 0$$

for all $x \in X$, and so

$$\frac{d}{dt}\bigg|_{t=0^+} P_u(\alpha(t)) = -\int_X \langle \exp_c^{-1}(u(x)), \alpha'(0)\rangle \, d\mu(x) < 0,$$

a contradiction to the fact that c is a minimum of P_u.

Similar computations imply that the gradient of P_u at an interior point $c \in B$ is given by

$$(\nabla P)_c = -\int_X \exp_c^{-1}(u(x)) \, d\mu(x).$$

Summarizing, we have the following result:

Proposition 7.1 *Let (X, μ) be a probability space and $u : X \to B \subset M$ a measurable map into an open convex ball B in a Riemannian manifold M where d^2 is strictly convex. Then there is a unique minimum $c(u) \in B$ of the function P_u defined in (7.1). Moreover, the point $c(u)$ satisfies*

$$\int_X \exp_{c(u)}^{-1}(u(x)) \, d\mu(x) = 0. \tag{7.2}$$

An estimate of the radius of a ball where d^2 is a convex function can be obtained from an upper bound on the sectional curvatures of M (e.g., Theorem 1.2.2 in Karcher [242]).

Usually, we shall take as $u : X \to B$ the inclusion of an open set or a hypersurface with its normalized Riemannian measure.

An important property is the compatibility of the center of mass with the action of the isometry group.

Proposition 7.2 *Let (X, μ) be a probability space and $u : X \to B \subset M$ a measurable map into an open convex ball B in a Riemannian manifold M where d^2 is strictly convex.. Let φ be an isometry of M. Then $\varphi \circ u : X \to \varphi(B)$ is measurable, $\varphi(B)$ is convex, and $\varphi(c(u)) = c(\varphi \circ u)$.*

Proof It follows from the equality $P_{\varphi \circ u}(\varphi(p)) = P_u(p)$ for all $p \in B$ since φ preserves the Riemannian distance. □

7.1.2 Blowing Up the Riemannian Metric at a Point

We start with a very useful Maclaurin-Taylor expansion of the squared modulus of a Jacobi field; see Proposition 2.7 in Chapter 5 of do Carmo [141] or Lemma 5.5 in Parker and Lee [264].

Lemma 7.3 *Let (M, g) be a Riemannian manifold, $p \in M$, and $v \in T_pM$, $w \in T_pM$ with $|w| = 1$. Consider the Jacobi field $J(r)$ along the geodesic $\gamma(r) = \exp_p(rv)$ with initial conditions $J(0) = 0$, $J'(0) = w$. Then we have*

$$|J(r)|^2 = r^2 - \frac{1}{3} R(w, v, v, w)\, r^4 - \frac{1}{6} (\nabla_v R)(v, w, w, v)\, r^5 + O(r^6), \tag{7.3}$$

where $O(r^k)$ denotes a function bounded in modulus by a constant times r^k.

Proof The proof is a simple calculation. Since $J(0) = 0$, $J'(0) = w$, we have, denoting by a prime both the derivative of a function with respect to r and the covariant derivative along $\gamma(r)$,

$$\begin{aligned}
\langle J, J\rangle(0) &= 0,\\
\langle J, J\rangle'(0) &= 2\,\langle J', J\rangle(0) = 0,\\
\langle J, J\rangle''(0) &= 2\,\langle J'', J\rangle(0) + 2\,\langle J', J'\rangle(0) = 2,\\
\langle J, J\rangle'''(0) &= 2\,\langle J''', J\rangle(0) + 6\,\langle J'', J\rangle(0) = 0.
\end{aligned}$$

Differentiating the Jacobi equation $J'' + R(J, \gamma')\gamma' = 0$ in (1.51)

$$J''' + (\nabla_{\gamma'} R)(J, \gamma')\gamma' + R(J', \gamma')\,\gamma' = 0$$

we have

$$\langle J, J\rangle^{\mathrm{iv}}(0) = 2\,\langle J^{\mathrm{iv}}, J\rangle(0) + 8\,\langle J''', J'\rangle(0) + 6\,\langle J'', J''\rangle(0) = -8\, R(w, v, v, w).$$

Differentiating twice the Jacobi equation

$$J^{\mathrm{iv}} + (\nabla^2_{\gamma'} R)(J, \gamma')\gamma' + 2(\nabla_{\gamma'} R)(J', \gamma')\gamma' + R(J'', \gamma')\gamma' = 0$$

we compute a last term

$$\begin{aligned}
\langle J, J\rangle^{\mathrm{v}}(0) &= 2\,\langle J^{\mathrm{v}}, J\rangle(0) + 10\langle J^{\mathrm{iv}}, J'\rangle(0) + 20\langle J''', J''\rangle(0)\\
&= -20\,(\nabla_v R)(w, v, v, w)
\end{aligned}$$

These equalities imply

$$|J(r)|^2 = |J'(0)|^2 r^2 - \frac{1}{3} R(J'(0), v, v, J'(0))\, r^4 \\ - \frac{1}{6} \nabla_v R(J'(0), v, v, J'(0))\, r^5 + O(r^6),$$

which is in turn (7.3). □

One of the first applications of Lemma 7.3 is the following asymptotic expansion for the volume and perimeter of a geodesic ball in a Riemannian manifold. Compare with Theorem 9.12 in Gray [188].

Lemma 7.4 *Let (M, g) be a Riemannian manifold, $p \in M$, and $r > 0$. Then we have the asymptotic expansions $A(p, r)$ and $V(p, r)$ for the perimeter and volume of the geodesic ball $B(p, r)$.*

$$A(p, r) = \mathbf{c}_{m-1} r^{m-1} \left(1 - \frac{1}{6m} \operatorname{scal}(p)\, r^2 + O(r^4)\right), \tag{7.4}$$

$$V(p, r) = \omega_m r^m \left(1 - \frac{1}{6\,(m+2)} \operatorname{scal}(p)\, r^2 + O(r^4)\right). \tag{7.5}$$

Proof Let UM be the unit sphere bundle. We parameterize the sphere $S(p, r)$ by the map $F_r : U_pM \to M$ defined by $F_r(v) = \exp_p(rv)$. We fix $v \in T_pM$ with $|v| = 1$, and we complete it to an orthonormal basis $e_1, \ldots, e_{m-1}, v$ of T_pM. We have

$$(dF_r)_v(e) = (d \exp_p)_{rv}(re) = J(r),$$

where J is the Jacobi field along the geodesic $\gamma_v(r) = \exp_p(rv)$ with initial conditions $J(0) = 0$, $J'(0) = e$. We consider the Jacobi fields $J_i(r)$ along the geodesic γ_r with initial conditions $J_i(0) = 0$, $J_i'(0) = e_i$. Then

$$\langle (dF_r)_v(e_i), (dF_r)_v(e_j) \rangle = \langle J_i(r), J_j(r) \rangle.$$

Formula $2 \langle J_i, J_j \rangle = |J_i + J_j|^2 - |J_i|^2 - |J_j|^2$ and (7.3) imply

$$\langle J_i(r), J_j(r) \rangle = r^2 \left(\delta_{ij} - \frac{1}{3} R(e_i, v, v, e_i)\, r^2 - \frac{1}{6} (\nabla_v R)(e_i, v, v, e_i)\, r^3 + O(r^4) \right).$$

We use the classical formula

$$\sqrt{\det(I + B)} = 1 + \tfrac{1}{2} \operatorname{trace}(B) + O(|B|^2).$$

By the area formula,

$$\begin{aligned}A(S(p,r)) &= \int_{U_pM} \operatorname{Jac} F_r \, d(U_pM) \\ &= \int_{U_pM} r^{m-1}\left(1 - \frac{1}{6}\operatorname{Ric}_p(v,v)\, r^2 - f(v)\, r^3 + O(r^4)\right) d(U_pM)(v),\end{aligned}$$

where

$$f(v) = -\frac{1}{12}\sum_{i=1}^{m-1}(\nabla_v R)(e_i, v, v, e_i).$$

The term in Ricci curvature is computed from

$$\int_{U_pM} \operatorname{Ric}_p(v,v)\, d\mathbb{S}^{m-1}(v) = \frac{\mathbf{c}_{m-1}}{m}\operatorname{scal}(p) \tag{7.6}$$

(see, e.g., p. 107 in do Carmo [141]). For the term in r^3, we have

$$\int_{U_pM} f(v)\, d(U_pM)(v) = 0$$

since $f(-v) = -f(v)$. This implies (7.4).

Finally, formula (7.5) for $V(p,r)$ is obtained by integrating $A(p,s)$ from 0 to r. □

Another application of Lemma 7.3 is the description of the blow-up of a Riemannian metric near a point in normal coordinates. Given a Riemannian manifold (M, g) and $p \in M$, recall that a *normal chart* is given by a set $V \subset T_pM$ containing 0 together with a basis $e_1, \ldots, e_m$ of T_pM such that $\exp_p : V \subset T_pM \to \exp_p(V)$ is a diffeomorphism.

Lemma 7.5 *Let (M, g) be a Riemannian manifold, $p \in M$, and $e_1, \ldots, e_m$ an orthonormal basis of T_pM. Assume that $\exp_p$ is a diffeomorphism on an open radial set $V \subset T_pM$. For $r \in \mathbb{R}$ and $v \in V_r = V/r$, we consider the map $F_r : V_r \to M$ defined by*

$$F_r(v) = \exp_p(rv).$$

Let $g_r = r^{-2}F_r^* g$. Then $\{g_r\}_{r\in\mathbb{R}}$ is a smooth family of Riemannian metrics on V_r, with $V_0 = T_pM$, such that g_0 is the Euclidean metric.

Proof The components of the Riemann tensor $F_r^* g$ at $v \in V_r$ are given by

$$\begin{aligned}
r^2 (g_r)_{ij}(v) &= \left(F_r^* g\right)_v \left(\frac{\partial}{\partial x_i}, \frac{\partial}{\partial x_j}\right) \\
&= g_{F_r(v)}\left((dF_r)_v\left(\frac{\partial}{\partial x_i}\right), (dF_r)_v\left(\frac{\partial}{\partial x_j}\right)\right) \\
&= g_{F_r(v)}\left((d\exp_p)_{rv}\left(r\frac{\partial}{\partial x_i}\right), (d\exp_p)_{rv}\left(r\frac{\partial}{\partial x_j}\right)\right) \\
&= g(J_i(r), J_j(r)),
\end{aligned}$$

where J_i is the Jacobi field along the geodesic $\gamma(r) = \exp_p(rv)$ with initial conditions $J_i(0) = 0$ and $J_i'(0) = (\partial/\partial x_i)_0 = e_i$. Using the Taylor expansion (7.3) together with the formula $2\langle J_i, J_j\rangle = |J_i + J_j|^2 - |J_i|^2 - |J_j|^2$, we get

$$g(J_i(r), J_j(r)) = r^2\left(\delta_{ij} - \frac{1}{3}R(e_i, v, v, e_j)\, r^2 + O(r^3)\right). \tag{7.7}$$

$$(g_r)_{ij}(v) = \delta_{ij} - \frac{1}{3}R(e_i, v, v, e_j)\, r^2 + O(r^3).$$

This proves the claim. □

Finally, as a consequence of Lemma 7.3 we obtain the following expression for the mean curvature of small geodesic spheres.

Lemma 7.6 *Let (M, g) be a Riemannian manifold, $p \in M$, and $v \in T_pM$ with $|v| = 1$. The mean curvature of the geodesic sphere $S(p, r)$ at $\exp_p(rv)$ is given by*

$$H_{S(p,r)}(\exp_p(rv)) = \frac{m-1}{r} - \frac{1}{3}\operatorname{Ric}_p(v, v)\, r + O(r^2). \tag{7.8}$$

Proof Let $e_1, \ldots, e_{m-1}, v$ be an orthonormal basis of T_pM. For $i = 1, \ldots, m-1$, consider the Jacobi field $J_i(r)$ along the geodesic $\gamma(r) = \exp_p(rv)$ with initial conditions $J_i(0) = 0, J_i'(0) = e_i$. We consider the square matrices $A(r) = \big(\langle J_i'(r), J_j(r)\rangle\big)$ and $B(r) = \big(\langle J_i(r), J_j(r)\rangle\big)$ or order $(m-1)$. Then

$$H_{S(p,r)}(\exp_p(rv)) = \operatorname{trace}\big(B(r)^{-1}A(r)\big).$$

Formula (7.7) implies the Taylor development

$$B(r) = I\, r^2 - \frac{1}{3}R\, r^4 + O(r^5), \tag{7.9}$$

where I is the identity matrix of order $(m-1)$ and $R = \big(R(e_i, v, v, e_j)\big)$. To obtain a similar development for $A(r)$, we take derivatives in (7.7) to obtain

$$\langle J_i'(r), J_j(r)\rangle + \langle J_i(r), J_j'(r)\rangle = 2I\,r - \frac{4}{3}\,R(e_i, v, v, e_i)\,r^3 + O(r^4).$$

Since

$$\langle J_i', J_j\rangle' - \langle J_i, J_j'\rangle' = -\langle R(J_i, \gamma')\gamma', J_j\rangle + \langle J_i, R(J_j, \gamma')\gamma'\rangle = 0,$$

and $\langle J_i', J_j\rangle(0) = \langle J_i, J_j'\rangle(0)$, we get

$$\langle J_i', J_j\rangle(r) = \langle J_i, J_j'\rangle(r)$$

for all r. We could have obtained this equality just by observing that $\langle J_i'(r), J_j(r)\rangle$ is the second fundamental form of $S(p, r)$ applied to the tangent vectors $J_i(r)$ and $J_j(r)$. So we get

$$A(r) = I\,r - \frac{2}{3}\,R\,r^3 + O(r^4). \tag{7.10}$$

Finally, we get from (7.9) and (7.10)

$$\begin{aligned} B(r)^{-1}A(r) &= \frac{1}{r^2}\Big(I + \frac{1}{3}\,R\,r^2 + O(r^3)\Big)\Big(I\,r - \frac{2}{3}\,R\,r^3 + O(r^4)\Big) \\ &= \frac{1}{r}\Big(I - \frac{1}{3}\,R\,r^2 + O(r^3)\Big). \end{aligned}$$

Taking traces, we obtain (7.8). □

7.1.3 *Pseudo Bubbles*

In this section, we consider a class of hypersurfaces containing the ones with constant mean curvature. For the most part, we follow Nardulli [317].

Let UM be the unit sphere bundle on the Riemannian manifold (M, g). For $p \in M$, $U_pM = \{v \in T_pM : |v| = 1\}$.

Definition 7.7 A hypersurface S in M is a *pseudo bubble* if:

1. There exist a point $p \in M$ and a smooth function u, defined on the unit sphere of T_pM, such that $S = \big\{\exp_p(u(v)\,v) : v \in U_pM\big\}$
2. The mean curvature of S is a constant plus an eigenfunction of eigenvalue 2 for the standard Laplacian on U_pM

Under conditions 1 and 2, we also say that S is a *pseudo bubble over the point* p. A hypersurface satisfying condition 1 in Definition 7.7 is called a radial graph. Of course, radial graphs with constant mean curvature are pseudo bubbles.

Now we prove the first result in this section.

Theorem 7.8 *Let* (M, g) *be a compact Riemannian manifold. Then there exist* $r_0 > 0$ *and a smooth function* $\Psi : (0, r_0) \times M \to UM$, *with* $\Psi(r, p) \in U_pM$, *such that*

$$S_{r,p} = \left\{ \exp_p \left(r(u + \Psi(r, p)(u))\, u \right) : u \in U_pM \right\}$$

is the unique pseudo bubble with center of mass p and mean curvature

$$\frac{m-1}{r} + H_r,$$

where H_r *is an eigenfunction of eigenvalue* 2 *of the standard Laplacian on* U_pM.

Moreover, for $r > 0$, *the volume of the domain* $\Omega_{r,p}$ *enclosed by the hypersurface* $S_{p,r}$ *is given by*

$$|\Omega_{r,p}| = \omega_m r^m h(r, p),$$

where $h(r, p)$ *is a smooth function with* $h(0, p) = 1$.

Proof We consider a local chart (V, φ) on M and let $\tau = \varphi^{-1}$. Let $E_1, \ldots, E_m$ be an orthonormal frame on V. We identify $\mathbb{R}^m$ with T_pM using the basis $(E_1)_p, \ldots, (E_m)_p$. To every $v = (v_1, \ldots, v_m) \in \mathbb{R}^m$, we associate the vector $\tilde{v} \in T_pM$ with coordinates $v_1, \ldots, v_m$ in the basis $(E_1)_p, \ldots, (E_m)_p$.

For any $x \in \mathbb{R}^m$, we define the map $F_{x,r} : \mathbb{R}^m \to M$ by

$$F_{x,r}(v) = \exp_{\tau(x)}(r\tilde{v}).$$

We take the Riemannian metric

$$g_{x,r} = r^{-2} F^*_{x,r} g \tag{7.11}$$

on $\mathbb{R}^m$. Observe that $g_{x,r}$ is a smooth family of metrics and that $g_{x,0}$ is the Euclidean metric by Lemma 7.5.

For a function $u \in C^2(\mathbb{S}^{m-1})$, the radial graph of u in $\mathbb{R}^m$ is

$$S_u = \left\{ v + u(v)\, v : v \in \mathbb{S}^{m-1} \right\}.$$

This is the normal graph over $\mathbb{S}^{m-1}$ of the function u.

We consider the function $H(r, x, u)(v)$, for $r \in \mathbb{R}$, $x \in \varphi(V)$, $u \in C^2(\mathbb{S}^{m-1})$, and $v \in \mathbb{S}^{m-1}$, defined as the mean curvature of S_u at the point $v + u(v)\, v$ with respect to the metric $g_{x,r}$. The function $H(r, x, u)$ is equal to r times the mean curvature of S_u with respect to the metric $F_{x,r}^* g$, which in turn coincides with the mean curvature of $S_{r,x,u} = F_{x,r}(S_u)$ with respect to the metric g. So we have

$$H(r, x, u)(v) = r \cdot H_{F_{x,r}(S_u)}(\exp_p(rv)).$$

Note that

$$F_{x,r}(S_u) = \left\{ \exp_{\tau(x)}(r\,(\tilde{v} + u(v)\,\tilde{v})) : v \in \mathbb{S}^{m-1} \right\}.$$

On the other hand, denote by $d\mathbb{S}_{x,r}^{m-1}$ the area element on $\mathbb{S}^{m-1}$ induced by the Riemannian metric $F_{x,r}^* g$. If we can properly define the center of mass of $S_{r,x,u}$, then $\tau(x)$ is the center of mass of $S_{r,x,u}$ if and only if

$$0 = \int_{\mathbb{S}^{m-1}} r\,(v + u(v)\,v)\, d\mathbb{S}_{x,r}^{m-1}(v).$$

We define

$$A(r, x, u) = \int_{\mathbb{S}^{m-1}} (v + u(v)\,v)\, d\mathbb{S}_{x,r}^{m-1}(v).$$

Consider on $\mathbb{S}^{m-1}$ the linear operator

$$L = -\left(\Delta_{\mathbb{S}^{m-1}} + (m-1)\right),$$

where $\Delta_{\mathbb{S}^{m-1}}$ is the standard Laplacian on $\mathbb{S}^{m-1}$. The operator L has an m-dimensional kernel given by the restrictions of the linear functions $L_a(x) = \langle x, a \rangle$, $x, a \in \mathbb{R}^m$, to $\mathbb{S}^{m-1}$:

$$\ker(L) = \left\{ L_a(x)|_{\mathbb{S}^{m-1}} : a \in \mathbb{R}^m \right\}.$$

Given the space $C^{k,\alpha}(\mathbb{S}^{m-1})$ of k differentiable functions whose k derivative satisfies an α-Hölder condition, $k \geqslant 0, \alpha \in (0,1)$, we consider the subspaces $C^{k,\alpha}(\mathbb{S}^{m-1})^\perp = C^{k,\alpha}(\mathbb{S}^{m-1}) \cap \ker(L)^\perp$. For any $k \geqslant 2$, the operator $L : C^{k,\alpha}(\mathbb{S}^{m-1})^\perp \to C^{k-2,\alpha}(\mathbb{S}^{m-1})^\perp$ is invertible.

Now consider an open subset $U \subset V$ and $r_0 > 0$ such that $B(q, r_0) \subset V$ for all $q \in U$, $B(q, r_0)$ is convex, and d_p^2 is strictly convex on $B(q, r_0)$ for all $p \in B(q, r_0)$. We define the map

$$\Phi : \left(-\tfrac{r_0}{2}, \tfrac{r_0}{2}\right) \times \varphi(U) \times C^{2,\alpha}(\mathbb{S}^{m-1}) \longrightarrow \mathbb{R}^m \times C^{0,\alpha}(\mathbb{S}^{m-1})^\perp \tag{7.12}$$

by

$$\Phi(r, x, u) = \left(A(r, x, u),\, Q(H(r, x, u) - (m-1))\right),$$

where Q is the L^2-orthogonal projection over $\ker(L)^{\perp}$. Note that Φ is just defined for functions u with $L_{\infty}(\mathbb{S}^{m-1})$-norm no larger than 1. This ensures that the center of mass of $S_{r,x,u}$ is well-defined. We compute

$$\frac{\partial \Phi}{\partial u}(0, x, 0) = \left(\frac{\partial A}{\partial u}(0, x, 0), Q\Big(\frac{\partial H}{\partial u}(0, x, 0)\Big)\right).$$

Since for $r = 0$ we have the flat metric on $\mathbb{R}^{m-1}$, Lemma 1.26 implies

$$\frac{\partial H}{\partial u}(0, x, 0) = L.$$

On the other hand,

$$\begin{aligned}\frac{\partial A}{\partial u}(0, x, 0)(w) &= \frac{d}{dt}\Big|_{t=0} \int_{\mathbb{S}^{m-1}} (v + tw(v)\, v)\, d\mathbb{S}^{m-1}(v)\\ &= \int_{\mathbb{S}^{m-1}} w(v)\, v\, d\mathbb{S}^{m-1}(v),\end{aligned}$$

where $d\mathbb{S}^{m-1}$ is the standard metric in $\mathbb{S}^{m-1}$.

Let us see that $(\partial\Phi/\partial u)(0, x, 0)$ is injective. If

$$\int_{\mathbb{S}^{m-1}} w(v)\, v\, d\mathbb{S}^{m-1}(v) = 0, \qquad Q(L(w)) = 0,$$

then the first condition implies that $w \in \ker(L)^{\perp}$ and the second one that $L(w)$ is a linear combination of functions in $\ker(L)$. Hence, $w = 0$. If $(c, w) \in \mathbb{R}^m \times C^{0,\alpha}(\mathbb{S}^m)^{\perp}$, then there exists a unique $v \in C^{2,\alpha}(\mathbb{S}^{m-1})^{\perp}$ such that $L(v) = w$ by Fredholm alternative. Note that $Q(L(v)) = w$. On the other hand, $v'(x) = v(x) + \langle x, a\rangle$ also satisfies the equation $Q(L(v')) = 0$, and a can be adjusted so that

$$\int_{\mathbb{S}^{m-1}} \langle x, a\rangle\, x\, d\mathbb{S}^m(x) = c - \int_{\mathbb{S}^{m-1}} v(x)\, x\, d\mathbb{S}^{m-1}(x).$$

Hence, the implicit function theorem ensures the existence of $r_0 > 0$, $U' \subset U$, and a function $u(r, x)$ with $u(0, x) = 0$ such that $S_{r,x,u(r,x)}$ has center of mass x and mean curvature

$$\frac{m-1}{r} + \text{first eigenfunction of } L$$

for $r \neq 0$. The result is proven by covering M by a finite number of sets U' taking as Ψ the function defined by $\Psi(r, p) = u(r, \varphi(p))$.

To compute the volume $|\Omega_{r,p}|$ of the domain enclosed by $S_{r,p}$, we take into account that this volume equals r^m times the volume of

$$\{(v + u(r, x)\, v)\, v : v \in \mathbb{S}^{m-1}\}$$

with respect to the metric $g_{x,r}$ for $x = \varphi^{-1}(p)$, which is equal to ω_m for $r = 0$. □

Let us see now that we can replace the radius r in Theorem 7.8 by the volume enclosed by the pseudo bubble.

Theorem 7.9 *Let (M, g) be a compact Riemannian manifold. Then there exist $v_0 > 0$ and a smooth function $\Theta : (0, v_0) \times M \to UM$, with $\Theta(v, p) \in U_pM$, such that*

$$S_{v,p} = \{\exp_p\left(r(u + \Theta(v, p)(u))\, u : u \in U_pM\right\}$$

is the unique pseudo bubble with center of mass p, positive mean curvature, and enclosing volume v.

Proof Let $\Omega_{r,p}$ be the domain enclosed by the pseudo bubble $S_{r,p}$. By Theorem 7.8, $|\Omega_{r,p}| = \omega_m r^m h(r, p)$, where $h(r, p)$ is a smooth function with $h(0, p) = 1$. Then

$$|\Omega_{r,p}|^{1/m} = r\, \omega_m^{1/m}\, h(r, p)^{1/m}.$$

We conclude that

$$\frac{d}{dr}\bigg|_{r=0} |\Omega_{r,p}|^{1/m} = \omega_m^{1/m}\, h(0, p)^{1/m} = \omega_m^{1/m},$$

so that for small $v > 0$, there is a unique $r > 0$ such that $|\Omega_{v(r),p}| = v$. □

7.1.4 Isoperimetric Sets of Small Volume

We prove in this section that isoperimetric sets of small volume in a compact manifold are indeed pseudo bubbles over their center of mass.

Theorem 7.10 *Let (M, g) be a compact Riemannian manifold. Then there exists $v_0 > 0$ such that any isoperimetric set of volume $v \in (0, v_0)$ is a pseudo bubble over its center of mass.*

Proof Let $E \subset M$ be an isoperimetric set of volume $0 < v < |M|$. We let S be the regular part of its boundary, S_0 its singular set, and H the mean curvature of S with respect to the outer normal. Assume that $\operatorname{Ric} \geqslant (m - 1)\,\kappa$ for some $\kappa \in \mathbb{R}$. Since every point in E is at minimum distance of some point in ∂E that belongs to the regular set S (its tangent cone is contained in a half-space), we can estimate the volume of E by

$$|E| \leqslant P(E) \int_0^r \left(\cos_\kappa(t) - \frac{H}{m-1}\sin_\kappa(t)\right)^{m-1} dt, \tag{7.13}$$

estimating the normal Jacobian by (1.62). Recall that the functions $\cos_\kappa$ and $\sin_\kappa$ are defined in (1.47) and (1.48). The value r is the inradius of E, the radius of the largest ball contained in E.

We apply (7.13) to the complementary region $M \setminus E$, bounded by S, with mean curvature $-H$ with respect to the outer normal to get

$$|M| - |E| \leqslant P(E) \int_0^{r'} \left(\cos_\kappa(t) + \frac{H}{m-1}\sin_\kappa(t)\right)^{m-1} dt.$$

We conclude that $H \to \infty$ when $|E| \to 0$ since $P(E) \to 0$.

Applying (7.13) to E, we get

$$|E| \leqslant C_1 r\, P(E),$$

for some constant C_1. On the other hand, $r \leqslant \cot_\kappa^{-1}(H/(m-1))$, where $\cot_\kappa^{-1}$ is the inverse function of $\cot_\kappa = \cos_\kappa / \sin_\kappa$, so that

$$r \leqslant \frac{C_2}{H}.$$

Hence, we get

$$H\,|E| \leqslant C_3 P(E). \tag{7.14}$$

We notice now that the boundary of E, and hence the boundary of E, must be connected. Since H is large, by (1.24), we have

$$\mathrm{Ric}(N,N) + |\sigma|^2 \geqslant \mathrm{Ric}(N,N) + \frac{H^2}{m-1}$$

and so $\mathrm{Ric}(N,N) + |\sigma|^2$ is strictly positive on every component of ∂E. We take a locally constant function non-zero function on every connected component of ∂E, and we multiply it by the functions obtained in Lemma 1.61. Adjusting the resulting function to have mean zero, we conclude from the second variation formula that ∂E must be connected. So E is connected.

Observe now that for small volumes, we have

$$C^{-1} v^{(m-1)/m} \leqslant I(v) \leqslant C\, v^{(m-1)/m}. \tag{7.15}$$

The first inequality follows from Theorem 3.7 and the second one by comparing with the perimeter of small balls of volume v centered at a given point of M.

Consider now a sequence of isoperimetric sets E_i with volumes $v_i \to 0$ and mean curvatures H_i. We consider the sequence $g_i = v_i^{-m} g$ of Riemannian metrics on M so that the sets E_i now have volume 1, perimeter equal to $v_i^{-(m-1)/m} P(E_i) \leqslant C$ by (7.15), and mean curvatures

$$v_i^m H_i \leqslant \frac{C_3 P(E_i)}{v_i} v_i^{(m-1)/m} \leqslant C_3 C$$

by (7.14) and (7.15).

By Nash embedding theorem [320], we assume that (M, g) is isometrically embedded in some Euclidean space $\mathbb{R}^k$. The manifolds (M, g_i) are just embedded in $\mathbb{R}$ as dilations of (M, g). The mean curvatures of the sets E_i are uniformly bounded and also the perimeters so that the sets E_i are contained inside balls of fixed radius by the monotonicity formula for varifolds; see 5.1(3) in Allard [11]. We may assume they are centered at the origin by translation and that each scaling of M is tangent to $\mathbb{R}^m \times \{0\}$. By compactness, the sets E_i converge to a solution in $\mathbb{R}^m \times \{0\}$, a ball of unit volume. Since the mean curvatures are bounded, $C^{1,\alpha}$ is obtained by Allard's regularity theorem; see Section 8 in [11]. The minimizers are then free of singular points and close in $C^{1,\alpha}$ topology to a round sphere.

Higher regularity convergence is obtained by expressing the difference of the immersions in non-parametric form (e.g., see Theorems 2.1 and 2.2 in Osserman [324]) as solutions of uniformly elliptic systems and applying Schauder's estimates; see Chapter 3, 1.2, and 1.3 in Ladyzhenskaya and Ural'tseva [257]. We refer the reader to page 1022 in Morgan and Johnson for a few more details [311].

It remains to prove that for large i, ∂E_i is a radial graph over its center of mass, but this follows easily from the higher-order convergence and the fact that the center of masses of the rescaled ∂E_i converges to the center of mass of the ball. □

7.1.5 *Proof of the Symmetry Result*

Finally, we have the following:

Theorem 7.11 *Let (M, g) be a compact Riemannian manifold. Then there exists $v_0 > 0$ such that any isoperimetric set of volume $v \in (0, v_0)$ is invariant by the isometries of M fixing its center of mass.*

Proof By Theorem 7.10, there exists $v_0 > 0$ such that the boundary S of an isoperimetric set $E \subset M$ of volume $v \in (0, v_0)$ is a pseudo bubble over its center of mass p. Reduce v_0 if necessary so that Theorem 7.9 applies. Let $f : M \to M$ be any isometry with $f(p) = p$. Then $f(S)$ is a pseudo bubble over $f(p) = p$, and $f(S)$ encloses the set $f(E)$ of volume v. By the uniqueness part of Theorem 7.9, we have $f(S) = S$ and so $F(E) = E$ as claimed. □

Although we have stated all the results in Sect. 7.9 for compact Riemannian manifolds, all proofs in the previous sections apply without changes to the case of a Riemannian manifold with a cocompact action of the isometry group. We only need to modify slightly the proof of Theorem 7.10 to show that isoperimetric sets of small volume have large constant mean curvature. We have the following:

Theorem 7.12 *Let (M, g) be a complete Riemannian manifold with a cocompact action of its isometry group. Then there exists $v_0 > 0$ such that any isoperimetric set of volume $v \in (0, v_0)$ is invariant by the isometries of M fixing its center of mass.*

7.1.6 Asymptotic Expansion of Area and Volume of Pseudo Bubbles

In this section, we prove an asymptotic expansion for the area of pseudo bubbles and the volume enclosed. Let us prove first the following:

Lemma 7.13 *Let $p \in M$, and, for $r > 0$ small enough, let $S_{r,p}$ be the unique pseudo bubble with center of mass p and mean curvature equal to $(m-1)/r$ plus a spherical harmonic. Let $\Omega_{r,p}$ be the region enclosed by $S_{r,p}$. Then we have*

$$A(S_{r,p}) = \mathfrak{c}_{m-1} r^{m-1} \left(1 - \frac{1}{2m} \operatorname{scal}(p)\, r^2 + O(r^4)\right), \tag{7.16}$$

$$|\Omega_{r,p}| = \omega_m r^m \left(1 - \frac{m+1}{2\,(m-1)(m+2)} \operatorname{scal}(p)\, r^2 + O(r^4)\right), \tag{7.17}$$

where scal *is the scalar curvature of (M, g).*

Moreover, from (7.16) and (7.17), we obtain the following expansion for the area of a pseudo bubble in terms of the volume

$$A(S_{p,v}) = \frac{\mathfrak{c}_{m-1}}{\omega_m^{(m-1)/m}}\, v^{(m-1)/m} \left(1 - \frac{1}{2n(n+2)} \operatorname{scal}(p)\, \frac{v^{2/m}}{\omega_m^{2/m}} + O(v^{4/m})\right). \tag{7.18}$$

Proof We use the same notation as in the proof of Theorem 7.8. The local chart (V, φ) contains p and $E_1, \ldots, E_m$ is a local orthonormal frame on V. The pseudo bubble $S_{r,p}$ is the radial graph of the function $u(r, \varphi(p))$, where $u(r, x)$ is obtained from an application of the implicit function theorem to the map Φ in (7.12) defined by

$$\Phi(r, x, u) = \big(A(r, x, u),\ Q(H(r, x, u) - (m-1))\big).$$

The function A is equal to

$$A(r, x, u) = \int_{\mathbb{S}^{m-1}} (v + u(v)\, v)\, d\mathbb{S}^{m-1}_{x,r}(v),$$

the area element $d\mathbb{S}^{m-1}_{x,r}$ on $\mathbb{S}^{m-1}$ is induced by the Riemannian metric $g_{x,r} = F^*_{x,r}\, g$ defined in (7.11), Q is the L^2-orthogonal projection over $\ker(L)^\perp$, and

$$H(x, r, u) \text{ is the mean curvature of } v \in \mathbb{S}^{m-1} \mapsto v + u(v)\, v$$

with respect to the metric $g_{x,r}$. Recall that the metric $g_{x,r}$ is defined on $\mathbb{R}^m$ via the exponential map $F_{x,r}((v_1, \ldots, v_m)) = \exp_{\varphi^{-1}(x)} \left(\sum_{i=1}^m v_i (E_i)_{\varphi^{-1}(x)} \right)$.

Since we have $u(0, x) = 0$ for all $x \in \varphi(V)$, we write

$$u(r, x) = r u_1 + r^2 u_2 + O(r^3). \tag{7.19}$$

Let us prove first

$$u(r, x) = r^2 u_2 + O(r^3), \qquad u_2 \in \ker(L)^\perp, \qquad L(u_2)(v) = -\frac{1}{3} \operatorname{Ric}_p(v, v). \tag{7.20}$$

The metric $g_{x,r}$ has asymptotic expansion

$$B(r)_{ij} = \left(g_{x,r}\right)_{ij} = \delta_{ij} - \frac{1}{3} R(e_i, v, v, e_j)\, r^2 + O(r^3)$$

at v (here, $e_1, \ldots, e_{m-1}$ are orthonormal and perpendicular to v). Since $B'(0) = 0$, we get

$$\begin{aligned} d\mathbb{S}^{m-1}_{x,r}(v) &= \sqrt{\det B(r)}\, d\mathbb{S}^{m-1} \\ &= \left(1 + \frac{1}{2} \operatorname{trace}(B''(0))\, r^2 + O(r^3)\right) d\mathbb{S}^{m-1} \\ &= \left(1 - \frac{1}{3} \operatorname{Ric}_p(v, v)\, r^2 + O(r^3)\right) d\mathbb{S}^{m-1}. \end{aligned} \tag{7.21}$$

From $A(r, x, u(r, x)) = 0$ and (7.19) and (7.21), we get

$$\begin{aligned} &\int_{\mathbb{S}^{m-1}} u_1(v)\, v\, d\mathbb{S}^{m-1}(v) = 0, \\ &\int_{\mathbb{S}^{m-1}} u_2(v)\, v\, d\mathbb{S}^{m-1}(v) = \frac{1}{3} \int_{\mathbb{S}^{m-1}} \operatorname{Ric}_p(v, v)\, v\, d\mathbb{S}^{m-1}(v) = 0. \end{aligned} \tag{7.22}$$

On the other hand, we have that $Q(\Phi(r, x, 0))$ is the projection to $\ker(L)^{\perp}$ of the mean curvature of the geodesic sphere $S(x, r)$ with respect to the metric $g_{x,r}$. This means

$$Q(\Psi(r, x, 0))(v) = (m-1) - \frac{1}{3}\operatorname{Ric}_p(v, v)\, r^2 + O(r^3).$$

Taking derivative with respect to r of the expression

$$0 = Q(\Psi(r, x, u(r, x)) = Q(\Psi(r, x, ru_1 r^2 u_2 + O(r^3))$$

and evaluating at $r = 0$, we get

$$0 = \frac{\partial(Q \circ \Psi)}{\partial r}(0, x, 0) + \frac{\partial(Q \circ \Psi)}{\partial u}(0, x, 0)\, u_1 = L(u_1).$$

Since $L(u_1) = 0$, $u_1 \in \ker(L)$, but the first equation in (7.22) implies $u \in (\ker L)^{\perp}$. So $u_1 = 0$. Taking a second derivative with respect to r and evaluating at $r = 0$, we get

$$0 = \frac{\partial^2(Q \circ \Psi)}{\partial r^2} + L(2u_2) = -\frac{2}{3}\operatorname{Ric}_p(v, v) + L(u_2).$$

We conclude that

$$u_2 \in \ker(L)^{\perp}, \qquad L(u_2) = \frac{1}{3}\operatorname{Ric}_p(v, v).$$

This proves (7.20). Since $L = -\Delta_{\mathbb{S}^{m-1}} - (m-1)$, integrating the last equality, we get

$$\int_{\mathbb{S}^{m-1}} (m-1)\, u_2\, d\mathbb{S}^{m-1} = -\frac{1}{3}\int_{\mathbb{S}^{m-1}} \operatorname{Ric}_p(v, v)\, d\mathbb{S}^{m-1}(v). \tag{7.23}$$

Let us now prove the asymptotic expressions (7.16) and (7.17) for area and volume. The pseudo bubble $S_{r,p}$ is parametrized by the map $F : U_pM \to M$ defined by

$$F(v) = \exp_p(t(v)\, v), \quad v \in U_pM, \quad t(v) = r\,(1 + u(r, x)(v)),$$

the composition of $v \mapsto t(v)\, v$ with $\exp_p$. From now on, we let $u = u(x, r)$. If $e \perp v$ and $\alpha(s)$ is a curve with $\alpha(0) = v$, $\alpha'(0) = e$, then

$$(dF)_{t(v)v}(e) = \left(d\exp_p\right)_{t(v)\, v}\left(t(v)\left(\frac{e(u)}{t(v)}\, v + e\right)\right) = J(t(v)),$$

where J is the Jacobi field with initial conditions $J(0) = 0$ and $J'(0) = (e(u)/t(v))\, v + e$.

We take $e_1, \ldots, e_{m-1}$ in $T_v U_p M$ an orthonormal set perpendicular to v and the Jacobi field J_i along the geodesic $\exp_p(s\, t(v)v)$ with initial conditions $J_i(0) = 0$, $J_i'(0) = (e_i(u)/t(v))\, v + e_i$. Then

$$\langle J_i(t), J_j(t)\rangle = t^2\Big(\delta_{ij} + \frac{e_i(u)e_j(u)}{(1+u)^2} - \frac{1}{3}\, R(e_i, v, ve_j)\, t^2 + O(t^3)\Big).$$

As $t = r(1+u) = r(1 + r^2 u_2 + O(r^3))$, we get

$$\begin{aligned}\det\big(\langle J_i(t), J_j(t)\rangle\big)^{1/2} &= r^{m-1}\big(1 + (m-1)u_2 r^2 + O(r^3)\big)\\ &\quad \big(1 - \frac{1}{6}\operatorname{Ric}_p(v,v) r^2 + O(r^3)\big)\\ &= r^{m-1}\big(1 + \big((m-1)\, u_2 - \frac{1}{3}\operatorname{Ric}_p(v,v)\big)\, r^2 + O(r^3)\big).\end{aligned}$$

Integrating over $U_p M \equiv \mathbb{S}^{m-1}$, using (7.23) and formula (7.6), we get (7.16).

As for the volume $|\Omega_{r,p}|$, we integrate the volume element of M in polar coordinates around p. Using $u = u_2 r^2 + O(r^3)$ and Eq. (7.23), we get

$$\begin{aligned}|\Omega_{r,p}| &= \int_{U_p M}\left\{\int_0^{r(1+u)} s^{m-1}\Big(1 - \frac{1}{6}\operatorname{Ric}_p(v,v)\, s^2 + O(s^4)\Big)\, ds\right\} d(U_p M)\\ &= \int_{U_p M}\left\{\frac{r^m}{m} + \big(m u_2 - \frac{1}{6(m+2)}\operatorname{Ric}_p(v,v)\big)\, r^{m+2} + O(r^{m+4})\right\} d(U_p M)\\ &= \omega_m r^m - \frac{m+1}{2(m-1)(m+2)}\Big(\int_{U_p M}\operatorname{Ric}_p(v,v)\, d(U_p M)\Big)\, r^{m+2} + O(r^{m+4}).\end{aligned}$$

Finally, Eq. (7.6) implies (7.17).

To prove the expansion (7.18) in terms of the volume, we use (7.16) and (7.17). We let $A(r) = A(S_{p,r})$ and $V(r) = |\Omega_{p,r}|$ and write

$$A(r) = \mathbf{c}_{m-1} r^{m-1}\big(1 - \alpha_m r^2 + O(r^4)\big), \quad V(r) = \omega_m r^m\big(1 - \gamma_m r^2 + O(r^4)\big), \tag{7.24}$$

where

$$\alpha_m = \frac{1}{2m}\operatorname{scal}(p), \quad \gamma_m = \frac{m+1}{2(m-1)(m+2)}\operatorname{scal}(p).$$

Using the expansion $(1+x)^{1/m} = 1 + \frac{1}{m}x + O(x^2)$, we get

$$\begin{aligned} w = v^{1/m} &= \omega_m^{1/m} r \left(1 - \alpha_m r^2 + O(r^4)\right)^{1/m} \\ &= \omega_m^{1/m} r \left(1 - \frac{\gamma_m}{m} r^2 + O(r^4)\right). \end{aligned}$$

Writing $r(w) = \alpha w + \beta w^2 + \delta w^3 + \rho w^4 + O(w^5)$ and replacing in the previous expression, we get

$$\begin{aligned} w &= \omega_m^{1/m} \left(\alpha w + \beta w^2 + \delta w^3 + \rho w^4 + O(w^5)\right)\left(1 - \frac{\gamma_m}{m} (\alpha w + \beta w^2)^2 + O(w^5)\right) \\ &= \omega_m^{1/m} \left(\alpha w + \beta w^2 + \left(\delta - \frac{\gamma_m}{m}\alpha^3\right) w^3 + \left(\rho - \frac{\gamma_m}{m} 2\alpha\beta\right) w^4 + O(w^5)\right). \end{aligned}$$

So we get

$$\alpha = \omega_m^{-1/m}, \quad \beta = 0, \quad \delta = \frac{\gamma_m}{m} \omega_m^{-3/m}, \quad \rho = 0,$$

and

$$r(w) = \omega_m^{-1/m} w \left(1 + \frac{\gamma_m}{m} \omega_m^{-2/m} w^2 + O(w^4)\right).$$

Finally, we get, since $w = v^{1/m}$ and $(1+x)^{m-1} = 1 + (m-1)x + O(x^2)$,

$$\begin{aligned} A(r(v)) &= \frac{\mathbf{c}_{m-1}}{\omega_m^{(m-1)/m}} v^{(m-1)/m} \left(1 + (m-1) \frac{\gamma_m}{m} \omega_m^{-2/m} v^{2/m} + O(v^{4/m})\right) \times \\ &\qquad \times \left(1 - \alpha_m \omega_m^{-2/m} v^{2/m} + O(v^{4/m})\right) \\ &= \frac{\mathbf{c}_{m-1}}{\omega_m^{(m-1)/m}} v^{(m-1)/m} \left(1 + \left(\frac{(m-1)\,\gamma_m}{m} - \alpha_m\right) \frac{v^{2/m}}{\omega_m^{2/m}}\right) + O(v^{4/m}). \end{aligned}$$

As

$$\frac{(m-1)\,\gamma_m}{m} - \alpha_m = -\frac{1}{2m\,(m+2)} \operatorname{scal}(p),$$

we get (7.18). □

Remark 7.14 The Taylor developments for the area and volume of geodesic spheres in terms of the radius were given in Lemma 7.4 and are equal to

$$A(p,r) = \mathbf{c}_{m-1} r^{m-1}\left(1 - \frac{1}{6m}\ \mathrm{scal}(p)\, r^2 + O(r^4)\right),$$

$$V(p,r) = \omega_m r^m\left(1 - \frac{1}{6\,(m+2)}\ \mathrm{scal}(p)\, r^2 + O(r^4)\right).$$

So formulas (7.24) are satisfied with coefficients

$$\alpha_m = \frac{1}{6m}\ \mathrm{scal}(p), \qquad \gamma_m = \frac{1}{6(m+2)}.$$

The arguments after formulas (7.24) show that

$$A(p,v) = \frac{\mathbf{c}_{m-1}}{\omega_m^{1/m}}\, v^{(m-1)/m}\left(1 + \left(\frac{(m-1)\,\gamma_m}{m} - \alpha_m\right)\frac{v^{2/m}}{\omega_m^{2/m}}\right) + O(v^{4/m}).$$

In this case,

$$\frac{(m-1)\,\gamma_m}{m} - \alpha_m = -\frac{1}{2m(m+2)}\ \mathrm{scal}(p).$$

So for geodesic spheres, we have formally the same expansion for the area in terms of the volume as the one in (7.18) for pseudo bubbles.

7.1.7 Isoperimetric Sets and Scalar Curvature

We prove in this section two results concerning the relation between isoperimetric sets of small volume and the scalar curvature. The first one indicates that isoperimetric solutions concentrate near the maxima of the scalar curvature. The second one is a comparison result for isoperimetric sets of small volume under an strict upper bound on the scalar curvature.

Theorem 7.15 *Let (M, g) be a compact manifold and $\{E_i\}_{i\in\mathbb{N}}$ a sequence of isoperimetric sets with $|E_i| \to 0$. For i large, let p_i be the center of mass of E_i. Then $\{p_i\}_{i\in\mathbb{N}}$ converges to a maximum of the scalar curvature of M.*

Proof For i large, we know that the boundaries ∂E_i are pseudo bubbles by Theorem 7.10. If p_i does not converge to a maximum of the scalar curvature, there exists a subsequence of $\{p_i\}_{i\in\mathbb{N}}$ which converges to a point $p \in M$ satisfying $\mathrm{scal}(p) < \sigma = \max_M \mathrm{scal}$. For simplicity, we assume that the whole sequence converges to p. Then $\mathrm{scal}(p_i) < \sigma$ for i large enough. Let $p_{\max}$ be a minimum of the scalar curvature and F_i be a pseudo bubble with center of mass $p_{\max}$ and volume v_i. By (7.18), we get

$$\begin{aligned} P(E_i) &= \frac{\mathbf{c}_{m-1}}{\omega_m^{(m-1)/m}}\, v_i^{(m-1)/m}\left(1 - \frac{1}{2n(n+2)}\ \mathrm{scal}(p)\,\frac{v_i^{2/m}}{\omega_m^{2/m}} + O(v_i^{4/m})\right) \\ &> \frac{\mathbf{c}_{m-1}}{\omega_m^{(m-1)/m}}\, v_i^{(m-1)/m}\left(1 - \frac{1}{2n(n+2)}\ \sigma\,\frac{v_i^{2/m}}{\omega_m^{2/m}} + O(v_i^{4/m})\right) \\ &= P(F_i) \end{aligned}$$

for i large enough. This contradiction shows that $\mathrm{scal}(p_i) \to \sigma$. □

Now we prove an isoperimetric comparison result for small volumes when scal $< m(m-1)\,K_0$. It was proven by Druet; see Theorem 2 in [142].

Theorem 7.16 *Let (M, g) be a compact Riemannian manifold with scalar curvature* scal *satisfying the inequality* scal $< m(m-1)\,K_0$. *Then there exists $v_0 > 0$ such that, for any set $E \subset M$ of finite perimeter with volume $|E| < v_0$, we have*

$$P(E) > P(B),$$

where B is the geodesic ball of volume $|E|$ in the space form of constant curvature K_0.

Proof Let $\sigma = \max_M$ scal. For any pseudo bubble S with center of mass p and volume v, we have

$$A(S) = \frac{\mathbf{c}_{m-1}}{\omega_m^{(m-1)/m}}\, v^{(m-1)/m}\left(1 - \frac{1}{2n(n+2)}\ \mathrm{scal}(p)\,\frac{v^{2/m}}{\omega_m^{2/m}} + O(v^{4/m})\right),$$

with a remainder $O(v^{4/m})$ uniform on M. On the other hand, for a geodesic ball of volume v in the space form $\mathbb{M}^m_{K_0}$ of constant sectional curvature K_0, we have

$$P(B) = \frac{\mathbf{c}_{m-1}}{\omega_m^{(m-1)/m}}\, v^{(m-1)/m}\left(1 - \frac{1}{2n(n+2)}\, m(m-1)\,K_0\,\frac{v^{2/m}}{\omega_m^{2/m}} + O(v^{4/m})\right).$$

Comparing the formulas for $A(S)$ and $P(B)$, we immediately obtain the existence of a constant $v_0 > 0$ such that $A(S) > P(B)$ for any pseudo bubble S in M enclosing a region of volume $v < v_0$ and a geodesic ball. In particular, since the boundaries of isoperimetric sets of small enough volume are pseudo bubbles, the result follows.

□

7.2 Large Isoperimetric Sets in $M \times \mathbb{R}$

Given a compact Riemannian manifold (M, g), we see in Theorem 7.19 in this section that isoperimetric sets of large volume in the Riemannian cylinder $M \times \mathbb{R}$ are slabs of the form $M \times [a, b]$, where $[a, b] \subset \mathbb{R}$ is a closed interval. The Riemannian metric in $M \times \mathbb{R}$ is the product metric $g \times g_0$, where g_0 is the standard metric on $\mathbb{R}$. We notice that the result is obviously false when M is not compact. We shall need the following preliminary results in the proof of Theorem 7.19.

Lemma 7.17 *Let (M, g) be a compact manifold. Then there exist constants $c_1, c_2 > 0$, only depending on M, such that, for any set $E \subset M$ of finite perimeter with $|E| \leqslant |M|/2$, we have:*

1. $P(E) \geqslant c_1|E|$
2. $P(E) \geqslant c_2|E|^{(m-1)/m}$

Proof It follows easily since the isoperimetric profile of (M, g) is strictly positive and asymptotic to the function $t \mapsto t^{(m-1)/m}$ for $t > 0$ small. □

A set $E \subset N = M \times \mathbb{R}$ is called *normalized* if the intersection

$$E_p = E \cap (\{p\} \times \mathbb{R})$$

is either empty or a vertical segment centered at $(p, 0)$ for all $p \in M$. Notice that a normalized set is invariant by the reflection σ in $M \times \mathbb{R}$ defined by $\sigma(p, t) = (p, -t)$, which is an isometry of $(M \times \mathbb{R}, g \times g_0)$.

Given any set $E \subset N$, and $t \in \mathbb{R}$, we denote the horizontal section E_t by

$$E_t = \{p \in M : (p, t) \in E\}.$$

Note that E_t is a subset of M, and also notice that for normalized sets, one has $E_t \subset E_s$ whenever $|s| \leqslant |t|$. In particular, E_0 is the largest horizontal section, and so $E_0 = \pi(E)$, where $\pi : M \times \mathbb{R} \to \mathbb{R}$ is the projection over the second factor. We denote by P and $|\cdot|$ the perimeter and volume in the manifold $(M \times \mathbb{R}, g \times g_0)$.

Let us prove first the following key linear isoperimetric inequality:

Lemma 7.18 *If $E \subset N$ is a normalized isoperimetric set and $|M \setminus E_0| > 0$, then there exists a constant $c > 0$ independent of $|E|$ such that*

$$P(E) \geqslant c\,|E|. \tag{7.25}$$

Proof As E is normalized, we can choose $\tau \geqslant 0$ so that $|E_t| \leqslant |M|/2$ for all $t \geqslant \tau$ and $|E_t| > |M|/2$ for all $t \in [0, \tau)$ in case $\tau > 0$.

Let us consider first the case $\tau > 0$.

We apply the coarea formula and Lemma 7.17(1) to obtain

$$\begin{aligned} P(E) \geqslant P(E, M \times (\tau, \infty)) &\geqslant \int_\tau^\infty P(E_s) ds \\ &\geqslant c_1 \int_\tau^\infty \mathcal{H}^m(E_s)\, ds \\ &= c_1 \left| E \cap (M \times (\tau, \infty)) \right|. \end{aligned} \tag{7.26}$$

On the other hand, for $t \in [0, \tau)$, we claim that inequality

$$\mathcal{H}^m(M \setminus E_t) \geqslant P(E, M \times (0, t)) \tag{7.27}$$

holds. Otherwise, we would have

$$\begin{aligned} \mathcal{H}^m(M) &= \mathcal{H}^m(M \setminus E_t) + \mathcal{H}^m(E_t) \\ &< P(E, M \times (0, t)) + P(E, M \times (t, \infty)) \\ &\leqslant P(E)/2. \end{aligned}$$

This is a contradiction since comparison of E with a slab $M \times [a, b]$ of the same volume implies that $P(E) \leqslant P(M \times [a, b]) = 2|M|$. This proves (7.27).

We consider the non-increasing function $y(t) = \mathcal{H}^m(M \setminus E_t)$. Using the coarea formula and Lemma 7.17(2), we may rewrite inequality (7.27) as

$$y(t) \geqslant c_2 \int_0^t y(s)^{(n-1)/n} ds.$$

As $y(t) > 0$ for all $t \in [0, \tau)$, we have

$$y(t) \geqslant \left(\frac{c_2}{m} \right)^m t^m.$$

In particular, taking limits when $t \to \tau^-$ and using that $y(t)$ is non-decreasing,

$$\mathcal{H}^m(M) \geqslant \mathcal{H}^m(M \setminus E_\tau) = y(\tau) \geqslant \left(\frac{c_2}{m} \right)^m \tau^m.$$

Hence,

$$\tau \leqslant \frac{m \mathcal{H}^m(M)^{1/m}}{c_2}$$

and so

$$\begin{aligned}\mathcal{H}^m(E \cap (M \times (0, \tau)) &= \int_0^\tau |E_s|\, ds \leqslant |E_0|\, \tau \\ &\leqslant |E_0| \frac{m\mathcal{H}^m(M)^{1/m}}{c_2} \\ &\leqslant \frac{m\mathcal{H}^m(M)^{1/m}}{c_2} \frac{P(E)}{2}.\end{aligned} \tag{7.28}$$

The last inequality follows from $2|E_0| \leqslant P(E)$, which holds since E is normalized and so $P(E)$ is the sum of a lateral area that projects to some set of measure zero on M and the area of the graphs of two C^1 functions u and $-u$ over some set $\Omega \subset M$ of full measure in E_0. So we have

$$|E_0| = |\Omega| = \int_\Omega dM \leqslant \int_\Omega \sqrt{1 + |\nabla u|^2}\, dM \leqslant \frac{P(E)}{2}.$$

This proves (7.28).

Now (7.25) follows from (7.26) and (7.28). This completes the proof in the case $\tau > 0$.

It remains to consider the case $\tau = 0$. In this case, Eq. (7.26) alone implies the linear isoperimetric inequality (7.25) since $|E \cap (M \times (0, +\infty))| = |E|/2$ as E is normalized. □

Theorem 7.19 *Let (M, g) be a compact manifold. For large volumes, isoperimetric regions in the Riemannian cylinder $(M \times \mathbb{R}, g \times g_0)$ are slabs of the form $M \times [a, b]$, where $[a, b] \subset \mathbb{R}$ is a closed bounded interval.*

Proof Existence of isoperimetric regions in $N = M \times \mathbb{R}$ is guaranteed by Theorem 4.25. Comparison with slabs $M \times [a, b]$ implies

$$I_{M\times\mathbb{R}}(v) \leqslant 2\mathcal{H}^m(M), \tag{7.29}$$

for any volume $v > 0$.

Seeing $M \times \mathbb{R}$ as a warped product with base M and fiber $\mathbb{R}$, we perform the symmetrization we defined in Sect. 5.1. For any set $F \subset M \times \mathbb{R}$, we replace $F_p = F \cap (\{p\} \times \mathbb{R})$ by the segment in $\{p\} \times \mathbb{R}$ centered at $(p, 0)$ of the same length as F_p. This is an analogous of Steiner symmetrization.

We take an isoperimetric set $E \subset M \times \mathbb{R}$ and let $\operatorname{sym}(E)$ be the symmetrized of E. Since $P(\operatorname{sym}(E)) \leqslant P(E)$ and $|\operatorname{sym}(E)| = |E|$, the set $\operatorname{sym}(E)$ is an isoperimetric set of the same volume as E which also satisfies $\pi(\operatorname{sym}(E)) = \pi(E)$. Observe that the set $\operatorname{sym}(E)$ is normalized.

If $|M \setminus \pi(\operatorname{sym}(E))| > 0$, then Lemma 7.18 provides a constant $c > 0$ such that $P(\operatorname{sym}(E)) \geqslant c\,|\operatorname{sym}(E)|$. But this is in contradiction to (7.29) whenever $|\operatorname{sym}(E)| = |E| > 2c^{-1}\mathcal{H}^m(M)$. Hence, $\pi(E)$ is a set of full measure in M, and

$\mathrm{sym}(E)$ is the region between the graph of a function $u : M \to \mathbb{R}$ and the one of $-u$. By regularity of isoperimetric regions, ∇u is defined and differentiable a.e. on M and

$$P(E) = 2\int_M \sqrt{1+|\nabla u|^2}\, dM \geqslant 2\int_M dM = 2\mathcal{H}^m(M).$$

Since $P(E) \leqslant 2\mathcal{H}^m(M)$ by (7.29), we should have equality in the above inequality. Hence, $\nabla u = 0$ and so $\mathrm{sym}(E)$ is a slab. Since $P(\mathrm{sym}(E)) = P(E)$, the arguments in Sect. 5.1.4 imply that, for a.e. $p \in M$, the set E_p is a segment. Hence, E is the set between the graphs of two functions u_1, u_2 and

$$P(E) = \sum_{i=1}^{2}\int_M \sqrt{1+|\nabla u_i|^2}\, dM \geqslant 2\mathcal{H}^m(M).$$

As $P(E) \leqslant 2\mathcal{H}^m(M)$, we obtain that u_1 and u_2 are constant and so E is also a slab. □

7.3 Large Isoperimetric Sets in $M \times \mathbb{R}^k$, $k > 1$

We prove in this section that isoperimetric sets of large volume in $N = M \times \mathbb{R}^k$, where $k \geqslant 2$ and M is a compact Riemannian manifold, are of the form $M \times B$, where $B \subset \mathbb{R}^k$ is a Euclidean ball. We follow the strategy of [373] for the most part.

7.3.1 Geometric Properties of $M \times \mathbb{R}^k$

We let $N = M \times \mathbb{R}^k$. With the usual convention on dimensions, we have $n = m + k$. A fundamental role in this section is played by the *anisotropic dilations* $\varphi_t : M \times \mathbb{R}^k \to M \times \mathbb{R}^k$ defined by

$$\varphi_t(p, x) = (p, tx), \qquad (p, x) \in M \times \mathbb{R}^k.$$

Since the Jacobian of the map φ_t is t^k, we have

$$|\varphi_t(E)| = t^k |E| \tag{7.30}$$

for any measurable set $E \subset M \times \mathbb{R}^k$.

Let $S \subset M \times \mathbb{R}^k$ be an $(n-1)$-rectifiable set. At a regular point $p \in S$, the unit normal ξ can be decomposed as $\xi = av + bw$, with v tangent to M, w tangent to $\mathbb{R}^k$, and $a^2 + b^2 = 1$. Then the Jacobian of $\varphi_t|S$ is equal to $t^{k-1}(t^2a^2 + b^2)^{1/2}$. For

$t \geqslant 1$, we get

$$t^k \mathcal{H}^{n-1}(S) \geqslant \mathcal{H}^{n-1}(\varphi_t(S)) \geqslant t^{k-1}\mathcal{H}^{n-1}(S) \tag{7.31}$$

and the reversed inequalities when $t \leqslant 1$. Similar properties hold for the perimeter. Equality holds on the right-hand side of (7.31) if and only if $a = 0$ or equivalently if and only if ξ is tangent to $\mathbb{R}^k$.

An open ball in $\mathbb{R}^k$ of radius $r > 0$ and center x will be denoted by $D(x, r)$. If it is centered at the origin, we set $D(r) = D(0, r)$. We shall also denote by $T(x, r)$ the set $M \times D(x, r)$ and by $T(r)$ the set $M \times D(r)$. Observe that $\varphi_t(T(x, r)) = T(tx, tr)$ and that $T(x, r)$ is the tubular neighborhood of radius $r > 0$ of $M \times \{x\}$.

A set $F \subset N$ is *normalized* if $F_p = F \cap (\{p\} \times \mathbb{R}^k)$ is either empty or $\{p\} \times D(r(p))$, where $D(r(p)$, with $r(p) \geqslant 0$, is the disk in $\mathbb{R}^k$ with the same $\mathcal{H}$ measure as F_p. Given any set $E \subset N$ of finite perimeter, we can replace it by a *normalized* set sym E by requiring $\operatorname{sym} E \cap (\{p\} \times \mathbb{R}^k) = \{p\} \times D(r(p))$, where $\mathcal{H}^k(D(r(p))$ is equal to the $\mathcal{H}^k$-measure of $E \cap (\{p\} \times \mathbb{R}^k)$. This is the symmetrization considered in Sect. 5.1.2 taking M as base manifold and $\mathbb{R}^k$ as fiber. By the results in Sect. 5.1.2 and Corollary 5.7, we get:

1. $|\operatorname{sym} E| = |E|$
2. $P(\operatorname{sym} E) \leqslant P(E)$

Given $E \subset N$, we denote by E^* its orthogonal projection onto M. If E is normalized, and $u : E^* \to \mathbb{R}^+$ measures the radius of the disk obtained projecting $E \cap (\{p\} \times \mathbb{R}^k)$ to $\mathbb{R}^k$, we get, assuming enough regularity of u, that

$$|E| = \omega_k \int_{E^*} u^k d\mathcal{H}^m,$$

$$\mathcal{H}^{n-1}(\partial E) = k\omega_k \int_{E^*} u^{k-1}\sqrt{1 + |\nabla u|^2} d\mathcal{H}^m,$$

where, as usual, $\omega_k = \mathcal{H}^k(D(1))$ and $k\omega_k = \mathbf{c}_{k-1} = \mathcal{H}^{k-1}(\mathbb{S}^{k-1})$. The above formulas imply

$$|T(r)| = \omega_k r^k \mathcal{H}^m(M),$$

$$P(T(r)) = k\omega_k r^{k-1}\mathcal{H}^m(M),$$

so that

$$P(T(r)) = k\left(\omega_k \mathcal{H}^m(M)\right)^{1/k} |T(r)|^{(k-1)/k}. \tag{7.32}$$

Existence of isoperimetric sets in $M \times \mathbb{R}^k$ is guaranteed by Theorem 4.25 since the quotient of $M \times \mathbb{R}^k$ by its isometry group is compact. Isoperimetric regions are also bounded by Lemma 4.27. From (7.32), we get

$$I(v) \leqslant k \left(\omega_k \mathcal{H}^m(M)\right)^{1/k} v^{(k-1)/k}, \tag{7.33}$$

for any $v > 0$. The regularity of the boundaries of isoperimetric sets follows from Theorem 1.60. In this case, the boundary is regular except for a singular set of vanishing $\mathcal{H}^{n-7}$ measure. The following properties of the isoperimetric profile hold:

Proposition 7.20 *The isoperimetric profile I of $M \times \mathbb{R}^k$ is non-decreasing and continuous.*

Proof Let $v_1 < v_2$ and $E \subset N$ an isoperimetric region of volume v_2. Let $0 < t < 1$ so that $|\varphi_t(E)| = v_1$. By (7.31), we have

$$I(v_1) \leqslant P(\varphi_t(E)) \leqslant P(E) = I(v_2).$$

This shows that I is non-decreasing.

Let us prove now the right continuity of I at v. Consider an isoperimetric region E of volume v. Take a smooth vector field Z with support in the regular part of the boundary of E such that $\int_E \operatorname{div} Z \neq 0$. The flow $\{\varphi_t\}_{t\in\mathbb{R}}$ of Z satisfies $(d/dt)|_{t=0}|\varphi_t(E)| \neq 0$. Using the inverse function theorem, we obtain a smooth family $\{E_w\}$, for w near v, with $|E_w| = w$ and $E_v = E$. The function $f(w) = P(E_w)$ satisfies $f \geqslant I$ and $I(v) = f(v)$. This implies that I is right-continuous at v since, for $v_i \downarrow v$, we have

$$I(v) = f(v) = \lim_{i\to\infty} f(v_i) \geqslant \lim_{i\to\infty} I(v_i) \geqslant I(v),$$

by the monotonicity of I.

To prove the left continuity of I at v, we take a sequence of isoperimetric regions E_i with $v_i = |E_i| \uparrow v$, and we consider balls B_i disjoint from E_i so that $|E_i \cup B_i| = |E_i| + |B_i|$. Then $I(v) \leqslant P(E_i \cup B_i) = I(v_i) + P(B_i) \leqslant I(v) + P(B_i)$ by the monotonicity of I, and the left continuity follows by taking limits since $\lim_{i\to\infty} P(B_i) = 0$. □

We also use the following well-known isoperimetric inequalities in M and $M \times \mathbb{R}^k$:

Lemma 7.21 *Given $0 < v_0 < \mathcal{H}^m(M)$, there exists a constant $a(v_0) > 0$ such that*

$$\mathcal{H}^{m-1}(\partial E) \geqslant a(v_0)\, \mathcal{H}^m(E)$$

for any set $E \subset M$ satisfying $0 < \mathcal{H}^m(E) < v_0$.

Lemma 7.22 *Given $v_0 > 0$, there exists a constant $c(v_0) > 0$ so that*

$$I(v) \geqslant c(v_0)\, v^{(n-1)/n} \tag{7.34}$$

for any $v \in (0, v_0)$.

Lemma 7.21 is a slight generalization of the first inequality in Lemma 7.17. Lemma 7.22 follows from the facts that $I(v)$ is strictly positive for $v > 0$ and asymptotic to the Euclidean isoperimetric profile when v approaches 0.

The strategy of the proof is to use symmetrization and show in Corollary 7.24 that anisotropic scaling of symmetrized isoperimetric regions of large volume L^1 converges to a tubular neighborhood of $M \times \{0\}$. This convergence is improved in Lemma 7.26 to Hausdorff convergence of the boundaries using the density estimates on tubes from Lemma 7.25, similar to the ones obtained by Ritoré and Vernadakis [371] for convex surfaces. Results of White [445] and Große-Brauckmann [201] on stable submanifolds then imply that the scaled boundaries are cylinders; see Theorem 7.28. For small dimensions, it is also possible to use a result by Morgan and Ros [314] to get the same conclusion only using L^1 convergence. Once it is shown that the symmetrized set is a tube, it is not difficult to prove that the original isoperimetric region is also a tube.

7.3.2 L^1 Convergence of Anisotropic Scalings

In this section, we shall prove that normalized isoperimetric regions of large volume, when scaled down to have constant volume v_0, have their boundaries L^1 close to the boundary of the normalized tube of volume v_0.

If $E \subset N$ is any finite perimeter set and $T(E)$ is the tube with the same volume as E, we define

$$E^- = E \cap T(E), \quad E^+ = E \setminus T(E).$$

Let $t > 0$ and $\Omega = \varphi_t(E)$. Since $\varphi_t(E^+) = \Omega^+$, (7.30) implies

$$\frac{|E^+|}{|E|} = \frac{|\Omega^+|}{|\Omega|}. \tag{7.35}$$

A similar equality holds replacing E^+ by E^-.

Proposition 7.23 *Let $\{E_i\}_{i\in\mathbb{N}}$ be a sequence of normalized sets with volumes $|E_i| \to \infty$. Let $v_0 > 0$ and $0 < t_i < 1$ so that $|\varphi_{t_i}(E_i)| = v_0$ for all $i \in \mathbb{N}$, and let T be the tube of volume v_0 around M_0.*

If $\varphi_{t_i}(E_i)$ does not converge to T in the L^1 topology, then there is a constant $c > 0$, only depending on $\{E_i\}_{i \in \mathbb{N}}$, so that, passing to a subsequence, there holds

$$\mathcal{H}^{n-1}(\partial E_i) \geqslant c|E_i|. \tag{7.36}$$

Proof Assume $T = M \times D(r)$, and set $\Omega_i = \varphi_{t_i}(E_i)$. As $|\Omega_i| = |T|$, we get $2\,|\Omega_i^+| = |\Omega_i \Delta T|$, and, since $|\Omega_i \Delta T|$ does not converge to 0, the sequence $|\Omega_i^+|$ does not converge to 0 either. Let $c_1 > 0$ be a constant so that $\limsup_{i\to\infty}(|\Omega_i^+|/|\Omega_i|) > c_1$. From (7.35), we obtain

$$\limsup_{i\to\infty} \frac{|E_i^+|}{|E_i|} > c_1. \tag{7.37}$$

Now we claim that

$$\liminf_{i\to\infty} \mathcal{H}^m((\Omega_i \cap \partial T)^*) < \mathcal{H}^m(M). \tag{7.38}$$

To prove (7.38), we argue by contradiction. Assume that $\liminf_{i\to\infty} \mathcal{H}^m((\Omega_i \cap \partial T)^*) = \mathcal{H}^m(M)$. As Ω_i is normalized, we have $(\Omega_i \cap \partial T)^* \subset (\Omega_i \cap T)^*$ and so $(T \setminus \Omega_i) \subset (M \setminus (\Omega_i \cap \partial T)^*) \times D(r)$. This implies $\limsup_{i\to\infty} |T \setminus \Omega_i| = 0$. Since $|\Omega_i| = |T|$, we get $\lim_{i\to\infty} |\Omega_i \Delta T| = 2\lim_{i\to\infty} |T \setminus \Omega_i| = 0$, a contradiction that proves the claim.

Hence, there exists $w \in (0, \mathcal{H}^m(M))$ so that

$$\liminf_{i\to\infty} \mathcal{H}^m((\Omega_i \cap \partial T)^*) < w. \tag{7.39}$$

Let $T(r_i)$ be the normalized tube with $|T(r_i)| = |E_i|$. As $\Omega_i \cap T = \varphi_{t_i}(E_i \cap T(r_i))$, we have $(E_i \cap \partial T(r_i))^* = (\Omega_i \cap \partial T)^*$; from (7.39), we get $\liminf_{i\to\infty} \mathcal{H}^m((E_i \cap \partial T(r_i))^*) < w$, and we obtain

$$\liminf_{i\to\infty} \mathcal{H}^m((E_i \cap \partial T(s))^*) < w, \qquad \forall s \geqslant r_i. \tag{7.40}$$

This last step to go from the particular r_i to every $s \geqslant r_i$ is easy to check as, for any normalized set $E = \bigcup_{p\in E^*}(\{p\} \times D(r(p)))$, we have $(E \cap \partial T(s))^* = \{p \in M : r(p) \geqslant s\}$; therefore, $(E \cap \partial T(s))^* \subset (E \cap \partial T(r))^*$ whenever $s \geqslant r$.

The above arguments imply, replacing the original sequence by a subsequence, that

$$|E_i^+| > c_1\,|E_i|, \qquad \mathcal{H}^m((E_i \cap \partial T(s))^*) < w, \qquad i \in \mathbb{N},\ s \geqslant r_i. \tag{7.41}$$

Let $a = a(w)$ be the constant in Lemma 7.21. For the elements of the subsequence satisfying (7.41), we have

$$\begin{aligned}
\mathcal{H}^{n-1}(\partial E_i) &\geqslant \mathcal{H}^{n-1}(\partial E_i \cap (N \setminus T(r_i))) \\
&\geqslant \int_{r_i}^{\infty} \mathcal{H}^{n-2}(\partial E_i \cap \partial T(s))\, ds \\
&\geqslant \int_{r_i}^{\infty} \mathcal{H}^{n-2}(\partial(E_i \cap \partial T(s)))\, ds \\
&= \int_{r_i}^{\infty} \mathcal{H}^{m-1}(\partial(E_i \cap \partial T(s))^*)\, \mathcal{H}^{k-1}(\partial D(s))\, ds \\
&\geqslant \int_{r_i}^{\infty} a\, \mathcal{H}^{m}((E_i \cap \partial T(s))^*)\, \mathcal{H}^{k-1}(\partial D(s))\, ds \\
&= a \int_{r_i}^{\infty} \mathcal{H}^{n-1}(E_i \cap \partial T(s))\, ds = a\, |E_i^+| > a\, c_1 |E_i|,
\end{aligned}$$

thus proving the result. In the previous inequalities, we have used the coarea formula for the distance function to $M \times \{0\}$; that $\partial(E_i \cap \partial T(s)) \subset \partial E_i \cap \partial T(s)$, where the first ∂ denotes the boundary operator in $\partial T(s)$; the fact that for an $O(k)$-invariant set F, we have $F \cap \partial T(s) = (F \cap \partial T(s))^* \times \partial D(s)$, and so $\mathcal{H}^{r+k-1}(F \cap \partial T(s)) = \mathcal{H}^{r}((F \cap \partial T(s))^*)\, \mathcal{H}^{k-1}(\partial D(s))$; that $(\partial(E_i \cap \partial T(s)))^* = \partial(E_i \cap \partial T(s))^*$; and the isoperimetric inequality on M given in Lemma 7.21. □

Corollary 7.24 *Let $\{E_i\}_{i \in \mathbb{N}}$ be a sequence of normalized isoperimetric sets with volumes $\lim_{i \to \infty} |E_i| = \infty$. Let $v_0 > 0$ and $0 < t_i < 1$ such that $\Omega_i = \varphi_{t_i}(E_i)$ has volume v_0 for all $i \in \mathbb{N}$. Then $\Omega_i \to T$ in the L^1 topology, where T is the tube of volume v_0.*

Proof Regularity results for isoperimetric regions imply that $P(E_i) = \mathcal{H}^{n-1}(\partial E_i)$, choosing as representative of every isoperimetric set the closure of the set of density one points. If Ω_i does not converge to T in the L^1 topology, then, using (7.36) in Proposition 7.23 and (7.33), we get

$$c\,|E_i| \leqslant P(E_i) \leqslant k\,\left(\omega_k \mathcal{H}^m(M)\right)^{1/k} |E_i|^{(k-1)/k}$$

for a subsequence, thus yielding a contradiction by letting $i \to \infty$ since $|E_i| \to \infty$. □

7.3.3 Density Estimates

Using density estimates, we show now that the L^1 convergence of the scaled isoperimetric regions can be improved to Hausdorff convergence.

In a similar way to Leonardi and Rigot [266, p. 18] (see also Ritoré and Vernadakis [371], David and Semmes [132], and the results in Sect. 3.7), given $E \subset N$, we define a function $h : \mathbb{R}^k \times (0, +\infty) \to \mathbb{R}^+$ by

$$h(x, R) = \frac{\min\{|E \cap T(x, R)|, |T(x, R) \setminus E|\}}{R^n},$$

for $x \in \mathbb{R}^k$ and $R > 0$. We remark that the quantity $h(x, R)$ is not homogeneous in the sense of being invariant by scaling since $h(x, R) \leqslant \frac{1}{2}(k\omega_k \mathcal{H}^m(M))\, R^{k-n}$, which goes to infinity when R goes to 0. When the set E should be explicitly mentioned, we shall write

$$h(E, x, R) = h(x, R).$$

Lemma 7.25 *Let $E \subset N$ be an isoperimetric region of volume $v > v_0$. Let $\tau > 1$ such that $\Omega = \varphi_\tau^{-1}(E)$ has volume v_0. Choose ε so that*

$$0 < \varepsilon < \min\left\{v_0, \left(\frac{c(v_0)\, v_0^{1/k}}{2k\omega_k \mathcal{H}^m(M)}\right)^n, \left(\frac{c(v_0)}{8n}\right)^n\right\}, \tag{7.42}$$

where $c(v_0)$ is as in (7.34).

Then, for any $x \in \mathbb{R}^k$ and $R \leqslant 1$ so that $h(\Omega, x, R) \leqslant \varepsilon$, we get

$$h(\Omega, x, R/2) = 0.$$

Moreover, in case $h(\Omega, x, R) = |\Omega \cap T(x, R))|\, R^{-n}$, we get $|\Omega \cap T(x, R/2)| = 0$, and, in case $h(\Omega, x, R) = |T(x, R) \setminus \Omega|\, R^{-n}$, we have $|T(x, R/2) \setminus \Omega| = 0$.

Proof Using Lemma 7.22, we get a positive constant $c(v_0)$ so that (7.34) is satisfied (i.e., $I(w) \geqslant c(v_0)\, w^{(n-1)/n}$, for all $0 \leqslant w \leqslant v_0$).

Assume first that

$$h(x, R) = h(\Omega, x, R) = \frac{|\Omega \cap T(x, R)|}{R^n}.$$

Define

$$\mu(r) = |\Omega \cap T(x, r)|, \quad 0 < r \leqslant R.$$

The function $\mu(r)$ is non-decreasing, and, for $r \leqslant R \leqslant 1$, we get

$$\mu(r) \leqslant \mu(R) \leqslant |\Omega \cap T(x, R)| \leqslant \varepsilon R^n \leqslant \varepsilon < v_0 \tag{7.43}$$

by (7.42). Hence, $v_0 - \mu(r) > 0$ for $0 < r \leqslant R$.

By the coarea formula, when $\mu'(r)$ exists, we get

$$\mu'(r) = \frac{d}{dr} \int_0^r \mathcal{H}^{n-1}(\Omega \cap \partial T(x, s))\, ds = \mathcal{H}^{n-1}(\Omega \cap \partial T(x, r)).$$

Now define

$$\lambda(r) = \frac{v_0^{1/k}}{(v_0 - \mu(r))^{1/k}} = \frac{v^{1/k}}{|E \setminus T(\tau x, \tau r)|^{1/k}} \geqslant 1,$$

and

$$\mu(r) = \varphi_{\lambda(r)}(\Omega \setminus T(x, r)),$$

so that $|\mu(r)| = |\Omega|$. Then

$$E(r) = \varphi_\tau(\mu(r)) = \varphi_{\lambda(r)}(E \setminus T(\tau x, \tau r)),$$

and $|E(r)| = |E|$. Then, using (7.31) for $\lambda(r) \geqslant 1$ and standard properties of finite perimeter sets [275, Lemmas 12.22 and 15.12], we have

$$\begin{aligned} I(v) &\leqslant P(E(r)) \leqslant \lambda(r)^k \big(P(E \setminus T(\tau x, \tau r))\big) \\ &\leqslant \frac{v_0}{v_0 - \mu(r)} \big(P(E) - P(E \cap T(\tau x, \tau r)) + 2\mathcal{H}^{n-1}(E \cap \partial T(\tau x, \tau r))\big). \end{aligned} \tag{7.44}$$

Since $\tau \geqslant 1$ and $E \cap \partial T(\tau x, \tau r)$ is part of a cylinder, using (7.31) again, we get

$$P(E \cap T(\tau x, \tau r) \geqslant \tau^{k-1} P(\Omega \cap T(x, r)) \geqslant \tau^{k-1} c(v_0)\, \mu(r)^{(n-1)/n},$$
$$\mathcal{H}^{n-1}(E \cap \partial T(\tau x, \tau r)) = \tau^{k-1} \mathcal{H}^{n-1}(\Omega \cap \partial T(x, r)) = \tau^{k-1} \mu'(r).$$

Replacing these expressions in (7.44), since $P(E) = I(v)$ and $\tau^k v_0 = v$, we have

$$\begin{aligned}
2\mu'(r) &\geqslant \mu(r)^{(n-1)/n}\left(c(v_0) - \frac{\mu(r)^{1/n}}{\tau^{k-1}v_0}\, I(v)\right)\\
&\geqslant \mu(r)^{(n-1)/n}\left(c(v_0) - \frac{\mu(r)^{1/n}}{v_0^{1/k}}\, \frac{I(v)}{v^{(k-1)/k}}\right)\\
&\geqslant \mu(r)^{(n-1)/n}\left(c(v_0) - \frac{\varepsilon^{1/n}}{v_0^{1/k}}\,(k\omega_k \mathcal{H}^m(M))\right)\\
&\geqslant \frac{c(v_0)}{2}\,\mu(r)^{(n-1)/n},
\end{aligned} \tag{7.45}$$

where we have also used $\mu(r) \leqslant \varepsilon$, (7.33), and (7.42)

If there is $r \in [R/2, R]$ such that $\mu(r) = 0$, then, by the monotonicity of the function $\mu(r)$, we would conclude $\mu(R/2) = 0$ as well. So we assume $\mu(r) > 0$ in $[R/2, R]$. Then by (7.45), we get

$$\frac{c(v_0)}{4} \leqslant \frac{\mu'(t)}{\mu(t)^{(n-1)/n}}, \qquad \mathcal{H}^1\text{-a.e.}$$

By (7.43), we get $\mu(r) \leqslant \varepsilon R^n$. Integrating between $R/2$ and R,

$$c(v_0)\, R/8 \leqslant n\,(\mu(R)^{1/n} - \mu(R/2)^{1/n}) \leqslant n\,\mu(R)^{1/n} \leqslant n\,\varepsilon^{1/n} R.$$

This is a contradiction, since $\varepsilon < (c(v_0)/8n)^n$ by (7.42). So the proof in case $h(x, R) = |\Omega \cap T(x, R)|\, R^{-n}$ is completed.

Now we deal with the case $h(x, R) = |T(x, R) \setminus \Omega|\, R^{-n}$. Define

$$\mu(r) = |T(x, r) \setminus \Omega|.$$

Then $\mu(r)$ is a non-decreasing function and

$$\mu'(r) = \mathcal{H}^{n-1}(\Omega^c \cap \partial T(x, r)) = \frac{1}{\tau^{k-1}}\,\mathcal{H}^{n-1}(E^c \cap \partial T(\tau x, \tau r)), \tag{7.46}$$

since $E^c \cap \partial T(\tau x, \tau r)$ is part of a tube. We also have $\mu(r) \leqslant \mu(R) \leqslant \varepsilon R^n \leqslant \varepsilon < v_0$ by (7.42). Observe that

$$P(E \cup T(\tau x, \tau r) \leqslant P(E) - P(T(\tau x, \tau r) \setminus E) + 2\mathcal{H}^{n-1}(E^c \cap \partial E(\tau x, \tau r)). \tag{7.47}$$

Since $\varphi_\tau(T(x,r)\setminus\Omega) = T(\tau x, \tau r)\setminus E$ and $\tau \geqslant 1$, we get

$$\begin{aligned} P(T(\tau x, \tau r)\setminus E) &= P(\varphi_\tau(T(x,r)\setminus\Omega)) \\ &\geqslant \tau^{k-1} P(T(x,r)\setminus\Omega) \geqslant \tau^{k-1}\, c(v_0)\, \mu(r)^{(n-1)/n}. \end{aligned} \tag{7.48}$$

Now, using that I is a non-decreasing function, we easily obtain $P(E) = I(v) \leqslant I(|E \cup T(\tau x, \tau r)|) \leqslant P(E \cup T(\tau x, \tau r))$. We estimate $P(E \cup T(\tau x, \tau r))$ from (7.47). Using (7.48) and (7.46), we get

$$\begin{aligned} I(v) = P(E) &\leqslant P(E \cup T(\tau x, \tau r)) \\ &\leqslant I(v) - \tau^{k-1} c(v_0)\, \mu(r)^{(k-1)/k} + 2\tau^{k-1}\mu'(r), \end{aligned} \tag{7.49}$$

and so

$$\frac{c(v_0)}{2} \leqslant \frac{\mu'(r)}{\mu(r)^{(n-1)/n}}, \qquad \mathcal{H}^1\text{-a.e.}$$

By (7.43), we get $\mu(R) \leqslant \varepsilon R^n$. Integrating between $R/2$ and R,

$$c(v_0)\, R/4 \leqslant n\, (\mu(R)^{1/n} - \mu(R/2)^{1/n}) \leqslant n\, \mu(R)^{1/n} \leqslant n\, \varepsilon^{1/n} R,$$

and we get a contradiction since by (7.42), we have $\varepsilon < (c(v_0)/(8n))^n < (c(v_0)/(4n))^n$. This concludes the proof. □

Let $F \subset N$; then we define $F_r = \{x \in N : d(x, F) \leqslant r\}$. We improve now the L^1 convergence of normalized isoperimetric regions obtained in Corollary 7.24 to Hausdorff convergence of their boundaries.

Lemma 7.26 *Let $\{E_i\}_{i\in\mathbb{N}}$ be a sequence of isoperimetric sets in N with $\lim_{i\to\infty} |E_i| = \infty$. Let $v_0 > 0$ and $\{t_i\}_{i\in\mathbb{N}}$ such that $\lim_{i\to\infty} t_i = 0$ and $|\Omega_i| = v_0$ for all $i \in \mathbb{N}$, where $\Omega_i = \varphi_{t_i}(E_i)$. Then for every $r > 0$, $\partial\Omega_i \subset (\partial T)_r$, for large enough $i \in \mathbb{N}$, where T is the tube of volume v_0.*

Proof Since $|\Omega_i| = v_0$, using (7.42), we can choose a uniform $\varepsilon > 0$ so that Lemma 7.25 holds with this ε for all Ω_i, $i \in \mathbb{N}$. This means that, for any $x \in N$ and $0 < r \leqslant 1$, whenever $h(\Omega_i, x, r) \leqslant \varepsilon$, we get $h(\Omega_i, x, r/2) = 0$.

As $\Omega_i \to T$ in $L^1(N)$ by Corollary 7.24, we can choose a sequence $r_i \to 0$ so that

$$|\Omega_i \,\triangle\, T| < r_i^{n+1}. \tag{7.50}$$

Now fix some $0 < r < 1$. We reason by contradiction assuming that, for some subsequence, there exist

$$x_i \in \partial\Omega_i \setminus (\partial T)_r. \tag{7.51}$$

We distinguish two cases.

First case: $x_i \in N \backslash T$, for a subsequence. Choosing i large enough, (7.51) implies $T(x_i, r_i) \cap T = \emptyset$, and (7.50) yields

$$|\Omega_i \cap T(x_i, r_i)| \leqslant |\Omega_i \setminus T| \leqslant |\Omega_i \Delta T| < r_i^{n+1}.$$

So, for i large enough, we get

$$h(\Omega_i, x_i, r_i) = \frac{|\Omega_i \cap T(x_i, r_i)|}{r_i^n} < r_i \leqslant \varepsilon.$$

By Lemma 7.25, we conclude that $|\Omega_i \cap T(x_i, r_i/2)| = 0$, a contradiction.

Second case: $x_i \in T$. Choosing i large enough, (7.51) implies $T(x_i, r_i) \subset T$ and so

$$|T(x_i, r_i) \setminus \Omega_i| \leqslant |T \setminus \Omega_i|, \quad \text{for every } r_i < r.$$

Then, by (7.50), we get

$$|T(x_i, r_i) \setminus \Omega_i| \leqslant |T \setminus \Omega_i| \leqslant |\Omega_i \Delta T| < r_i^{n+1}.$$

So, for i large enough, we get

$$h(\Omega_i, x_i, r_i) = \frac{|T(x_i, r_i) \setminus \Omega_i|}{r_i^n} < r_i \leqslant \varepsilon.$$

By Lemma 7.25, we conclude that $|T(x_i, r_i/2) \setminus \Omega_i| = 0$, and we get again a contradiction. This proves the result. □

7.3.4 Strict $O(k)$ Stability of Tubes with Large Radius

In this section, we consider the orthogonal group $O(k)$ acting on the product $M \times \mathbb{R}^k$ through the second factor.

A compact hypersurface $S \subset M \times \mathbb{R}^k$ with constant mean curvature is a critical point of the area functional under volume-preserving deformations. By the second variation formula (1.27), S is a second-order minimum of the area under volume-preserving variations if and only if

$$\int_S \left(|\nabla u|^2 - q\,u^2\right) dS \geqslant 0, \tag{7.52}$$

for any smooth function $u : S \to \mathbb{R}$ with mean zero on S. In the above formula, ∇ is the gradient on S, and q is the function $\operatorname{Ric}(\xi,\xi) + |\sigma|^2$, where $|\sigma|^2$ is the sum of the squared principal curvatures in S, ξ is a unit vector field normal to S, and Ric is the Ricci curvature on N.

We have called these surfaces *stable* in and have referred to (7.52) as the *stability condition* in Sect. 6.1. In case S is $O(k)$-invariant, we can consider an *equivariant* stability condition: we say that S is *strictly* $O(k)$*-stable* if there exists a positive constant $\lambda > 0$ such that

$$\int_S \left(|\nabla u|^2 - q\,u^2\right) dS \geqslant \lambda \int_S u^2\, dS$$

for any $O(k)$-invariant function $u : S \to \mathbb{R}$ with mean zero.

We consider now the tube $T(r) = M \times D(r)$. The boundary of $T(r)$ is the $O(k)$-invariant cylinder $S(r) = M \times \partial D(r)$, with $(k-1)$ principal curvatures equal to $1/r$. Hence, its mean curvature is equal to $(k-1)/r$, and the squared norm of the second fundamental form satisfies $|\sigma|^2 = (k-1)/r^2$. The inner unit normal to $S(r)$ is the normal to $\partial D(r)$ in $\mathbb{R}^k$ (it is tangent to the factor $\mathbb{R}^k$). This implies $\operatorname{Ric}(\xi,\xi) = 0$.

We have the following result:

Lemma 7.27 *The cylinder $S(r)$ is strictly $O(k)$-stable if and only if*

$$r^2 > \frac{k-1}{\lambda_1(M)},$$

where $\lambda_1(M)$ is the first positive eigenvalue of the Laplacian in M.

Proof Let $S = S(r) = M \times D(r)$. Observe that an $O(k)$-invariant function with mean zero on S is determined by a function $u : M \to \mathbb{R}$ with $\int_M u\, dM = 0$. Hence,

$$\begin{aligned}
\int_S \left(|\nabla u|^2 - q\,u^2\right) dS &= k\omega_k r^{k-1} \int_M \left(|\nabla_M u|^2 - \frac{k-1}{r^2}\,u^2\right) dM \\
&\geqslant k\omega_k r^{k-1}\left(\lambda_1(M) - \frac{k-1}{r^2}\right) \int_M u^2\, dM \\
&= \left(\lambda_1(M) - \frac{k-1}{r^2}\right) \int_S u^2\, dS.
\end{aligned}$$

This proves the lemma. □

Using results by White [445] and Große-Brauckmann [201], we get the following:

Theorem 7.28 *Let T be a normalized tube so that $S = \partial T$ is a strictly $O(k)$-stable cylinder. Then there exists $r > 0$ so that any $O(k)$-invariant finite perimeter set E with $|E| = |T|$ and $\partial E \subset T_r$ has larger perimeter than T unless $E = T$.*

Proof Since S is strictly $O(k)$-stable, Lemma 5 in Große-Brauckmann [201] (see also Sect. 6.1.3) imply that, for some $C > 0$, S has strictly positive second variation for the functional

$$F_C = \text{area} + H\,\text{vol} + \frac{C}{2}\,(\text{vol} - \text{vol}(T))^2,$$

in the sense that the second variation of F_C in the normal direction of a function u satisfies

$$\delta_u^2 F_C = \int_S (|\nabla u|^2 - q\,u^2)\,dS + C\left(\int_S u\,dS\right)^2 \geqslant \lambda \int_S u^2\,dS,$$

for any smooth $O(k)$-invariant function u (see the discussion in the proof of Theorem 2 in Morgan and Ros [314]). In White's proof of Theorem 3 in [445], it is observed that a sequence of minimizers of F_C in tubular neighborhoods of radius $1/i$ of S are *almost minimizing* and hence $C^{1,\alpha}$ submanifolds that converge Hölder differentiably to S, contradicting the positivity of the second variation of S. The symmetrization of each of these minimizers is again a minimizer. Thus, we get a family of $O(k)$ minimizers of F_C converging Hölder differentiably to S, thus contradicting the strict $O(k)$ stability of S. □

7.3.5 Statement and Proof of the Main Result

Theorem 7.29 *Let M be a compact Riemannian manifold. There exists a constant $v_0 > 0$ such that any isoperimetric region in $M \times \mathbb{R}^k$ of volume $v \geqslant v_0$ is a tubular neighborhood of $M \times \{x\}$, with $x \in \mathbb{R}^k$.*

Proof First, we claim that there exists $v_0 > 0$ such that, for any isoperimetric region E of volume $|E| \geqslant v_0$, the set sym E is a tube.

To prove this, consider a sequence of isoperimetric regions $\{E_i\}_{i\in\mathbb{N}}$ with $\lim_{i\to\infty} |E_i| = \infty$. We know that $\{\text{sym}\,E_i\}_{i\in\mathbb{N}}$ are also isoperimetric regions. Let $T = M \times D$ be a strictly $O(k)$-stable tube, which exists by Lemma 7.27. For large i, we scale down the sets sym E_i so that $\Omega_i = \varphi_{t_i}^{-1}(\text{sym}\,E_i)$ has the same volume as T. As sym E_i is isoperimetric and $t_i > 1$, we get from (7.33) and (7.31) that $P(\Omega_i) \leqslant P(T)$. By Corollary 7.24, the sets $\{\partial\Omega_i\}_{i\in\mathbb{N}}$ converge to ∂T in Hausdorff distance. By Theorem 7.28, $\Omega_i = T$ for large i and so sym E_i is a tube. This proves the claim. In particular, $\mathcal{H}^m(E \cap (\{p\} \times \mathbb{R}^k)) = \mathcal{H}^m(D)$ for any $p \in M$.

Hence, the isoperimetric profile satisfies $I(v) = C\,v^{(k-1)/k}$ for the constant C in (7.32) and any $v \geqslant v_0$. We conclude that

$$I(t^k v) = t^{k-1} I(v), \quad \text{whenever } t^k v \geqslant v_0. \tag{7.53}$$

Let E be an isoperimetric region with volume $|E| > v_0$ and $t < 1$ so that $t^k|E| = v_0$. Then

$$I(t^k|E|) \leqslant P(\varphi_t(E)) \leqslant t^{k-1} P(E) = t^{k-1} I(|E|)$$

by the inequality corresponding to (7.31) when $t \leqslant 1$. By (7.53), equality holds, and the unit normal ξ to $\operatorname{reg}(\partial E)$, the regular part of ∂E, is tangent to the $\mathbb{R}^k$ factor. This implies that the m-Jacobian of the restriction f of the projection $\pi_1 : M \times \mathbb{R}^k \to M$ to the regular part of ∂E is equal to 1. By Federer's coarea formula for rectifiable sets [154, 3.2.22], we get

$$\mathcal{H}^{n-1}(\partial E) = \int_M \mathcal{H}^{k-1}(f^{-1}(p))\, d\mathcal{H}^m(p).$$

Assume that $\operatorname{sym} E$ is the tube $T(E) = M \times D$. The Euclidean isoperimetric inequality implies $\mathcal{H}^{k-1}(f^{-1}(p)) \geqslant \mathcal{H}^{k-1}(\{p\} \times \partial D)$ and so $\mathcal{H}^{n-1}(\partial E) \geqslant \mathcal{H}^{n-1}(\partial T(E))$, again by the coarea formula. As $P(E) = P(\operatorname{sym} E) = P(T(E))$, we get $\mathcal{H}^{k-1}(f^{-1}(p)) = \mathcal{H}^{k-1}(\partial D)$ for $\mathcal{H}^m$-a.e. $p \in M$, and so $\pi_1^{-1}(p)$ is equal to a disk $\{p\} \times D_p$ for $\mathcal{H}^m$- a.e. $p \in M$.

The fact that ξ is tangent to $\mathbb{R}^k$ in $\operatorname{reg}(\partial E)$ implies that $\operatorname{reg}(\partial E)$ is locally a cylinder of the form $U \times S$, where $U \subset M$ is an open set and $S \subset \mathbb{R}^k$ is a smooth hypersurface. Hence, the disks D_p are centered at the same point (i.e., E is the translation of a normalized tube, which proves the theorem). □

7.4 Notes

Notes for Sect. 7.1 Theorem 7.11 seems to have been obtained by Kleiner in the early 1990s. In a 1993 paper by Tomter [423] on constant mean curvature surfaces in the Heisenberg group, an unpublished result by Kleiner is cited in page 493: "Kleiner can now use the implicit function theorem for Banach spaces to conclude that for any point x and for sufficiently small volume V, there exists a unique closed hypersurface $M_V(x)$, whose area $A(V, x)$ is stationary with respect to diffeomorphisms $DV(x)$." Here, $DV(x)$ is the group of volume-preserving diffeomorphism which fix x. In the introduction to Nardulli's paper [317], it is said explicitly that "B. Kleiner convinced us that he had a proof of Theorems 1 and 2 along the same lines."

Assuming that the scalar curvature of a compact manifold is a Morse function, Ye [451, 452] obtained a foliation by constant mean curvature hypersurfaces of punctured neighborhoods of the isolated critical points of the scalar curvature. It can be proven that these hypersurfaces are pseudo bubbles, so that Theorem 7.11 applies. It can also be proven (see page 130 in [317]) that the sheets of Ye's foliation near the maxima of the scalar curvature are isoperimetric sets.

Pacard and Xu [327] obtained existence results of small constant mean curvature radial hypersurfaces without assuming any non-degeneracy condition on the scalar curvature of the manifold.

Notes for Sect. 7.2 Theorem 7.19 was proven by Duzaar and Steffen; see Proposition 2.11 in [143]. It was extended to cylinders with density by Castro [96]. We have followed her proof in Sect. 7.2. Morgan observed that one can also give a proof of Theorem 7.19 by using the monotonicity formula or lower density estimates for isoperimetric sets (e.g., see Corollary 4.12 in [370]).

Notes for Sect. 7.3 The key tool to prove Theorem 7.29 is Proposition 7.23, where it is shown that rescaling large isoperimetric regions, we have L^1 convergence to tubes. The use of density estimates allows us to improve the convergence to Hausdorff convergence and use the local minimality results of Große-Brauckmann [201].

Theorem 7.29 was proven independently by Gonzalo [187].

There are many results concerning the existence and uniqueness of isoperimetric foliations under assumptions on the asymptotic geometry of the manifold. Let us cite the result by Eichmeir and Metzger (see Theorem 1.1 in [148]): let M be an m-dimensional manifold M which is an initial data set and asymptotically C^0 to a Schwarzschild manifold of mass $m > 0$. Then isoperimetric regions exists for large volume. If, in addition, M is asymptotically C^2 to a Schwarzschild manifold, then the boundaries of large isoperimetric sets are the sheets of the unique foliation near infinity by constant mean curvature spheres. The reader is referred to the references in [148] for more information. Chodosh, Eichmair, and Vollmann [111] treated the case of manifolds asymptotically conical to cones over manifolds with non-negative Ricci curvature. They prove existence of a foliation by constant mean curvature hypersurfaces outside a compact set (Theorem 1) and existence and uniqueness of large isoperimetric regions (Theorem 3). The case of a purely conical manifold was considered by Morgan and Ritoré [313].

Chapter 8
Isoperimetric Comparison for Sectional Curvature

We consider in this chapter the validity of the Euclidean isoperimetric inequality in a Cartan–Hadamard manifold, a complete, simply connected manifold with non-negative sectional curvature K_{sec}. The following conjecture appeared in Aubin; see Conjecture 1 in p. 581 in [30].

Conjecture 8.1 (Aubin) For a measurable set E contained in a ball B (where the curvature is no larger than K) in a manifold M, we have that $\text{Mink}_-(E)$ is larger than or equal to the area of the boundary of a ball, in a manifold of constant curvature K, of volume $|E|$.

A similar conjecture for Cartan–Hadamard manifolds appears in Burago and Zalgaller (see §36.5.10 in [83]) and Gromov (see §6.28 in [198]). These conjectures read as follows:

Conjecture 8.2 Let M be a Cartan–Hadamard manifold. Then

$$I_M \geqslant I_{\mathbb{R}^m}. \tag{8.1}$$

Moreover, if a bounded set $E \subset M$ satisfies $P(E) = I_{\mathbb{R}^m}(|E|)$, then E is isometric to a flat geodesic ball.

Burago-Zalgaller and Gromov also mention the possibility that the lower bound on the profile I_M could be improved assuming a more strict bound, either on the Ricci curvature of M or on the sectional curvature of M. This leads to a generalized form of Conjecture 8.2.

Conjecture 8.3 Let M be a Cartan–Hadamard manifold with $K_{\text{sec}} \leqslant c \leqslant 0$, for some constant $c \leqslant 0$. Then

$$I_M \geqslant I_{\mathbb{M}_c^m}, \tag{8.2}$$

M. Ritoré, *Isoperimetric Inequalities in Riemannian Manifolds*, Progress in Mathematics 348, https://doi.org/10.1007/978-3-031-37901-7_8

where $\mathbb{M}_c^m$ is m-dimensional Euclidean or hyperbolic space of constant sectional curvature c.

Moreover, if there exists a bounded set $E \subset M$ such that $P(E) = I_{\mathbb{M}_c}(|E|)$, then E is isometric to the geodesic ball of volume $|E|$ in $\mathbb{M}_c$.

It is usual referring to Conjecture 8.2 as the *Cartan–Hadamard conjecture* (see pp. 235–236 in Hebey [223]) and to Conjecture 8.3 as the *generalized Cartan–Hadamard conjecture*. Morgan, §17.3 in [307], also refers to Conjecture 8.3 as the *Aubin conjecture*.

For two-dimensional Cartan–Hadamard manifolds, Conjecture 8.2 follows from the results by Weil [439] and Beckenbach and Radó [49]; see Theorem 2.3. The generalized Conjecture 8.3 follows from Bol-Fiala's inequality (2.10); see [68] and [158].

It is not difficult to see that balls in Cartan–Hadamard manifolds with $K_{\sec} \leqslant c \leqslant 0$ satisfy the isoperimetric inequality $P(B) \geqslant I_c(|B|)$, where I_c is the isoperimetric profile of $\mathbb{M}_c^m$. In order to check it, we fix $p \in M$, and we let $H(a)$ be the mean curvature of the ball centered at p of perimeter a. Let $r(a)$ be the radius of such ball. Theorem 1.29 implies that $H(a) \geqslant (H_c)_{S_c(r(a))}$, where $(H_c)_{S_c(r(a))}$ is the mean curvature in $\mathbb{M}_c^m$ of a geodesic sphere $S_c(r(a))$ of radius $r(a)$. We also know from Theorem 1.29 that $a = A(S(p,r(a)) \geqslant A_c(S_c(r(a)))$. Since H_c is a decreasing function of a, we conclude

$$H(a) \geqslant H_c(a). \tag{8.3}$$

Let now $a(r) = P(B(p,r))$ and $v(r) = |B(p,r)|$. As $v(r)$ is smooth and increasing for $r > 0$, the function a can be expressed in terms of v. We take the derivative of $I_c^{-1}(a(v))$ with respect to v, and we get

$$\frac{dI_c^{-1}(a(v))}{dv} = \frac{dI_c^{-1}(a(v(r)))/dr}{dv/dr} = \frac{(1/H_c(a)) \int_{S(p,r)} H(a)\, dS}{a} \geqslant 1$$

by (8.3). Integrating between 0 and any $v > 0$, we have $I_c^{-1}(a(v)) \geqslant v$ and so

$$a(v) \geqslant I_c(v),$$

which is the desired result.

Another evidence of the validity of Conjecture 8.2 comes from the result below by Yau; see Proposition 3 in page 498 of [450] and §34.2.6 in [83].

Proposition 8.4 *Let E be a subset with finite perimeter and volume in a Cartan–Hadamard manifold with $K_{\sec} \leqslant c$. Then*

$$P(E) \geqslant (m-1)\sqrt{-c}\,|E|. \tag{8.4}$$

Proof Let d be the distance to a fixed point $p \in M$. The Laplacian of d is $(m-1)$ times the mean curvature of the sphere $S(p,d)$. Hence, $\Delta d \geqslant (m-1)\sqrt{-c}\coth(\sqrt{-c}d) \geqslant (m-1)\sqrt{-c}$ and so

$$(m-1)\sqrt{-c}\,|E| \leqslant \int_E \Delta d\, dM = \int_{\partial^* E} \langle \nu_E, \nabla d\rangle\, dP \leqslant P(E).$$

□

Nota that inequality (8.4) implies

$$P(E) \geqslant \mathbf{c}(m)\,|E|^{(m-1)/m}$$

for any set E of volume

$$|E| \geqslant \left(\frac{\mathbf{c}(m)}{(m-1)\sqrt{-c}}\right)^m.$$

One more reason to believe that the Conjecture 8.2 should hold is that inequality

$$P(E) \geqslant \varepsilon_m I_{\mathbb{R}^m}(|E|)$$

holds, where $\varepsilon_m < 1$ are constants depending only on the dimension m of the manifold; see Croke [126], Hoffman–Spruck [231, 232] (also Michael-Simon [283]), and Burago–Zalgaller [83].

Apart from the two-dimensional case, the generalized Conjecture 8.3 has been solved in case $n = 3$ by Kleiner [247] (for different proofs, see Ritoré [362] and Schultze [401] for the case $c = 0$). In addition, the Cartan–Hadamard Conjecture 8.3 has only been settled when $n = 4$ by Croke [127].

It is generally believed that Croke's proof would hardly provide a solution of the Cartan–Hadamard conjecture in dimensions different from 4, but that Kleiner's of the generalized Cartan–Hadamard conjecture could be extended to higher dimensions provided that an estimate on the total curvature of an isoperimetric set in Cartan–Hadamard manifolds is obtained.

We review Kleiner's proof in Sect. 8.3, the alternative proof in Sect. 8.4, and Croke's in Sect. 8.5.

8.1 A Proof of the Euclidean Isoperimetric Inequality Involving Total Curvature

Most of the proofs of the classical isoperimetric inequality in $\mathbb{R}^m$ cannot be adapted to a Cartan–Hadamard manifold. There is however a strategy, introduced by Almgren [12], that could be extended. See also Chavel, §II.1.2 in [101].

Assume that $E \subset \mathbb{R}^m$ is an isoperimetric set. We know from the results in Chap. 4 that isoperimetric sets exist for every volume $v > 0$ and that they are bounded. Moreover, its boundary is composed of a regular part S and a singular set S_0 of small Hausdorff dimension.

Let GK denote the *Gauss-Kronecker curvature* of S, defined as the product of the principal curvatures of S, and let

$$S^+ = \{q \in S : GK(q) \geqslant 0\}.$$

Pick any unit vector $v \in \mathbb{S}^{m-1}$. The height function $x \mapsto \langle x, v \rangle$ achieves a maximum at some $p \in \partial E$. At this point, the set E is contained in a half-space, and so it is its tangent cone at p. This implies that $p \in S$ and, since the principal curvatures of S are non-negative at p, that $p \in S^+$. Moreover, the Gauss map N satisfies $N(p) = v$. An argument first given by B.-Y. Chen (see Theorems 1 and 2 in [108]) implies

$$\int_{S^+} \left(\frac{H}{m-1} \right)^{m-1} dS \geqslant \int_{S^+} GK\, dS \geqslant \mathbf{c}_{m-1} \tag{8.5}$$

by the arithmetic-geometric mean inequality and the facts that $N : S^+ \to \mathbb{S}^{m-1}$ is surjective and GK is the Jacobian of the Gauss map on S^+. Equality holds if and only if S^+ is totally umbilical. Since E is an isoperimetric region, H is constant (and non-negative) and

$$\left(\frac{H}{m-1} \right)^{m-1} P(E) \geqslant \mathbf{c}_{m-1}.$$

Observe that for any geodesic ball $B(p, r) \subset \mathbb{R}^{m-1}$,

$$\left(\frac{H_{S(p,r)}}{m-1} \right)^{m-1} P(B(p,r)) = \mathbf{c}_{m-1},$$

and so

$$H \geqslant H_{S(p,r)}, \tag{8.6}$$

where $S(p, r)$ is a sphere with area equal to $P(E)$.

The isoperimetric profile $I_{\mathbb{R}^m}$ of $\mathbb{R}^m$ is a continuous increasing function. This follows from Theorem 4.17 or simply from the fact that $I_{\mathbb{R}^m}(v) = C\, v^{(m-1)/m}$, for some $C > 0$, as dilations in $\mathbb{R}^m$ modify the volume and perimeter as $P(\lambda E) = \lambda^{m-1} P(E)$, $|\lambda E| = \lambda^m |E|$ for any set $E \subset \mathbb{R}^m$. Hence, $I_{\mathbb{R}^m}$ is differentiable almost everywhere.

Let us denote by I_{balls} the function that assigns to every $v > 0$ the perimeter of the ball of volume v. The function I_{balls} is strictly increasing and smooth. Its inverse

I_{balls}^{-1} in $(0, +\infty)$ is also smooth and continuous. The composition $I_{\text{balls}}^{-1} \circ I_{\mathbb{R}^m}$ is non-decreasing and continuous. Taking $v > 0$ so that $I_{\mathbb{R}^m}$ is differentiable at v, we have

$$\left(I_{\text{balls}}^{-1} \circ I_{\mathbb{R}^m}\right)'(v) = (I_{\text{balls}}^{-1})'(I_{\mathbb{R}^m}(v)) \cdot I_{\mathbb{R}^m}'(v) = \frac{H}{H_{S(p,r)}},$$

where H is the mean curvature of any isoperimetric region of volume v and $H_{S(p,r)}$ is the mean curvature of a ball of area $I_{\mathbb{R}^m}(v) = P(E)$. By (8.6),

$$\left(I_{\text{balls}}^{-1} \circ I_{\mathbb{R}^m}\right)' \geqslant 1$$

almost everywhere in $(0, +\infty)$. Since $I_{\text{balls}}^{-1} \circ I_{\mathbb{R}^m}$ is monotone non-decreasing and $(I_{\text{balls}}^{-1} \circ I_{\mathbb{R}^m})(0) = 0$, integrating from 0 to v,

$$\left(I_{\text{balls}}^{-1} \circ I_{\mathbb{R}^m}\right)(v) \geqslant v,$$

and so

$$I_{\text{balls}}(v) \geqslant I_{\mathbb{R}^m}(v).$$

As $I_{\text{balls}} \leqslant I_{\mathbb{R}^m}$, we finally get

$$I_{\text{balls}} = I_{\mathbb{R}^m}.$$

This inequality holds almost everywhere but, since both functions are continuous, there holds everywhere. Hence, equality holds in (8.5), and so any isoperimetric region is a geodesic ball.

8.2 A Mean Curvature Estimate

Conjecture 8.3 was proven by Kleiner [247] for any $c \leqslant 0$ in dimension 3 by considering ideas similar to the ones in Sect. 8.1. In his proof, there is a scheme common to any dimension. The hypothesis on the dimension is only used to prove the following proposition. Its proof will be given in the next section.

Proposition 8.5 *Let M be a Cartan–Hadamard manifold with sectional curvatures bounded above by a constant $c \leqslant 0$. Let $E \subset M$ be a compact set with $C^{1,1}$ boundary S. Then*

$$\max_{\partial E} H_{\partial E} \geqslant H_c(P(E)), \tag{8.7}$$

where H_c is the mean curvature in the model space $\mathbb{M}_c^3$ of the geodesic ball of area $P(E)$.

If equality holds in (8.7), then E is isometric to a geodesic ball in $\mathbb{M}_c^3$.

Let us see that Conjecture 8.3 is true in any dimension if the analogous of Proposition 8.5 is valid. One of the main difficulties to apply the strategy of Sect. 8.1 is that isoperimetric domains may not exist in a non-compact manifold. To solve this problem, geodesic balls in M will be considered.

We fix a point in M and consider a geodesic ball B in M centered at p. Since M is a Cartan–Hadamard manifold, ∂B is a smooth hypersurface in M with mean curvature $H_B \geqslant 1/\text{radius}(B) > 0$ by classical comparison results; see Theorem 1.29.

By Theorem 1.62, for every $0 < v < |B|$, there exists an isoperimetric region E of volume v in Ω whose boundary satisfies the following properties:

1. ∂E is of class $C^{1,1}$ near $\partial\Omega$.
2. $\partial E \cap \Omega$ in the union of a smooth hypersurface S with constant mean curvature H and a closed singular set S_0 of Hausdorff dimension at most $m - 8$.

In addition, the isoperimetric profile I_B of B is non-decreasing by Lemma 4.12, and when the radius of B goes to ∞, the profile I_B decreases to I_M by Lemma 4.9. At regular points of I_B, its derivative is the constant mean curvature of the regular part of the isoperimetric boundary in B. This is easily proven taking a variation supported in B moving S while keeping constant the enclosed volume. Moreover, the mean curvature $H_{\partial E}$ of S satisfies

$$H = \max_{\partial E} H_{\partial E}.$$

This property is proven like in Theorem 3.7 in [412].

Since I_B is a non-decreasing function, it is differentiable almost everywhere. For a given regular point $0 < v < |B|$ of I_B, let $E \subset \overline{B}$ be an isoperimetric set in B of volume v, and let H be the (constant) mean curvature of the regular part of $S \cap B$. Assuming Proposition 8.5 holds, then $H = \max_{\partial E} H_{\partial E} \geqslant H_c(P(E))$, and equality holds when E is isometric to a geodesic ball of area $P(E)$ in $\mathbb{M}_c^m$.

We consider the function $I_c^{-1} \circ I_B$, defined on $(0, |B|)$. It is a continuous non-decreasing function, differentiable at points where I_B is differentiable. At the regular point v, there holds

$$\left(I_c^{-1} \circ I_B\right)'(v) = (I_c^{-1})'(I_B(v)) \cdot I_B'(v) = \frac{H}{H_c(P(E))} \geqslant 1$$

by (8.7) since $H = \max_{\partial E} H_{\partial E}$. Integrating from 0 to any positive value, we get

$$I_B \geqslant I_c \quad \text{on } (0, |B|).$$

Letting the radius of B go to $+\infty$, we have

$$I_M \geqslant I_c \tag{8.8}$$

on $(0, +\infty)$.

Assume now that there exists a bounded set $E \subset M$ such that $P(E) = I_c(|E|)$. The set E is contained in a large ball $\overline{B}$ and so $P(E) = I_B(|E|)$. Then equality holds on (8.8) at $|E|$. This implies $I_B = I_c$ on $(0, |E|)$. Hence, $H = \max_{\partial E} H_{\partial E} = H_c(P(E))$, and E is isometric to a geodesic ball in $\mathbb{M}_c^m$.

8.3 The Three-Dimensional Case: Kleiner's Proof

This section is devoted to proving Proposition 8.5 in the three-dimensional case following Kleiner [247]. A key tool is the use of the Gauss-Bonnet formula on the two-dimensional boundary of an isoperimetric set. Thus, the proof does not generalize directly to higher dimensions. Note that in the three-dimensional case, an isoperimetric set E inside a boundary has $C^{1,1}$ boundary (C^∞ in the interior of the ball) since there are no singularities at the boundary of E when $m \leqslant 7$. Let $S = \partial E$ and let $H = \max_{\partial E} H_{\partial E}$. Inequality (8.7) is trivial when S is a sphere since

$$4\pi = \int_S K\, dA = \int_S (K_{\mathrm{sec}} + \kappa_1\kappa_2)\, dA \leqslant \int_S (c + \kappa_1\kappa_2)\, dA \leqslant (c + H^2)A(S),$$

and equality holds if and only if $K_{\mathrm{sec}} \equiv c$ over S and the surface is totally umbilical. In case S is a geodesic sphere in a space of constant curvature c, equality holds in the above inequality. This shows

$$H \geqslant H_c(A(S)).$$

If equality holds in the above inequality, then S is a totally umbilical sphere so that the sectional curvature of the tangent plane equals c and S has the same second fundamental form as of the sphere of area $A(S)$ in $\mathbb{M}_c^3$. It follows from Theorem 7 in [400] that the domain enclosed by S is a geodesic ball in $\mathbb{M}_c^3$.

Let us assume now that S is any $C^{1,1}$ surface which encloses a domain E. Consider the closed convex hull D_0 of E. The set D_0 is convex and compact since it is contained in a (convex) ball of M. Using ideas of Federer [153], we consider the domains (Fig. 8.1)

$$D_s = \{x \in M : d(x, D_0) \leqslant s\}.$$

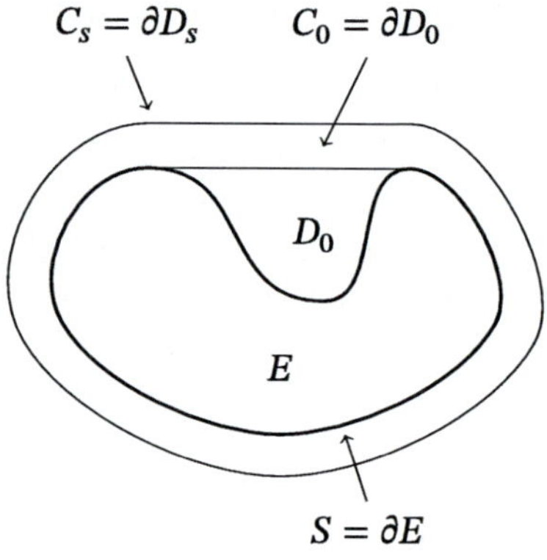

Fig. 8.1 Closed convex hull D_0 of E

We know that (see Proposition 8.6):

- D_s is convex
- The metric projection $r : M \setminus \operatorname{int} D_0 \to S$ is well-defined, and it is distance non-increasing
- $C_s = \partial D_s$ is of class $C^{1,1}$ and homeomorphic to $\mathbb{S}^2$

Let us call $r_s = r|_{C_s}$. As C_s is a $C^{1,1}$ surface, its Gauss curvature and its Gauss-Kronecker curvature GK (product of principal curvatures) exist by Rademacher's theorem. As C_s is a sphere, the total curvature of C_s equals 4π. Then

$$\begin{aligned} 4\pi = \int_{C_s} K_{C_s} &= \int_{C_s} (K_{\text{sec}} + GK_{C_s}) \leqslant \int_{C_s} (c + GK_{C_s}) \\ &= \int_{r_s^{-1}(S)} (c + GK_{C_s}) + \int_{C_s \setminus r_s^{-1}(S)} (c + GK_{C_s}) \\ &\leqslant \int_{r_s^{-1}(S)} (c + (\frac{H_s}{2})^2) + c\,A(C_s \setminus r_s^{-1}(S)) + \int_{C_s \setminus r_s^{-1}(S)} GK_{C_s}. \end{aligned} \tag{8.9}$$

In the first inequality, we have bounded K_{sec} by c, and in the second one, GK_{C_s} by the mean curvature of C_s. Equality holds in the above inequality if and only if $K_{\text{sec}} = c$ along C_s and $r_s^{-1}(S)$ is totally umbilical.

We treat now the terms in the last line of (8.9).

Let us see that

$$\lim_{s \to 0} \int_{r_s^{-1}(S)} (c + (\frac{H_s}{2})^2) \leqslant (c + (\frac{H}{2})^2)\, A(S \cap C_0). \tag{8.10}$$

We only have to take into account that if C_s is C^2 in p (this happens for almost every $p \in C_s$) and $p \in C_s$ and $r_s(p) \in C_0 \cap S$, then

$$H_s(p) \leqslant H - s\,(\mathrm{Ric}_-),$$

where Ric_- is the infimum of the Ricci curvatures in the unit sphere over the union of C_s and H_s is the mean curvature of C_s at p. Passing to the limit when $s \to 0$, we have (8.10). The way of getting the above inequality is to apply the formula

$$\frac{dH_t}{dt}(p) = -(\mathrm{Ric}(N, N) + |\sigma|^2) \leqslant -\mathrm{Ric}(N, N)$$

and integrate with respect to t between 0 and s. If equality holds in (8.10), then $A(C_0 \cap S) = A(S)$, so that $|D_0| = |E|$, from where we conclude $D_0 = E$. It follows that S is convex.

Let us see now

$$\lim_{s\to 0} A(r_s^{-1}(S)) = A(C_0 \cap S). \tag{8.11}$$

We use area formula for Lipschitz maps, and we get

$$\int_{r_s^{-1}(S)} \mathrm{Jac}(r_s)\, dC_s = A(C_0 \cap S),$$

so that

$$\begin{aligned} A(r_s^{-1}(S)) &= \int_{r_s^{-1}(S)} \mathrm{Jac}(r_s)\, dC_s + \int_{r_s^{-1}(S)} (1 - \mathrm{Jac}(r_s))\, dC_s \\ &= A(C_0 \cap S) + \int_{r_s^{-1}(S)} (1 - \mathrm{Jac}(r_s))\, dC_s \to A(C_0 \cap S), \end{aligned}$$

since $\mathrm{Jac}(r_s)$ converges uniformly to 1.

We finally see

$$\lim_{s\to 0} \int_{C_s \setminus r_s^{-1}(S)} GK_{C_s} = 0. \tag{8.12}$$

Note first that if $p \in C_0 \setminus S$ is a smooth point, then $GK_{C_0}(p) = 0$: otherwise, we could push D_0 near p toward the interior of D_0 to contradict the convex hull property of D_0.

Fix now $s_0 > 0$, and write

$$\int_{C_s \setminus r_s^{-1}(S)} GK_{C_s}\, dC_s = \int_{C_{s_0} \setminus r_{s_0}^{-1}(S)} (GK_{C_s} \circ r_{s_0 s})\, \mathrm{Jac}(r_{s_0 s})\, dC_{s_0},$$

where $r_{s_0 s} = r_s^{-1} \circ r_{s_0}$. The right integral is uniformly bounded because the second fundamental form of C_s in $q = r_{s_0 s}(p)$ applied to a unit vector $e \in T_q C_s$ equals to

$$\langle A_s(e), e \rangle = \frac{\langle J', J \rangle}{|J|^2},$$

where J is a Jacobi field along the geodesic $\gamma(s) = r_{s_0 s}(p)$ that has modulus 1 over C_{s_0}, orthogonal to C_{s_0}, and such that $J/|J| = e$. J is induced by a family of orthogonal geodesics leaving from C_{s_0}. The quantity $\langle A_s(e), e \rangle$ is bounded by the classical comparison theorems for geodesics starting from a submanifold with sectional curvature bounded from below. We remark that the second fundamental form of C_s is uniformly bounded from above, since every point in C_s is supported by a ball of radius s.

By the above discussion, $GK_{C_s} \circ r_{s_0 s}$ converges to 0 for almost every point of $C_{s_0} \setminus r_{s_0}^{-1}(S)$, so that the integral of GK_{C_s} converges to 0 in $C_{s_0} \setminus r_{s_0}^{-1}(S)$.

Then we conclude from (8.10), (8.11), and (8.12)

$$4\pi \leqslant (c + H^2) A(S),$$

which implies (8.7).

Should equality hold in (8.7), we would get:

- ∂E has constant mean curvature equal to the one of the geodesic ball of the same area in $\mathbb{M}_c^3$.
- The sectional curvature of the tangent plane to ∂E equals c.
- ∂E is totally umbilical.

Then a result by Schroeder and Ziller [400] implies that ∂E is a geodesic ball in $\mathbb{M}_c^m$. This concludes the proof.

To finish this section, we prove the properties of convex sets in Cartan–Hadamard manifolds we have used in the proof.

Proposition 8.6 *Let M be a Cartan–Hadamard manifold and D_0 a compact convex set in M. Then:*

1. *The metric projection $r : M \to \partial D_0$ is well-defined, and it is distance non-increasing.*
2. *The gradient of the distance function $d(\cdot, C)$ exists at any point $p \in M \setminus D_0$, and it is equal to the tangent vector to the unique unit speed minimizing geodesic joining $r(p)$ to p.*
3. *The hypersurfaces $(\partial D_0)_s$ are of class $C^{1,1}$ for all $s > 0$.*

In particular, the hypersurfaces ∂D_s are of class $C^{1,1}$ for $s > 0$.

Proof Let us check 1. Take a point $p \in M \setminus D_0$, and assume there are two different points $p_1, p_2 \in D_0$ such that $d(p, p_1) = d(p, p_2) = d(p, C)$. Let $\gamma : [0, L] \to M$ be the unit speed geodesic satisfying $\gamma(0) = p_1$ and $\gamma(L) = p_2$, where $L = d(p_1, p_2)$. By comparison for sectional curvature, the function $f(t) = (d^2/2)(\gamma(t), p)$ is strictly convex. This implies

$$\begin{aligned} d(\gamma(t), p) < \min\{d(p, \gamma(0)), d(\gamma(L), p)\} &= \min\{d(p, p_1), d(p, p_2)\} \\ &= d(p, C). \end{aligned}$$

Since $\gamma(t) \in D_0$ for all $t \in [0, L]$ by the convexity of D_0, the above displayed inequality yields a contradiction. That $d(\cdot, C)$ is distance non-increasing is well-known. From now on, we denote the point $r(p)$, sometimes called the footpoint of p in D_0, by p^*.

To prove assertion 2, we notice first that $d(\cdot, C)$ is Lipschitz and, by Rademacher's theorem, differentiable almost everywhere. Let $p \in M \setminus D_0$ be a regular point of $d(\cdot, C)$, let $R = d(p, C)$, and let $\gamma : [0, R] \to M$ be the unit speed geodesic connecting p^* and p. Then $d(\gamma(R + \varepsilon, C) = R + \varepsilon$ for $\varepsilon > 0$, which implies $\langle (\nabla d)_p, \gamma'(R)\rangle = 1$. On the other hand, if $v \in T_pM$ is orthogonal to $\gamma'(R)$, we consider a curve $\alpha(s)$ so that $\alpha(0) = p$ and $\alpha'(0) = v$. Then $d(\alpha(s), C) \leqslant d(\alpha(s), p^*)$, and both functions are differentiable and coincide at $s = 0$. Build a one-parameter family of geodesics defined on $[0, R]$ joining p^* and $\alpha(s)$, whose length is $d(\alpha(s), p^*)$. The first variation formula for the length then implies

$$\langle (\nabla d(\cdot, C))_p, v\rangle = \frac{d}{ds}\bigg|_{s=0} d(\alpha(s), C) = \frac{d}{ds}\bigg|_{s=0} d(\alpha(s), p^*) = \langle \gamma'(R), v\rangle = 0.$$

Hence, $(\nabla d(\cdot, C))_p = \gamma'(R)$ as claimed.

For every point $p \in M \setminus D_0$, let $N(p)$ be the tangent vector at p to the unit speed geodesic γ_p connecting p^* and p. Then N is continuous in $M \setminus D_0$: if $\{p_i\}_{i\in\mathbb{N}}$ converges to p, then the footpoints p_i^* converge to p^* (otherwise, we would have two different minimizing geodesics connecting D_0 and p), and also the unit tangent vectors $\gamma_{p_i}'(0)$ converge to $\gamma_p'(0)$. This implies that $N(p_i) \to N(p)$. Lemma 4.7 in Federer [153] implies that $d(\cdot, C)$ is C^1 on $M \setminus D_0$ and its gradient is N.

Finally, to prove assertion 3, we notice that the hypersurfaces $(\partial D_0)_s$ admit supporting balls of radius s at any point from both sides. Taking a local chart, the same property holds for the corresponding Euclidean hypersurface perhaps with a positive radius smaller than s. Then the Euclidean hypersurface is $C^{1,1}$; see Theorem 1.8 in Dalphin [128] or Lemma 2.6 in Ghomi and Spruck [180]. Hence, $(\partial D_0)_s$ is also $C^{1,1}$. □

8.4 The Three-Dimensional Case: A Proof Using the Willmore Functional

A different proof of Proposition 8.5 was given in [362]. It is based on the following estimate of Willmore type.

Proposition 8.7 *Let M be a Cartan–Hadamard manifold with sectional curvatures bounded above by a constant $c \leqslant 0$. Let $E \subset M$ be a compact set with $C^{1,1}$ boundary S and $H_{\partial E} \geqslant 0$. Then*

$$\int_{\partial E} \left(c + \left(\frac{H_{\partial E}}{2}\right)^2 \right) d(\partial E) \geqslant 4\pi, \tag{8.13}$$

and so

$$\left(c + \max_{\partial E} \left(\frac{H_{\partial E}}{2}\right)^2 \right) A(\partial E) \geqslant 4\pi. \tag{8.14}$$

If equality holds in (8.14), then E is isometric to a geodesic ball in $\mathbb{M}^3_c$.

Estimate (8.13) implies (8.7) since for any geodesic ball B in $\mathbb{M}^3_c$, we have

$$\left(c + \left(\frac{H_B}{2}\right)^2 \right) A(\partial B) = 4\pi,$$

and so (8.14) implies $\max_{\partial E} H_{\partial E} \geqslant H_B$ when $A(\partial E) = A(\partial B)$. Moreover, if $\max_{\partial E} H_{\partial E} = H_B$ when $A(\partial E) = A(\partial B)$, then equality holds in (8.14), and, by Proposition 8.7, the set E is isometric to a geodesic ball in $\mathbb{M}^3_c$. Hence, Proposition 8.7 implies the Cartan–Hadamard conjecture in dimension 3 by Sect. 8.2.

The proof of Proposition 8.7 is given in the following two subsections, where we consider separately the cases $c = 0$ and $c < 0$.

8.4.1 The Euclidean Case

Assume $K_{\text{sec}} \leqslant 0$. Let $S = \partial E$ and $H = H_{\partial E}$. Inequality (8.13) reads

$$\int_S \left(\frac{H}{2}\right)^2 dS \geqslant 4\pi. \tag{8.15}$$

Let $p \in S$ and let $d(\cdot) = d(\cdot, p)$ be the distance function to p. For every $\varepsilon > 0$, we consider the conformal metric

$$g_\varepsilon = \rho_\varepsilon^2\, g = e^{2u_\varepsilon}\, g,$$

where

$$\rho_\varepsilon = \frac{2\varepsilon}{1+\varepsilon^2 d^2}, \qquad u_\varepsilon = \log\left(\frac{2\varepsilon}{1+\varepsilon^2 d^2}\right).$$

In case M is the Euclidean space, this metric is obtained by applying a conformal transformation to the metric of the sphere and projecting this metric orthogonally to the Euclidean space by means of the stereographic projection.

Taking into account the well-known relation between the sectional curvatures of conformal metrics, we get

$$\begin{aligned} e^{2u_\varepsilon}(K_\varepsilon)_{\text{sec}} &= K_{\text{sec}} - \left(\frac{\varepsilon^2}{1+\varepsilon^2 d^2}\right)^2 4d^2 + \left(\frac{\varepsilon^2}{1+\varepsilon^2 d^2}\right)(\nabla^2 d^2(e,e) + \nabla^2 d^2(v,v)) \\ &= K_{\text{sec}} + \left(\frac{\varepsilon^2}{1+\varepsilon^2 d^2}\right)(\nabla^2 d^2(e,e) + \nabla^2 d^2(v,v) - 4) + e^{2u_\varepsilon}, \end{aligned} \tag{8.16}$$

where $(K_\varepsilon)_{\text{sec}}$ and K_{sec} are the sectional curvatures of the tangent plane generated by the unitary orthogonal vectors e, v (e.g., §5 in Chapter 1 of Chen [109]). Since on the Cartan–Hadamard manifold M we have $\nabla d^2(w,w) \geqslant 2$ for any unit vector w, (8.16) implies

$$e^{2u_\varepsilon}(K_\varepsilon)_{\text{sec}} \geqslant K_{\text{sec}} + e^{2u_\varepsilon}. \tag{8.17}$$

From now on, we assume that e, v generate the tangent plane to S. Hence,

$$\begin{aligned} \int_S H^2\, dS &= \int_S (H^2 + 4K_{\text{sec}})\, dS - \int_S 4K_{\text{sec}}\, dS \\ &= \int_S (H_\varepsilon^2 + 4(K_\varepsilon)_{\text{sec}})\, dS_\varepsilon - \int_S 4K_{\text{sec}}\, dS \\ &\geqslant \int_S H_\varepsilon^2\, dS_\varepsilon + 4\int_S dS_\varepsilon \\ &\geqslant 4\int_S dS_\varepsilon, \end{aligned}$$

where in the second equality we have used the conformal invariance of $\int_S (H^2 + 4K_{\text{sec}})\, dS$ and in the first inequality we have used (8.17). The area element dS_ε is computed with respect to the metric g_ε, and H_ε is the mean curvature of S in (M, g_ε). The limit of the last integral can be computed by passing to polar (ambient) coordinates or taking into account that it corresponds geometrically to blowing up the surface S at the point p with a spherical metric, so that

$$\lim_{\varepsilon\to\infty}\int_S dS_\varepsilon = 4\pi.$$

So we get (8.15) and

$$\max_{\partial E}\Big(\frac{H_{\partial E}}{2}\Big)^2 A(\partial E) \geqslant 4\pi. \tag{8.18}$$

To analyze what happens when equality holds in (8.13), it is necessary to rewrite the estimate of $\int_S H^2\,dS$ as follows:

$$\begin{aligned}
4\pi &= \int_S H^2\,dS\\
&= \int_S (H^2+4K_{\text{sec}})\,dS - \int_S 4K_{\text{sec}}\,dS\\
&= \int_S (H_\varepsilon^2+4(K_\varepsilon)_{\text{sec}})\,dS_\varepsilon - \int_S 4K_{\text{sec}}\,dS\\
&= 4\int_S dS_\varepsilon + 4\int_S\Big(\frac{\varepsilon^2}{1+\varepsilon^2 d^2}\Big)\big(\nabla^2 d^2(e,e)+\nabla^2 d^2(v,v)-4\big)\,dS\\
&\quad + \int_S H_\varepsilon^2\,dS_\varepsilon.
\end{aligned}$$

We already know that $\lim_{\varepsilon\to 0}\int_S dS = 4\pi$. Since

$$\nabla^2 d^2(e,e)+\nabla^2 d^2(v,v)-4 \geqslant 0$$

and $H_\varepsilon^2 \geqslant 0$, the limit of both integrands is 0. In particular,

$$\int_S\Big(\frac{1}{d^2}\Big)\big(\nabla^2 d^2(e,e)+\nabla^2 d^2(v,v)-4\big)\,dS = 0.$$

Hence, we have

$$\nabla^2 d^2(e,e) = \nabla^2 d^2(v,v) = 2.$$

Standard comparison theorems in the Riemannian geometry then show that, if the geodesic starting from p leaves the enclosed domain E in a non-tangential way, then $\nabla^2 d^2 = 2\,g$ at the leaving point. Standard comparison also shows that $\nabla d^2 \equiv 2\,g$ along the geodesic. Changing slightly the initial speed of the geodesic, we get a cone so that $\nabla d^2 \equiv 2\,g$ inside this cone. Since every point in the interior of E can be connected with S by a minimizing geodesic hitting S orthogonally, we conclude that every point inside S is flat and so E is flat. Indeed, this argument shows that the convex envelope of E (i.e., the intersection of all convex subsets in M containing E) is flat.

If equality holds in (8.18), then the mean curvature H of ∂E is constant. As we already know that E is flat, Corollary 4.33 or Theorem 5.33 implies

$$3\,|E| \leqslant \frac{2}{H}\,A(S),$$

and equality holds if and only if E is isometric to a geodesic ball in Euclidean space. But the classical Minkowski formula

$$3\,|E| = \frac{2}{H}\,A(S)$$

holds in E since the function $(1/2)\,d^2$ has Hessian on E proportional to twice the identity matrix. From this, we conclude our proof of Proposition 8.7 in the flat case.

Remark 8.8 If equality holds in (8.15) and we assume that $H_{\partial E} > 0$, then Theorem 5.33 implies that E is isometric to a geodesic ball in $\mathbb{R}^3$. For isoperimetric mimimizers in balls of Cartan–Hadamard manifolds, we know that their mean curvature is strictly positive and bounded below by the one of the ball.

8.4.2 The Hyperbolic Case

By scaling the metric on M, we assume $c = -1$. Letting $H = H_{\partial E}$, the estimate to be obtained in this case is

$$\int_S \left(-1 + \left(\frac{H}{2}\right)^2 \right) dS \geqslant 4\pi. \tag{8.19}$$

We consider the following family of conformal metrics on M.

$$g_\varepsilon = \left(\frac{2\varepsilon}{(1-\varepsilon^2) + (1+\varepsilon^2)\cosh(d)} \right) g, \qquad \varepsilon > 1.$$

When M is the hyperbolic space, this family is obtained by writing the spherical metric in a disk D of $\mathbb{R}^m$ via stereographical projection in terms of the hyperbolic metric of constant curvature -1 in D and applying a family of conformal transformations.

Expressing the relation between the sectional curvatures of a plane generated by the orthonormal vectors e, v

$$e^{2u_\varepsilon}(K_\varepsilon)_{\mathrm{sec}} = K_{\mathrm{sec}} + 1 + e^{2u_\varepsilon} + \left(\frac{1+\varepsilon^2}{(1-\varepsilon^2)+(1+\varepsilon^2)\cosh(d)}\right)\times$$
$$\times\left(\nabla^2\cosh(d)(e,e) + \nabla^2\cosh(d)(v,v) - 2\cosh(d)\right).$$

Since $\nabla^2\cosh(d)(w,w) \geqslant \cosh(d)$ for a unit vector w, we obtain

$$e^{2u_\varepsilon}(K_\varepsilon)_{\mathrm{sec}} \geqslant K_{\mathrm{sec}} + e^{2u_\varepsilon} + 1.$$

From now on, we consider the tangent plane to S. Taking into account the conformal invariance of $\int_S (H^2 + 4K_{\mathrm{sec}})\,dS$ and the fact that

$$\lim_{\varepsilon\to\infty}\int_S dA = 4\pi,$$

we get

$$\begin{aligned}\int_S(-4+H^2)\,dS &= \int_S (H^2+4K_{\mathrm{sec}})\,dS - \int_S 4\,(K_{\mathrm{sec}}+1)\,dS\\ &= \int_S (H_\varepsilon^2 + 4(K_\varepsilon)_{\mathrm{sec}})\,dS_\varepsilon - \int_S 4\,(K_{\mathrm{sec}}+1)\,dS\\ &\geqslant \int_S H_\varepsilon^2\,dS_\varepsilon + 4\int_S dS_\varepsilon\\ &\geqslant 4\int_S dS_\varepsilon,\end{aligned}$$

which yields (8.19) and

$$\left(-1+\max_{\partial E}\left(\frac{H}{2}\right)^2\right)A(S) \geqslant 4\pi. \tag{8.20}$$

To analyze the equality in (8.19), it is more convenient to rewrite the estimate on $\int_S(-4+H^2)\,dS$ as

$$16\pi = \int_S(-4+H^2)\,dA = 4\int_S dS_\varepsilon + \int_S H_\varepsilon^2\,dS_\varepsilon$$
$$+4\int_S\left(\frac{1+\varepsilon^2}{(1-\varepsilon^2)+(1+\varepsilon^2)\cosh(d)}\right)\times$$
$$\times\left(\nabla^2\cosh(d)(e,e)+\nabla^2\cosh(d)(v,v)-2\cosh(d)\right)dA.$$

Letting $\varepsilon \to \infty$ and taking into account that $\lim_{\varepsilon\to\infty} \int_S dA_\varepsilon = 4\pi$, we deduce that the remaining positive integrals tend to 0 when $\varepsilon \to \infty$. In particular,

$$\nabla^2 \cosh(d)(e, e) = \nabla^2 \cosh(d)(v, v) = \cosh(d).$$

By standard comparison theorems, and arguing as in the Euclidean case, we conclude that the metric in the convex envelope of E is hyperbolic.

If, in addition, equality holds in (8.20), then H is constant. Moreover, from Theorem 4.38, we conclude

$$\int_S \left(\cosh(d) + \frac{H}{2} \sinh(d)\langle \partial/\partial d, N\rangle\right) dA \geqslant 0,$$

and equality holds only when S is a geodesic sphere. But since the metric in E is hyperbolic, we have $\nabla^2 \cosh(d) = 2\langle\, ,\rangle$, so that

$$\int_S \left(\cosh(d) + \frac{H}{2} \sinh(d)\langle \partial/\partial d, N\rangle\right) dA = 0,$$

by Corollary 4.37. Hence, Proposition 8.7 also follows in the hyperbolic case.

8.5 The Four-Dimensional Case: Croke's Proof

We show in this section that an isoperimetric inequality of Euclidean type holds on any Cartan–Hadamard manifold of dimension $m \geqslant 3$. This inequality coincides with the Euclidean isoperimetric inequality only when $m = 4$. This result was proven by Croke in [127]. We closely follow his proof.

Theorem 8.9 *Let M be a Cartan–Hadamard manifold with dimension $m \geqslant 3$ and $E \subset M$ a bounded set with smooth boundary. Then*

$$P(E) \geqslant C(m)|E|^{(m-1)/m}, \tag{8.21}$$

where

$$C(m)^m = \frac{\mathbf{c}_{m-1}^{m-1}}{\mathbf{c}_{m-2}^{m-2}\left\{\int_0^{\pi/2} \cos(t)^{m/(m-2)} \sin(t)^{m-2}\, dt\right\}^{m-2}}. \tag{8.22}$$

Equality holds in (8.21) if and only if $n = 4$ and E is isometric to a flat geodesic ball.

The proof is based on techniques of geometric probability. We consider the unit sphere bundle $\pi : UM \to M$ with its canonical *kinematic density* or *Liouville measure* dv, defined locally as the product of the Riemannian measure with the standard measure of spheres. For any $v \in UM$, we denote by $\gamma_v(t)$ the geodesic with initial conditions $\gamma(0) = \pi(v)$ and $\gamma_v'(0) = v$. The *geodesic flow* ξ^t on UM assigns to any $v \in UM$ the unit vector $\gamma_v'(t)$. It is well-known that the kinematic density is invariant by the geodesic flow (e.g., see Theorem VII.1.3 in Chavel [102]).

Let $E \subset M$ be a compact set with smooth boundary S and inner unit normal N, and let U^+S be the bundle of inwardly pointing unit vectors, that is,

$$U^+S = \{u \in UM : \pi(u) \in S, \langle u, N_{\pi(u)}\rangle > 0\}.$$

For every $p \in S$, we let $U_p^+S = U_pM \cap U^+S$. The measure du on U^+S is locally the product of the measure of S and of the half-sphere $\{u \in U_pM : \langle u, N_{\pi(u)}\rangle > 0\}$. We represent $\langle u, N_{\pi(u)}\rangle$ by $\cos(u)$. For $u \in \pi^{-1}(E)$, let

$$\ell(u) = \max\{t : \gamma_v(t) \in E\}.$$

Note that $\gamma_u(\ell(u)) \in S$. We also define $\mathrm{ant}(u) = -\gamma_u'(\ell(u))$. Observe that $\ell(\mathrm{ant}(u)) = \ell(u)$.

A main ingredient in the proof of Theorem 8.9 is the following result, a proof of which can be found in Santaló; see pp. 336–338 in [392].

Proposition 8.10 *Let $f : \pi^{-1}(E) \to \mathbb{R}$ be an integrable function. Then*

$$\int_{\pi^{-1}(E)} f(v)\,dv = \int_{U^+S}\int_0^{\ell(u)} f(\xi^t(u))\cos(u)\,dt\,du. \tag{8.23}$$

The proof of Theorem 8.9 is based on the following two lemmas:

Lemma 8.11

1. $|E| = \mathbf{c}_{m-1}^{-1}\int_{U^+S}\ell(u)\cos(u)\,du$.
2. *For any integrable function g on U^+S, we have*

$$\int_{U^+S} g(u)\cos(u)\,du = \int_{U^+S} g(ant(u))\cos(u)\,du.$$

Proof Part 1 follows from (8.23) taking $f = 1$ since $\int_{\pi^{-1}(E)} dv = \mathbf{c}_{m-1}\,|E|$.

To prove part 2, we first observe that (8.23) implies that the geodesic flow ξ is a volume-preserving map from Q to $\pi^{-1}(E)$, where $Q = \{(u, t) : u \in U^+S, 0 \leqslant t \leqslant \ell(u)\}$ has the measure $\cos(u)\,dt\,du$. The map ξ has an inverse which is smooth almost everywhere since

$$\xi^{-1}(v) = (-\gamma_v'(\ell(-v)), \ell(-v)).$$

As the antipodal map $-\mathbf{1}$ is volume preserving on $\pi^{-1}(E)$, the map $\xi^{-1}\circ(-\mathbf{1})\circ\xi : Q \to Q$ is volume preserving, but clearly

$$\left(\xi^{-1}\circ(-\mathbf{1})\circ\xi\right)(u,t) = (\mathrm{ant}(u), \ell(u)-t).$$

So we get, for any measurable function $G : Q \to \mathbb{R}$,

$$\int_{U^+S}\int_0^{\ell(u)} G(u,t)\cos(u)\,dt\,du = \int_{U^+S}\int_0^{\ell(u)} G(\mathrm{ant}(u), \ell(u)-t)\cos(u)\,dt\,du.$$

To obtain 2, we simply take $G(u,t) = g(u)/\ell(u)$ in the previous formula and integrate with respect to t. □

Lemma 8.12

1. *We have*

$$\int_{U^+S} \frac{\ell(u)^{m-1}}{\cos(ant(u))}\,du \leqslant A(S)^2.$$

   Equality holds if and only if E is flat and convex.
2. *We have*

$$\int_{U^+S} (\cos(ant(u)))^{1/(m-2)}(\cos(u))^{(m-1)/(m-2)}\,du \leqslant C_2(m)\,A(S),$$

   where

$$C_2(m) = \mathbf{c}_{m-2}\int_0^{\pi/2} \cos^{m/(m-2)}(t)\,\sin^{m-2}(t)\,dt.$$

   Equality holds if and only if $\cos(u) = \cos(ant(u))$ *almost everywhere.*

Proof Let $q \in S$. In normal polar coordinates based on q, we have $dM = F(u,r)\,du\,dr$ in the region

$$\left\{\exp_q(tu) : 0 \leqslant t \leqslant \ell(u)\right\}.$$

Here, $F(u,r)$ is the Jacobian of the parameterization of $S(q,r)$ by normal coordinates. By the discussion in Sect. 1.3.5, we know that $F(u,r) \geqslant r^{m-1}$. Let

$$B = \exp_q\{\ell(u)\,u : \langle u, N_{\pi(u)}\rangle \geqslant 0\}.$$

Then $B \subset S$, and the measure dS on B is equal to

$$dS = \frac{F(u,\ell(u))}{\cos(\mathrm{ant}(u))}\,du.$$

So we have

$$\int_{U_q^+S} \frac{F(u, \ell(u))}{\cos(\mathrm{ant}(u))}\, du = A(B) \leqslant A(S).$$

Integrating over $q \in S$, we have

$$\int_{U^+S} \frac{F(u, \ell(u))}{\cos(\mathrm{ant}(u))}\, du \leqslant A(S)^2.$$

Finally, estimating $F(u, \ell(u)) \geqslant \ell(u)^{m-1}$, we get 1.

To prove 2, we apply the Cauchy-Schwarz inequality and Lemma 8.11(1) to obtain

$$\begin{aligned}
&\int_{U^+S} \cos(\mathrm{ant}(u))^{1/(m-2)} \cos(u)^{(m-1)/(m-2)} du \\
&= \int_{U^+S} \cos(\mathrm{ant}(u))^{1/(m-2)} \cos(u)^{1/(m-2)} \cos(u)\, du \\
&\leqslant \left(\int_{U^+S} \cos(\mathrm{ant}(u))^{2/(m-2)} \cos(u)\, du\right)^{1/2} \left(\int_{U^+S} \cos(u)^{2/(m-2)} \cos(u)\, du\right)^{1/2} \\
&= \int_{U^+S} \cos(u)^{m/(m-2)} du \\
&= \int_S \left\{\int_{U_q^+S} \cos(u)^{m/(m-2)}\, du\right\} dM \\
&= A(S) \cdot C_2(m).
\end{aligned}$$

Equality holds if and only if $\cos(\mathrm{ant}(u))$ and $\cos(u)$ are proportional almost everywhere. Since the maximum value of both functions is equal to 1, it is clear that $\cos(\mathrm{ant}(u)) = \cos(u)$. □

Using Lemmas 8.11 and 8.12, we now give the proof of the main result.

Proof of Theorem 8.9 By Lemma 8.11(1) and Hölder's inequality,

$$\begin{aligned}
|E| &= \frac{1}{\mathbf{c}_{m-1}} \int_{U^+S} \ell(u) \cos(u)\, du \\
&= \frac{1}{\mathbf{c}_{m-1}} \int_{U^+S} \frac{\ell(u)}{\cos(\mathrm{ant}(u))^{1/(m-1)}} \cos(\mathrm{ant}(u))^{1/(m-1)} \cos(u)\, du \\
&\leqslant \frac{1}{\mathbf{c}_{m-1}} \left(\int_{S^+U} \frac{\ell(u)^{m-1}}{\cos(\mathrm{ant}(u))}\right)^{1/(m-1)} \cdot \\
&\quad \cdot \left(\int_{U^+S} \cos(\mathrm{ant}(u))^{1/(m-2)} \cos(u)^{(m-1)/(m-2)} du\right)^{(m-2)/(m-1)}.
\end{aligned}$$

Lemma 8.12 implies

$$|E| \leqslant \frac{1}{\mathbf{c}_{m-1}} A(S)^{2/(m-1)} A(S)^{(m-2)/(m-1)} C_2(m)^{(m-2)/(m-1)}.$$

Hence, we get

$$C(m)\,|E|^{m-1} = \frac{\mathbf{c}_{m-1}}{C_2(m)^{m-2}}\,|E|^{m-1} \leqslant A(S)^m.$$

This implies (8.21).

Equality is obtained in (8.21) if equality holds in Lemma 8.12 and in the above Hölder inequality. In particular, E must be flat and convex, and hence, the equality follows from the classical isoperimetric inequality in $\mathbb{R}^m$. But $C(4)$ is the Euclidean isoperimetric constant and $C(m)$, for $m \neq 4$, is not. □

8.6 Notes

Notes for Sect. 8.2 Apart from the two-dimensional results by Weil [439], Beckenbach and Radó [49], Bol [68], and Fiala [158], the generalized Conjecture 8.3 has only been proven in dimension 3 by Kleiner [247] and Conjecture 8.2 in case $K_{\text{sec}} \leqslant 0$ by Croke [127]. Some other proofs in the three-dimensional case were given by Ritoré [362] (of the generalized conjecture) and Schulze [401]. Croke's results have been revisited by Kloeckner and Kuperberg [248].

The conjecture has been proven for small volumes when the dimension is odd by Morgan and Johnson (see Theorem 4.4 in [311]) and also for all dimensions and small volume under the stronger assumption $K_{\text{sec}} < c \leqslant 0$.

Cao and Escobar obtained in [93] an optimal isoperimetric inequality in simply connected piecewise flat manifolds with non-positive curvature.

Non-optimal isoperimetric inequalities of Euclidean type in Cartan–Hadamard manifolds have been obtained by Hoffman and Spruck [231, 232] and Croke [126, 127].

Notes for Sect. 8.3 The strategy of Kleiner [247] could be extended to an m-dimensional Cartan–Hadamard manifold provided the total Gauss-Kronecker curvature of a bounded convex set with $C^{1,1}$ boundary can be estimated from below by the volume of $\mathbb{S}^{m-1}$; see Problem 1.1 in page 2 of Ghomi and Spruck [180]. In this paper, the authors provide detailed proofs of the validity of Kleiner's strategy in higher-dimensional Cartan–Hadamard manifolds.

Some work has been done to estimate the total Gauss-Kronecker curvature of a bounded convex set in a Cartan–Hadamard manifold. Willmore and Saleemi [446] stated Problem 1.1 in [180], but their proof contained a mistake. Borbély [73] has proven the estimate for some convex sets in rotationally symmetric Cartan–Hadamard manifold and also that the total curvature of convex sets is monotone with respect to the inclusion in hyperbolic spaces [74]. Unfortunately, this property is no longer true in general Cartan–Hadamard manifolds; see Dekster [136].

Notes for Sect. 8.4 The proof in [362] only employs a lower bound on the Willmore functional inspired on techniques developed by Li and Yau [270]. The same technique was used by Choe and Ritoré [120] to provide an optimal isoperimetric inequality outside a convex set in a three-dimensional Cartan–Hadamard manifold.

Notes for Sect. 8.5 Although we have discussed Croke's Theorem 8.9 for a subset of a Cartan–Hadamard manifold, the result in [127] is stated and proved for a manifold with boundary, assuming that every geodesic ray minimizes length until it hits the boundary. Such a manifold is not necessarily contained in a manifold with boundary as shown by Hass [220], who constructed a negatively curved ball with totally concave boundary. Kloeckner and Kuperberg have shown in Theorem 1.9 in [248] the existence, for every $\varepsilon > 0$, of a ball with sectional curvature $-1 \leqslant K \leqslant -(1-\varepsilon)$ with arbitrarily large volume and bounded area depending only on ε. Indeed, they obtain in Corollary 1.10 that, for every $v, a > 0$, there exists a Riemannian 3-ball Ω with $K \leqslant -1$, $|\Omega| = v$, $A(\partial\Omega)$, thus preventing the existence of any possible isoperimetric relation in a negatively curved ball. The reader is referred to the interesting paper by Kloeckner and Kuperberg [248] for a deeper insight in Croke's result.

Chapter 9
Relative Isoperimetric Inequalities

In this chapter, isoperimetric inequalities on domains with smooth boundary in complete Riemannian manifolds are considered. The perimeter here is the one *relative* to the domain Ω. If a smooth interface S separates Ω into two sets, the relative perimeter of each one of these sets is the area of the interface. This means that there are no contributions to the relative perimeter coming from pieces in $\partial\Omega$. While many of the techniques are adapted from the boundaryless case, some subtle differences appear. The geometry of the boundary, in particular its second fundamental form, will play a determinant role.

We follow in this chapter a similar scheme to the boundaryless case in Chap. 3. We start by computing the first and second variation formulas of the perimeter and volume. This leads to the definition of *stable* hypersurface in this setting: a second-order minimum of the relative perimeter under a volume constraint. As expected, terms involving the boundary of the interface appear in these formulas. We see that the isoperimetric profile for small volumes is asymptotic to the isoperimetric profile of a Euclidean half-space. Then a differential inequality for sets with *convex boundary* in manifolds with Ricci curvature satisfying the inequality $\text{Ric} \geqslant (m-1)\delta$ is obtained. Convexity of the boundary is defined infinitesimally in terms of the non-negativity of the second fundamental form of $\partial\Omega$. This differential inequality implies some comparison results, including a version of the Lévy-Gromov isoperimetric inequality. The differential inequality for convex domains in a Riemannian manifold was obtained by Bayle and Rosales [48] after previous work by Bayle [46] for manifolds without boundary; see also Chap. 3.

Afterward, a classification of *stable* hypersurfaces inside and outside Euclidean balls, following a recent result by Wang and Xia [436], is obtained. These results are also valid on spheres and hyperbolic spaces and for isoperimetric boundaries with a small singular set.

One of the consequences of the Bayle-Rosales differential inequality is the concavity of the function I_C^{α}, for $1 \leqslant \alpha \leqslant m/(m-1)$, where I_C of the isoperimetric profile of a convex body C with smooth boundary in the Euclidean space $\mathbb{R}^m$.

M. Ritoré, *Isoperimetric Inequalities in Riemannian Manifolds*,
Progress in Mathematics 348, https://doi.org/10.1007/978-3-031-37901-7_9

By approximation on Hausdorff distance of any convex set by convex sets with smooth boundary, we are able to extend this result to arbitrary convex bodies and to extract some geometric and topological consequences for isoperimetric sets. Density estimates are obtained, and it is proven that the isoperimetric profile for small volumes is asymptotic to the one of the tangent cone to the convex body with the smallest aperture. In particular, isoperimetric sets of small volumes in polytopes are characterized. They are small spheres centered at the vertex with the largest aperture in the boundary of the polytope. The concavity of the profile for arbitrary convex sets in $\mathbb{R}^m$ was first proven by E. Milman [284]. For most of the results in this last section, we follow Ritoré and Vernadakis [371].

9.1 The Isoperimetric Profile of a Domain

Given an open set $\Omega \subset M$ and a measurable set $E \subset M$, the *relative perimeter* of E in Ω is defined by

$$P(E, \Omega) = \sup\left\{ \int_S \operatorname{div} X \, dM : X \in \mathfrak{X}_0(\Omega), |X| \leqslant 1 \right\},$$

where $\mathfrak{X}_0(\Omega)$ is the set of smooth vector fields with compact support in Ω.

If $S = \partial E \cap \Omega$ is a C^1 hypersurface, then $P(E, \Omega)$ is the area of S inside Ω by the divergence theorem.

Definition 9.1 Given a domain Ω in a Riemannian manifold (M, g), we define the *isoperimetric profile* I_Ω of Ω by

$$I_\Omega(v) = \inf\left\{P(E, \Omega) : E \subset \Omega, |E| = v\right\}. \tag{9.1}$$

When the volume of Ω is finite, the *normalized isoperimetric profile* of Ω is the function $h_\Omega : (0, 1) \to \mathbb{R}^+$ defined by

$$h_\Omega(\lambda) = \frac{I_\Omega(\lambda|\Omega|)}{|\Omega|}. \tag{9.2}$$

9.1.1 Isoperimetric Sets: Boundary Regularity of Isoperimetric Sets

Definition 9.2 Given a domain Ω in a Riemannian manifold (M, g), an *isoperimetric set* in Ω is a measurable set $E \subset \Omega$ satisfying

$$P(E, \Omega) = I_\Omega(|E|).$$

In case Ω is relatively compact with Lipschitz boundary, the compactness property in Theorem 1.42 and the lower semicontinuity of perimeter in Proposition 1.41 imply the existence of isoperimetric sets for any given volume. The lower semicontinuity of the perimeter and the existence of local deformations also imply the continuity of the isoperimetric profile as in Remark 3.6. The positivity of the profile follows from a local relative isoperimetric inequality in balls inside Ω.

The problem of finding perimeter minimizers under a volume constraint in a set is often named as the *partitioning problem.*

Theorem 9.3 *Let Ω be a relatively compact domain with Lipschitz boundary in a Riemannian manifold M. Then:*

1. *Isoperimetric sets exist on Ω for any volume $0 < v < |\Omega|$.*
2. *The isoperimetric profile I_Ω is continuous.*
3. *$I_\Omega > 0$ on $(0, |\Omega|)$.*

As for the regularity, we have the following result, which follows from Giusti [184]; Gonzalez et al. [186]; Bombieri [69]; Grüter [203]; and Morgan [302]; see Proposition 2.3 in Bayle and Rosales [48].

Theorem 9.4 *Let E be a measurable set of finite volume minimizing perimeter under a volume constraint in a domain Ω with smooth boundary of a M. Then:*

1. *If $m \leqslant 7$, then the boundary $S = \mathrm{cl}(\partial E \cap \Omega)$ of E is a smooth hypersurface.*
2. *If $m > 7$, then the boundary of $\mathrm{cl}(\partial E \cap \Omega)$ is the union of a smooth hypersurface S and a closed singular set S_0 of Hausdorff dimension at most $m - 8$ (i.e., $\mathcal{H}^{m-8+\gamma}(S_0) = 0$ for all $\gamma > 0$).*

Theorem 9.7 then implies that the regular part S has constant scalar mean curvature and meets $\partial\Omega$ in an orthogonal way. Moreover, S is stable in the sense that the second derivative of the area of S under volume-preserving variations is non-negative.

A hypersurface $S \subset \Omega$ satisfying $\partial S \subset \partial\Omega$ which separates Ω into two sets is called an *interface.*

9.1.2 *The Isoperimetric Profile of a Half-Space*

Let $\mathbb{M}^m_\kappa$, for $m \geqslant 2$, be the complete simply connected manifold with constant sectional curvatures equal to κ. A complete totally geodesic hypersurface S in $\mathbb{M}^m_\kappa$ separates the manifold into two connected open sets Ω^+_S and Ω^-_S, which are clearly isometric since they are applied one into the other by the orthogonal reflection on S. The sets $\Omega^\pm$ are called *open half-spaces* in $\mathbb{M}^m_\kappa$. Given another complete totally geodesic hypersurface S', there is always an isometry of $\mathbb{M}^m_\kappa$ taking S into S' and so taking Ω^+_S onto $\Omega^\pm_{S'}$. We denote by $\mathbb{H}^m_\kappa$ any *open half-space* in $\mathbb{M}^m_\kappa$.

The optimal isoperimetric inequality in $\mathbb{H}^m_\kappa$ can be obtained from the one in $\mathbb{M}^m_\kappa$.

Lemma 9.5 *For every $\kappa \in \mathbb{R}$ and $m \in \mathbb{N}$, $m \geqslant 2$,*

$$I_{\mathbb{M}^m_\kappa}(v) = 2\, I_{\mathbb{H}^m_\kappa}\left(\tfrac{v}{2}\right). \tag{9.3}$$

Moreover, the isoperimetric sets in $\mathbb{H}^m_\kappa$ are the intersection with $\mathbb{H}^m_\kappa$ of geodesic balls in $\mathbb{H}^m_\kappa$ centered at points in $\partial\mathbb{H}^m_\kappa$.

Proof Let $E \subset \mathbb{M}^m_\kappa$ be an isoperimetric set of volume $v > 0$ and S a totally geodesic submanifold separating $\mathbb{M}^m_\kappa$ into two open half-spaces Ω^+ and Ω^- such that $|E \cap \Omega^+| = |E \cap \Omega^-|$. Then

$$I_{\mathbb{M}^m_\kappa}(v) = P(E) \geqslant P(E, \Omega^+) + P(E, \Omega^-) \geqslant 2 I_{\mathbb{H}^m_\kappa}\left(\tfrac{v}{2}\right). \tag{9.4}$$

To prove the opposite inequality, let $F \subset \mathbb{H}^m_\kappa$ be an isoperimetric set of volume $v/2$ in a half-space Ω of $\mathbb{M}^m_\kappa$, and let f be the mirror symmetry with respect to $\partial\Omega$. Let $F' = f(F)$ and $\Omega' = f(\Omega)$. Then $|F \cup F'| = v$ and

$$I_{\mathbb{M}^m_\kappa}(v) \leqslant P(F \cup F') = P(F, \Omega) + P(F', \Omega') = 2 I_{\Omega^m_\kappa}\left(\tfrac{v}{2}\right). \tag{9.5}$$

Inequalities (9.4) and (9.5) imply (9.3). Moreover, if F is an isoperimetric set in a half-space, then (9.4) and (9.5) imply that $F \cup F'$ is an isoperimetric set in $\mathbb{M}^m_\kappa$. Hence, $F \cup F'$ is a ball and F is a half-ball. □

9.1.3 The Isoperimetric Profile for Small Volumes

For compact manifolds without boundary, Theorem 3.7 by Berard and Meyer [52] describes the behavior of the isoperimetric profile for small volumes. In the case of a domain with smooth boundary in a Riemannian manifold, the proof follows without relevant changes; see Lemme de localisation II in Appendix C of Berard and Meyer [52] or Proposition 2.1 in Bayle and Rosales [48].

Theorem 9.6 *Let Ω be a domain with smooth boundary and compact closure in a Riemannian manifold (M, g). Then there exists a positive constant $v_0 = v_0(\Omega, g, \varepsilon)$ such that any set $E \subset M$ of volume $0 < v \leqslant v_0$ satisfies*

$$P(E, \Omega) \geqslant (1 - \varepsilon)\, \frac{\mathbf{c}(m)}{2}\, |E|^{(m-1)/m}, \tag{9.6}$$

where $\mathbf{c}(m)$ is the isoperimetric constant in $\mathbb{R}^m$.

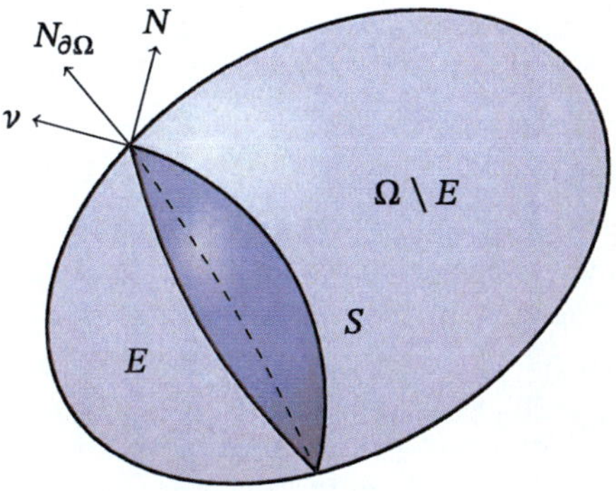

Fig. 9.1 Geometric configuration of an interface S separating Ω into two sets

9.2 Variation Formulas in Sets with Boundary

In this section, the first variation formula for the area of *interfaces*, hypersurfaces separating a domain into two open sets, is computed. Critical points of the area are hypersurfaces with constant mean curvature meeting $\partial\Omega$ along ∂S in an orthogonal way. The second derivative of the functional $(\text{area} - H \text{ vol})$ is also computed.

9.2.1 First Variation

The first variation of perimeter and volume of an interface is computed in the following result (Fig. 9.1).

Theorem 9.7 *Let $\Omega \subset M$ be an open set with smooth boundary and S be a smooth hypersurface with $\partial S \subset \partial\Omega$ separating Ω into two regions E and $\Omega \setminus E$. We assume that E has finite volume. Let $N_{\partial\Omega}$ be the outer unit normal to $\partial\Omega$, N the unit normal to S pointing into $\Omega \setminus E$, H the mean curvature of S with respect to N, and ν the unit outer conormal to ∂S inside S.*

Let $\{\varphi_t\}_{t\in\mathbb{R}}$ be a one-parameter family of diffeomorphisms with compact support such that $\varphi_t(\partial\Omega) \subset \partial\Omega$ for all t.

Then we have

$$\frac{d}{dt}\bigg|_{t=0} A(\varphi_t(S)) = \int_S H \langle X, N\rangle \, dS + \int_{\partial S} \langle X, \nu\rangle \, d(\partial S) \tag{9.7}$$

and

$$\frac{d}{dt}\bigg|_{t=0} |\varphi_t(E)| = \int_S \langle X, N\rangle \, dS. \tag{9.8}$$

Assume that, for any one-parameter family of diffeomorphisms $\{\varphi_t\}_{t\in\mathbb{R}}$ *such that* $(d/dt)_{t=0}|\varphi_t(E)| = 0$*, we have* $(d/dt)_{t=0}A(\varphi_t(S)) = 0$*. Then* H *is constant and* $\nu = N_{\partial\Omega}$ *on* ∂S*.*

Proof Formula (9.7) coincides with formula (1.19) for hypersurfaces. Equation (9.8) follows from (1.25) taking into account that $\partial E = S \cup (\overline{E} \cap \partial\Omega)$ and $\langle X, N_{\partial\Omega}\rangle = 0$, where $N_{\partial\Omega}$ is the outer unit normal to $\partial\Omega$, since $\varphi_t(\partial\Omega) \subset \partial\Omega$.

Assume that for any one-parameter family of diffeomorphisms $\{\varphi_t\}_{t\in\mathbb{R}}$ such that $(d/dt)_{t=0}|\varphi_t(E)| = 0$, we have $(d/dt)_{t=0}A(\varphi_t(S)) = 0$. Then it follows that H is constant on $S \cap \Omega$ (by using variations supported in Ω) and that $\langle v, \nu_p\rangle = 0$ for any $p \in \partial S \cap \partial\Omega$ and any $v \in T_p(\partial\Omega)$ (by extending $v \in T_p(\partial\Omega)$ to a vector field X with compact support tangent to $\partial\Omega$). Hence, ν is proportional to $N_{\partial\Omega}$, and, since both vectors have modulus 1 and point outside Ω, we obtain $\nu = N_{\partial\Omega}$. □

We made in Remark 1.15 the observation that the first variation formulas at any t_0 can be computed by considering the one-parameter family of diffeomorphisms $\psi_t = \varphi_{t+t_0} \circ \varphi_{t_0}^{-1}$ and using the formulas for the derivatives at $t = 0$. The vector field associated with this one-parameter family is denoted by X_{t_0} and is equal to

$$(X_{t_0})_p = \frac{d}{dt}\bigg|_{t=0} (\varphi_{t+t_0} \circ \varphi_{t_0}^{-1})(p).$$

This allows us to compute the first variation of area and volume at any instant t

$$\frac{d}{dt}A(\varphi_t(S)) = \int_{S_t} H_t\,\langle X_t, N_t\rangle\, dS_t + \int_{\partial S_t} \langle X_t, \nu_t\rangle\, d(\partial S_t), \tag{9.9}$$

$$\frac{d}{dt}|\varphi_t(E)| = \int_{S_t} \langle X_t, N_t\rangle\, dS_t, \tag{9.10}$$

where $S_t = \varphi_t(S)$, $E_t = \varphi_t(E)$, N_t is the unit normal to S_t pointing into $\Omega \setminus E_t$, and ν_t is the unit conormal to ∂S_t pointing outside S_t. Recall that, in case $\{\varphi_t\}_{t\in\mathbb{R}}$ is a one-parameter *group* of diffeomorphisms, equality $X_t = X$ holds for all t.

9.2.2 Second Variation

For the second variation, we reason as in Theorem 1.27. We have the following:

Theorem 9.8 *Let* $\Omega \subset M$ *be an open set with smooth boundary and* S *be a smooth hypersurface with* $\partial S \subset \partial\Omega$ *separating* Ω *into two regions* E *and* $\Omega \setminus E$*. We assume that* E *has finite volume. Let* $N_{\partial\Omega}$ *be the outer unit normal to* $\partial\Omega$*,* N *the unit normal to* S *pointing into* $\Omega \setminus E$*,* H *the mean curvature of* S *with respect to* N*, and* ν *the unit outer conormal to* ∂S *inside* S*.*

Let $\{\varphi_t\}_{t\in\mathbb{R}}$ *be a one-parameter family of diffeomorphisms with compact support such that* $\varphi_t(\partial\Omega) \subset \partial\Omega$ *for all* t*.*

If H is constant and $\nu = N_{\partial\Omega}$, then

$$\frac{d^2}{dt^2}\bigg|_{t=0} A(\varphi_t(S)) - H \frac{d^2}{dt^2}\bigg|_{t=0} |\varphi_t(E)|$$

is equal to

$$-\int_S u\,\big(\Delta_S u + \big(\operatorname{Ric}(N,N) + |\sigma|^2\big)\,u\big)\,dS + \int_{\partial S} u\left(\frac{\partial u}{\partial \nu} - \mathrm{II}(N,N)\,u\right) d(\partial S), \tag{9.11}$$

where II *is the second fundamental form of $\partial\Omega$ with respect to the outer unit normal $N_{\partial\Omega}$ and $u = \langle X, N\rangle$. Using the divergence theorem, formula* (9.11) *can be expressed as*

$$\int_S \big(|\nabla_S u|^2 - \big(\operatorname{Ric}(N,N) + |\sigma|^2\big)\,u^2\big)\,dS - \int_{\partial S} \mathrm{II}(N,N)\,u^2\,d(\partial S). \tag{9.12}$$

Proof We use formulas (9.9) and (9.10) for the first variation of perimeter and volume. Let $\psi_t = \varphi_{t+t_0} \circ \varphi_{t_0}^{-1}$ be the one-parameter family of diffeomorphisms with associated vector field X_{t_0}. Then

$$\begin{aligned}
\frac{d}{dt}\bigg|_{t=t_0} A(\varphi_t(S)) &= \frac{d}{dt}\bigg|_{t=0} A(\varphi_{t+t_0}(S)) \\
&= \frac{d}{dt}\bigg|_{t=t_0} A(\psi_t(S_{t_0})) \\
&= \int_{S_{t_0}} H_{t_0} \langle X_{t_0}, N_{t_0}\rangle\, dS_{t_0} + \int_{\partial S_{t_0}} \langle X_{t_0}, \nu_{t_0}\rangle\, d(\partial S_{t_0}),
\end{aligned}$$

where H_t is the mean curvature of $S_t = \varphi_t(S)$. The second derivative of this function, evaluated at $t = 0$, is equal to

$$\begin{aligned}
\frac{d^2}{dt^2}\bigg|_{t=0} A(\varphi_t(S)) &= \frac{d}{dt}\bigg|_{t=0} \int_S \big[H_t \langle X_t, N_t\rangle) \circ \varphi_t\big] \operatorname{Jac}(\varphi_t)\, dS \\
&\quad + \frac{d}{dt}\bigg|_{t=0} \int_{\partial S} \big[\langle X_t, \nu_t\rangle \circ \varphi_t\big] \operatorname{Jac}(\varphi_t)\, d(\partial S) \\
&= \int_S \frac{d}{dt}\bigg|_{t=0} (H_t \circ \varphi_t)\, u\, dS \\
&\quad + H \int_S \frac{d}{dt}\bigg|_{t=0} \big[(\langle X_t, N_t\rangle \circ \varphi_t) \operatorname{Jac}(\varphi_t)\big]\, dS
\end{aligned}$$

$$+\int_{\partial S} \frac{d}{dt}\bigg|_{t=0} \big[\langle X_t, \nu_t\rangle \circ \varphi_t\big]\, d(\partial S)$$

$$+\int_{\partial S} \langle X, \nu\rangle \frac{d}{dt}\bigg|_{t=0} \operatorname{Jac}(\varphi_t)\, d(\partial S).$$

As H is constant on S, the derivative of the mean curvature $(d/dt)_{t=0}(H_t \circ \varphi_t)$ is computed from (1.41) and equals $-\Delta_S u - (\operatorname{Ric}(N,N) + |\sigma|^2)\, u$. The derivative $(d/dt)_{t=0} \int_{S_t} \langle X_t, N_t\rangle\, dS_t$ is computed from (1.38), and it is equal to $(d^2/dt^2)_{t=0} |\varphi_t(E)|$. As $\langle X, \nu\rangle = \langle X, N_{\partial\Omega}\rangle = 0$ on ∂S, we get that $(A(\varphi_t(S)) - H\, |\varphi_t(E)|)''_{t=0}$ is equal to

$$-\int_S u\, \big(\Delta_S u + \big(\operatorname{Ric}(N,N) + |\sigma|^2\big)\, u\big)\, dS + \int_{\partial S} \frac{d}{dt}\bigg|_{t=0} \big[\langle X_t, \nu_t\rangle \circ \varphi_t\big]\, d(\partial S).$$

It remains to compute the last term. Note that X_t is tangent to $\partial\Omega$ for all t. Fix $p \in \partial S$, and let $\gamma(t) = \varphi_t(p)$. Let D/dt be the covariant derivative along γ. We have at $t = 0$

$$\frac{d}{dt}\langle X_t, \nu_t\rangle = \langle \frac{D}{dt} X_t, \nu_t\rangle + \langle X_t, \frac{D}{dt}\nu_t\rangle.$$

The first summand is

$$\langle \frac{D}{dt} X_t, \nu_t\rangle(0) = \langle \frac{D}{dt}\bigg|_{t=0} X_t, (N_{\partial\Omega})_p\rangle = -\langle X_p, \frac{D}{dt}\bigg|_{t=0} N_{\partial\Omega}\rangle = -\mathrm{II}(X_p, X_p).$$

For the second one, we decompose $X = X^\top + \langle X, N\rangle\, N$, where $X^\top$ is the orthogonal projection over ∂S. Letting $Z(t) = (d\varphi_t)_p(X_p^\top)$, Lemma 1.13 implies

$$\langle X_p^\top, \frac{D}{dt}\bigg|_{t=0} \nu_t\rangle = \langle Z_p, \frac{D}{dt}\bigg|_{t=0} \nu_t\rangle = -\langle D_{X_p^\top} X, (N_{\partial\Omega})_p\rangle = \mathrm{II}(X_p^\top, X_p).$$

Moreover, by (1.40),

$$\begin{aligned}\langle N_p, \frac{D}{dt}\bigg|_{t=0} \nu_t\rangle &= -\langle \frac{D}{dt}\bigg|_{t=0} N, N_{\partial\Omega}\rangle \\ &= -\langle \nabla_{X_p^\top} N, (N_{\partial\Omega})_p\rangle - \langle (\nabla_S u)_p, (N_{\partial\Omega})_p\rangle \\ &= \mathrm{II}(X_p^\top, N_p) + \left(\frac{\partial u}{\partial \nu}\right)_p.\end{aligned}$$

Hence,

$$\begin{aligned}\frac{d}{dt}\langle X_t, \nu_t\rangle(0) &= -\mathrm{II}(X_p, X_p) + \mathrm{II}(X_p^\top, X_p) + u(p)\left(\mathrm{II}(X_p^\top, N_p) + \left(\frac{\partial u}{\partial \nu}\right)_p\right) \\ &= -\mathrm{II}(N_p, N_p)\, u(p)^2 + u(p)\left(\frac{\partial u}{\partial \nu}\right)_p .\end{aligned}$$

This implies (9.11). □

Remark 9.9 Given a smooth function $u : S \to \mathbb{R}$ with compact support in S such that $\int_S u\, dS = 0$, a one-parameter family of diffeomorphisms $\{\psi_t\}_{t\in\mathbb{R}}$ in M can be constructed so that:

- $|\psi_t(E)| = |E|$ for small t
- $\langle Y, N\rangle = u$, where $Y_p = \frac{d}{dt}\Big|_{t=0} \psi_t(p)$ is the velocity vector field

This is done following the strategy of Lemma 1.24. First, uN is extended to a vector field X with compact support in M such that $X_p \in T_p(\partial\Omega)$ for all $p \in \partial\Omega$. Then we take a function $v : S \to \mathbb{R}$ with compact support in Ω such that $\int_S v\, dS \neq 0$, and the vector field vN on S is extended to a vector field Z with compact support in Ω which satisfies $\int_E \operatorname{div}(Z)\, dM = \int_S v\, dS \neq 0$. If $\{\xi_t\}_{t\in\mathbb{R}}$ is the flow associated with Z, then an application of the implicit function theorem to the function $(s,t) \mapsto |(\xi_s \circ \varphi_t)(E)|$ as in Lemma 1.24 provides the one-parameter family $\{\psi_t\}_{t\in\mathbb{R}}$.

Definition 9.10 We say that a hypersurface $S \subset \Omega$ such that $\partial S \subset \partial\Omega$ is:

- *Stationary* if S has constant mean curvature and meets $\partial\Omega$ orthogonally along ∂S
- *Stable* if S is stationary and the quantity in (9.12) is non-negative for any function $u : S \to \mathbb{R}$ with compact support such that $\int_S u\, dS = 0$

9.3 A Differential Inequality for the Isoperimetric Profile of a Bounded Set

The main result in this section is a differential inequality in weak sense (see Definition 3.12), satisfied by the isoperimetric profile of a set with convex boundary in a Riemannian manifold with $\mathrm{Ric} \geqslant (m-1)\,\delta$.

Definition 9.11 We say that a domain Ω with smooth boundary has *convex boundary* if the principal curvatures of $\partial\Omega$ with respect to the outer unit normal are non-negative.

Given a domain Ω with smooth boundary in a Riemannian manifold with the property that two points can be connected by a minimizing geodesic contained in Ω, it can be proven that any geodesic tangent to $\partial\Omega$ is locally outside Ω. It was established by Bishop that this property implies that the second fundamental form of $\partial\Omega$ with respect to the outer unit normal is non-negative; see [59]. Hence, Ω has convex boundary.

9.3.1 The Differential Inequality

Now we prove the following differential inequality for the profile. This is the analogous of Theorem 3.13 for compact manifolds without boundary. This result was proven as Theorem 3.2 in Bayle and Rosales [48].

Theorem 9.12 *Let Ω be a bounded domain with smooth convex boundary in a Riemannian manifold. Assume that* $\operatorname{Ric} \geqslant (m-1)\,\delta$ *on Ω. Let $1 \leqslant \alpha \leqslant m/(m-1)$. Then I_Ω^α satisfies in weak sense the differential inequality*

$$\left(I_\Omega^\alpha\right)'' \leqslant -\alpha(m-1)\delta\left(I_\Omega^\alpha\right)^{(\alpha-2)/\alpha}. \tag{9.13}$$

Moreover, equality holds in (9.13) for some $v \in (0, |\Omega|)$ if $\alpha = m/(m-1)$, the boundary $S = \operatorname{cl}(\partial E \cap \Omega)$ of any isoperimetric set E of volume v is totally umbilical, and

$$\operatorname{Ric}(N, N) = (m-1)\,\delta \text{ on } S, \qquad \mathrm{II}(N, N) = 0 \text{ on } S \cap \partial\Omega,$$

where II *is the second fundamental form of $\partial\Omega$ with respect to its outer unit normal.*

In addition, the function I_Ω^α is locally the sum of a concave function and a smooth function for any $1 \leqslant \alpha \leqslant m/(m-1)$.

Remark 9.13 In particular, if $\alpha = m/(m-1)$, then $I_\Omega^{m/(m-1)}$ satisfies in weak sense the differential inequality

$$\left(I_\Omega^{m/(m-1)}\right)'' \leqslant -m\delta\left(I_\Omega^{m/(m-1)}\right)^{(2-m)/m} \tag{9.14}$$

with equality for some $v \in (0, |\Omega|)$ if the boundary $S = \operatorname{cl}(\partial E \cap \Omega)$ of any isoperimetric set E of volume v is totally umbilical, $\operatorname{Ric}(N, N) = (m-1)\,\delta$ on S, and $\mathrm{II}(N, N) = 0$ on ∂S. Note also that in the half-space $\mathbb{H}_\delta^m$, the differential equality

$$\left(I_{\mathbb{H}_\delta^m}^{m/(m-1)}\right)'' = -m\delta\left(I_{\mathbb{H}_\delta^m}^{m/(m-1)}\right)^{(2-m)/m}$$

is satisfied in a strong sense since the isoperimetric sets in $\mathbb{H}_\delta^m$ are half-balls by Lemma 9.5.

Proof of Theorem 9.12 We fix some volume $0 < v_0 < |\Omega|$ and consider an isoperimetric set $E \subset \Omega$ of volume v_0. Let S be the regular part of $\operatorname{cl}(\partial E \cap \Omega)$ and

S_0 be the singular set. Lemma 1.61 yields a sequence of smooth functions $\{f_i\}_{i\in\mathbb{N}}$ with compact support in S satisfying:

1. $0 \leqslant f_i \leqslant 1$ for all i
2. The sequence $\{f_i\}_{i\in\mathbb{N}}$ converges pointwise to the constant function 1 on S
3. $\lim_{i\to\infty} \int_S |\nabla f_i|^2 dS = 0$

For any i, we take a vector field X_i with compact support on M so that:

- $X_i = f_i N$ on S, where N is the outer unit normal to E in S
- X_i is tangent to $\partial\Omega$ at points of $\partial\Omega$

The vector field X_i can be chosen to satisfy the second condition because S meets $\partial\Omega$ orthogonally. Let $\{\varphi_t^i\}_{t\in\mathbb{R}}$ be the flow associated with X_i. We have

$$\left.\frac{d}{dt}\right|_{t=0} |\varphi_t^i(E)| = \int_S f_i \, dS > 0,$$

and so we can take the volume as a parameter of the deformation $\varphi_t^i(E)$ for v close to v_0 and write $A_i(v) = P(\varphi_{t(v)}^i(E))$. We trivially have $A_i^\alpha(v) \geqslant I_\Omega^\alpha(v)$, with equality at v_0.

Arguing as in the proof of Theorem 3.13 using (9.12), there follows

$$\begin{aligned}\frac{d^2}{dv^2} A_i^\alpha(v_0) = \alpha A(S)^{\alpha-1}\Bigg((\alpha-1)\frac{H^2}{A(S)} &+ \frac{1}{\left(\int_S f_i dS\right)^2}\bigg(\int_S \left(|\nabla_S f_i|^2 - (\mathrm{Ric}(N,N) + |\sigma|^2\right) f_i^2 \, dS \\ &- \int_{\partial S} \mathrm{II}(N,N)\, f_i^2 \, d(\partial S)\bigg)\Bigg).\end{aligned}$$

By the properties of the sequence $\{f_i\}_{i\in\mathbb{N}}$ and the fact that $\mathrm{II}(N,N) \geqslant 0$ by the convexity of $\partial\Omega$, we obtain

$$\begin{aligned}\limsup_{i\to\infty} \frac{d^2}{dv^2} A_i^\alpha(v_0) &= -\alpha A(S)^{\alpha-3} \int_S \left(\mathrm{Ric}(N,N) + \left(|\sigma|^2 - (\alpha-1)H^2\right)\right) dS \\ &\leqslant -\alpha(m-1)\delta A(S)^{\alpha-2} = \alpha(m-1)\delta(I(v_0)^\alpha)^{(\alpha-2)/\alpha}\end{aligned}$$

since $\alpha - 1 \leqslant 1/(m-1)$ and $|\sigma|^2 - H^2/(m-1) \geqslant 0$, with equality if and only if $\mathrm{cl}(\partial E \cap \Omega)$ is totally umbilical, $\mathrm{Ric}(N,N) = (m-1)\,\delta$ on S, and $\mathrm{II}(N,N) = 0$ on ∂S. Hence, inequality (9.13) holds weakly in the sense of Definition 3.12.

Finally, to prove that I_Ω^α is locally the sum of a concave function and a smooth function, we take $0 < v_0 < |\Omega|$ and a small closed interval $J \subset (0, |\Omega|)$. Because

I_Ω is continuous and positive on J, there exists a constant $C > 0$ such that

$$-\alpha(m-1)\delta\left(I_\Omega^\alpha\right)^{(\alpha-2)/\alpha} \leqslant C$$

on J. Then the function $I_\Omega^\alpha(v) - Cv^2$ satisfies the inequality

$$\left(I_\Omega^\alpha(v) - Cv^2\right)'' \leqslant 0$$

in weak sense and so is concave on J by Lemma 3.11. □

A direct consequence of Theorem 9.12 is the concavity of some powers of the profile and the normalized profile when the Ricci curvature is non-negative on Ω.

Corollary 9.14 *Let Ω be a bounded domain with smooth convex boundary in a Riemannian manifold. Assume that* Ric $\geqslant (m-1)\,\delta \geqslant 0$ *on Ω. Then I_Ω^α, h_Ω^α are concave functions when $1 \leqslant \alpha \leqslant m/(m-1)$.*

Another of the consequences of the differential inequality (9.13) are the following regularity properties of the isoperimetric profile I_Ω.

Corollary 9.15 *Let Ω be a bounded domain with smooth convex boundary in a Riemannian manifold. Assume that* Ric $\geqslant (m-1)\,\delta$ *on Ω. Then I_Ω^α is locally the sum of a concave function and a smooth function for all $1 \leqslant \alpha \leqslant m/(m-1)$. In particular:*

1. *Left and right derivatives $(I_\Omega)'_-$, $(I_\Omega)'_+$ exist everywhere, and $(I_\Omega)'_-(v) \geqslant (I_\Omega)'_+(v)$. Moreover, I_Ω is differentiable except on a countable set.*
2. *If $v, w \in (0, |M|)$, then*

$$I_\Omega(w) - I_\Omega(v) = \int_v^w (I_\Omega)'_+(\xi)\,d\xi = \int_v^w (I_\Omega)'_-(\xi)\,d\xi.$$

3. *If $E \subset \Omega$ is an isoperimetric region of volume v_0 with boundary mean curvature H, then*

$$(I_\Omega)'_+(v_0) \leqslant H \leqslant (I_\Omega)'_-(v_0).$$

   In particular, if I_Ω is regular at v_0, then all isoperimetric regions have boundary mean curvature $H = I'_\Omega(v_0)$.
4. *If I_Ω is not regular at v_0, then there exist two isoperimetric regions in Ω of volume v_0 with boundary mean curvatures $(I_\Omega)'_+(v_0)$ and $(I_\Omega)'_-(v_0)$.*
5. *If v is a regular value of I_Ω, let $H(v)$ be the boundary mean curvature of any isoperimetric set of volume v. Then*

$$I_\Omega(w) - I_\Omega(v) = \int_v^w H(\xi)\,d\xi$$

   for all $0 < v < w < |\Omega|$.

Proof It is similar to the ones of Lemma 3.19 and Theorem 3.20. □

9.3.2 Geometrical and Topological Restrictions for Isoperimetric Sets

In case the Ricci curvature of M is non-negative on Ω, we have several restrictions on the geometry and topology of isoperimetric sets.

Theorem 9.16 *Let Ω be a bounded domain with smooth boundary in a complete Riemannian manifold. Assume that I_Ω is strictly concave. Then the isoperimetric regions in Ω are connected. In particular, the hypothesis is satisfied when the Ricci curvature of M is non-negative on Ω.*

Proof Let $E \subset \Omega$ be an isoperimetric region of volume $v = v_1 + v_2$. Assume that E has two components E_1 and E_2 of volumes v_1, v_2, respectively. Then

$$I_\Omega(v_1) + I_\Omega(v_2) \leqslant P(E_1) + P(E_2) = P(E) = I_\Omega(v). \tag{9.15}$$

But the strict concavity of the function I_Ω implies, since $I_\Omega(0) = 0$,

$$\frac{I_\Omega(v_1) - I_\Omega(0)}{v_1 - 0} > \frac{I_\Omega(v_1 + v_2) - I_\Omega(v_2)}{(v_1 + v_2) - v_2},$$

which implies $I_\Omega(v) = I_\Omega(v_1 + v_2) < I_\Omega(v_1) + I_\Omega(v_2)$, a contradiction to (9.15). □

As for the isoperimetric boundary, we have the following result:

Theorem 9.17 *Let Ω be a bounded domain with smooth convex boundary in a Riemannian manifold M such that* $\operatorname{Ric} \geqslant (m-1)\delta$. *Let* II *be the second fundamental form of $\partial\Omega$ with respect to the outer unit normal. Let $E \subset \Omega$ be an isoperimetric set and S the regular part of* $\operatorname{cl}(\partial E \cap \Omega)$ *and N a unit normal to S. Then:*

1. *If $\delta > 0$, then M is connected.*
2. *If $\delta = 0$ and S has more than one connected component, then* $\operatorname{Ric}(N, N) \equiv 0$ *on S and* $\mathrm{II}(N, N) \equiv 0$ *on ∂S. In particular, if S is disconnected and Ω is strictly convex (has positive principal curvatures), then $S \cap \partial\Omega = \emptyset$.*

Proof Let S_1 and S_2 be two connected components of the regular part S of $\operatorname{cl}(\partial E \cap \Omega)$. Lemma 1.61 provides two families of functions $\varphi_\varepsilon^i : S_i \to \mathbb{R}$, $i = 1, 2$, $\varepsilon > 0$, so that:

- φ_ε^i has compact support on S_i
- φ_ε^i converges to the function 1 on each S_i
- $\int_{S_i} |\nabla_{S_i} \varphi_\varepsilon^i|^2 \, dS_i \to 0$ when $\varepsilon \to 0$

For every $\varepsilon > 0$, let $a_\varepsilon \in \mathbb{R}$ such that $\int_S(\varphi_\varepsilon^1 - a_\varepsilon\varphi_\varepsilon^2)\,dS = 0$. Note that $a_\varepsilon \to a_\infty = A(S_1)/A(S_2)$ when $\varepsilon \to 0$. Inserting the test function $(\varphi_\varepsilon^1 - a_\varepsilon\varphi_\varepsilon^2)$ in the second variation formula (9.12), taking into account that this variation is non-negative, and taking limits when $\varepsilon \to 0$, we get

$$\sum_{i=1}^{2}(a_\infty^i)^2\left\{\int_{S_i}\big(\mathrm{Ric}(N,N)+|\sigma|^2\big)\,dS_i + \int_{\partial S_i}\mathrm{II}(N,N)\,d(\partial S_i)\right\} \leqslant 0,$$

where $a_\infty^1 = 1$, $a_\infty^2 = a_\infty$, N is the outer unit normal to E, and σ is the second fundamental form of S. Hence, $\mathrm{Ric}(N,N) \equiv 0$ and $|\sigma|^2 \equiv 0$ on $S_1 \cup S_2$ and $\mathrm{II}(N,N) \equiv 0$ on $\partial S_1 \cup \partial S_2$. From these equalities, the result follows. □

9.3.3 *Comparison Results*

Perhaps the most interesting consequence of Theorem 9.12 is a version for sets with smooth boundary of the Lévy-Gromov inequality; see Sect. 3.5.1 for the corresponding inequality when M is a compact manifold without boundary. This result was proven by Bayle and Rosales [48].

Theorem 9.18 (Lévy-Gromov Inequality for Domains with Convex Boundary) *Let Ω be a bounded domain with smooth convex boundary in a Riemannian manifold. Assume that* Ric $\geqslant (m-1)\,\delta > 0$ *on Ω. Then*

$$h_\Omega \geqslant h_{\mathbb{H}_\delta^m}. \tag{9.16}$$

Moreover, equality holds in (9.16) if and only if Ω is isometric to $\mathbb{H}_\delta^m$ and $\partial\Omega$ is totally geodesic in M.

Proof A direct computation using (9.14) shows that $f = h_\Omega^{m/(m-1)}$ satisfies in weak sense the differential inequality

$$f'' \leqslant \Lambda(f), \tag{9.17}$$

where

$$\Lambda(x) = -m\delta\, x^{(2-m)/m}.$$

On the other hand, the function $f_\delta = h_{\mathbb{H}_\delta^m}^{m/(m-1)}$ satisfies the differential equation

$$f_\delta'' = \Lambda(f_\delta).$$

To prove that $f \geqslant f_\delta$, we assume the existence of $0 < \lambda_0 < 1$ such that $f(\lambda_0) < f_\delta(\lambda_0)$. As f, f_δ are continuous and coincide at 0, 1, there exists a maximal interval $[\lambda_1, \lambda_2] \subset [0, 1]$ containing λ_0 so that $f < f_\delta$ on (λ_1, λ_2). Then $f = f_\delta$ at λ_1, λ_2. On the interval $[\lambda_1, \lambda_2]$, the inequality

$$(f - f_\delta)'' \leqslant \Lambda(f) - \Lambda(f_\delta) \leqslant 0$$

is satisfied since $f < f_\delta$ and Λ is an increasing function. This implies that $f - f_\delta$ is a non-positive concave function on $[\lambda_1, \lambda_2]$ vanishing at the endpoints of the interval, and so $f = f_\delta$ on $[\lambda_1, \lambda_2]$. This contradiction proves $f \geqslant f_\delta$ and (9.18).

Assume now that there exists some $\lambda_0 \in (0, 1)$ such that $f(\lambda_0) = f_\delta(\lambda_0)$. By symmetry of the normalized profiles with respect to $\lambda = 1/2$, we may assume that $0 < \lambda_0 \leqslant 1/2$. The same arguments as in the proof of the rigidity part in Sect. 3.5.1 imply $f = f_\delta$ on $[0, \lambda_0]$. The asymptotic expansion of the normalized profile at 0 then implies $|\Omega| = |\mathbb{H}^m_\delta|$. The Bishop rigidity (see Theorem 2.7(ii) in [48]) implies that Ω is isometric to $\mathbb{H}^m_\delta$ and that $\partial\Omega$ is totally geodesic in M. □

To finish this section, we give a version of the Toponogov-Cheng theorem for compact Riemannian manifolds, which characterizes the equality case in the Bonnet-Myers theorem. Let us get first an estimate of the diameter of a set in terms of the normalized isoperimetric profile of the set; see [48] and compare with Proposition 6.0 in Gallot [176].

Lemma 9.19 *The diameter of a bounded domain Ω with smooth boundary in a complete Riemannian manifold M satisfies*

$$\operatorname{diam}(\Omega) \leqslant \int_0^1 \frac{d\xi}{h_\Omega(\xi)},$$

with equality when Ω coincides with some $\mathbb{H}^m_\delta$.

Proof We take two points $p_0, p_1 \in \overline{\Omega}$ so that $d(p_0, p_1) = \operatorname{diam}(\Omega)$. By the coarea formula,

$$|E| = \int_0^\infty \mathcal{H}^{m-1}(E \cap S(p_0, t))\, dt.$$

Hence, the function $\xi(t) = |\Omega \cap B(0, t)|/|\Omega|$ is absolutely continuous on $[0, \operatorname{diam}(\Omega)]$ and

$$\xi'(t) = \frac{\mathcal{H}^{m-1}(\Omega \cap S(p_0, t))}{|\Omega|} \geqslant \frac{P(\Omega \cap B(p_0, t), \Omega)}{|\Omega|} \geqslant h_\Omega(\xi(t)),$$

for a.e. $t \in [0, \operatorname{diam}(\Omega)]$. Equality holds when Ω coincides with some $\mathbb{H}^m_\delta$. Integrating the inequality $\xi'(t)/h_\Omega(\xi(t)) \geqslant 1$ between 0 and $\operatorname{diam}(\Omega)$, we get the estimate. □

A consequence of the Lévy-Gromov inequality (9.16) and Lemma 9.19 is the analogous of the Toponogov-Cheng theorem.

Theorem 9.20 *Let* Ω *be a bounded domain with smooth convex boundary in a Riemannian manifold* M. *If* $\mathrm{Ric} \geqslant (m-1)\,\delta > 0$ *on* Ω, *then*

$$\mathrm{diam}(\Omega) \leqslant \frac{\pi}{\sqrt{\delta}}.$$

Equality holds if and only if Ω *is isometric to a half-space* $\mathbb{H}^m_\delta$.

Proof By Lemma 9.19 and the Lévy-Gromov inequality (9.16), we have

$$\mathrm{diam}(\Omega) \leqslant \int_0^1 \frac{d\xi}{h_\Omega(\xi)} \leqslant \int_0^1 \frac{d\xi}{h_{\mathbb{H}^m_\delta}(\xi)} = \mathrm{diam}(\mathbb{H}^m_\delta) = \frac{\pi}{\sqrt{\delta}}.$$

Equality holds when Ω is isometric to $\mathbb{H}^m_\delta$. □

9.4 The Isoperimetric Profile of the Ball

In this section, we prove that stable interfaces separating a Euclidean ball into two regions of prescribed volumes are spherical caps, a result proven by Wang and Xia [436]. The result makes use of the second variation formula (9.11) and a special type of conformal vector fields in $\mathbb{R}^m$.

9.4.1 *A Family of Conformal Deformations*

In Sect. 6.2.2, we studied the stability of compact embedded hypersurfaces $S \subset \mathbb{R}^m$ using the test function

$$u = (m-1) - H\,\langle W, N\rangle,$$

where

$$W = \nabla\Big(\frac{d^2}{2}\Big)$$

is the radial vector field in $\mathbb{R}^m$ and d is the distance to 0. This test function is the normal component of a deformation consisting of taking parallels to S and then correcting the enclosed volume by a family of dilations.

In this section, we use a different family of deformations, better adapted to the boundary of the ball $\mathbb{B} = B(0, 1)$. We take $a \in \mathbb{R}^m$. By abuse of notation, we also denote by a the parallel vector field on $\mathbb{R}^m$ whose components are the coordinates of a. We consider the vector field

$$X_a = \langle W, a\rangle\, W - \frac{1}{2}(1 + d^2)\, a. \tag{9.18}$$

The vector field X_a has the following properties:

- $X_0 = 0$ and X_a vanishes only at the points $\pm a/|a|$ when $a \neq 0$.
- If $p \in \partial\mathbb{B}$, then $(X_a)_p$ is tangent to $\partial\mathbb{B}$.

The first property follows easily since, when $a \neq 0$, the vector field X_a vanishes at p if and only if $W_p = p$ and a are proportional. Since $|W| = d$, this implies $p = \pm a/|a|$. For the second property, we notice that $W|_{\partial\mathbb{B}}$ is the outer unit normal to $\partial\mathbb{B}$, and, for $p \in \partial\mathbb{B}$, we have $d(p) = 1$ and so $\langle (X_a)_p, W_0\rangle = 0$.

We also have the property

$$\langle \nabla_e X_a, e\rangle = \langle W_p, a\rangle\, |e|^2, \qquad e \in T_p\mathbb{R}^m,\ p \in \mathbb{R}^m. \tag{9.19}$$

This is obtained simply by computing

$$\nabla_e X_a = \langle e, a\rangle\, W_p + \langle W_p, a\rangle\, e - \langle W_p, e\rangle\, a$$

since $\nabla_e W = e$ and $W = \nabla(d^2/2)$. Taking scalar product with e, we get (9.19).

The vector field X_a is the push forward of the radial vector field in $\mathbb{R}^m$ by a conformal transformation $\varphi : \mathbb{R}^m \setminus \{\text{point}\} \to \mathbb{R}^m$ taking the boundary of the hyperplane $x_m = 0$ to $\partial\mathbb{B}$; see pp. 1854–1855 in Wang and Xia [436].

For $a \in \mathbb{R}^m$ and a hypersurface $S \subset \mathbb{R}^m$ with possibly non-empty boundary, we define the function

$$u_a = (m-1)\,\langle W, a\rangle - H\,\langle X_a, N\rangle, \tag{9.20}$$

where N is a unit normal to S and H is the mean curvature of S computed with respect to this normal.

The functions u_a will be used in the proof of the characterization of stable hypersurfaces in a Euclidean ball. We now establish some basic properties of these functions.

Lemma 9.21 (First Minkowski Formula for X_a) *Let $S \subset \mathbb{R}^m$ be a hypersurface meeting $\partial\mathbb{B}$ orthogonally along ∂S. Then u_a has mean zero on S for all $a \in \mathbb{R}^m$.*

Proof We consider the tangential projection $X_a^\top$ of X_a to S. If ν is the outer conormal to S, the divergence theorem implies

$$\int_S \operatorname{div}_S(X_a^\top)\, dS = \int_{\partial S} \langle X_a^\top, \nu\rangle\, d(\partial S) = 0.$$

The last equality since $\langle X_a^\top, \nu\rangle = \langle X_a, \nu\rangle$, as $X_a^\top$, ν are tangent to S, and $\langle X_a, \nu\rangle = 0$ since X_a is tangent to $\partial\mathbb{B}$ and ν is perpendicular to $\partial\mathbb{B}$. On the other hand, if $e_1, \ldots, e_{m-1}$ is an orthonormal basis of T_pS, then (9.19) implies

$$\begin{aligned}(\operatorname{div}_S X_a^\top)(p) &= (\operatorname{div}_S X_a)(p) - H\,\langle (X_a)_p, N_p\rangle \\ &= (m-1)\,\langle W_p, a\rangle - H\,\langle (X_a)_p, N_p\rangle = u_a.\end{aligned}$$

Hence, $\int_S u_a dS = 0$. □

Lemma 9.22 *Let $S \subset \mathbb{R}^m$ be a hypersurface with unit normal N and constant mean curvature H with respect to N. Let $\mathcal{L} = \Delta_S + |\sigma|^2$. Then*

$$\begin{aligned}\Delta_S\langle W, a\rangle &= -H\,\langle N, a\rangle, \\ \Delta_S\big(d^2/2\big) &= (m-1) - H\,\langle W, N\rangle, \\ \mathcal{L}\langle N, a\rangle &= 0, \\ \mathcal{L}\langle W, N\rangle &= H, \\ \mathcal{L}\langle X_a, N\rangle &= \langle W, a\rangle\, H - (m-1)\langle N, a\rangle.\end{aligned} \tag{9.21}$$

Proof The first formula is trivial taking into account that $\nabla_e W = e$ for all $e \in T_pM$.

The second formula is just an application of Lemma 4.31 since $\nabla(d^2/2)(e, e) = 1$ for any unit vector e and $W = \nabla(d^2/2)$.

The third one is obtained since $-\mathcal{L}\langle N, a\rangle$ is the derivative of the mean curvature of $\varphi_t(S)$, where $\{\varphi_t\}_{t\in\mathbb{R}}$ is the flow associated with the vector field a, by Lemma 1.26. Since φ_t is a translation in $\mathbb{R}^m$ for all $t \in \mathbb{R}$, the mean curvature is preserved and so $\mathcal{L}\langle N, a\rangle = 0$.

The fourth formula is obtained in a similar way. The flow $\{\varphi_t\}_{t\in\mathbb{R}}$ associated with W is given by $\varphi_t(x) = e^t x$, the dilation of ratio e^{-t}, for all $t \in \mathbb{R}$. Hence, the mean curvature of $\varphi_t(S)$ is $e^{-t}H$ and so $\mathcal{L}\langle W, N\rangle = H$.

The last formula is obtained from the previous ones. From the expression (9.18) for X_a, we get

$$\mathcal{L}\langle X_a, N\rangle = \mathcal{L}(\langle W, a\rangle\langle W, N\rangle) - \mathcal{L}(\tfrac{1}{2}(1+d^2)\langle a, N\rangle).$$

Using formula $\mathcal{L}(uv) = u\mathcal{L}(v) + v\Delta_S u + 2\langle\nabla_S u, \nabla_S v\rangle$ and the previous ones, we obtain

$$\begin{aligned}\mathcal{L}\big(\langle W, a\rangle\langle W, N\rangle\big) &= H\,\langle W, a\rangle - H\langle W, N\rangle\langle N, a\rangle + 2\langle\nabla_S\langle W, a\rangle, \nabla_S\langle W, N\rangle\rangle, \\ \mathcal{L}\big(\tfrac{1}{2}(1+d^2)\,\langle a, N\rangle\big) &= \langle a, N\rangle\big((m-1) - H\langle W, N\rangle\big) + 2\langle\nabla_S\langle a, N\rangle, \nabla_S\big(\tfrac{1}{2}\,d^2\big)\rangle.\end{aligned}$$

Taking an orthonormal basis $e_1, \dots, e_{m-1}$ of T_pS

$$\langle \nabla_S \langle W, a \rangle, \nabla_S \langle W, N \rangle \rangle = \sum_{i=1}^{m-1} \langle e_i, a \rangle \langle W, \nabla_{e_i} N \rangle,$$

$$\langle \nabla_S \langle a, N \rangle, \nabla_S \big(\tfrac{1}{2} d^2\big) \rangle = \sum_{i=1}^{m-1} \langle a, \nabla_{e_i} N \rangle \langle W, e_i \rangle.$$

If $e_1, \dots, e_{m-1}$ is composed of principal vectors, we immediately see that both expressions coincide. Hence, the last formula is proven. □

Proposition 9.23 *Let $S \subset \mathbb{B}$ be a hypersurface with constant mean curvature so that $\partial S \subset \partial \mathbb{B}$ and S and $\partial \mathbb{B}$ meet orthogonally along ∂S. Then the function u_a defined in (9.20) satisfies*

$$\int_S u_a dS = 0, \tag{9.22}$$

and

$$\mathcal{L}(u_a) = \big((m-1)|\sigma|^2 - H^2\big) \langle W, a \rangle, \qquad \text{on } S, \tag{9.23}$$

$$\nu(u_a) = u_a, \qquad \text{on } \partial S. \tag{9.24}$$

Proof Equation (9.22) follows from the first Minkowski formula for X_a in Lemma 9.21. The expression for $\mathcal{L}(u_a)$ in (9.23) is a consequence of the values of $\Delta_S \langle W, a \rangle$ y $\mathcal{L}\langle X_a, N \rangle$ obtained in Lemma 9.22.

To prove (9.24), we recall that $u_a = (m-1) \langle W, a \rangle - H \langle X_a, N \rangle$ and notice

$$\nu \langle W, a \rangle = \langle \nu, a \rangle.$$

On the other hand,

$$\nu \langle X_a, N \rangle = \langle \nabla_\nu X_a, N \rangle + \langle X_a, \nabla_\nu N \rangle.$$

The second summand is 0 because ν is a principal direction of S since $\partial \mathbb{B}$ is totally umbilical. To prove this, we take e tangent to S and orthogonal to ν, and we have

$$\langle \nabla_\nu N, e \rangle = \sigma(\nu, e) = \sigma(W, \nabla_e N) = -\mathrm{II}(e, N) = 0.$$

Hence, $\nabla_\nu N = \lambda \nu$ and $\langle X_a, \lambda \nu \rangle = 0$ as ν is orthogonal to $\partial \mathbb{B}$ and X_a is tangent. As for the first summand, we have

$$\langle \nabla_\nu X_a, N \rangle = \langle \nabla_\nu \big(\langle W, a \rangle W - \frac{1}{2}(1 + d^2) a\big), N \rangle = -\nu\big(\tfrac{1}{2} d^2\big) \langle a, N \rangle = \langle X_a, N \rangle,$$

since $\langle \nu, N\rangle = \langle W, N\rangle = 0$ on $\partial\mathbb{B}$. Hence,

$$\nu(u_a) = (m-1)\langle \nu, a\rangle - H\nu\big(\tfrac{1}{2}\,d^2\big)\,\langle X_a, N\rangle. \tag{9.25}$$

As $\nu = W$ and $\nu(\frac{1}{2}\,d^2) = 1$, Eq. (9.25) implies (9.24). □

9.4.2 *Stable Hypersurfaces Inside a Euclidean Ball*

We prove in this section that stable hypersurfaces inside a Euclidean ball are indeed $(m-1)$-dimensional disks or spherical caps.

Theorem 9.24 *Let $S \subset \mathbb{B}$ be a hypersurface with constant mean curvature meeting $\partial\mathbb{B}$ orthogonally along ∂S. If S is stable, then it is either totally geodesic or a spherical cap.*

Proof For any $a \in \mathbb{R}^m$, we consider the functions u_a defined in (9.20). They have mean zero by (9.22). Inserting this function in the second variation formula (9.11) using (9.24) and (9.23), the stability condition provides the inequality

$$\int_S \big((m-1)|\sigma|^2 - H^2\big)\,\langle W, a\rangle\big((m-1)\,\langle W, a\rangle - H\,\langle X_a, N\rangle\big)\,dS \leqslant 0.$$

Choosing a in an orthonormal basis $e_1, \ldots, e_m$ of $\mathbb{R}^m$ and adding up the corresponding inequalities, we get

$$\int_S \big((m-1)|\sigma|^2 - H^2\big)\,\big((m-1)\,|W|^2 - H\frac{1}{2}(d^2-1)\,\langle W, N\rangle\big)\,dS \leqslant 0 \tag{9.26}$$

since

$$\sum_{i=1}^{m}\langle W, e_i\rangle^2 = |W|^2, \qquad \sum_{i=1}^{m}\langle W, e_i\rangle\,X_{e_i} = \frac{1}{2}(d^2-1)\,W.$$

We introduce the function

$$\Phi = \frac{1}{2}\,(d^2-1)\,H - (m-1)\langle W, N\rangle.$$

Using Lemma 9.22, we have

$$\Delta_S\Phi = \big((m-1)|\sigma|^2 - H^2\big)\,\langle W, N\rangle. \tag{9.27}$$

As $d = 1$ and $\langle W, N\rangle = 0$ on $\partial S \subset \partial\mathbb{B}$, we have $\Phi = 0$ on ∂S. Hence,

$$\int_S \Delta_S\Big(\frac{\Phi^2}{2}\Big)\, dS = \int_{\partial S} \Phi\nu(\Phi)\, d(\partial S) = 0.$$

Adding $\int_S \Delta_S(\Phi^2/2)\, dS = \int_S(\Phi\Delta_S\Phi + |\nabla_S\Phi|^2)\, dS$ to (9.26) and using (9.27), we get

$$\int_S (m-1)|W^\top|^2\big((m-1)|\sigma|^2 - H^2\big)\, dA + \int_S |\nabla_S\Phi|^2 dS \leqslant 0.$$

Since the integrals in this inequality are non-negative, we get $|W^\top|((m-1|\sigma|^2 - H^2) = |\nabla_S\Phi| = 0$. In particular, Φ is constant on S and so $\Delta_S\Phi = 0$ on S. By (9.27), we obtain $\big((m-1)|\sigma|^2 - H^2\big)\langle W, N\rangle = 0$. So we get $(m-1)|\sigma|^2 - H^2 = 0$ on S, which means that S is totally umbilical. Hence, it is a totally geodesic hypersurface or a spherical cap.

Finally, S must be connected since $\partial\mathbb{B}$ is strictly convex. □

9.4.3 *Stable Hypersurfaces Outside a Euclidean Ball*

Now we replace the open ball $\mathbb{B}$ by the complement of its closure $\mathbb{R}^m \setminus \overline{\mathbb{B}}$, the exterior of the unit ball in Euclidean space.

Theorem 9.25 *Let $S \subset \mathbb{R}^m \setminus \overline{\mathbb{B}}$ be a compact hypersurface with constant mean curvature meeting $\partial\mathbb{B}$ orthogonally along ∂S. If S is stable and connected, then it is a spherical cap.*

Proof The same proof as that of Theorem 9.24 works. The only difference is that $\nu = -W$ on $\partial\mathbb{B}$. Hence, (9.24) must be replaced by

$$\nu(u_a) = -u_a.$$

This follows immediately from Eq. (9.25). As $\mathrm{II}(e, e) = -|e|^2$ in our setting, we get $\nu(u_a) = \mathrm{II}(N, N)\, u_a$, and we proceed as in the proof of Theorem 9.24. Notice also that the case $H = 0$ cannot hold by the maximum principle. □

9.4.4 *Isoperimetric Sets Inside a Euclidean Ball*

As a consequence of Theorem 9.24, the classification of isoperimetric sets in $\mathbb{B}$ follows.

Theorem 9.26 *Let $E \subset \mathbb{B}$ be an isoperimetric set. Then ∂E is either an $(m-1)$-dimensional disk or a spherical cap meeting $\partial\mathbb{B}$ in an orthogonal way.*

Proof Since the boundary of E is composed of a regular part S and a singular set S_0 with $\mathcal{H}^{m-8+\gamma}(S_0) = 0$, we can use the functions $\{\varphi_i\}_{i\in\mathbb{N}}$ of Lemma 1.61 and consider the test functions $\varphi_i u_a$. The arguments in Sect. 6.2.4 apply, and so we conclude that S is either a totally geodesic disk or a spherical cap. □

9.4.5 Isoperimetric Sets Outside a Euclidean Ball

Theorem 9.25 allows us to classify the isoperimetric sets in $\mathbb{R}^m \setminus \overline{\mathbb{B}}$ as follows.

Theorem 9.27 *Let $E \subset \mathbb{R}^m \setminus \overline{\mathbb{B}}$ be an isoperimetric set. Then ∂E is a spherical cap meeting $\partial\mathbb{B}$ in an orthogonal way.*

Proof We need to consider first the problem of existence of isoperimetric sets since the characterization is done as in the proof of Theorem 9.26. Consider a minimizing sequence $\{E_i\}_{i\in\mathbb{N}}$ of sets of volume v in $\mathbb{R}^m \setminus \overline{\mathbb{B}}$ so that

$$\lim_{i\to\infty} P(E_i, \mathbb{R}^m \setminus \overline{\mathbb{B}}) = I_{\mathbb{R}^m\setminus\overline{\mathbb{B}}}(v).$$

Theorem 4.21 applies so that we can find sequences $\{E_i^c\}_{i\in\mathbb{N}}$ and $\{E_i^d\}_{i\in\mathbb{N}}$ and a measurable set $E \subset \mathbb{R}^m \setminus \overline{\mathbb{B}}$ with the properties $E_i^c \to E$ and E_i^d diverges. The set E is bounded by the proof of the boundedness Lemma 4.27. If $|E| < v$, then the divergent part of the sequence can be replaced by a sequence of Euclidean balls B_i with $|B_i| = |E_i^d| \to v - |E| > 0$ by the Euclidean isoperimetric inequality. Then $E \cup B$, where B is a ball of volume $v - |E|$ disjoint from E, is an isoperimetric set. Indeed, $|E|$ should be v since moving the ball B until there is a first contact with either E of $\mathbb{B}$ would provide a non-permitted singularity at the boundary of $E \cup B$.

Now Theorem 9.25 implies that each connected component of $S = \text{cl}(\partial E \cap (\mathbb{R}^m \setminus \mathbb{B})$ is a spherical cap. But using rotations about the origin, we could move two connected components of S until they touch, producing a non-permitted singularity of the isoperimetric set. Hence, S is connected and it is a spherical cap. □

Remark 9.28 Note that the proof of Theorem 9.27 shows that any minimizing sequence of sets of volume v converges to an isoperimetric set of the same volume.

9.5 The Isoperimetric Profile of a General Convex Body

A *convex body* $C \subset \mathbb{R}^m$ is a compact convex set with *non-empty* interior; see Schneider [398]. By abuse of terminology, we define the isoperimetric profile I_C of a convex body C as the isoperimetric profile of its interior. This is

$$I_C = I_{\text{int}(C)}$$

by definition. The normalized profile of $\text{int}(C)$ is denoted by h_C.

Note that the isoperimetric profile of a convex body C is defined regardless the regularity of the boundary ∂C. When no assumption is made on the regularity of ∂C, we keep the interior regularity properties of isoperimetric interfaces, while we have no knowledge of the regularity properties of such interfaces at ∂C.

We prove in this section that the power I_C^α, $1 \leqslant \alpha \leqslant m/(m-1)$, is a concave function for any convex body C. The strategy of the proof is the approximation of any convex set by convex sets with smooth boundary in Hausdorff distance.

9.5.1 Concavity of the Profile

Let us define first the Hausdorff distance for a pair of compact sets in $\mathbb{R}^m$.

Definition 9.29 Given two compact sets $A, B \subset\in \mathbb{R}^m$, their *Hausdorff distance* $\delta(A, B)$ is the infimum of the set of non-negative r such that

$$A \subseteq B_r, \qquad B \subseteq A_r,$$

where, for any set $E \subset \mathbb{R}^m$, $E_r = \big\{x \in \mathbb{R}^m : d(x, E) \leqslant r\big\}$ is the closed tubular neighborhood of radius $r > 0$.

It can be easily proven that δ is a distance in the space of compact subsets of $\mathbb{R}^m$; see §1.8 in [398].

Definition 9.30 A sequence $\{K_i\}_{i\in\mathbb{N}}$ of compact sets in $\mathbb{R}^m$ converges in *Hausdorff distance* to a compact set $K \subset \mathbb{R}^m$ if, for every $\varepsilon > 0$, there exists $i_0 \in \mathbb{N}$ such that $\delta(K_i, K) < \varepsilon$ for all $i \geqslant i_0$.

The Hausdorff limit C of a sequence of convex bodies $\{C_i\}_{i\in\mathbb{N}}$ is a compact convex set, although it may have empty interior. The following property of Hausdorff convergence for convex bodies is well-known; see Theorem 1.2.8 in [398].

Theorem 9.31 *Let $\{C_i\}_{i\in\mathbb{N}}$ be a sequence of compact convex sets converging in Hausdorff distance to a compact convex set C. Then* $\lim_{i\to\infty} |C_i| = |C|$.

Let us prove the main technical result in this section.

Theorem 9.32 *Let $\{C_i\}_{i\in\mathbb{N}}$ be a sequence of convex bodies converging in Hausdorff distance to a convex body C. Let $\{\lambda_i\}_{i\in\mathbb{N}} \subset (0, 1)$ converging to $\lambda \in (0, 1)$ and $\{E_i\}_{i\in\mathbb{N}}$ be a sequence of isoperimetric sets of volume $\lambda_i|C_i|$. Then there exists a subsequence of $\{E_i\}_{i\in\mathbb{N}}$ converging in $L^1_{loc}(\mathbb{R}^m)$ to an isoperimetric set $E \subset C$. Moreover,*

$$\lim_{i\to\infty} h_{C_i}(\lambda_i) = h_C(\lambda). \tag{9.28}$$

Proof As the sequence $\{C_i\}_{i\in\mathbb{N}}$ converges to C in Hausdorff distance, the sets C_i and C are contained in a fixed closed ball B of radius r_B. Note that this implies that the isoperimetric profiles of all sets C_i and C are uniformly bounded. To prove this, we simply take a hyperplane Π dividing C into two subsets of volumes v and $|C| - v$. Then

$$I_C(v) \leqslant \mathcal{H}^{m-1}(\Pi \cap C) \leqslant \mathcal{H}^{m-1}(\Pi \cap B) \leqslant \omega_{m-1} r_B^{m-1}. \tag{9.29}$$

The same inequality holds for C_i. Since the volumes $|C_i|$ are uniformly bounded below (as they converge to $|C| > 0$), the normalized profiles h_{C_i} and h_C are also uniformly bounded above.

Observe also that

$$\mathcal{H}^{m-1}(C_i) \leqslant \mathcal{H}^{m-1}(B) = \mathbf{c}_{m-1} r_B^{m-1} \tag{9.30}$$

for all $i \in \mathbb{N}$ since the metric projection of $\mathbb{R}^m$ onto C_i is a distance non-increasing mapping.

For each $i \in \mathbb{N}$, let $E_i \subset C_i$ be an isoperimetric set of volume $\lambda_i |\Omega_i|$. The perimeter $P(E_i)$ of E_i in $\mathbb{R}^m$ satisfies

$$P(E_i) \leqslant P(E_i, \operatorname{int}(C_i)) + \mathcal{H}^{m-1}(C_i),$$

which is uniformly bounded above by (9.29) and (9.30). Theorem 1.42 allows us to extract a non-relabeled convergent subsequence to some set $E \subset \mathbb{R}^m$ of finite perimeter in $\mathbb{R}^m$. Since the characteristic functions $\mathbf{1}_{E_i}$ of $E_i \subset C_i$ converge almost everywhere to the characteristic function $\mathbf{1}_E$ of E, and C_i converges to C in Hausdorff distance, we get $E \subset C$. Let X be a vector field with compact support in $\operatorname{int}(C)$ and $|X| \leqslant 1$. As $\{C_i\}_{i \in \mathbb{N}}$ converges to C in Hausdorff distance, $\operatorname{supp}(X) \subset \operatorname{int}(C_i)$ for i large. Then we have

$$\int_{\mathbb{R}^m} \mathbf{1}_E \operatorname{div}(X)\, d\mathbb{R}^m = \lim_{i \to \infty} \int_{\mathbb{R}^m} \mathbf{1}_{E_i} \operatorname{div}(X)\, d\mathbb{R}^m \leqslant \liminf_{i \to \infty} P(E_i, C_i).$$

Taking supremum over the set of vector fields with compact support in $\operatorname{int}(C)$ and modulus no larger than 1, we get

$$P(E, \operatorname{int}(C)) \leqslant \liminf_{i \to \infty} P(E_i, \operatorname{int}(C_i))$$

and so

$$I_C(\lambda |C|) \leqslant P(E, \operatorname{int}(C)) \leqslant \liminf_{i \to \infty} P(E_i, \operatorname{int}(C_i)) = I_{C_i}(\lambda_i |C_i|).$$

Hence, we get

$$h_C(\lambda) \leqslant \liminf_{i \to \infty} h_{C_i}(\lambda_i). \tag{9.31}$$

Now we apply the following construction: since C is a convex body, it contains some interior point that we assume to be 0. As C_i converges to C in Hausdorff distance, 0 is also an interior point of C_i for i large. As $C_i \to C$ in Hausdorff distance, we can take a sequence μ_i converging to 1 so that $C_i \subset \mu_i C$ (see Lemma 1.8.18 and the proof of Theorem 1.8.20 in [398]). Fixing an isoperimetric region $E \subset C$ of volume $\lambda|C|$ and considering the sets $\mu_i E \subset \mu_i C$, we restrict them to C_i, and we add or remove a small ball $B_i \subset C_i$ so that we get a set $E_i^* \subset C_i$ with volume $\lambda_i|C_i|$. Since $|B_i| \to 0$, we have

$$\begin{aligned}
\limsup_{i\to\infty} I_{C_i}(\lambda_i|C_i|) &\leqslant \limsup_{i\to\infty} P(E_i^*, \operatorname{int}(C_i)) \\
&\leqslant \limsup_{i\to\infty} \big(P(\mu_i E, \operatorname{int}(C_i)) + P(B_i)\big) \\
&\leqslant \limsup_{i\to\infty} \big(P(\mu_i E, \operatorname{int}(\mu_i C)) + P(B_i)\big) \\
&= \limsup_{i\to\infty} \big(\mu_i^{n-1} P(E, \operatorname{int}(C)) + P(B_i)\big) \\
&= P(E, C) = I_C(\lambda|C|).
\end{aligned}$$

Hence,

$$\limsup_{i\to\infty} h_{C_i}(\lambda_i) \leqslant h_C(\lambda). \tag{9.32}$$

Inequalities (9.31) and (9.32) then imply (9.28) and that E is an isoperimetric set in C of volume $\lambda|C|$. □

Theorem 9.33 *Let $C \subset \mathbb{R}^m$ be a convex body and $1 \leqslant \alpha \leqslant m/(m-1)$. Then the functions h_C^α and I_C^α are concave.*

Proof Let $\{C_i\}_{i\in\mathbb{N}}$ be a sequence of convex sets with smooth boundary converging to C in Hausdorff distance (c.f., §5.2 in [267]). The functions $h_{C_i}^\alpha$ are concave by Corollary 9.14. Since $\lim_{\to\infty} h_{C_i}^\alpha(\lambda) = h_C^\alpha(\lambda)$ for all $0 < \lambda < 1$ by Theorem 9.32, the function h_C^α is concave as a pointwise limit of concave functions.

Since I_C^α is the composition of h_C^α with affine functions, it is also a concave function. □

As in Theorem 3.17, the strict concavity of the profile I_C, which follows from the concavity of $I_C^{m/(m-1)}$, implies the connectedness of isoperimetric sets in C.

Theorem 9.34 *Let $C \subset \mathbb{R}^m$ be a convex body and $C \subset E$ an isoperimetric set. Then E is connected.*

We also have the following generalization of Lemma 9.15 for general convex bodies in $\mathbb{R}^m$:

Theorem 9.35 *Let $C \subset \mathbb{R}^m$ be a convex body. Then I_C^α is concave for $1 \leqslant \alpha \leqslant m/(m-1)$. In particular:*

1. *Left and right derivatives $(I_C)'_-, (I_C)'_+$ exist everywhere, and $(I_C)'_-(v) \geqslant (I_C)'_+(v)$. Moreover, I_C is differentiable except on a countable set.*
2. *If $v, w \in (0, |M|)$, then*

$$I_C(w) - I_C(v) = \int_v^w (I_C)'_+(\xi)\, d\xi = \int_v^w (I_C)'_-(\xi)\, d\xi.$$

3. *If $E \subset C$ is an isoperimetric region of volume v_0 with boundary mean curvature H, then*

$$(I_C)'_+(v_0) \leqslant H \leqslant (I_C)'_-(v_0).$$

   In particular, if I_C is regular at v_0, then all isoperimetric regions have boundary mean curvature $H = I'_C(v_0)$.
4. *If I_C is not regular at v_0, then there exist two isoperimetric regions in Ω of volume v_0 with boundary mean curvatures $(I_C)'_+(v_0)$ and $(I_C)'_-(v_0)$.*
5. *If v is a regular value of I_C, let $H(v)$ be the boundary mean curvature of any isoperimetric set of volume v. Then*

$$I_C(w) - I_C(v) = \int_v^w H(\xi)\, d\xi$$

   for all $0 < v < w < |\Omega|$.

9.5.2 The Ahlfors Property and a Relative Isoperimetric Inequality for Small Volumes

In this section, we prove that a convex body satisfies an Ahlfors inequality for the volume of relative balls and a relative isoperimetric inequality depending on geometric properties of the convex body.

Given a convex body C, a point $x \in C$, and $r > 0$, we define the *intrinsic ball* $B_C(x, r)$ by

$$B_C(x, r) = B(x, r) \cap C,$$

where $B(x, r)$ is the ball in $\mathbb{R}^m$ centered at x of radius r. The set $\overline{B}_C(x, r)$ is the intersection of the closed ball $\overline{B}(x, r)$ with C.

Recall that the *inradius* of a set E, $\operatorname{inr}(E)$, is the supremum of the radii of balls contained in E.

Lemma 9.36 (Ahlfors Property) *Let $C \subset \mathbb{R}^m$ be a convex body. Given $r_0 > 0$, there exists a positive constant $\ell_1 > 0$, only depending on δ/r_0, where $\delta > 0$ is a lower bound on the inradius of the sets $\overline{B}_C(x, r_0)$, $x \in C$, such that*

$$\ell_1 r^m \leqslant |\overline{B}_C(x, r)| \leqslant \omega_m r^m. \tag{9.33}$$

Proof Inequality $|\overline{B}_C(x, r)| \leqslant \omega_m r^m$ follows trivially for any $r > 0$ since $\overline{B}_C(x, r) \subset \overline{B}(x, r)$. To prove the lower estimate on $|\overline{B}_C(x, r)|$, we consider

$$\inf_{x \in C} \operatorname{inr}(\overline{B}_C(x, r_0)),$$

which is positive by the compactness of C. We take $0 < \delta < \inf_{x\in C} \operatorname{inr}(\overline{B}_C(x, r_0))$. Then, for every $x \in C$, there exists $y(x) \in \overline{B}_C(x, r_0)$ such that

$$\overline{B}(y(x), \delta) \subset \overline{B}_C(x, r_0) \subset \overline{B}(x, r_0).$$

Now take $0 < r \leqslant r_0$. Let $0 < \lambda \leqslant 1$ so that $r = \lambda r_0$, and denote by $h_{x,\lambda}$ the dilation of center x and ratio λ. Then $h_{x,\lambda}(\overline{B}(y(x), \delta)) \subset h_{x,\lambda}(\overline{B}_C(x, r_0))$ and so

$$\overline{B}(h_{x,\lambda}(y(x)), \lambda\delta) \subset \overline{B}_{h_{x,\lambda}(C)}(x, \lambda r_0) \subset \overline{B}_C(x, \lambda r_0),$$

since, by the convexity of C, $h_{x,\lambda}(C) \subset C$ when $x \in C, 0 < \lambda \leqslant 1$. So we have

$$\begin{aligned} |\overline{B}(x, r) \cap C| &= |\overline{B}(x, \lambda r_0) \cap C| \geqslant |h_{x,\lambda}(\overline{B}(x, r_0) \cap C)| \\ &= \lambda^m |\overline{B}(x, r_0) \cap C| \geqslant \lambda^m |\overline{B}(y(x), \delta)| \\ &= \omega_m (\delta/r_0)^m \, r^m, \end{aligned}$$

and (9.33) is satisfied taking $\ell_1 = \omega_m (\delta/r_0)^m$. □

The strategy to prove a relative isoperimetric inequality in a convex set is to transfer the relative isoperimetric inequality of the ball obtained in Sect. 9.4 to the convex set by means of a bilipschitz map. By definition, a map $f : (X, d) \to (X', d')$ between two metric spaces is *bilipschitz* if f is bijective and both f, f^{-1} are Lipschitz maps. As usual, we denote by $\operatorname{Lip}(f)$ and $\operatorname{Lip}(f^{-1})$ the smallest Lipschitz constants of f and f^{-1}, respectively.

Remark 9.37 (Basic Properties of Lip) We refer the reader to Lemma 2.1 in [371] for the following basic properties of $\operatorname{Lip}(f)$. Let (X, d) be an arbitrary metric space.

- Let f be a Lipschitz function on (X, d) so that $|f| \geqslant M > 0$. Then $1/f$ is a Lipschitz function and $\operatorname{Lip}(1/f) \leqslant \operatorname{Lip}(f)/M^2$.

- Let f_1, f_2 be Lipschitz functions on (X, d). Then $f_1 + f_2$ is a Lipschitz function and $\mathrm{Lip}(f_1 + f_2) \leqslant \mathrm{Lip}(f_1) + \mathrm{Lip}(f_2)$.
- Let f_1, f_2 be Lipschitz functions on (X, d) so that $|f_i| \leqslant M_i$, $i = 1, 2$. Then $f_1 f_2$ is a Lipschitz function and $\mathrm{Lip}(f_1 f_2) \leqslant M_1 \mathrm{Lip}(f_2) + M_2 \mathrm{Lip}(f_1)$.
- If $\lambda : (X, d) \to \mathbb{R}$ is Lipschitz with $|\lambda| \leqslant L'$, and $f : (X, d) \to \mathbb{R}^m$ is Lipschitz with $|f| < M'$, then $\mathrm{Lip}(\lambda f) \leqslant M' \mathrm{Lip}(\lambda) + L' \mathrm{Lip}(f)$.

Given a convex body $C \subset \mathbb{R}^m$ containing 0 in its interior, its *radial function* $\rho(C, \cdot) : \mathbb{S}^{m-1} \to \mathbb{R}$ is defined by

$$\rho(C, u) = \max \{\lambda \geqslant 0 : \lambda u \in C\}.$$

From this definition, it follows that $\rho(C, u)u \in \partial C$ for all $u \in \mathbb{S}^{m-1}$.

Lemma 9.38 *Let $C \subset \mathbb{R}^m$ be a convex body so that $B(0, r) \subset C \subset B(0, R)$. Then the radial function $\rho(C, \cdot) : \mathbb{S}^{m-1} \to \mathbb{R}$ is R^2/r-lipschitz.*

Proof Let C^* be the polar body of C, [398, §1.6]. Theorem 1.6.1 in [398] implies that $(C^*)^* = C$ and that $B(0, 1/R) \subset C^* \subset B(0, 1/r)$. Let $h(C^*, \cdot)$ be the support function of C^*. Using $(C^*)^* = C$, Remark 1.7.7 in [398] implies

$$\rho(C, u) = \frac{1}{h(C^*, u)}.$$

By Lemma 1.8.10 in [398], the function $h(C^*, \cdot)$ is $1/r$-lipschitz. Since $h(C^*, \cdot) \geqslant 1/R$, we conclude by the properties in Remark 9.37 that $\rho(C, \cdot)$ is an R^2/r-lipschitz function. □

The next result provides a construction of bilipschitz maps between two convex subsets.

Proposition 9.39 *Let C, $C' \subset \mathbb{R}^m$ be convex bodies with radial functions ρ, ρ'. Assume that $\overline{B}(0, 2r) \subset C \cap C'$ and $C \cup C' \subset \overline{B}(0, R) \subset \mathbb{R}^m$. Let $f : C \to C'$ be the bilipschitz map defined by*

$$f(x) = \begin{cases} x, & |x| \leqslant r, \\ r\,\dfrac{x}{|x|} + (|x| - r)\,\dfrac{\rho'\big(\frac{x}{|x|}\big) - r}{\rho\big(\frac{x}{|x|}\big) - r}\,\dfrac{x}{|x|}, & |x| \geqslant r. \end{cases} \tag{9.34}$$

Then we have

$$1 \leqslant \mathrm{Lip}(f), \mathrm{Lip}(f^{-1}) \leqslant 1 + \frac{R}{r}\Big(\frac{R}{r} - 1\Big)\Big(\frac{R^2}{r^2} + 1\Big). \tag{9.35}$$

Proof We trivially have $\mathrm{Lip}(f) \geqslant \mathrm{Lip}(f|_{\{|x| \leqslant r\}}) = 1$ and also $\mathrm{Lip}(f^{-1}) \geqslant 1$. Let us compute $\mathrm{Lip}(f|_{\{|x| \geqslant r\}}$. Observe that the map $F_r : \mathbb{R}^m \setminus B(0, r) \to \overline{B}(0, r)$ defined by $F_r(x) = rx/|x|$ is the metric projection on the convex set $\overline{B}(0, r)$ and so $\mathrm{Lip}(F_r) \leqslant 1$. Hence, the map $x \in \mathbb{R}^m \setminus B(0,r) \mapsto x/|x| \in \mathbb{R}^m$ has Lipschitz constant no larger than $1/r$. By Lemma 9.38 and the properties in Remark 9.37, for the map Λ defined by

$$x \in \mathbb{R}^m \setminus B(0,r) \mapsto \frac{\rho'\big(\frac{x}{|x|}\big) - r}{\rho\big(\frac{x}{|x|}\big) - r}$$

we have

$$\mathrm{Lip}(\Lambda) \leqslant \frac{1}{r}\frac{R^2}{r}\frac{1}{r} + (R-r)\frac{R^2}{r}\frac{1}{r}\frac{1}{r} = \frac{R^2}{r^3} + (R-r)\frac{R^2}{r^4}.$$

As the function Λ is bounded above by $(R-r)/r$, and $x \mapsto x/|x|$ is bounded above by 1, having Lipschitz constant no larger than $1/r$, we get

$$\mathrm{Lip}\Big(\Lambda\frac{F_r}{r}\Big) \leqslant \frac{R^2}{r^3} + (R-r)\frac{R^2}{r^4} + \frac{R-r}{r}\frac{1}{r}.$$

Thus, as the function $(\Lambda F_r/r)$ is bounded above by $(R-r)/r$, and $x \mapsto |x| - r$ is bounded from above by $R - r$, having Lipschitz constant no larger than 1, from Remark 9.37, we get

$$\begin{aligned}
\mathrm{Lip}(f) &\leqslant 1 + (R-r)\Big(\frac{R^2}{r^3} + (R-r)\frac{R^2}{r^4} + \frac{R-r}{r^2}\Big) + \frac{R-r}{r} \\
&\leqslant 1 + \Big(\frac{R-r}{r}\Big)\Big(\frac{R^2}{r^2} + \Big(\frac{R-r}{r}\Big)\frac{R^2}{r^2} + \frac{R-r}{r} + 1\Big) \\
&\leqslant 1 + \Big(\frac{R}{r} - 1\Big)\Big(\frac{R^3}{r^3} + \frac{R}{r}\Big).
\end{aligned}$$

This implies (9.35) for f. Since f^{-1} is of the form (9.34), the same estimates obtained for f are valid for f^{-1}. □

Using Proposition 9.39, we get the following relative isoperimetric inequality for convex sets.

Theorem 9.40 (Relative Isoperimetric Inequality for Convex Sets) *Let $C \subset \mathbb{R}^m$ be a convex body. Let $x, y \in C$, $0 < r < R$ such that $\overline{B}(y, 2r) \subset C \subset \overline{B}(x, R)$. Then there exists a constant $K > 0$, depending only on R/r and m, such that*

$$I_C(v) \geqslant K \min\big\{v, |C| - v\big\}^{(m-1)/m}, \tag{9.36}$$

for all $0 < v < |C|$.

Proof We observe first that in the closed unit ball $\overline{B} = \overline{B}(0, 1)$, the non-optimal isoperimetric inequality

$$I_{\overline{B}} \geqslant K_0 \min\left\{v, |\overline{B}| - v\right\}^{(m-1)/m} \tag{9.37}$$

is satisfied. This follows since $I_{\overline{B}}$ is continuous, positive, and symmetric with respect to $|\overline{B}|/2$ and asymptotic to $v \mapsto v^{(m-1)/m}$ near $v = 0$. Inequality (9.37) is invariant by dilations, and so it holds for any Euclidean ball of any radius.

Since $\overline{B}(y, 2r) \subset C \subset \overline{B}(x, R) \subset \overline{B}(y, 2R)$, Proposition 9.39 provides a bilipschitz map $f : C \to \overline{B}(x, R)$ so that $\mathrm{Lip}(f)$ and $\mathrm{Lip}(f^{-1})$ only depend on r/R.

Let $F \subset C$ be any measurable set. Then we have

$$P_{\overline{B}}(f(F)) \leqslant \big(\mathrm{Lip}(f)\big)^m P_C(F), \qquad |F| \leqslant \big(\mathrm{Lip}(f^{-1})\big)^m |f(E)|.$$

Using these inequalities and (9.37), if $E \subset C$ is an isoperimetric set of volume v, letting $\overline{B} = \overline{B}(x, R)$, we have

$$\begin{aligned}
I_C(v) &= P_C(E) \\
&\geqslant \big(\mathrm{Lip}(f)\big)^{-m} P_{\overline{B}}(f(E)) \\
&\geqslant \big(\mathrm{Lip}(f)\big)^{-m} K_0 \min\left\{|f(E)|, |\overline{B} \setminus f(E)|\right\}^{(m-1)/m} \\
&\geqslant \big(\mathrm{Lip}(f)\,\mathrm{Lip}(f^{-1})\big)^{-m} K_0 \min\left\{|E|, |C \setminus E|\right\}^{(m-1)/m} \\
&= \big(\mathrm{Lip}(f)\,\mathrm{Lip}(f^{-1})\big)^{-m} K_0 \min\left\{v, |C| - v\right\}^{(m-1)/m}.
\end{aligned}$$

The inequalities in (9.35) imply that the constant $\big(\mathrm{Lip}(f)\,\mathrm{Lip}(f^{-1})\big)^{-m} K_0$ is bounded below by a constant depending only on r/R and m, since K_0 only depends on m. We take as K such a lower bound. □

Corollary 9.41 (Relative Isoperimetric Inequality for Relative Balls) *Let $C \subset \mathbb{R}^m$ be a convex body and $r_0 > 0$. Let $0 < \delta < \min_{x \in C} \mathrm{inr}(\overline{B}_C(x, r_0))$. Then there exists a constant K, only depending on δ/r_0 and m, such that*

$$I_{\overline{B}_C(x,r)}(v) \geqslant K \min\left\{v, |C| - v\right\}^{(m-1)/m} \tag{9.38}$$

for all $x \in C$, $0 < r \leqslant r_0$, and $0 < v < |\overline{B}_C(x, r)|$.

Proof The existence of δ satisfying $0 < \delta < \min_{x \in C} \mathrm{inr}(\overline{B}_C(x, r_0))$ follows as in the proof of Lemma 9.36 by the continuity of the inradius of $\overline{B}_C(x, r_0)$ as a function of $x \in C$ and the compactness of C.

For every $x \in C$, there exists $y(x) \in \overline{B}_C(x, r_0)$ such that

$$\overline{B}(y(x), 2\,(\delta/2)) \subset \overline{B}_C(x, r_0) \subset \overline{B}(x, r_0).$$

For any $0 < r \leqslant r_0$, taking λ such that $r = \lambda r_0$, the convexity of C implies

$$\overline{B}(h_{x,\lambda}(y(x)), 2\,(\lambda\delta/2)) \subset \overline{B}_C(x, \lambda r_0) \subset \overline{B}(x, \lambda r_0),$$

where $h_{x,\lambda}$ is the dilation of center x and ratio λ. Theorem 9.40 then implies that (9.38) is satisfied with a uniform constant that only depends on $2\delta/r_0$ and m.

□

9.5.3 Density Estimates

In this section, we closely follow the paper by Leonardi and Rigot [266]. Our intention is to prove a density estimate for isoperimetric sets in the spirit of David and Semmes for quasi-minimizing sets [132].

To simplify the notation, when $C \subset \mathbb{R}^m$ is a convex body, we denote the perimeter of P in $\mathrm{int}(C)$ by P_C. This means

$$P_C(E) = P(E, \mathrm{int}(C))$$

by definition for any measurable set E.

We make first the following observation. If $C \subset \mathbb{R}^m$ is a convex subset, $\lambda > 0$, and $E \subset C$ is a measurable set, then $|\lambda E| = \lambda^m|E|$, and $P_{\lambda C}(\lambda E) = \lambda^{m-1} P_C(E)$. This implies

$$I_{\lambda C}(v) = \lambda^{m-1} I_C\left(\frac{v}{\lambda^m}\right). \tag{9.39}$$

In particular, if $\lambda \geqslant 1$, then $v/\lambda^m \leqslant v$, and the concavity of I_C implies $I_C(v/\lambda^m) \geqslant \lambda^m I_C(v)$. From (9.39), we conclude

$$I_{\lambda C}(v) \geqslant I_C(v), \quad \lambda \geqslant 1, \quad v \in (0, |C|). \tag{9.40}$$

Given a measurable set $E \subset C$, the function $h : C \times (0, +\infty) \to (0, \frac{1}{2})$ is defined by

$$h(E, C, x, R) = \frac{\min\left\{|E \cap B_C(x, R)|, |B_C(x, R) \setminus E|\right\}}{|B_C(x, R)|},$$

for $x \in C$ and $R > 0$. When E and C are fixed, we shall simply denote

$$h(x, R) = h(E, C, x, R).$$

Before stating the main result in this section, let us prove the following technical lemma.

Lemma 9.42 *For any $v > 0$, consider the function $f_v : (0, v] \to \mathbb{R}$ defined by*

$$f_v(s) = s^{-(m-1)/m}\left(\left(\frac{v-s}{v}\right)^{(m-1)/m} - 1\right).$$

Then there is a constant $0 < c_2 < 1$ that does not depend on v so that $f_v(s) \geqslant -(1/2)\, v^{-(m-1)/m}$ for all $0 \leqslant s \leqslant c_2 v$.

Proof Let $\alpha = (m-1)/m$. Define $f_1 : (0, 1] \to \mathbb{R}$ by $f_v(sv) = f_1(s)\, v^{-\alpha}$. Hence,

$$f_1(s) = s^{-\alpha}\big((1-s)^{\alpha} - 1\big)$$

The function f_1 extends continuously to $s = 0$ by setting $f_1(0) = 0$. We also have $f_1(1) = -1$. By continuity, there exists $0 < c_2 < 1$ such that $f_1(s) \geqslant -1/2$ for all $s \in [0, c_2]$. This implies $f_v(s) = f_1(s/v)\, v^{-\alpha} \geqslant -(1/2)\, v^{-\alpha}$ for all $s \in [0, c_2 v]$.

□

The key density estimate in this section is the following:

Theorem 9.43 *Let $C \subset \mathbb{R}^m$ be a convex body and $E \subset C$ an isoperimetric region of volume $0 < v < |C|$. Choose ε so that*

$$\varepsilon < \min\left\{\frac{v}{\omega_m}, \frac{|C|-v}{\omega_m}, c_2 v, c_2(|C|-v), \frac{I_C(v)^m}{\omega_m 8^m v^{m-1}}, \frac{I_C(v)^m}{\omega_m 8^m (|C|-v)^{m-1}}\right\}, \tag{9.41}$$

where c_2 is the constant in Lemma 9.42.

Then, for any $x \in C$ and $R \leqslant 1$ so that $h(x, R) \leqslant \varepsilon$, there holds

$$h(x, R/2) = 0.$$

Moreover, in case $h(x, R) = |E \cap B_C(x, R)||B_C(x, R)|^{-1}$, we get $|E \cap B_C(x, R/2)| = 0$, and, in case $h(x, R) = |B_C(x, R) \setminus E||B_C(x, R)|^{-1}$, we have $|B_C(x, R/2) \setminus E| = 0$.

Proof Since $I_C(0) = 0$ and $I_C^{m/(m-1)}$ is concave, we get, for all $0 \leqslant w \leqslant v$,

$$I_C(w) \geqslant c_1 w^{(m-1)/m}, \qquad c_1 = \frac{I_C(v)}{v^{(m-1)/m}}, \tag{9.42}$$

for all $0 \leqslant w \leqslant v$.

Assume first that

$$h(x, R) = \frac{|E \cap B_C(x, R)|}{|B_C(x, R)|}.$$

As $\mu(t) = |E \cap B_C(x, t)|$, defined for $0 < t \leqslant R$, is a non-decreasing function, we have

$$\begin{aligned}\mu(t) &\leqslant m(R) = |E \cap B_C(x, R)| \\ &= h(x, R)\,|B_C(x, R)| \leqslant h(x, R)\,\omega_m R^m \leqslant \varepsilon\omega_m < v,\end{aligned} \tag{9.43}$$

for $t \leqslant R \leqslant 1$ by (9.41). Hence, $v - \mu(t) > 0$.

By the coarea formula, we get

$$m'(t) = \frac{d}{dt}\int_0^t \mathcal{H}^{m-1}(E \cap \partial B_C(x, s))ds = \mathcal{H}^{m-1}(E \cap \partial B_C(x, t)), \tag{9.44}$$

for a.e. t. In this equation, we have denoted $\partial B(x, t) \cap \text{int}(C)$ by $\partial B_C(x, t)$. Define

$$\lambda(t) = \frac{v^{1/m}}{(v - \mu(t))^{1/m}}, \qquad E(t) = \lambda(t)(E \setminus B_C(x, t)). \tag{9.45}$$

Then $E(t) \subset \lambda(t)C$ and $|E(t)| = |E| = v$. As $\lambda(t) \geqslant 1$, we get $I_{\lambda(t)C} \geqslant I_C$ by (9.40). Combining this with [455, Cor. 5.5.3], Eq. (9.44), and elementary properties of the perimeter functional, we get

$$\begin{aligned}I_C(v) &\leqslant I_{\lambda(t)C}(v) \leqslant P_{\lambda(t)C}(E(t)) = \lambda^{m-1}(t)\, P_C(E \setminus B_C(x, t)) \\ &\leqslant \lambda^{m-1}(t)\big(P_C(E) - P(E, B_C(x, t)) + \mathcal{H}^{m-1}(E \cap \partial B_C(x, t))\big) \\ &\leqslant \lambda^{m-1}(t)\big(P_C(E) - P_C(E \cap B_C(x, t)) + 2\mathcal{H}^{m-1}(E \cap \partial B_C(x, t))\big) \\ &\leqslant \lambda^{m-1}(t)\big(I_C(v) - c_1\mu(t)^{(m-1)/m} + 2\mu'(t)\big),\end{aligned}$$

where c_1 is the constant in (9.42). Multiplying both sides by $I_C(v)^{-1}\lambda(t)^{-(m-1)}$

$$\lambda(t)^{-(m-1)} - 1 + \frac{c_1}{I_C(v)}\mu(t)^{(m-1)/m} \leqslant \frac{2}{I_C(v)}\mu'(t).$$

Set

$$a = \frac{2}{I_C(v)}, \qquad b = \frac{c_1}{I_C(v)} = \frac{1}{v^{(m-1)/m}}.$$

From the definition (9.45) of $\lambda(t)$, we get

$$f(\mu(t)) \leqslant a\mu'(t) \qquad \mathcal{H}^1\text{-a.e,} \tag{9.46}$$

where

$$\frac{f(s)}{s^{(m-1)/m}} = b + \frac{\left(\frac{v-s}{v}\right)^{(m-1)/m} - 1}{s^{(m-1)/m}}.$$

By Lemma 9.42, there exists a universal constant $0 < c_2 < 1$, not depending on v, so that

$$\frac{f(s)}{s^{(m-1)/m}} \geqslant b/2 \qquad \text{whenever} \qquad 0 < s \leqslant c_2 v. \tag{9.47}$$

Since $\varepsilon \leqslant c_2 v$ by (9.41), Eq. (9.47) holds in the interval $[0, \varepsilon]$. If there were $t \in [R/2, R]$ such that $\mu(t) = 0$, then, by monotonicity of $\mu(t)$, we would conclude $\mu(R/2) = 0$ as well. So we assume $\mu(t) > 0$ in $[R/2, R]$. Then by (9.46) and (9.47), we get

$$b/2a \leqslant \frac{\mu'(t)}{\mu(t)^{(m-1)/m}}, \qquad \mathcal{H}^1\text{-a.e.}$$

Integrating between $R/2$ and R, we get by (9.43)

$$\begin{aligned} bR/4a &\leqslant \mu(R)^{1/m} - \mu(R/2)^{1/m} \\ &\leqslant \mu(R)^{1/m} \leqslant (\varepsilon\omega_m)^{1/m} R \leqslant (\varepsilon\omega_m)^{1/m}. \end{aligned}$$

This is a contradiction, since $\varepsilon\omega_m < (b/4a)^m = I_C(v)^m/(8^m v^{m-1})$ by (9.41). So the proof in case $h(x, R) = |E \cap B_C(x, R)|\,(|B_C(x, R))|^{-1}$ is completed.

For the remaining case, when

$$h(x, R) = \frac{|B_C(x, R) \setminus E|}{|B_C(x, R)|},$$

we replace E by $C \setminus E$, which is also an isoperimetric region, and we are reduced to the previous case. □

Remark 9.44 The case $h(x, R) = |B_C(x, R)|^{-1}|B_C(x, R) \setminus E|$ in the proof of the density estimate in Theorem 9.43 is treated in Leonardi and Rigot [266] in a completely different way using the monotonicity of the isoperimetric profile in Carnot groups.

We define the sets

$$\begin{aligned} E_1 &= \{x \in C : \exists r > 0 \text{ such that } |B_C(x, r) \setminus E| = 0\}, \\ E_0 &= \{x \in C : \exists r > 0 \text{ such that } |B_C(x, r) \cap E| = 0\}, \\ S &= \{x \in C : h(x, r) > \varepsilon \text{ for all } r \leqslant 1\}. \end{aligned}$$

In the same way as in Theorem 4.3 of [266], we get the following:

Proposition 9.45 *Let ε be as in Theorem 9.43. Then we have:*

1. *E_0, E_1, and S form a partition of C.*
2. *E_0 and E_1 are open in C.*
3. *$E_0 = E(0)$ and $E_1 = E(1)$.*
4. *$S = \partial E_0 = \partial E_1$, where the boundary is taken relative to C.*

As a consequence, we get the following two corollaries:

Corollary 9.46 (Lower Density Bound) *Let $C \subset \mathbb{R}^m$ be a convex body and $E \subset C$ an isoperimetric region of volume v. Then there exists a constant $K > 0$, only depending on ε, on the constant $K' > 0$ in the isoperimetric inequality for balls (9.38), and on an Ahlfors constant ℓ_1, such that*

$$P(E, B_C(x, r)) \geqslant K r^{m-1}, \tag{9.48}$$

for all $x \in \partial E_1$ and $r \leqslant 1$.

Proof If $x \in \partial E_1$, the choice of ε and the relative isoperimetric inequality (9.36) give

$$\begin{aligned} P(E, B_C(x, r)) &\geqslant K' \min\{|E \cap B_C(x, r)|, |B_C(x, r) \setminus E|\}^{(m-1)/m} \\ &= K' \left(|B_C(x, r)|\, h(x, r)\right)^{(m-1)/m} \geqslant M(|B_C(x, r)|\, \varepsilon)^{(m-1)/m} \\ &\geqslant K' \left(\ell_1 \varepsilon\right)^{(m-1)/m} r. \end{aligned}$$

This implies the desired inequality with $K = K' \left(\ell_1 \varepsilon\right)^{(m-1)/m}$. □

Remark 9.47 If C_i is a sequence of convex bodies converging to a convex body C in Hausdorff distance, and $E_i \subset C_i$ is a sequence of isoperimetric regions converging weakly to an isoperimetric region $E \subset C$ of volume $0 < v < |C|$, then a constant $K > 0$ in (9.48) can be chosen independent of $i \in \mathbb{N}$. In fact, by (9.41), the constant ε only depends on $|E_i|$, $|C_i| - |E_i|$, and $I_{C_i}(|E_i|)$, which are uniformly bounded since $|C_i| \to |C|$ and $|E_i| \to |E|$. By the convergence in Hausdorff distance of C_i to C, both a lower Ahlfors constant ℓ_1 and a Poincaré constant can be chosen uniformly for all $i \in \mathbb{N}$.

Now we prove that isoperimetric regions also converge in Hausdorff distance to their $L^1_{loc}(\mathbb{R}^m)$ limits, which are also isoperimetric regions by Theorem 9.32. It is necessary to choose a representative of the isoperimetric regions in the class of finite perimeter so that Hausdorff convergence makes sense: we simply consider the closure of the set E_1 of points of density 1.

Theorem 9.48 *Let $\{C_i\}_{i \in \mathbb{N}}$ be a sequence of convex bodies converging in Hausdorff distance to a convex body C. Let $E_i \subset C_i$ be a sequence of isoperimetric sets of* *volumes* $v_i \to v \in (0, |C|)$ *converging in* $L^1(\mathbb{R}^m)$ *to an isoperimetric set* $E \subset C$ *of volume* v. *Then* E_i *converges to* E *in Hausdorff distance.*

Proof By Remark 9.47, we can choose $\varepsilon > 0$ so that Theorem 9.43 holds with this ε for all $i \in \mathbb{N}$.

We now prove that $E_i \to E$ in Hausdorff distance. As $\mathbf{1}_{E_i} \to \mathbf{1}_E$ in $L^1(\mathbb{R}^m)$, we can choose a sequence $r_i \to 0$ so that

$$|E_i \triangle E| < r_i^{m+2}. \tag{9.49}$$

Now fix some $0 < r < 1$, and assume that, for some subsequence, there exist $x_i \in E_i \setminus E_r$. Choose i large enough so that $r_i < \min\{\ell_1, r\}$. Then, by (9.49),

$$|E_i \cap B_{C_i}(x_i, r_i)| \leqslant |E_i \setminus E| \leqslant |E_i \Delta E| < r_i^{m+2} < \ell_1 r_i^{m+1} \leqslant |B_{C_i}(x_i, r_i)|\, r_i. \tag{9.50}$$

So, for i large enough, we get

$$h(E_i, C_i, x_i, r_i) = \frac{|E_i \cap B_{C_i}(x_i, r_i)|}{|B_{C_i}(x_i, r_i)|} < r_i \leqslant \varepsilon.$$

By Theorem 9.43, we conclude that $|E_i \cap B_{C_i}(x_i, r_i/2)| = 0$ for i large enough. As we have chosen as representative of E_i the closure of the set of points of E_i of density 1, we get a contradiction that shows that $E_i \subset (E)_r$ for i large enough. In a similar way, we get that $E \subset (E_i)_r$, which proves that the Hausdorff distance between E and E_i is less than an arbitrary $r > 0$. So $E_i \to E$ in Hausdorff distance. □

9.5.4 *The Isoperimetric Profile for Small Volumes*

In this section, we show that isoperimetric sets of small volume inside a convex body are close to boundary points which tangent cone has minimum solid angle. This will be proven by rescaling the isoperimetric regions and then studying their convergence, as in Morgan and Johnson [311]. We recall first some results on closed convex cones.

A *cone* $K \subset \mathbb{R}^m$ of vertex p is, by definition, a set invariant by all dilations of center p and positive ratio. Let $K \subset \mathbb{R}^m$ be a closed convex cone with vertex p. Define

$$\alpha(K) = \mathcal{H}^{m-1}(\partial B(p, 1) \cap \operatorname{int}(K))$$

as the *solid angle* of K.

It is known that the geodesic balls centered at the vertex are isoperimetric regions in K (see Lions and Pacella [272] or Ritoré and Rosales [367]) and that they are the only ones (see Figalli and Indrei [159] for general convex cones) without assuming any regularity assumption on the boundary of the cone. The isoperimetric profile of K is then given by

$$I_K(v) = \alpha(K)^{1/m}\, m^{(m-1)/m} v^{(m-1)/m}. \tag{9.51}$$

Consequently, the isoperimetric profile of a convex cone is completely described by its solid angle.

We define the tangent cone C_p of a convex body C at a given boundary point $p \in \partial C$ as the closure of the set

$$\bigcup_{\lambda>0} h_{p,\lambda}(C),$$

where $h_{p,\lambda}$ denotes the dilation of center p and factor λ. The solid angle $\alpha(C_p)$ of C_p will be denoted by $\alpha(p)$. Tangent cones to convex bodies have been intensively studied in convex geometry under the name of supporting cones [398, §2.2] or projection cones [72]. In the following result, we prove the lower semicontinuity of the solid angle of tangent cones in convex sets.

Lemma 9.49 *Let $C \subset \mathbb{R}^m$ be a convex body, $\{p_i\}_{i\in\mathbb{N}} \subset \partial C$ so that $p = \lim_{i\to\infty} p_i$. Then*

$$\alpha(p) \leqslant \liminf_{i\to\infty} \alpha(p_i). \tag{9.52}$$

In particular, this implies the existence of points in ∂C whose tangent cones are minima of the solid angle function.

Proof We may assume that $\alpha(p_i)$ converges to $\liminf_{i\to\infty} \alpha(p_i)$ passing to a subsequence if necessary. Since the sequence $C_{p_i} \cap \overline{B}(p_i, 1)$ is bounded in Hausdorff distance, we can extract a non-relabeled subsequence converging to a convex body $C_\infty \subset \overline{B}(p, 1)$. It is easy to check that C_∞ is the intersection of a closed convex cone K_∞ of vertex p with $\overline{B}(p, 1)$ and that $C_p \subset K_\infty$. By the continuity of the volume with respect to the Hausdorff distance, we have

$$m^{-1}\alpha(p) = |C_p \cap \overline{B}(p, 1)| \leqslant |C_\infty| = \lim_{i\to\infty} |C_{p_i} \cap \overline{B}(p_i, 1)| = \lim_{i\to\infty} m^{-1}\alpha(p_i),$$

thus providing (9.52).

To show the existence of tangent cones with the smallest solid angle, we simply take a sequence $\{p_i\}_{i\in\mathbb{N}}$ of points at the boundary of C so that $\alpha(p_i)$ converges to $\inf\{\alpha(p) : p \in \partial C\}$, we extract a convergent subsequence, and we apply the lower semicontinuity of the solid angle function. □

The isoperimetric profiles of tangent cones which are minima of the solid angle function coincide. The common profile will be denoted by $I_{C_{\min}}$.

Proposition 9.50 *Let $C \subset \mathbb{R}^m$ be a convex body. Then*

$$I_C(v) \leqslant I_{C_{\min}}(v) \leqslant \mathbb{H}_0^m(v), \tag{9.53}$$

for all $0 \leqslant v \leqslant |C|$.

Remark 9.51 Proposition 9.50 provides an alternative proof of the fact that the intersection $E \cap \partial C$ is not empty when $E \subset C$ is isoperimetric since if $E \cap \partial C$ is empty, then $P(E) \geqslant I_{\mathbb{R}^m}(|E|)$.

Proof of Proposition 9.50 Fix $0 < v < |C|$, and let $p \in \partial C$ such that $I_{C_p} = I_{C_{\min}}$. Let $r > 0$ such that $|B_C(p,r)| = v$. The closure of the set $\partial B(p,r) \cap \operatorname{int}(C)$ is a geodesic sphere of the closed cone K_p of vertex p subtended by the closure of $\partial B(p,r) \cap \operatorname{int}(C)$. If $S = \partial B(p,r) \cap \operatorname{int}(C)$, then $S = \partial B(p,r) \cap \operatorname{int}(K_p)$ as well. By the convexity of C,

$$B(p,r) \cap \operatorname{int}(K_p) \subset B(p,r) \cap \operatorname{int}(C)$$

and so $v_0 = \mathcal{H}^m(B(p,r) \cap \operatorname{int}(K_p)) \leqslant v$. Since $K_p \subset C_p$, (9.51) implies $\mathcal{H}^{m-1}(S) \leqslant I_{C_{\min}}(v_0)$. So we have

$$I_C(v) \leqslant P_C(B_C(p,r)) = \mathcal{H}^n(S) \leqslant I_{C_{\min}}(v_0) \leqslant I_{C_{\min}}(v),$$

as $I_{C_{\min}}$ is an increasing function. This proves (9.53). □

We now describe the asymptotic behavior of the isoperimetric profile of an arbitrary convex cone.

Theorem 9.52 *Let $C \subset \mathbb{R}^m$ be a convex body. Then*

$$\lim_{v \to 0} \frac{I_C(v)}{I_{C_{\min}}(v)} = 1. \tag{9.54}$$

Moreover, a rescaling of a sequence of isoperimetric regions of volumes approaching 0 has a convergent subsequence in Hausdorff distance to a geodesic ball centered at some vertex in a tangent cone with the smallest solid angle.

Proof To prove (9.54), we notice that the tangent cones C_p and $(\lambda C)_{\lambda p}$ coincide up to a translation. Hence, (9.53) holds replacing C by λC for any $\lambda > 0$. We consider a sequence $\{\lambda_i\}_{i \in \mathbb{N}}$ of positive numbers such that $\lambda_i \to \infty$ and take limits to get

$$\limsup_{i \to \infty} I_{\lambda_i C}(v) \leqslant I_{C_{\min}}(v), \tag{9.55}$$

for any $v > 0$.

Consider now a sequence $\{E_i\}_{i\in\mathbb{N}} \subset C$ of isoperimetric sets in C of volumes $v_i \to 0$. We take $p_i \in E_i \cap \partial C$. Translating C and passing to a subsequence, we may assume that $p_i \to 0 \in \partial C$. Let $\lambda_i = v_i^{-1/(n+1)}$. Then $\lambda_i \to \infty$ and $\lambda_i E_i$ are isoperimetric regions in $\lambda_i C$ of volume 1. By Theorem 9.34, the sets $\lambda_i E_i$ are connected. We claim that

$$\sup_{i\in\mathbb{N}} \operatorname{diam}(\lambda_i E_i) < \infty. \tag{9.56}$$

Let us complete the proof assuming that (9.56) holds. We pick a large closed ball $\overline{B}(0, 2R)$ so that $|C_0 \cap \overline{B}(0, 2R)| > 1$ and $\lambda_i E_i$ is contained in the smaller ball $\overline{B}(0, R)$. The last property yields

$$P_{\lambda_i C}(\lambda_i E_i) = P(\lambda_i E, \operatorname{int}(\lambda_i C \cap \overline{B}(0, 2R))). \tag{9.57}$$

Since $\lambda_i C \cap \overline{B}(0, 2R)$ converges in Hausdorff distance to $C_0 \cap \overline{B}(0, 2R)$, Theorem 9.32 implies that a subsequence of $\lambda_i E_i$ converges in $L^1_{loc}(\mathbb{R}^m)$ to an isoperimetric set $E \subset C_0 \cap \overline{B}(0, 2R)$. But, since $\lambda_i E \subset \overline{B}(0, R)$, we also have $E \subset \overline{B}(0, R)$ and so

$$P_{C_0}(E) = P(E, \operatorname{int}(C_0 \cap \overline{B}(0, 2R))). \tag{9.58}$$

Since we have chosen R to satisfy $|C_0 \cap \overline{B}(0, 2R)| > 1$, the ball B centered at 0 of volume 1 is strictly contained in $C_0 \cap \overline{B}(0, 2R)$ and so $P_{C_0}(B) = P(B, \operatorname{int}(C_0 \cap \overline{B}(0, 2R)))$. As E is isoperimetric in $C_0 \cap \overline{B}(0, 2R)$ and B is isoperimetric in C_0, Eqs. (9.57) and (9.58) imply

$$P_{C_0}(E) \leqslant P_{C_0}(B) = I_{C_0}(1).$$

This implies that E is an isoperimetric set in C_0.

By the property $P_{C_0}(E) \liminf_{i\to\infty} P_{\lambda_i C}(\lambda_i E_i)$, we obtain

$$I_{C_0}(1) \leqslant \liminf_{i\to\infty} I_{\lambda_i C}(1)$$

and, by (9.55),

$$\limsup_{i\to\infty} I_{\lambda_i C}(1) \leqslant I_{C_{\min}}(1) \leqslant I_{C_0}(1) \leqslant \liminf_{i\to\infty} I_{\lambda_i C}(1).$$

Thus,

$$I_{C_0}(1) = I_{C_{\min}}(1) = \lim_{i\to\infty} I_{\lambda_i C}(1). \tag{9.59}$$

From (9.51), we deduce that C_0 has minimum solid angle. Finally, from (9.59), (9.39), and the fact that $\lambda C_0 = C_0$, we deduce

$$1 = \lim_{i\to\infty} \frac{I_{\lambda_i C}(1)}{I_{C_0}(1)} = \lim_{i\to\infty} \frac{\lambda_i^n I_C(1/\lambda_i^m)}{\lambda_i^n I_{C_0}(1/\lambda_i^m)} = \lim_{i\to\infty} \frac{I_C(v_i)}{I_{C_0}(v_i)}.$$

So it remains to prove (9.56) to conclude the proof. For this, it is enough to prove

$$P_{\lambda_i C}(F_i, B_{\lambda_i C}(x,r)) \geqslant K r^{m-1}, \tag{9.60}$$

for any $0 < r \leqslant 1$, $x \in \partial F_i$, and any isoperimetric region $F_i \subset \lambda_i C$ of volume 1. The constant $K > 0$ is independent of i.

To prove (9.60), observe first that the constant in the relative isoperimetric inequality (9.36) is invariant by dilations and, if the factor of dilation is chosen larger than 1, then the estimate $r \leqslant r_0$ is uniform. The same argument can be applied to a lower Ahlfors constant ℓ_1. The constant $\omega_m = |\overline{B}(0,1)|$ is universal and does not depend on the convex set.

Now we modify the proof of Theorem 9.43 to show that there exists some $\varepsilon > 0$, independent of i, so that if $h(\lambda_i E_i, \lambda_i C, x, r) \leqslant \varepsilon$, then $h(\lambda_i E_i, \lambda_i C, x, r/2) = 0$, for $0 < r \leqslant 1$.

First, we treat the case

$$h(F_i, \lambda_i C, x, R) = \frac{|F_i \cap B_{\lambda_i C}(x,R)|}{|B_{\lambda_i C}(x,R)|}.$$

By Theorem 9.43, since $I_C(1) \leqslant I_{\lambda_i C}(1)$ for all $i \in \mathbb{N}$, it is enough to take

$$0 < \varepsilon \leqslant \min\left\{\frac{1}{\omega_m}, c_2, \frac{I_C(1)^m}{\omega_m 8^m}\right\}.$$

Now when

$$h(F_i, \lambda_i C, x, R) = \frac{|B_{\lambda_i C}(x,R) \setminus F_i|}{|B_{\lambda C}(x,R)|},$$

we proceed as in the proof of Case 1 of Lemma 5.2 in [266]. For λ_i large enough, we have $1 + \omega_m = |\lambda_i E_i| + \omega_m < |\lambda_i C|/2$. As $I_{\lambda_i C}$ is increasing in the interval $(0, |\lambda_i C|/2]$, the proof of Case 1 in Lemma 5.2 of [266] provides an $\varepsilon > 0$ independent of i.

As in Remark 9.47, we conclude the existence of $K > 0$ independent of i so that (9.60) holds.

Now, if $\operatorname{diam}(\lambda_i E_i)$ is not uniformly bounded, (9.60) implies that $P_{\lambda_i C}(\lambda_i E_i)$ is unbounded. But this contradicts the fact that $P_{\lambda_i C}(\lambda_i E_i) = I_{\lambda_i C}(1) \leqslant I_{C_{\min}}(1)$ for all i. □

Theorem 9.53 *Let $P \subset \mathbb{R}^m$ be a convex polytope. For small volumes, the isoperimetric regions in P are relative balls $B_P(x, r)$ centered at vertexes x with $I_{C_x} = I_{C_{\min}}$ (vertexes with the smallest possible solid angle of its tangent cone).*

Proof Let $E_i \subset P$ be a sequence of isoperimetric sets in P with $|E_i| \to 0$. By Theorem 9.52, the sets E_i are contained, for i large, in an arbitrarily small ball B centered at a point $x \in \partial P$ with $I_{C_x} = I_{C_{\min}}$. We choose B small enough so that $B \cap P = B \cap P_x$. Then E_i are also isoperimetric sets in P_x, and, by the results of Figalli and Indrei [159], they are relative balls centered at x. □

9.6 Notes

Notes for Sect. 9.1 The key argument to provide regularity of perimeter minimizers under a volume constraint in a set with smooth boundary was given by Grüter [203]. His method consists of reflecting an isoperimetric set inside Ω locally across the boundary $\partial\Omega$. The reflected set satisfies a quasi-minimization property. See also the paper by Grüter and Jost [209] on varifolds with free boundary; Grüter et al. [208]; and Grüter [204–207].

Notes for Sect. 9.2 The first and second variation formulas in this section are well-known for flows associated with a vector field; see Simon [404]. For one-parameter families of diffeomorphisms, we have followed an approach similar to the one in Sect. 1.3.

Notes for Sect. 9.3 Most of the results of this section are taken from the paper by Bayle and Rosales [48]. To obtain the comparison results, they followed the approach in Bayle [46, 47]. They defined the upper second derivative of a function f at a point x by

$$\overline{D^2 f}(x) = \limsup_{h\to 0} \frac{f(x+h) - f(x-h) - 2f(x)}{h^2}$$

and use condition $\overline{D^2 f}(x) \leqslant C$ instead of the weak condition we introduced in Definition 3.12 following Morgan and Johnson [311].

The first proof of the concavity of the isoperimetric profile of a Euclidean convex set was given by Sternberg and Zumbrun [411]. The concavity of the power $m/(m-1)$ of the profile was shown by Kuwert [256].

It is possible to give a proof of the Lévy-Gromov inequality in Theorem 9.18 following the classical argument using Heintze-Karcher inequalities. Using this strategy, Morgan [305] proved a general Lévy-Gromov inequality (9.16) for convex sets with non-smooth boundary in manifolds with Ricci curvature satisfying the inequality $\operatorname{Ric} \geqslant (m-1)\,\delta > 0$.

Comparison results for the isoperimetric profile outside Euclidean convex sets have been obtained by Choe [114, 115, 117] (see also [116]) and Choe et al. [118,

119]. Outside convex sets in Cartan-Hadamard manifolds, an optimal isoperimetric inequality has been obtained by Choe and Ritoré [120].

Notes for Sect. 9.4 This section is based on the results of Wang and Xia [436] for capillary surfaces in geodesic balls of space forms. Previous results in dimension 3 for the free boundary problem, with capillary angle $\pi/2$, have been obtained by Ros and Vergasta [383], Bürger and Kuwert [84], and Nunes [322], who completed the classification of stable surfaces in a *three*-dimensional ball.

The vector field X_a defined in (9.18) generates a one-parameter family of conformal automorphisms of $\mathbb{B} \subset \mathbb{R}^m$; see p. 274 in Li and Yau [270]. It has been used by Fraser and Schoen [167] to obtain information about the first Steklov eigenvalue in manifolds with boundary. The family of deformations induced by the vector fields X_a has been used by Marinov [276] in dimension 2 and by Li and Xiong [269] in higher dimensions in their study of capillary surfaces. The problem we have considered in this chapter is the one of capillary angle $\pi/2$.

The isoperimetric problem for the relative perimeter inside a Euclidean ball can be solved by symmetrization; see the discussion in §18 in Burago and Zalgaller [83] and also Burago and Maz'ya [82], Bokowski and Sperner [67], and Almgren [13].

Wang and Xia's results in [436] are much more general than the ones included in Sect. 9.4. They prove that any immersed stable capillary hypersurface in a ball of a simply connected space forms is totally umbilical. They also obtain a Heintze-Karcher-type inequality for hypersurfaces with free boundary in a ball, which together with the new Minkowski formula yields a new proof of Alexandrov's theorem for embedded hypersurfaces with constant mean curvature with free boundary inside a ball using Reilly-Ros' techniques.

Notes for Sect. 9.5 The concavity of the isoperimetric profile of open, bounded, strictly convex sets in Euclidean space was proven by Sternberg and Zumbrun in Theorem 2.8 in [411]. Kuwert first proved the concavity of the function $I_\Omega^{m/(m-1)}$ in [256]. Milman [284] first established this property when Ω is a bounded open set in $\mathbb{R}^m$ without any hypothesis on the regularity of $\partial\Omega$. For the proofs in this section, we have followed Ritoré and Vernadakis [371]. While Theorem 4.3 in [371] is not correct, all proofs in [371] depending on such result can be obtained from Theorem 9.32. The concavity of $I_\Omega^{m/(m-1)}$ for arbitrary open convex sets $\Omega \subset \mathbb{R}^m$ was established in Leonardi et al. [267].

In convex cones in Euclidean spaces, balls about the vertex are isoperimetric sets. This result was proven by Lions and Pacella [272]. Uniqueness of these isoperimetric sets when the convex cone has smooth boundary out of the vertex was proven by Ritoré and Rosales [367]. For general boundaries, it was established by Figalli and Indrei [159]. A classification of isoperimetric sets in cones over compact manifolds with non-negative Ricci curvature was provided by Morgan and Ritoré [313].

The proof of the density estimates for isoperimetric sets is taken from [371] following Leonardi and Rigot [266]; see also David and Semmes [132]. It is well-known that the classical monotonicity formula for rectifiable varifolds [404] can be applied to get the lower density bounds similar to (9.48) for small r. Assuming C^2

regularity of the boundary of C (convexity is no longer needed), a monotonicity formula for varifolds with free boundary under boundedness condition on the mean curvature has been obtained by Grüter and Jost [209]. This monotonicity formula implies the lower density bound (9.48).

Another proof of the fact that in convex Euclidean polytopes, balls about a vertex uniquely solve the partitioning problem for small volumes is given by Morgan [305].

When Ω is a bounded convex domain with smooth boundary, Fall [150] has proven that isoperimetric sets of small volume in Ω accumulate near the maxima of the mean curvature function of $\partial\Omega$. He also showed in [151] the existence of a foliation of hypersurfaces with constant mean curvature meeting orthogonally $\partial\Omega$ near non-degenerate critical points of the mean curvature of $\partial\Omega$. He also obtained an asymptotic expansion of the isoperimetric profile of an open bounded set. His techniques are quite similar to the ones used by Nardulli [317] in the boundaryless case.

References

1. U. Abresch, J. Langer, The normalized curve shortening flow and homothetic solutions. J. Differ. Geom. **23**(2), 175–196 (1986)
2. U. Abresch, H. Rosenberg, A Hopf differential for constant mean curvature surfaces in $\mathbf{S}^2 \times \mathbf{R}$ and $\mathbf{H}^2 \times \mathbf{R}$. Acta Math. **193**(2), 141–174 (2004)
3. U. Abresch, H. Rosenberg, Generalized Hopf differentials. Mat. Contemp. **28**, 1–28 (2005)
4. E. Acerbi, N. Fusco, M. Morini, Minimality via second variation for a nonlocal isoperimetric problem. Commun. Math. Phys. **322**(2), 515–557 (2013)
5. C. Adams, F. Morgan, Isoperimetric curves on hyperbolic surfaces. Proc. Am. Math. Soc. **127**(5), 1347–1356 (1999)
6. C. Adams, F. Morgan, S. Nardulli, Isoperimetric profile continuous? Frank Morgan's blog entry, 26 July 2013
7. A.D. Aleksandrov, Uniqueness theorems for surfaces in the large. V. Vestnik Leningrad. Univ. **13**(19), 5–8 (1958)
8. A.D. Aleksandrov, Uniqueness theorems for surfaces in the large. V. Am. Math. Soc. Transl. (2) **21**, 412–416 (1962)
9. S.B. Alexander, R.L. Bishop, Warped products of Hadamard spaces. Manuscripta Math. **96**(4), 487–505 (1998)
10. L.J. Alías, A. Brasil, Jr., O. Perdomo, Stable constant mean curvature hypersurfaces in the real projective space. Manuscripta Math. **121**(3), 329–338 (2006)
11. W.K. Allard, On the first variation of a varifold. Ann. Math. (2) **95**, 417–491 (1972)
12. F. Almgren, Optimal isoperimetric inequalities. Indiana Univ. Math. J. **35**(3), 451–547 (1986)
13. F. Almgren, Spherical symmetrization, in *Proceedings of the International Workshop on Integral Functionals in the Calculus of Variations (Trieste, 1985)*, vol. 15, pp. 11–25 (1987)
14. J.F.J. Almgren, Existence and regularity almost everywhere of solutions to elliptic variational problems with constraints. Mem. Am. Math. Soc. **4**(165), viii+199 (1976)
15. F. Alter, V. Caselles, Uniqueness of the Cheeger set of a convex body. Nonlinear Anal. **70**(1), 32–44 (2009)
16. F. Alter, V. Caselles, A. Chambolle, A characterization of convex calibrable sets in $\mathbb{R}^N$. Math. Ann. **332**(2), 329–366 (2005)
17. L. Ambrosio, Some fine properties of sets of finite perimeter in Ahlfors regular metric measure spaces. Adv. Math. **159**(1), 51–67 (2001)
18. L. Ambrosio, Calculus, heat flow and curvature-dimension bounds in metric measure spaces, in *Proceedings of the International Congress of Mathematicians—Rio de Janeiro 2018. Vol. I. Plenary lectures* (World Scientific Publishing, Hackensack, NJ, 2018), pp. 301–340

M. Ritoré, *Isoperimetric Inequalities in Riemannian Manifolds*,
Progress in Mathematics 348, https://doi.org/10.1007/978-3-031-37901-7

19. L. Ambrosio, V. Caselles, S. Masnou, J.-M. Morel, Connected components of sets of finite perimeter and applications to image processing. J. Eur. Math. Soc. (JEMS) **3**(1), 39–92 (2001)
20. L. Ambrosio, S. Di Marino, N. Gigli, Perimeter as relaxed Minkowski content in metric measure spaces. Nonlinear Anal. **153**, 78–88 (2017)
21. L. Ambrosio, N. Fusco, D. Pallara. *Functions of Bounded Variation and Free Discontinuity Problems*. Oxford Mathematical Monographs (The Clarendon Press, Oxford University Press, New York, 2000)
22. L. Ambrosio, R. Ghezzi, Sobolev and bounded variation functions on metric measure spaces, in *Geometry, Analysis and Dynamics on Sub-Riemannian Manifolds, Vol. II* (Eur. Math. Soc., Zürich, 2016), pp. 211–273
23. S. Angenent, Nodal properties of solutions of parabolic equations. Rocky Mountain J. Math. **21**(2), 585–592 (1991) Current directions in nonlinear partial differential equations (Provo, UT, 1987)
24. G. Antonelli, E. Bruè, M. Fogagnolo, M. Pozzetta, On the existence of isoperimetric regions in manifolds with nonnegative Ricci curvature and Euclidean volume growth. Calc. Var. Part. Differ. Equ. **61**(2), Paper No. 77, 40 (2022)
25. G. Antonelli, M. Fogagnolo, M. Pozzetta, *The Isoperimetric Problem on Riemannian Manifolds via Gromov-Hausdorff Asymptotic Analysis* (2021). https://doi.org/10.48550/arXiv.2101.12711
26. G. Antonelli, S. Nardulli, M. Pozzetta, *The Isoperimetric Problem via Direct Method in Noncompact Metric Measure Spaces with Lower Ricci Bounds*. ESAIM Control Optim. Calc. Var. **28**, Paper No. 57, 32 (2022). https://doi.org/10.1051/cocv/2022052
27. G. Antonelli, E. Pasqualetto, M. Pozzetta, Isoperimetric sets in spaces with lower bounds on the Ricci curvature. Nonlinear Anal. **220**, Paper No. 112839, 59 (2022)
28. G. Antonelli, E. Pasqualetto, M. Pozzetta, D. Semola, *Sharp Isoperimetric Comparison and Asymptotic Isoperimetry on Non Collapsed Spaces with Lower Ricci Bounds* (2022). https://doi.org/10.48550/arXiv.2201.04916
29. M. Athanassenas, A variational problem for constant mean curvature surfaces with free boundary. J. Reine Angew. Math. **377**, 97–107 (1987)
30. T. Aubin, Problèmes isopérimétriques et espaces de Sobolev. J. Differ. Geom. **11**(4), 573–598 (1976)
31. T. Aubin, *Nonlinear analysis on manifolds. Monge-Ampère equations*, vol. 252. *Grundlehren der mathematischen Wissenschaften [Fundamental Principles of Mathematical Sciences]* (Springer, New York, 1982)
32. A. Baernstein, II, *Symmetrization in Analysis*, vol. 36. *New Mathematical Monographs* (Cambridge University Press, Cambridge, 2019). With David Drasin and Richard S. Laugesen, With a foreword by Walter Hayman
33. D. Bakry, M. Ledoux, Lévy-Gromov's isoperimetric inequality for an infinite-dimensional diffusion generator. Invent. Math. **123**(2), 259–281 (1996)
34. W. Ballmann, M. Gromov, V. Schroeder, *Manifolds of Nonpositive Curvature*, vol. 61. *Progress in Mathematics* (Birkhäuser Boston, Boston, MA, 1985)
35. C. Bandle, On a differential inequality and its applications to geometry. Math. Z. **147**(3), 253–261 (1976)
36. C. Bandle, *Isoperimetric Inequalities and Applications*, vol. 7. *Monographs and Studies in Mathematics* (Pitman (Advanced Publishing Program), Boston, MA, London, 1980)
37. J.L. Barbosa, M. do Carmo, On the size of a stable minimal surface in R^3. Am. J. Math. **98**(2), 515–528 (1976)
38. J.L. Barbosa, M. do Carmo, Stability of hypersurfaces with constant mean curvature. Math. Z. **185**(3), 339–353 (1984)
39. J.L. Barbosa, M. do Carmo, J. Eschenburg, Stability of hypersurfaces of constant mean curvature in Riemannian manifolds. Math. Z. **197**(1), 123–138 (1988)
40. L. Barbosa, P. Bérard, Eigenvalue and "twisted" eigenvalue problems, applications to CMC surfaces. J. Math. Pures Appl. (9) **79**(5), 427–450 (2000)

41. F. Barthe, Extremal properties of central half-spaces for product measures. J. Funct. Anal. **182**(1), 81–107 (2001)
42. F. Barthe, An isoperimetric result for the Gaussian measure and unconditional sets. Bull. Lond. Math. Soc. **33**(4), 408–416 (2001)
43. F. Barthe, Isoperimetric inequalities, probability measures and convex geometry, in *European Congress of Mathematics* (Eur. Math. Soc., Zürich, 2005), pp. 811–826
44. F. Barthe, B. Maurey, Some remarks on isoperimetry of Gaussian type. Ann. Inst. H. Poincaré Probab. Stat. **36**(4), 419–434 (2000)
45. C. Bavard, P. Pansu, Sur le volume minimal de $\mathbf{R}^2$. Ann. Sci. École Norm. Sup. (4) **19**(4), 479–490 (1986)
46. V. Bayle, A differential inequality for the isoperimetric profile. Int. Math. Res. Not. **7**, 311–342 (2004)
47. V. Bayle, *Propriétés de Concavité du Profil Isopérimétrique et Applications*, PhD thesis, Institut Fourier (2004)
48. V. Bayle, C. Rosales, Some isoperimetric comparison theorems for convex bodies in Riemannian manifolds. Indiana Univ. Math. J. **54**(5), 1371–1394 (2005)
49. E.F. Beckenbach, T. Rado, Subharmonic functions and surfaces of negative curvature. Trans. Am. Math. Soc. **35**, 662–674 (1933)
50. I. Benjamini, J. Cao, A new isoperimetric comparison theorem for surfaces of variable curvature. Duke Math. J. **85**(2), 359–396 (1996)
51. P. Bérard, G. Besson, S. Gallot, Sur une inégalité isopérimétrique qui généralise celle de Paul Lévy-Gromov. Invent. Math. **80**(2), 295–308 (1985)
52. P. Bérard, D. Meyer, Inégalités isopérimétriques et applications. Ann. Sci. École Norm. Sup. (4) **15**(3), 513–541 (1982)
53. P.H. Bérard, *Lectures on Spectral Geometry*, 15° Colóquio Brasileiro de Matemática [15th Brazilian Mathematics Colloquium] (Instituto de Matemática Pura e Aplicada (IMPA), Rio de Janeiro, 1985). Some aspects of direct problems in spectral geometry, With an appendix by Gérard Besson, With a bibliography by Bérard and M. Berger
54. P.H. Bérard, From vanishing theorems to estimating theorems: the Bochner technique revisited. Bull. Am. Math. Soc. (N.S.) **19**(2), 371–406 (1988)
55. M. Berger, *Geometry I*. Universitext (Springer, Berlin, 2009). Translated from the 1977 French original by M. Cole, S. Levy, Fourth printing of the 1987 English translation [MR0882541]
56. M. Berger, P. Gauduchon, E. Mazet, *Le spectre d'une variété riemannienne*. *Lecture Notes in Mathematics*, vol. 194 (Springer, Berlin, New York, 1971)
57. F. Bernstein, Über die isoperimetrische Eigenschaft des Kreises auf der Kugeloberfläche und in der Ebene. Math. Ann. **60**(1), 117–136 (1905)
58. E. Bierstone, P.D. Milman, Semianalytic and subanalytic sets. Inst. Hautes Études Sci. Publ. Math. **67**, 5–42 (1988)
59. R.L. Bishop, Infinitesimal convexity implies local convexity. Indiana Univ. Math. J. **24**, 169–172 (1974/75)
60. R.L. Bishop, R.J. Crittenden, *Geometry of Manifolds* (AMS Chelsea Publishing, Providence, RI, 2001). Reprint of the 1964 original
61. R.L. Bishop, B. O'Neill, Manifolds of negative curvature. Trans. Am. Math. Soc. **145**, 1–49 (1969)
62. W. Blaschke, *Vorlesungen über Differentialgeometrie und geometrische Grundlagen von Einsteins Relativitätstheorie. Band I. Elementare Differentialgeometrie*, 3d edn. (Dover Publications, New York, NY, 1945)
63. W. Blaschke, *Kreis und Kugel* (Chelsea Publishing Co., New York, 1949)
64. V. Blåsjö, The isoperimetric problem. Am. Math. Mon. **112**(6), 526–566 (2005)
65. S.G. Bobkov, An isoperimetric inequality on the discrete cube, and an elementary proof of the isoperimetric inequality in Gauss space. Ann. Probab. **25**(1), 206–214 (1997)
66. S.G. Bobkov, C. Houdré, Some connections between isoperimetric and Sobolev-type inequalities. Mem. Am. Math. Soc. **129**(616), viii+111 (1997)

67. J. Bokowski, J.E. Sperner, Zerlegung konvexer Körper durch minimale Trennflächen. J. Reine Angew. Math. **311/312**, 80–100 (1979)
68. G. Bol, Isoperimetrische Ungleichungen für Bereiche auf Flächen. Jber. Deutsch. Math.-Verein. **51**, 219–257 (1941)
69. E. Bombieri, Regularity theory for almost minimal currents. Arch. Ration. Mech. Anal. **78**(2), 99–130 (1982)
70. T. Bonnesen, Über das isoperimetrische Defizit ebener Figuren. Math. Ann. **91**(3-4), 252–268 (1924)
71. T. Bonnesen, *Les problèmes des isopérimètres et des isépiphanes, Collection de monographies sur la théorie des fonctions* (Gauthier-Villars, Paris, 1929), 175 p.
72. T. Bonnesen, W. Fenchel, *Theory of Convex Bodies* (BCS Associates, Moscow, ID, 1987) Translated from the German and edited by L. Boron, C. Christenson, B. Smith
73. A. Borbély, On the total curvature of hypersurfaces in negatively curved Riemannian manifolds. Publ. Math. Debrecen **59**(1-2), 17–23 (2001)
74. A. Borbély, On the total curvature of convex hypersurfaces in hyperbolic spaces. Proc. Am. Math. Soc. **130**(3), 849–854 (2002)
75. C. Borell, The Brunn-Minkowski inequality in Gauss space. Invent. Math. **30**(2), 207–216 (1975)
76. S. Brendle, The isoperimetric inequality for a minimal submanifold in Euclidean space. J. Am. Math. Soc. **34**(2), 595–603 (2021)
77. Y. Brenier, Polar factorization and monotone rearrangement of vector-valued functions. Commun. Pure Appl. Math. **44**(4), 375–417 (1991)
78. H. Brezis, *Functional Analysis, Sobolev Spaces and Partial Differential Equations*. Universitext (Springer, New York, 2011)
79. M.R. Bridson, A. Haefliger, *Metric Spaces of Non-Positive Curvature*, vol. 319. *Grundlehren der mathematischen Wissenschaften [Fundamental Principles of Mathematical Sciences]* (Springer, Berlin, 1999)
80. J.E. Brothers, F. Morgan, The isoperimetric theorem for general integrands. Mich. Math. J. **41**(3), 419–431 (1994)
81. D. Burago, Y. Burago, S. Ivanov, *A Course in Metric Geometry*, vol. 33. *Graduate Studies in Mathematics* (American Mathematical Society, Providence, RI, 2001)
82. Y.D. Burago, V.G. Maz'ya, *Potential Theory and Function Theory for Irregular Rregions*. Seminars in Mathematics, V. A. Steklov Mathematical Institute, Leningrad, vol. 3 (Consultants Bureau, New York, 1969). Translated from Russian
83. Y.D. Burago, V.A. Zalgaller, *Geometric Inequalities*, vol. 285. *Grundlehren der Mathematischen Wissenschaften [Fundamental Principles of Mathematical Sciences]* (Springer, Berlin, 1988). Translated from the Russian by A. B. Sosinskiĭ, Springer Series in Soviet Mathematics
84. W. Bürger, E. Kuwert, Area-minimizing disks with free boundary and prescribed enclosed volume. J. Reine Angew. Math. **621**, 1–27 (2008)
85. P. Buser, A note on the isoperimetric constant. Ann. Sci. École Norm. Sup. (4) **15**(2), 213–230 (1982)
86. A. Cañete, M. Ritoré, The isoperimetric problem in complete annuli of revolution with increasing Gauss curvature. Proc. R. Soc. Edinburgh Sect. A **138**(5), 989–1003 (2008)
87. X. Cabré, Elliptic PDE's in probability and geometry: symmetry and regularity of solutions. Discrete Contin. Dyn. Syst. **20**(3), 425–457 (2008)
88. X. Cabré, Isoperimetric, Sobolev, and eigenvalue inequalities via the Alexandroff-Bakelman-Pucci method: a survey. Chin. Ann. Math. Ser. B **38**(1), 201–214 (2017)
89. X. Cabré, X. Ros-Oton, J. Serra, Sharp isoperimetric inequalities via the ABP method. J. Eur. Math. Soc. (JEMS) **18**(12), 2971–2998 (2016)
90. L. Cadeddu, M.A. Farina, A brief note on the coarea formula. Abh. Math. Semin. Univ. Hambg. **88**(1), 193–199 (2018)
91. L.A. Caffarelli, The regularity of mappings with a convex potential. J. Am. Math. Soc. **5**(1), 99–104 (1992)

92. A. Cañete, Stable and isoperimetric regions in rotationally symmetric tori with decreasing Gauss curvature. Indiana Univ. Math. J. **56**(4), 1629–1659 (2007)
93. J. Cao, J.F. Escobar, A new 3-dimensional curvature integral formula for PL-manifolds of non-positive curvature. Comm. Anal. Geom. **11**(3), 489–551 (2003)
94. C. Carathéodory, E. Study, Zwei Beweise des Satzes, daßder Kreis unter allen Figuren gleichen Umfanges den größten Inhalt hat. Math. Ann. **68**(1), 133–140 (1909)
95. T. Carleman, Zur Theorie der Minimalflächen. Math. Z. **9**, 154–160 (1921)
96. K. Castro, Isoperimetric inequalities in cylinders with density. Nonlinear Anal. **217**, Paper No. 112726 (2022)
97. F. Cavalletti, A. Mondino, Sharp and rigid isoperimetric inequalities in metric-measure spaces with lower Ricci curvature bounds. Invent. Math. **208**(3), 803–849 (2017)
98. G.R. Chambers, Proof of the log-convex density conjecture. J. Eur. Math. Soc. (JEMS) **21**(8), 2301–2332 (2019)
99. I. Chavel, On A. Hurwitz' method in isoperimetric inequalities. Proc. Am. Math. Soc. **71**(2), 275–279 (1978)
100. I. Chavel, *Eigenvalues in Riemannian Geometry*, vol. 115. *Pure and Applied Mathematics* (Academic Press, Orlando, FL, 1984). Including a chapter by Burton Randol, With an appendix by Jozef Dodziuk
101. I. Chavel, *Isoperimetric Inequalities*, vol. 145 . *Cambridge Tracts in Mathematics* (Cambridge University Press, Cambridge, 2001). Differential geometric and analytic perspectives
102. I. Chavel, *Riemannian Geometry*, vol. 98. *Cambridge Studies in Advanced Mathematics*, 2nd edn. (Cambridge University Press, Cambridge, 2006). A modern introduction
103. J. Cheeger, A lower bound for the smallest eigenvalue of the Laplacian, in *Problems in Analysis (Sympos. In Honor of Salomon Bochner, Princeton University, Princeton, NJ, 1969)* (Princeton University Press, Princeton, NJ, 1970), pp. 195–199
104. J. Cheeger, D.G. Ebin, *Comparison Theorems in Riemannian Geometry*. North-Holland Mathematical Library, vol. 9 (North-Holland Publishing, Amsterdam-Oxford; American Elsevier Publishing, New York, 1975)
105. J. Cheeger, D. Gromoll, The splitting theorem for manifolds of nonnegative Ricci curvature. J. Differ. Geom. **6**, 119–128 (1971/72)
106. J. Cheeger, D. Gromoll, On the structure of complete manifolds of nonnegative curvature. Ann. Math. (2) **96**, 413–443 (1972)
107. J. Cheeger, M. Gromov, M. Taylor, Finite propagation speed, kernel estimates for functions of the Laplace operator, and the geometry of complete Riemannian manifolds. J. Differ. Geom. **17**(1), 15–53 (1982)
108. B.-Y. Chen, On the total curvature of immersed manifolds. I. An inequality of Fenchel-Borsuk-Willmore. Am. J. Math. **93**, 148–162 (1971)
109. B.-y. Chen, *Geometry of Submanifolds*. Pure and Applied Mathematics, vol. 22 (Marcel Dekker, Inc., New York, 1973)
110. M. Chlebík, A. Cianchi, N. Fusco, The perimeter inequality under Steiner symmetrization: cases of equality. Ann. Math. (2) **162**(1), 525–555 (2005)
111. O. Chodosh, M. Eichmair, A. Volkmann, Isoperimetric structure of asymptotically conical manifolds. J. Differ. Geom. **105**(1), 1–19 (2017)
112. O. Chodosh, M. Engelstein, L. Spolaor, The Riemannian quantitative isoperimetric inequality. J. Eur. Math. Soc. **25**(5), 1711–1741 (2023). https://www.doi.org/10.4171/JEMS/1223
113. J. Choe, Index, vision number and stability of complete minimal surfaces. Arch. Ration. Mech. Anal. **109**(3), 195–212 (1990)
114. J. Choe, Relative isoperimetric inequality for domains outside a convex set. Arch. Inequal. Appl. **1**(2), 241–250 (2003)
115. J. Choe, A note on the three and four-dimensional relative isoperimetric inequality, in *Proceedings of the Eighth International Workshop on Differential Geometry*, pp. 93–100 (Kyungpook National University, Taegu, 2004)
116. J. Choe, Isoperimetric inequalities of minimal submanifolds, in *Global Theory of Minimal Surfaces*, vol. 2. *Clay Math. Proc.* (Amer. Math. Soc., Providence, RI, 2005), pp. 325–369

117. J. Choe, The double cover relative to a convex domain and the relative isoperimetric inequality. J. Aust. Math. Soc. **80**(3), 375–382 (2006)
118. J. Choe, M. Ghomi, M. Ritoré, Total positive curvature of hypersurfaces with convex boundary. J. Differ. Geom. **72**(1), 129–147 (2006)
119. J. Choe, M. Ghomi, M. Ritoré, The relative isoperimetric inequality outside convex domains in $\mathbf{R}^n$. Calc. Var. Part. Differ. Equ. **29**(4), 421–429 (2007)
120. J. Choe, M. Ritoré, The relative isoperimetric inequality in Cartan-Hadamard 3-manifolds. J. Reine Angew. Math. **605**, 179–191 (2007)
121. M. Cicalese, G.P. Leonardi, A selection principle for the sharp quantitative isoperimetric inequality. Arch. Ration. Mech. Anal. **206**(2), 617–643 (2012)
122. E. Cinti, A. Pratelli, Regularity of isoperimetric sets in $\mathbb{R}^2$ with density. Math. Ann. **368**(1-2), 419–432 (2017)
123. S. Cohn-Vossen, Kürzeste Wege und Totalkrümmung auf Flächen. Compos. Math. **2**, 69–133 (1935)
124. A. Cotton, D. Freeman, A. Gnepp, T. Ng, J. Spivack, C. Yoder, The isoperimetric problem on some singular surfaces. J. Aust. Math. Soc. **78**(2), 167–197 (2005)
125. T. Coulhon, L. Saloff-Coste, Isopérimétrie pour les groupes et les variétés. Rev. Mat. Iberoamericana **9**(2), 293–314 (1993)
126. C.B. Croke, Some isoperimetric inequalities and eigenvalue estimates. Ann. Sci. École Norm. Sup. (4) **13**(4), 419–435 (1980)
127. C.B. Croke, A sharp four-dimensional isoperimetric inequality. Comment. Math. Helv. **59**(2), 187–192 (1984)
128. J. Dalphin, Some characterizations of a uniform ball property, in *Congrès SMAI 2013*, vol. 45. *ESAIM Proc. Surveys* (EDP Sci., Les Ulis, 2014), pp. 437–446
129. B. Daniel, Isometric immersions into 3-dimensional homogeneous manifolds. Comment. Math. Helv. **82**(1), 87–131 (2007)
130. B. Daniel, P. Mira, Existence and uniqueness of constant mean curvature spheres in Sol_3. J. Reine Angew. Math. **685**, 1–32 (2013)
131. G. David, S. Semmes, Quantitative rectifiability and Lipschitz mappings. Trans. Am. Math. Soc. **337**(2), 855–889 (1993)
132. G. David, S. Semmes, Quasiminimal surfaces of codimension 1 and John domains. Pac. J. Math. **183**(2), 213–277 (1998)
133. E. De Giorgi, Nuovi teoremi relativi alle misure $(r-1)$-dimensionali in uno spazio ad r dimensioni. Ricerche Mat. **4**, 95–113 (1955)
134. E. De Giorgi, *Selected Papers*. Springer Collected Works in Mathematics (Springer, Heidelberg, 2013) [Author name on title page: Ennio Giorgi], ed. by L. Ambrosio, G. Dal Maso, M. Forti, M. Miranda, S. Spagnolo. Reprint of the 2006 edition [MR2229237]
135. G. De Philippis, G. Franzina, A. Pratelli, Existence of isoperimetric sets with densities "converging from below" on $\mathbb{R}^N$. J. Geom. Anal. **27**(2), 1086–1105 (2017)
136. B.V. Dekster, Monotonicity of integral Gauss curvature. J. Differ. Geom. **16**(2), 281–291 (1981)
137. C. Delaunay, Sur la surface de revolution dont la courbure moyenne est constante. J. Math. Pures Appl. **16**, 309–321 (1841)
138. C. Delaunay, Sur la surface de revolution dont la courbure moyenne est constante. J. Math. Pure et App. **16**, 309–321 (1841)
139. M. do Carmo, C.K. Peng, Stable complete minimal surfaces in $\mathbf{R}^3$ are planes. Bull. Am. Math. Soc. (N.S.) **1**(6), 903–906 (1979)
140. M. do Carmo, M. Ritoré, A. Ros, Compact minimal hypersurfaces with index one in the real projective space. Comment. Math. Helv. **75**(2), 247–254 (2000)
141. M.P. do Carmo, *Riemannian Geometry*. Mathematics: Theory & Applications (Birkhäuser Boston, Boston, MA, 1992). Translated from the second Portuguese edition by Francis Flaherty
142. O. Druet, Sharp local isoperimetric inequalities involving the scalar curvature. Proc. Am. Math. Soc. **130**(8), 2351–2361 (2002)

143. F. Duzaar, K. Steffen, Existence of hypersurfaces with prescribed mean curvature in Riemannian manifolds. Indiana Univ. Math. J. **45**(4), 1045–1093 (1996)
144. F. Edler, Completion for Steiner's basic geometric proofs of the Theorem that a circle possesses a greater area than each other plane figure with same circuit. Gött. Nachr. **1882**, 73–80 (1882)
145. J. Eells, L. Lemaire, A report on harmonic maps. Bull. Lond. Math. Soc. **10**(1), 1–68 (1978)
146. J. Eells, L. Lemaire, Another report on harmonic maps. Bull. Lond. Math. Soc. **20**(5), 385–524 (1988)
147. A. Ehrhard, Symétrisation dans l'espace de Gauss. Math. Scand. **53**(2), 281–301 (1983)
148. M. Eichmair, J. Metzger, Unique isoperimetric foliations of asymptotically flat manifolds in all dimensions. Invent. Math. **194**(3), 591–630 (2013)
149. L.C. Evans, R.F. Gariepy, *Measure Theory and Fine Properties of Functions*. Studies in Advanced Mathematics (CRC Press, Boca Raton, FL, 1992)
150. M.M. Fall, Area-minimizing regions with small volume in Riemannian manifolds with boundary. Pac. J. Math. **244**(2), 235–260 (2010)
151. M.M. Fall, Embedded disc-type surfaces with large constant mean curvature and free boundaries. Commun. Contemp. Math. **14**(6), 1250037, 35 (2012)
152. H.M. Farkas, I. Kra, *Riemann Surfaces*, vol. 71. *Graduate Texts in Mathematics*, 2nd edn. (Springer, New York 1992)
153. H. Federer, Curvature measures. Trans. Am. Math. Soc. **93**, 418–491 (1959)
154. H. Federer, *Geometric Measure Theory*. Die Grundlehren der mathematischen Wissenschaften, Band 153 (Springer New York Inc., New York, 1969)
155. I. Fernández, P. Mira, A characterization of constant mean curvature surfaces in homogeneous 3-manifolds. Differ. Geom. Appl. **25**(3), 281–289 (2007)
156. I. Fernández, P. Mira, Holomorphic quadratic differentials and the Bernstein problem in Heisenberg space. Trans. Am. Math. Soc. **361**(11), 5737–5752 (2009)
157. F. Fiala, Le problème des isopérimètres sur les surfaces ouvertes à courbure positive. Comment. Math. Helv. **13**, 293–346 (1941)
158. F. Fiala, Le problème des isopérimètres dans les plans de Riemann à courbure de signe constant. Comment. Math. Helv. **15**, 249–264 (1943)
159. A. Figalli, E. Indrei, A sharp stability result for the relative isoperimetric inequality inside convex cones. J. Geom. Anal. **23**(2), 938–969 (2013)
160. A. Figalli, F. Maggi, A. Pratelli, A mass transportation approach to quantitative isoperimetric inequalities. Invent. Math. **182**(1), 167–211 (2010)
161. R. Finn, *Equilibrium Capillary Surfaces*, vol. 284. *Grundlehren der mathematischen Wissenschaften [Fundamental Principles of Mathematical Sciences]* (Springer, New York, 1986)
162. D. Fischer-Colbrie, On complete minimal surfaces with finite Morse index in three-manifolds. Invent. Math. **82**(1), 121–132 (1985)
163. D. Fischer-Colbrie, R. Schoen, The structure of complete stable minimal surfaces in 3-manifolds of nonnegative scalar curvature. Commun. Pure Appl. Math. **33**(2), 199–211 (1980)
164. I. Fonseca, The Wulff theorem revisited. Proc. Roy. Soc. Lond. Ser. A **432**(1884), 125–145 (1991)
165. I. Fonseca, S. Müller, A uniqueness proof for the Wulff theorem. Proc. R. Soc. Edinburgh Sect. A **119**(1-2), 125–136 (1991)
166. V. Franceschi, F. Montefalcone, R. Monti, CMC spheres in the Heisenberg group. Anal. Geom. Metr. Spaces **7**(1), 109–129 (2019)
167. A. Fraser, R. Schoen, The first Steklov eigenvalue, conformal geometry, and minimal surfaces. Adv. Math. **226**(5), 4011–4030 (2011)
168. B. Fuglede, Stability in the isoperimetric problem. Bull. Lond. Math. Soc. **18**(6), 599–605 (1986)
169. B. Fuglede, Stability in the isoperimetric problem for convex or nearly spherical domains in $\mathbf{R}^n$. Trans. Am. Math. Soc. **314**(2), 619–638 (1989)
170. N. Fusco, The quantitative isoperimetric inequality and related topics. Bull. Math. Sci. **5**(3), 517–607 (2015)

171. N. Fusco, F. Maggi, A. Pratelli, The sharp quantitative isoperimetric inequality. Ann. Math. (2) **168**(3), 941–980 (2008)
172. M. Gage, R.S. Hamilton, The heat equation shrinking convex plane curves. J. Differ. Geom. **23**(1), 69–96 (1986)
173. M.E. Gage, Curve shortening makes convex curves circular. Invent. Math. **76**(2), 357–364 (1984)
174. M.E. Gage, Curve shortening on surfaces. Ann. Sci. École Norm. Sup. (4) **23**(2), 229–256 (1990)
175. M. Galli, M. Ritoré, Existence of isoperimetric regions in contact sub-Riemannian manifolds. J. Math. Anal. Appl. **397**(2), 697–714 (2013)
176. S. Gallot, Inégalités isopérimétriques et analytiques sur les variétés riemanniennes. *Astérisque* (163-164), 5–6, 31–91, 281 (1988/1989). On the geometry of differentiable manifolds (Rome, 1986)
177. N. Garofalo, F.-H. Lin, Monotonicity properties of variational integrals, A_p weights and unique continuation. Indiana Univ. Math. J. **35**(2), 245–268 (1986)
178. N. Garofalo, F.-H. Lin, Unique continuation for elliptic operators: a geometric-variational approach. Commun. Pure Appl. Math. **40**(3), 347–366 (1987)
179. N. Garofalo, D.-M. Nhieu, Isoperimetric and Sobolev inequalities for Carnot-Carathéodory spaces and the existence of minimal surfaces. Commun. Pure Appl. Math. **49**(10), 1081–1144 (1996)
180. M. Ghomi, J. Spruck, Total curvature and the isoperimetric inequality in Cartan-Hadamard manifolds. J. Geom. Anal. **32**(2), Paper No. 50, 54 (2022)
181. M. Giaquinta, S. Hildebrandt, *Calculus of Variations. I*, vol. 310. *Grundlehren der Mathematischen Wissenschaften [Fundamental Principles of Mathematical Sciences]* (Springer, Berlin, 1996). The Lagrangian formalism
182. N. Gigli, On the differential structure of metric measure spaces and applications. Mem. Am. Math. Soc. **236**(1113), vi+91 (2015)
183. D. Gilbarg, N.S. Trudinger, *Elliptic Partial Differential Equations of Second Order*. Classics in Mathematics (Springer, Berlin, 2001). Reprint of the 1998 edition
184. E. Giusti, *Minimal Surfaces and Functions of Bounded Variation*, vol. 80. *Monographs in Mathematics* (Birkhäuser Verlag, Basel, 1984)
185. E. Gonzalez, U. Massari, I. Tamanini, Minimal boundaries enclosing a given volume. Manuscripta Math. **34**(2-3), 381–395 (1981)
186. E. Gonzalez, U. Massari, I. Tamanini, On the regularity of boundaries of sets minimizing perimeter with a volume constraint. Indiana Univ. Math. J. **32**(1), 25–37 (1983)
187. J. Gonzalo, Soap bubbles and isoperimetric regions in the product of a closed manifold with Euclidean space. arXiv:1312.6311, 21 Dec 2013
188. A. Gray, *Tubes*, vol. 221. *Progress in Mathematics*, 2nd edn. (Birkhäuser Verlag, Basel, 2004). With a preface by Vicente Miquel
189. M.A. Grayson, The heat equation shrinks embedded plane curves to round points. J. Differ. Geom. **26**(2), 285–314 (1987)
190. M.A. Grayson, Shortening embedded curves. Ann. Math. (2) **129**(1), 71–111 (1989)
191. R.E. Greene, H. Wu, C^∞ convex functions and manifolds of positive curvature. Acta Math. **137**(3-4), 209–245 (1976)
192. R.E. Greene, H. Wu, Lipschitz convergence of Riemannian manifolds. Pac. J. Math. **131**(1), 119–141 (1988)
193. P. Griffiths, J. Harris, *Principles of Algebraic Geometry*. Pure and Applied Mathematics (Wiley-Interscience [Wiley], New York, 1978)
194. R. Grimaldi, S. Nardulli, P. Pansu, Differentiability of the isoperimetric profile and topology of analytic Riemannian manifolds. C. R. Math. Acad. Sci. Paris **347**(5-6), 293–297 (2009)
195. R. Grimaldi, S. Nardulli, P. Pansu, Semianalyticity of isoperimetric profiles. Differ. Geom. Appl. **27**(3), 393–398 (2009)
196. M. Gromov, Groups of polynomial growth and expanding maps. Inst. Hautes Études Sci. Publ. Math. **53**, 53–73 (1981)

197. M. Gromov, Isoperimetry of waists and concentration of maps. Geom. Funct. Anal. **13**(1), 178–215 (2003)
198. M. Gromov, *Metric Structures for Riemannian and Non-Riemannian spaces*. Modern Birkhäuser Classics (Birkhäuser Boston, Boston, MA, 2007). english edition. Based on the 1981 French original, With appendices by M. Katz, P. Pansu, S. Semmes, Translated from the French by Sean Michael Bates
199. M. Gromov, Erratum to: Isoperimetry of waists and concentration of maps GAFA 13:1 (2003), pp. 178–215 [mr1978494]. Geom. Funct. Anal. **18**(5), 1786 (2009)
200. W. Groß, Die Minimaleigenschaft der Kugel. Monatsh. Math. Phys. **28**(1), 77–97 (1917)
201. K. Grosse-Brauckmann, Stable constant mean curvature surfaces minimize area. Pac. J. Math. **175**(2), 527–534 (1996)
202. P.M. Gruber, *Convex and Discrete Geometry*, vol. 336. *Grundlehren der mathematischen Wissenschaften [Fundamental Principles of Mathematical Sciences]* (Springer, Berlin, 2007)
203. M. Grüter, Boundary regularity for solutions of a partitioning problem. Arch. Ration. Mech. Anal. **97**(3), 261–270 (1987)
204. M. Grüter, Free boundaries in geometric measure theory and applications, in *Variational Methods for Free Surface Interfaces (Menlo Park, Calif., 1985)* (Springer, New York, 1987), pp. 77–83
205. M. Grüter, Optimal regularity for codimension one minimal surfaces with a free boundary. Manuscripta Math. **58**(3), 295–343 (1987)
206. M. Grüter, Regularity results for minimizing currents with a free boundary. J. Reine Angew. Math. **375/376**, 307–325 (1987)
207. M. Grüter, The monotonicity formula in geometric measure theory, and an application to a partially free boundary problem, in *Partial Differential Equations and Calculus of Variations*, vol. 1357. *Lecture Notes in Math.* (Springer, Berlin, 1988), pp. 238–255
208. M. Grüter, S. Hildebrandt, J.C.C. Nitsche, Regularity for stationary surfaces of constant mean curvature with free boundaries. Acta Math. **156**(1-2), 119–152 (1986)
209. M. Grüter, J. Jost, Allard type regularity results for varifolds with free boundaries. Ann. Scuola Norm. Sup. Pisa Cl. Sci. (4) **13**(1), 129–169 (1986)
210. P. Günther, Einige Sätze über das Volumenelement eines Riemannschen Raumes. Publ. Math. Debrecen **7**, 78–93 (1960)
211. H. Hadwiger. *Vorlesungen über Inhalt, Oberfläche und Isoperimetrie* (Springer, Berlin-Göttingen-Heidelberg, 1957)
212. H. Hadwiger, Gitterperiodische Punktmengen und Isoperimetrie. Monatsh. Math. **76**, 410–418 (1972)
213. P. Hajłasz, P. Koskela, Sobolev met Poincaré. Mem. Am. Math. Soc. **145**(688), x+101 (2000)
214. T.C. Hales, Jordan's proof of the Jordan curve theorem. Stud. Logic Grammar Rhetoric **10**(23), 45–60 (2007)
215. R.R. Hall, A quantitative isoperimetric inequality in n-dimensional space. J. Reine Angew. Math. **428**, 161–176 (1992)
216. R. Hardt, M. Hoffmann-Ostenhof, T. Hoffmann-Ostenhof, N. Nadirashvili, Critical sets of solutions to elliptic equations. J. Differ. Geom. **51**(2), 359–373 (1999)
217. G.H. Hardy, J.E. Littlewood, G. Pólya, *Inequalities* (Cambridge University Press, Cambridge, 1934), xii, 314 p.
218. P. Hartman, Geodesic parallel coordinates in the large. Am. J. Math. **86**, 705–727 (1964)
219. P. Hartman, *Ordinary Differential Equations* (S. M. Hartman, Baltimore, MD, 1973). Corrected reprint
220. J. Hass, Bounded 3-manifolds admit negatively curved metrics with concave boundary. J. Differ. Geom. **40**(3), 449–459 (1994)
221. J. Hass, Isoperimetric regions in nonpositively curved manifolds. arXiv:1604.02768, 11 Apr 2016
222. L. Hauswirth, J. Pérez, P. Romon, A. Ros, The periodic isoperimetric problem. Trans. Am. Math. Soc. **356**(5), 2025–2047 (2004)

223. E. Hebey, *Nonlinear Analysis on Manifolds: Sobolev Spaces and Inequalities*, vol. 5. *Courant Lecture Notes in Mathematics* (New York University, Courant Institute of Mathematical Sciences, New York; American Mathematical Society, Providence, RI, 1999)
224. J. Heinonen, P. Koskela, N. Shanmugalingam, J.T. Tyson, *Sobolev Spaces on Metric Measure Spaces*, vol. 27. *New Mathematical Monographs* (Cambridge University Press, Cambridge, 2015). An approach based on upper gradients
225. E. Heintze, H. Karcher, A general comparison theorem with applications to volume estimates for submanifolds. Ann. Sci. École Norm. Sup. (4) **11**(4), 451–470 (1978)
226. F. Hélein, Inégalité isopérimétrique et calibration. Ann. Inst. Fourier (Grenoble) **44**(4), 1211–1218 (1994)
227. F. Hélein, Isoperimetric inequalities and calibrations, in *Progress in Partial Differential Equations: the Metz Surveys, 4*, vol. 345. *Pitman Res. Notes Math. Ser.* (Longman, Harlow, 1996), pp. 92–105
228. G. Henry, J. Petean, On Yamabe constants of products with hyperbolic spaces. J. Geom. Anal. **25**(2), 1387–1400 (2015)
229. J. Hersch, Quatre propriétés isopérimétriques de membranes sphériques homogènes. C. R. Acad. Sci. Paris Sér. A-B **270**, A1645–A1648 (1970)
230. M.W. Hirsch, *Differential Topology*, vol. 33. *Graduate Texts in Mathematics* (Springer, New York, 1994). Corrected reprint of the 1976 original
231. D. Hoffman, J. Spruck, Sobolev and isoperimetric inequalities for Riemannian submanifolds. Commun. Pure Appl. Math. **27**, 715–727 (1974)
232. D. Hoffman, J. Spruck, A correction to: "Sobolev and isoperimetric inequalities for Riemannian submanifolds" (Commun. Pure Appl. Math. **27**, 715–725 (1974)). Commun. Pure Appl. Math. **28**(6), 765–766 (1975)
233. H. Hopf, *Differential Geometry in the Large*, vol. 1000. *Lecture Notes in Mathematics* (Springer, Berlin, 1983). Notes taken by Peter Lax, John Gray, With a preface by S. S. Chern
234. W.-T. Hsiang, W.-Y. Hsiang, On the uniqueness of isoperimetric solutions and imbedded soap bubbles in noncompact symmetric spaces. I. Invent. Math. **98**(1), 39–58 (1989)
235. W.-Y. Hsiang, A symmetry theorem on isoperimetric regions (1988) PAM-409, UC Berkeley
236. W.-Y. Hsiang, Isoperimetric regions and soap bubbles, in *Differential Geometry*, vol. 52. *Pitman Monogr. Surveys Pure Appl. Math.* (Longman Sci. Tech., Harlow, 1991), pp. 229–240
237. W.-Y. Hsiang, On soap bubbles and isoperimetric regions in noncompact symmetric spaces. I. Tohoku Math. J. (2) **44**(2), 151–175 (1992)
238. W.-H. Huang, C.-C. Lin, Negatively curved sets on surfaces of constant mean curvature in $\mathbf{R}^3$ are large. Arch. Ration. Mech. Anal. **141**(2), 105–116 (1998)
239. A. Huber, On the isoperimetric inequality on surfaces of variable Gaussian curvature. Ann. Math. (2) **60**, 237–247 (1954)
240. A. Hurwitz, Sur le problème des isopérimètres. C. R. Acad. Sci. Paris **132**, 401–403 (1901)
241. N. Jacobson, *Basic Algebra. I* (W. H. Freeman and Co., San Francisco, CA, 1974)
242. H. Karcher, Riemannian center of mass and mollifier smoothing. Commun. Pure Appl. Math. **30**(5), 509–541 (1977)
243. B. Kawohl, *Rearrangements and Convexity of Level Sets in PDE*, vol. 1150. *Lecture Notes in Mathematics* (Springer, Berlin, 1985)
244. J.L. Kazdan, Unique continuation in geometry. Commun. Pure Appl. Math. **41**(5), 667–681 (1988)
245. J.L. Kazdan, F.W. Warner, Curvature functions for compact 2-manifolds. Ann. Math. (2) **99**, 14–47 (1974)
246. B. Klartag, Needle decompositions in Riemannian geometry. Mem. Am. Math. Soc. **249**(1180), v+77 (2017)
247. B. Kleiner, An isoperimetric comparison theorem. Invent. Math. **108**(1), 37–47 (1992)
248. B.R. Kloeckner, G. Kuperberg, The Cartan-Hadamard conjecture and the Little Prince. Rev. Mat. Iberoam. **35**(4), 1195–1258 (2019)

249. W.R. Knorr, *Textual Studies in Ancient and Medieval Geometry* (Birkhäuser Boston, Boston, MA, 1989)
250. W.R. Knorr, *The Ancient Tradition of Geometric Problems* (Dover Publications, New York, 1993). Corrected reprint of the 1986 original
251. H. Knothe, Contributions to the theory of convex bodies. Mich. Math. J. **4**, 39–52 (1957)
252. S. Kobayashi, K. Nomizu, *Foundations of Differential Geometry. Vol. I*. Wiley Classics Library (Wiley, New York, 1996). Reprint of the 1963 original, A Wiley-Interscience Publication
253. S. Kobayashi, K. Nomizu, *Foundations of Differential Geometry. Vol. II*. Wiley Classics Library (Wiley, New York, 1996). Reprint of the 1969 original, A Wiley-Interscience Publication
254. M. Koiso, Deformation and stability of surfaces with constant mean curvature. Tohoku Math. J. (2) **54**(1), 145–159 (2002)
255. N.J. Korevaar, R. Kusner, B. Solomon, The structure of complete embedded surfaces with constant mean curvature. J. Differ. Geom. **30**(2), 465–503 (1989)
256. E. Kuwert, Note on the isoperimetric profile of a convex body, in *Geometric Analysis and Nonlinear Partial Differential Equations* (Springer, Berlin, 2003), pp. 195–200
257. O.A. Ladyzhenskaya, N.N. Ural'tseva, *Linear and Quasilinear Elliptic Equations* (Academic Press, New York-London, 1968). Translated from the Russian by Scripta Technica, Inc, Translation editor: Leon Ehrenpreis
258. J.L. Lagrange, Essai d'une nouvelle methode pour determiner les maxima et les minima des formules integrales indefinies. Miscellanea Taurinensia **325**(1), 173–199 (1760)
259. G. Lawlor, A new minimization proof for the brachistochrone. Am. Math. Mon. **103**(3), 242–249 (1996)
260. G. Lawlor, Proving area minimization by directed slicing. Indiana Univ. Math. J. **47**(4), 1547–1592 (1998)
261. G. Lawlor, F. Morgan, Paired calibrations applied to soap films, immiscible fluids, and surfaces or networks minimizing other norms. Pac. J. Math. **166**(1), 55–83 (1994)
262. G. Lawlor, F. Morgan, Curvy slicing proves that triple junctions locally minimize area. J. Differ. Geom. **44**(3), 514–528 (1996)
263. J.M. Lee, *Introduction to Riemannian Manifolds*, vol. 176. *Graduate Texts in Mathematics* (Springer, Cham, 2018). Second edition of [MR1468735]
264. J.M. Lee, T.H. Parker, The Yamabe problem. Bull. Am. Math. Soc. (N.S.) **17**(1), 37–91 (1987)
265. L. Lemaire, Applications harmoniques de surfaces riemanniennes. J. Differ. Geom. **13**(1), 51–78 (1978)
266. G.P. Leonardi, S. Rigot, Isoperimetric sets on Carnot groups. Houst. J. Math. **29**(3), 609–637 (2003)
267. G.P. Leonardi, M. Ritoré, E. Vernadakis, Isoperimetric inequalities in unbounded convex bodies. Mem. Am. Math. Soc. **276**(1354), 1–86 (2022)
268. P. Levy, *Problèmes concrets d'analyse fonctionalle* (Gauthier-Villars, Paris, 1951)
269. H. Li, C. Xiong, Stability of capillary hypersurfaces in a Euclidean ball. Pac. J. Math. **297**(1), 131–146 (2018)
270. P. Li, S.T. Yau, A new conformal invariant and its applications to the Willmore conjecture and the first eigenvalue of compact surfaces. Invent. Math. **69**(2), 269–291 (1982)
271. X. Li-Jost, Uniqueness of minimal surfaces in Euclidean and hyperbolic 3-space. Math. Z. **217**(2), 275–285 (1994)
272. P.-L. Lions, F. Pacella, Isoperimetric inequalities for convex cones. Proc. Am. Math. Soc. **109**(2), 477–485 (1990)
273. J. Lott, C. Villani, Ricci curvature for metric-measure spaces via optimal transport. Ann. Math. (2) **169**(3), 903–991 (2009)
274. A. Lytchak, S. Wenger, Isoperimetric characterization of upper curvature bounds. Acta Math. **221**(1), 159–202 (2018)

275. F. Maggi, *Sets of Finite Perimeter and Geometric Variational Problems*, vol. 135. *Cambridge Studies in Advanced Mathematics* (Cambridge University Press, Cambridge, 2012). An introduction to geometric measure theory
276. P.I. Marinov, *Stability Analysis of Capillary Surfaces with Planar or Spherical Boundary in the Absence of Gravity* (ProQuest LLC, Ann Arbor, MI, 2010). Thesis (Ph.D.)–The University of Toledo
277. F.C. Marques, A. Neves, Min-max theory and the Willmore conjecture. Ann. Math. (2) **179**(2), 683–782 (2014)
278. Y. Matsushima, *Differentiable Manifolds*, vol. 9. *Pure and Applied Mathematics* (Marcel Dekker, Inc., New York, 1972). Translated from the Japanese by E. T. Kobayashi
279. Q. Maurmann, F. Morgan, Isoperimetric comparison theorems for manifolds with density. Calc. Var. Part. Differ. Equ. **36**(1), 1–5 (2009)
280. R.J. McCann, Existence and uniqueness of monotone measure-preserving maps. Duke Math. J. **80**(2), 309–323 (1995)
281. J. McCuan, The stability of cylindrical pendant drops. Mem. Am. Math. Soc. **250**(1189), v+109 (2017)
282. W.H. Meeks, III, P. Mira, J. Pérez, A. Ros, Isoperimetric domains of large volume in homogeneous three-manifolds. Adv. Math. **264**, 546–592 (2014)
283. J.H. Michael, L.M. Simon, Sobolev and mean-value inequalities on generalized submanifolds of R^n. Commun. Pure Appl. Math. **26**, 361–379 (1973)
284. E. Milman, On the role of convexity in isoperimetry, spectral gap and concentration. Invent. Math. **177**(1), 1–43 (2009)
285. E. Milman, Model spaces for sharp isoperimetric inequalities. C. R. Math. Acad. Sci. Paris **350**(19-20), 897–902 (2012)
286. E. Milman, Sharp isoperimetric inequalities and model spaces for the curvature-dimension-diameter condition. J. Eur. Math. Soc. (JEMS) **17**(5), 1041–1078 (2015)
287. E. Milman, J. Neeman, The Gaussian double-bubble and multi-bubble conjectures. Ann. Math. (2) **195**(1), 89–206 (2022)
288. V.D. Milman, G. Schechtman, *Asymptotic Theory of Finite-Dimensional Normed Spaces*, vol. 1200. *Lecture Notes in Mathematics* (Springer, Berlin, 1986). With an appendix by M. Gromov
289. H. Minkowski, Über die Begriffe Länge, Oberfläche und Volumen. Jahresber. Dtsch. Math.-Ver. **9**(1), 115–121 (1901)
290. H. Minkowski, Volumen und Oberfläche. Math. Ann. **57**, 447–495 (1903)
291. C. Miranda, *Partial Differential Equations of Elliptic Type*. Ergebnisse der Mathematik und ihrer Grenzgebiete, Band 2 (Springer, New York-Berlin, 1970). Second revised edition. Translated from the Italian by Zane C. Motteler
292. J.M. Miranda, Functions of bounded variation on "good" metric spaces. J. Math. Pures Appl. (9) **82**(8), 975–1004 (2003)
293. M. Miranda, Jr., D. Pallara, F. Paronetto, M. Preunkert, Heat semigroup and functions of bounded variation on Riemannian manifolds. J. Reine Angew. Math. **613**, 99–119 (2007)
294. A. Mondino, S. Nardulli, Existence of isoperimetric regions in non-compact Riemannian manifolds under Ricci or scalar curvature conditions. Comm. Anal. Geom. **24**(1), 115–138 (2016)
295. A. Mondino, E. Spadaro, On an isoperimetric-isodiametric inequality. Anal. PDE **10**(1), 95–126 (2017)
296. R. Montgomery, *A Tour of Subriemannian Geometries, Their Geodesics and Applications*, vol. 91. *Mathematical Surveys and Monographs* (American Mathematical Society, Providence, RI, 2002)
297. S. Montiel, Stable constant mean curvature hypersurfaces in some Riemannian manifolds. Comment. Math. Helv. **73**(4), 584–602 (1998)
298. S. Montiel, Unicity of constant mean curvature hypersurfaces in some Riemannian manifolds. Indiana Univ. Math. J. **48**(2), 711–748 (1999)

299. S. Montiel, A. Ros, Minimal immersions of surfaces by the first eigenfunctions and conformal area. Invent. Math. **83**(1), 153–166 (1986)
300. S. Montiel, A. Ros, Compact hypersurfaces: the Alexandrov theorem for higher order mean curvatures, in *Differential Geometry*, vol. 52. *Pitman Monogr. Surveys Pure Appl. Math.* (Longman Sci. Tech., Harlow, 1991), pp. 279–296
301. F. Morgan, Clusters minimizing area plus length of singular curves. Math. Ann. **299**(4), 697–714 (1994)
302. F. Morgan, Regularity of isoperimetric hypersurfaces in Riemannian manifolds. Trans. Am. Math. Soc. **355**(12), 5041–5052 (2003)
303. F. Morgan, Manifolds with density. Not. Am. Math. Soc. **52**(8), 853–858 (2005)
304. F. Morgan, Isoperimetric estimates in products. Ann. Global Anal. Geom. **30**(1), 73–79 (2006)
305. F. Morgan, The Levy-Gromov isoperimetric inequality in convex manifolds with boundary. J. Geom. Anal. **18**(4), 1053–1057 (2008)
306. F. Morgan, In orbifolds, small isoperimetric regions are small balls. Proc. Am. Math. Soc. **137**(6), 1997–2004 (2009)
307. F. Morgan, *Geometric Measure Theory*, 5th edn. (Elsevier/Academic Press, Amsterdam, 2016). A beginner's guide, Illustrated by James F. Bredt
308. F. Morgan, The isoperimetric problem with density. Math. Intell. **39**(4), 2–8 (2017)
309. F. Morgan, S. Howe, N. Harman, Steiner and Schwarz symmetrization in warped products and fiber bundles with density. Rev. Mat. Iberoam. **27**(3), 909–918 (2011)
310. F. Morgan, M. Hutchings, H. Howards, The isoperimetric problem on surfaces of revolution of decreasing Gauss curvature. Trans. Am. Math. Soc. **352**(11), 4889–4909 (2000)
311. F. Morgan, D.L. Johnson, Some sharp isoperimetric theorems for Riemannian manifolds. Indiana Univ. Math. J. **49**(3), 1017–1041 (2000)
312. F. Morgan, A. Pratelli, Existence of isoperimetric regions in $\mathbb{R}^n$ with density. Ann. Global Anal. Geom. **43**(4), 331–365 (2013)
313. F. Morgan, M. Ritoré, Isoperimetric regions in cones. Trans. Am. Math. Soc. **354**(6), 2327–2339 (2002)
314. F. Morgan, A. Ros, Stable constant-mean-curvature hypersurfaces are area minimizing in small L^1 neighborhoods. Interfaces Free Bound. **12**(2), 151–155 (2010)
315. A. Muñoz Flores, S. Nardulli, Continuity and differentiability properties of the isoperimetric profile in complete noncompact Riemannian manifolds with bounded geometry. arXiv:1404.3245, 11 Apr 2014
316. A.E. Muñoz Flores, S. Nardulli, Local Hölder continuity of the isoperimetric profile in complete noncompact Riemannian manifolds with bounded geometry. Geom. Dedicata **201**, 1–12 (2019)
317. S. Nardulli, The isoperimetric profile of a smooth Riemannian manifold for small volumes. Ann. Global Anal. Geom. **36**(2), 111–131 (2009)
318. S. Nardulli, Generalized existence of isoperimetric regions in non-compact Riemannian manifolds and applications to the isoperimetric profile. Asian J. Math. **18**(1), 1–28 (2014)
319. S. Nardulli, P. Pansu, A complete Riemannian manifold whose isoperimetric profile is discontinuous. Ann. Sc. Norm. Super. Pisa Cl. Sci. (5) **18**(2), 537–549 (2018)
320. J. Nash, The imbedding problem for Riemannian manifolds. Ann. Math. (2) **63**, 20–63 (1956)
321. J.C.C. Nitsche, *Lectures on Minimal Surfaces. Vol. 1* (Cambridge University Press, Cambridge, 1989). Introduction, fundamentals, geometry and basic boundary value problems, Translated from the German by Jerry M. Feinberg, With a German foreword
322. I. Nunes, On stable constant mean curvature surfaces with free boundary. Math. Z. **287**(1-2), 473–479 (2017)
323. B. O'Neill, *Semi-Riemannian Geometry*, vol. 103. *Pure and Applied Mathematics* (Academic Press, Inc. [Harcourt Brace Jovanovich, Publishers], New York, 1983). With applications to relativity
324. R. Osserman, Minimal varieties. Bull. Am. Math. Soc. **75**, 1092–1120 (1969)
325. R. Osserman, The isoperimetric inequality. Bull. Am. Math. Soc. **84**(6), 1182–1238 (1978)
326. R. Osserman, Bonnesen-style isoperimetric inequalities. Am. Math. Mon. **86**(1), 1–29 (1979)

327. F. Pacard, X. Xu, Constant mean curvature spheres in Riemannian manifolds. Manuscripta Math. **128**(3), 275–295 (2009)
328. B. Palmer, Index and stability of harmonic Gauss maps. Math. Z. **206**(4), 563–566 (1991)
329. B. Palmer, Stability of the Wulff shape. Proc. Am. Math. Soc. **126**(12), 3661–3667 (1998)
330. P. Pansu, Une inégalité isopérimétrique sur le groupe de Heisenberg. C. R. Acad. Sci. Paris Sér. I Math. **295**(2), 127–130 (1982)
331. P. Pansu, An isoperimetric inequality on the Heisenberg group, Number Special Issue, pp. 159–174 (1984), in *Conference on Differential Geometry on Homogeneous Spaces* (Turin, 1983)
332. P. Pansu, Sur la régularité du profil isopérimétrique des surfaces riemanniennes compactes. Ann. Inst. Fourier (Grenoble) **48**(1), 247–264 (1998)
333. P. Papasoglu, E. Swenson, A surface with discontinuous isoperimetric profile and expander manifolds. Geom. Dedicata **206**, 43–54 (2020)
334. T.H. Parker, Bubble tree convergence for harmonic maps. J. Differ. Geom. **44**(3), 595–633 (1996)
335. L.E. Payne, Isoperimetric inequalities and their applications. SIAM Rev. **9**, 453–488 (1967)
336. R.H.d.L. Pedrosa, *On the Uniqueness of Isoperimetric Regions in Cylindrical Spaces* (ProQuest LLC, Ann Arbor, MI, 1988). Thesis (Ph.D.)–University of California, Berkeley
337. R.H.L. Pedrosa, The isoperimetric problem in spherical cylinders. Ann. Global Anal. Geom. **26**(4), 333–354 (2004)
338. R.H.L. Pedrosa, M. Ritoré, Isoperimetric domains in the Riemannian product of a circle with a simply connected space form and applications to free boundary problems. Indiana Univ. Math. J. **48**(4), 1357–1394 (1999)
339. O. Perdomo, Low index minimal hypersurfaces of spheres. Asian J. Math. **5**(4), 741–749 (2001)
340. J. Petean, Isoperimetric regions in spherical cones and Yamabe constants of $M \times S^1$. Geom. Dedicata **143**, 37–48 (2009)
341. J. Petean, J.M. Ruiz, Isoperimetric profile comparisons and Yamabe constants. Ann. Global Anal. Geom. **40**(2), 177–189 (2011)
342. J. Petean, J.M. Ruiz, On the Yamabe constants of $S^2 \times \mathbb{R}^3$ and $S^3 \times \mathbb{R}^2$. Differ. Geom. Appl. **31**(2), 308–319 (2013)
343. S. Peters, Convergence of Riemannian manifolds. Compos. Math. **62**(1), 3–16 (1987)
344. P. Petersen, *Riemannian Geometry*, vol. 171. *Graduate Texts in Mathematics*, 2nd edn. (Springer, New York, 2006)
345. P. Petersen, *Riemannian Geometry*, vol. 171. *Graduate Texts in Mathematics*, 3rd edn. (Springer, Cham, 2016)
346. C. Pittet, The isoperimetric profile of homogeneous Riemannian manifolds. J. Differ. Geom. **54**(2), 255–302 (2000)
347. G. Pólya, G. Szegö, *Isoperimetric Inequalities in Mathematical Physics*. Annals of Mathematics Studies, vol. 27 (Princeton University Press, Princeton, NJ, 1951)
348. J. Pozuelo, M. Ritoré, Pansu-Wull shapes in $\mathbb{H}^1$. Adv. Calc. Var. **16**(1), 69–98 (2023). https://doi.org/10.1515/acv-2020-0093
349. A. Pratelli, G. Saracco, On the isoperimetric problem with double density. Nonlinear Anal. **177**(part B), 733–752 (2018)
350. A. Pratelli, G. Saracco, The $\varepsilon - \varepsilon^\beta$ property in the isoperimetric problem with double density, and the regularity of isoperimetric sets. Adv. Nonlinear Stud. **20**(3), 539–555 (2020)
351. M.H. Protter, H.F. Weinberger, *Maximum Principles in Differential Equations* (Springer, New York, 1984). Corrected reprint of the 1967 original
352. T. Rado, The isoperimetric inequality on the sphere. Am. J. Math. **57**(4), 765–770 (1935)
353. M. Reed, B. Simon, *Methods of Modern Mathematical Physics. IV. Analysis of Operators* (Academic Press [Harcourt Brace Jovanovich, Publishers], New York-London, 1978)
354. R.C. Reilly, Applications of the Hessian operator in a Riemannian manifold. Indiana Univ. Math. J. **26**(3), 459–472 (1977)

355. R.C. Reilly, Geometric applications of the solvability of Neumann problems on a Riemannian manifold. Arch. Ration. Mech. Anal. **75**(1), 23–29 (1980)
356. R.C. Reilly, Mean curvature, the Laplacian, and soap bubbles. Am. Math. Mon. **89**(3), 180–188, 197–198 (1982)
357. M. Ritoré, Applications of compactness results for harmonic maps to stable constant mean curvature surfaces. Math. Z. **226**(3), 465–481 (1997)
358. M. Ritoré, Examples of constant mean curvature surfaces obtained from harmonic maps to the two sphere. Math. Z. **226**(1), 127–146 (1997)
359. M. Ritoré, Index one minimal surfaces in flat three space forms. Indiana Univ. Math. J. **46**(4), 1137–1153 (1997)
360. M. Ritoré, Constant geodesic curvature curves and isoperimetric domains in rotationally symmetric surfaces. Comm. Anal. Geom. **9**(5), 1093–1138 (2001)
361. M. Ritoré, The isoperimetric problem in complete surfaces of nonnegative curvature. J. Geom. Anal. **11**(3), 509–517 (2001)
362. M. Ritoré, Optimal isoperimetric inequalities for three-dimensional Cartan-Hadamard manifolds, in *Global Theory of Minimal Surfaces*, vol. 2. *Clay Math. Proc.* (Amer. Math. Soc., Providence, RI, 2005), pp. 395–404
363. M. Ritoré, A proof by calibration of an isoperimetric inequality in the Heisenberg group $\mathbb{H}^n$. Calc. Var. Part. Differ. Equ. **44**(1-2), 47–60 (2012)
364. M. Ritoré, Continuity of the isoperimetric profile of a complete Riemannian manifold under sectional curvature conditions. Rev. Mat. Iberoam. **33**(1), 239–250 (2017)
365. M. Ritoré, A. Ros, Stable constant mean curvature tori and the isoperimetric problem in three space forms. Comment. Math. Helv. **67**(2), 293–305 (1992)
366. M. Ritoré, A. Ros, The spaces of index one minimal surfaces and stable constant mean curvature surfaces embedded in flat three manifolds. Trans. Am. Math. Soc. **348**(1), 391–410 (1996)
367. M. Ritoré, C. Rosales, Existence and characterization of regions minimizing perimeter under a volume constraint inside Euclidean cones. Trans. Am. Math. Soc. **356**(11), 4601–4622 (2004)
368. M. Ritoré, C. Rosales, Area-stationary surfaces in the Heisenberg group $\mathbb{H}^1$. Adv. Math. **219**(2), 633–671 (2008)
369. M. Ritoré, C. Sinestrari, *Mean Curvature Flow and Isoperimetric Inequalities*. Advanced Courses in Mathematics. CRM Barcelona (Birkhäuser Verlag, Basel, 2010). Edited by Vicente Miquel and Joan Porti
370. M. Ritoré, E. Vernadakis, Isoperimetric inequalities in convex cylinders and cylindrically bounded convex bodies. Calc. Var. Part. Differ. Equ. **54**(1), 643–663 (2015)
371. M. Ritoré, E. Vernadakis, Isoperimetric inequalities in Euclidean convex bodies. Trans. Am. Math. Soc. **367**(7), 4983–5014 (2015)
372. M. Ritoré, E. Vernadakis, Isoperimetric inequalities in conically bounded convex bodies. J. Geom. Anal. **26**(1), 474–498 (2016)
373. M. Ritoré, E. Vernadakis, Large isoperimetric regions in the product of a compact manifold with Euclidean space. Adv. Math. **306**, 958–972 (2017)
374. M. Ritoré, J. Yepes Nicolás, Brunn-Minkowski inequalities in product metric measure spaces. Adv. Math. **325**, 824–863 (2018)
375. M. Ritoré, Superficies con curvatura media constante. Ph. D. Thesis, Universidad de Granada (1994)
376. R.T. Rockafellar, *Convex Analysis*. Princeton Landmarks in Mathematics (Princeton University Press, Princeton, NJ, 1997). Reprint of the 1970 original, Princeton Paperbacks
377. A. Ros, Compact hypersurfaces with constant higher order mean curvatures. Rev. Mat. Iberoamericana **3**(3-4), 447–453 (1987)
378. A. Ros, The Willmore conjecture in the real projective space. Math. Res. Lett. **6**(5-6), 487–493 (1999)
379. A. Ros, Isoperimetric inequalities in crystallography. J. Am. Math. Soc. **17**(2), 373–388 (2004)

380. A. Ros, The isoperimetric problem, in *Global Theory of Minimal Surfaces*, vol. 2. *Clay Math. Proc.* (Amer. Math. Soc., Providence, RI, 2005), pp. 175–209
381. A. Ros, One-sided complete stable minimal surfaces. J. Differ. Geom. **74**(1), 69–92 (2006)
382. A. Ros, Stable periodic constant mean curvature surfaces and mesoscopic phase separation. Interfaces Free Bound. **9**(3), 355–365 (2007)
383. A. Ros, E. Vergasta, Stability for hypersurfaces of constant mean curvature with free boundary. Geom. Dedicata **56**(1), 19–33 (1995)
384. C. Rosales, Isoperimetric regions in rotationally symmetric convex bodies. Indiana Univ. Math. J. **52**(5), 1201–1214 (2003)
385. C. Rosales, A. Cañete, V. Bayle, F. Morgan, On the isoperimetric problem in Euclidean space with density. Calc. Var. Part. Differ. Equ. **31**(1), 27–46 (2008)
386. M. Ross, Schwarz' P and D surfaces are stable. Differ. Geom. Appl. **2**(2), 179–195 (1992)
387. W. Rudin, *Real and Complex Analysis*, 3rd edn. (McGraw-Hill Book Co., New York, 1987)
388. E.A. Ruh, Asymptotic behaviour of non-parametric minimal hypersurfaces. J. Differ. Geom., 4:509–513 (1970)
389. L.A. Santaló, A demonstration of the isoperimetric property of the circle. Publ. Inst. Mat. Univ. Nac. Litoral **2**, 37–46 (1940)
390. L.A. Santaló, On the isoperimetric inequality for surfaces of constant negative curvature. Univ. Nac. Tucumán. Revista A. **3**, 243–259 (1942)
391. L.A. Santaló, *Introduction to Integral Ggeometry*. Publ. Inst. Math. Univ. Nancago, II (Hermann & Cie, Paris, 1953)
392. L.A. Santaló, *Integral Geometry and Geometric Probability*. Encyclopedia of Mathematics and Its Applications, Vol. 1 (Addison-Wesley Publishing Co., Reading, MA-London-Amsterdam, 1976). With a foreword by Mark Kac
393. E. Schmidt, Über das isoperimetrische Problem im Raum von n Dimensionen. Math. Z. **44**(1), 689–788 (1939)
394. E. Schmidt, Die isoperimetrischen Ungleichungen auf der gewöhnlichen Kugel und für Rotationskörper im n-dimensionalen sphärischen Raum. Math. Z. **46**, 743–794 (1940)
395. E. Schmidt, Über die isoperimetrische Aufgabe im n-dimensionalen Raum konstanter negativer Kuümmung. I. Die isoperimetrischen Ungleichungen in der hyperbolischen Ebene und für Rotationskörper im n-dimensionalen hyperbolischen Raum. Math. Z. **46**, 204–230 (1940)
396. E. Schmidt, Über eine neue Methode zur Behandlung einer Klasse isoperimetrischer Aufgaben im Grossen. Math. Z. **47**, 489–642 (1942)
397. E. Schmidt, Beweis der isoperimetrischen Eigenschaft der Kugel im hyperbolischen und sphärischen Raum jeder Dimensionenzahl. Math. Z. **49**, 1–109 (1943)
398. R. Schneider, *Convex Bodies: The Brunn-Minkowski Theory*, vol. 151. *Encyclopedia of Mathematics and Its Applications*, expanded edn. (Cambridge University Press, Cambridge, 2014)
399. R. Schoen, L. Simon, A new proof of the regularity theorem for rectifiable currents which minimize parametric elliptic functionals. Indiana Univ. Math. J. **31**(3), 415–434 (1982)
400. V. Schroeder, W. Ziller, Local rigidity of symmetric spaces. Trans. Am. Math. Soc. **320**(1), 145–160 (1990)
401. F. Schulze, Nonlinear evolution by mean curvature and isoperimetric inequalities. J. Differ. Geom. **79**(2), 197–241 (2008)
402. H.A. Schwarz, Beweis des Satzes, dass die Kugel kleinere Oberfläche besitzt, als jeder andere Körper gleichen Volumens. Gött. Nachr. **1884**, 1–13 (1884)
403. K. Shiohama, T. Shioya, M. Tanaka, *The Geometry of Total Curvature on Complete Open Surfaces*, vol. 159. *Cambridge Tracts in Mathematics* (Cambridge University Press, Cambridge, 2003)
404. L. Simon, *Lectures on Geometric Measure Theory*, vol. 3. *Proceedings of the Centre for Mathematical Analysis, Australian National University* (Australian National University, Centre for Mathematical Analysis, Canberra, 1983)
405. J. Simons, Minimal varieties in Riemannian manifolds. Ann. Math. (2) **88**, 62–105 (1968)

406. M. Spivak, *A Comprehensive Introduction to Differential Geometry. Vol. IV*, 2nd edn. (Publish or Perish, Inc., Wilmington, De, 1979)
407. J. Steiner, Sur le maximum et le minimum des figures dans le plan, sur la sphère, et dans l'espace en général. J. Reine Angew. Math. **24**, 190–250 (1842)
408. J. Steiner, Sur le maximum et le minimum des figures dans le plan, sur la sphère, et dans l'espace en général. J. Reine Angew. Math. **24**, 93–152 (1842)
409. J. Steiner, Elementare Lösung einer Aufgabe über das ebene und sphärische Dreieck. J. Reine Angew. Math. **28**, 375–379 (1844)
410. J. Steiner, *Jacob Steiner's gesammelte Werke. Band 1. Herausgegeben von K. Weierstrass* (AMS Chelsea Publishing, Providence, RI, 1997). Republication of the 1971 2nd edn.
411. P. Sternberg, K. Zumbrun, On the connectivity of boundaries of sets minimizing perimeter subject to a volume constraint. Comm. Anal. Geom. **7**(1), 199–220 (1999)
412. E. Stredulinsky, W.P. Ziemer, Area minimizing sets subject to a volume constraint in a convex set. J. Geom. Anal. **7**(4), 653–677 (1997)
413. K.-T. Sturm, On the geometry of metric measure spaces. I. Acta Math. **196**(1), 65–131 (2006)
414. K.-T. Sturm, On the geometry of metric measure spaces. II. Acta Math. **196**(1), 133–177 (2006)
415. V.N. Sudakov, B.S. Cirel'son, Extremal properties of half-spaces for spherically invariant measures. Zap. Naučn. Sem. Leningrad. Otdel. Mat. Inst. Steklov. (LOMI) **41**, 14–24, 165 (1974). Problems in the theory of probability distributions, II
416. B. Süssmann, Curve shortening flow and the Banchoff-Pohl inequality on surfaces of nonpositive curvature. Beiträge Algebra Geom. **40**(1), 203–215 (1999)
417. G. Szegö, Inequalities for certain eigenvalues of a membrane of given area. J. Ration. Mech. Anal. **3**, 343–356 (1954)
418. M. Talagrand, Isoperimetry, logarithmic Sobolev inequalities on the discrete cube, and Margulis' graph connectivity theorem. Geom. Funct. Anal. **3**(3), 295–314 (1993)
419. G. Talenti, The standard isoperimetric theorem, in *Handbook of Convex Geometry, Vol. A, B* (North-Holland, Amsterdam, 1993), pp. 73–123
420. I. Tamanini, *Regularity Results for Almost Minimal Oriented Hypersurfaces in* $\mathbb{R}^N$, vol. 1. Quaderni del Dipartimento di Matematica dell'Universitá di Lecce (Universitá di Lecce, Lecce, 1984)
421. M. Tamm, Subanalytic sets in the calculus of variation. Acta Math. **146**(3-4), 167–199 (1981)
422. J.E. Taylor, Crystalline variational problems. Bull. Am. Math. Soc. **84**(4), 568–588 (1978)
423. P. Tomter, Constant mean curvature surfaces in the Heisenberg group, in *Differential Geometry: Partial Differential Equations on Manifolds (Los Angeles, CA, 1990)*, vol. 54. *Proc. Sympos. Pure Math.* (Amer. Math. Soc., Providence, RI, 1993), pp. 485–495
424. L. Tonelli, Sulla proprietà di minimo della sfera. Rend. Circ. Mat. Palermo **39**, 109–138 (1915)
425. P. Topping, Mean curvature flow and geometric inequalities. J. Reine Angew. Math. **503**, 47–61 (1998)
426. P. Topping, The isoperimetric inequality on a surface. Manuscripta Math. **100**(1), 23–33 (1999)
427. M. Troyanov, Metrics of constant curvature on a sphere with two conical singularities, in *Differential Geometry (Peñíscola, 1988)*, vol. 1410. *Lecture Notes in Math.* (Springer, Berlin, 1989), pp. 296–306
428. M. Troyanov, Prescribing curvature on compact surfaces with conical singularities. Trans. Am. Math. Soc. **324**(2), 793–821 (1991)
429. A.R. Veeravalli, Stability of compact constant mean curvature hypersurfaces in a wide class of Riemannian manifolds. Geom. Dedicata **159**, 1–9 (2012)
430. C. Viana, The isoperimetric problem for lens spaces. Math. Ann. **374**(1-2), 475–497 (2019)
431. C. Viana, Isoperimetry and volume preserving stability in real projective spaces. arXiv:1907.09445, 22 Jul 2019
432. C. Villani, *Topics in Optimal Transportation*, vol. 58. *Graduate Studies in Mathematics* (American Mathematical Society, Providence, RI, 2003)

433. T.I. Vogel, Stability of a liquid drop trapped between two parallel planes. SIAM J. Appl. Math. **47**(3), 516–525 (1987)
434. T.I. Vogel, Uniqueness for certain surfaces of prescribed mean curvature. Pac. J. Math. **134**(1), 197–207 (1988)
435. V.A. Šarafutdinov, Convex sets in a manifold of nonnegative curvature. Mat. Zametki **26**(1), 129–136, 159 (1979)
436. G. Wang, C. Xia, Uniqueness of stable capillary hypersurfaces in a ball. Math. Ann. **374**(3-4), 1845–1882 (2019)
437. F.W. Warner, *Foundations of Differentiable Manifolds and Lie Groups* (Scott, Foresman and Co., Glenview, IL-London, 1971)
438. F.W. Warner, *Foundations of Differentiable Manifolds and Lie Groups*, vol. 94. *Graduate Texts in Mathematics* (Springer, New York-Berlin, 1983). Corrected reprint of the 1971 edition
439. A. Weil, Sur les surfaces à courbure négative. C. R. Acad. Sci., Paris **182**, 1069–1071 (1926)
440. H.C. Wente, The stability of the axially symmetric pendent drop. Pac. J. Math. **88**(2), 421–470 (1980)
441. H.C. Wente, The symmetry of sessile and pendent drops. Pac. J. Math. **88**(2), 387–397 (1980)
442. H.C. Wente, A note on the stability theorem of J. L. Barbosa and M. Do Carmo for closed surfaces of constant mean curvature. Pac. J. Math. **147**(2), 375–379 (1991)
443. B. White, Complete surfaces of finite total curvature. J. Differ. Geom. **26**(2), 315–326 (1987)
444. B. White, Existence of smooth embedded surfaces of prescribed genus that minimize parametric even elliptic functionals on 3-manifolds. J. Differ. Geom. **33**(2), 413–443 (1991)
445. B. White, A strong minimax property of nondegenerate minimal submanifolds. J. Reine Angew. Math. **457**, 203–218 (1994)
446. T.J. Willmore, B.A. Saleemi, The total absolute curvature of immersed manifolds. J. Lond. Math. Soc. **41**, 153–160 (1966)
447. S. Winklmann, Estimates for stable hypersurfaces of prescribed F-mean curvature. Manuscripta Math. **118**(4), 485–499 (2005)
448. S. Winklmann, A note on the stability of the Wulff shape. Arch. Math. (Basel) **87**(3), 272–279 (2006)
449. J.A. Wolf, *Spaces of Constant Curvature*, 6th edn. (AMS Chelsea Publishing, Providence, RI, 2011)
450. S.T. Yau, Isoperimetric constants and the first eigenvalue of a compact Riemannian manifold. Ann. Sci. École Norm. Sup. (4) **8**(4), 487–507 (1975)
451. R. Ye, Foliation by constant mean curvature spheres. Pac. J. Math. **147**(2), 381–396 (1991)
452. R. Ye, Constant mean curvature foliation: singularity structure and curvature estimate. Pac. J. Math. **174**(2), 569–587 (1996)
453. J.-W. Yim, Distance nonincreasing retraction on a complete open manifold of nonnegative sectional curvature. Ann. Global Anal. Geom. **6**(2), 191–206 (1988)
454. Y. Zheng, Uniform Lipschitz continuity of the isoperimetric profile of compact surfaces under normalized Ricci flow. Proc. Am. Math. Soc. **149**(5), 2105–2119 (2021)
455. W.P. Ziemer, *Weakly Differentiable Functions*, vol. 120. *Graduate Texts in Mathematics* (Springer, New York, 1989). Sobolev spaces and functions of bounded variation

Index

M. Ritoré, *Isoperimetric Inequalities in Riemannian Manifolds*,
Progress in Mathematics 348, https://doi.org/10.1007/978-3-031-37901-7

Notation index

M. Ritoré, *Isoperimetric Inequalities in Riemannian Manifolds*,
Progress in Mathematics 348, https://doi.org/10.1007/978-3-031-37901-7